Timothy C. Frank, Bruce S. Holden
Industrial Chemical Separation

Also of interest

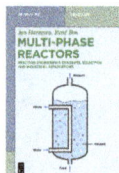

Multiphase Reactors.
Reaction Engineering Concepts, Selection, and Industrial Applications
Harmsen, Bos, 2023
ISBN 978-3-11-071376-3, e-ISBN (PDF) 978-3-11-071377-0

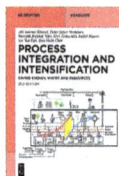

Sustainable Process Integration and Intensification.
Saving Energy, Water and Resources
Klemeš, Varbanov, Wan Alwi, Manan, Fan, Chin, 2023
ISBN 978-3-11-078283-7, e-ISBN (PDF) 978-3-11-078298-1

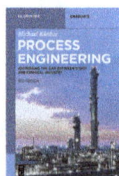

Process Engineering.
Addressing the Gap between Study and Chemical Industry
Kleiber 2023
ISBN 978-3-11-102811-8, e-ISBN (PDF) 978-3-11-102814-9

Process Technology.
An Introduction
De Haan, Paading, 2022
ISBN 978-3-11-071243-8, e-ISBN (PDF) 978-3-11-071244-5

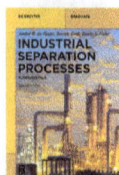

Industrial Separation Processes.
Fundamentals
Haan, Eral, Schuur, 2020
ISBN 978-3-11-065473-8, e-ISBN (PDF) 978-3-11-065480-6

Timothy C. Frank, Bruce S. Holden

Industrial Chemical Separation

Historical Perspective, Fundamentals, and Engineering
Practice

DE GRUYTER

Authors
Dr. Timothy C. Frank
Serious Separations LLC
Beaverton, Michigan
United States of America
E-mail: seriousseparationsllc@gmail.com

Bruce S. Holden
Midland, Michigan
United States of America
E-mail: bsh0@hotmail.com

ISBN 978-3-11-069502-1
e-ISBN (PDF) 978-3-11-069505-2
e-ISBN (EPUB) 978-3-11-069513-7

Library of Congress Control Number: 2023936596

Bibliographic information published by the Deutsche Nationalbibliothek
The Deutsche Nationalbibliothek lists this publication in the Deutsche Nationalbibliografie;
detailed bibliographic data are available on the Internet at http://dnb.dnb.de.

© 2023 Walter de Gruyter GmbH, Berlin/Boston
Cover image: Abani Biswas/iStock/Getty Images Plus. More than decoration, the image on the cover signals what this book is all about: bringing together two distinct material phases in close contact to transfer specific components from one phase to the other. The image suggests liquid-liquid extraction, showing droplets of one liquid suspended in another, or high-pressure distillation, showing liquid-like bubbles in a liquid near its critical point.
Typesetting: Integra Software Services Pvt. Ltd.
Printing and binding: CPI books GmbH, Leck

www.degruyter.com

To our wives, Martha and Kathy, for their love, encouragement, and patience

Why Separations?

A single molecule, all alone, is not a thing we know of.
It's always surrounded by many another, friend or foe of.
And yet, its physical journey
from liquid to vapor or solid and such,
matters much
to the company it keeps
and the purpose it serves.
So let us distill, extract, and crystallize
adsorb, permeate, and organize
to refine the matter
as needed of.

T. C. Frank

About the authors

Tim Frank retired as Fellow and Associate R&D Director at The Dow Chemical Company after 34 years of industrial experience. He holds a B.S. degree in chemical engineering from Montana State University, Bozeman (1978), and M.S. and Ph.D. degrees in chemical engineering from the University of Colorado, Boulder (1981, 1984). At Dow, Dr. Frank held a variety of positions in technology development and technical management within R&D, focusing on the development of distillation, extraction, crystallization, and adsorption-based processes for commercial implementation. He is co-author of 32 publications on various aspects of separation and reaction technology and is co-inventor on 20 granted patent families. Dr. Frank is editor and co-author of Section 15, "Liquid-Liquid Extraction and Other Liquid-Liquid Operations and Equipment," in *Perry's Chemical Engineers' Handbook*, 8th and 9th eds. (McGraw-Hill, 2007, 2018), and co-editor/co-author of Section 4, "Thermodynamics," in *Perry's Chemical Engineers' Handbook*, 9th ed. He has also been active in the American Institute of Chemical Engineers (AIChE) as Chair (2010–2011) and Director (2012–2015) of the local Mid-Michigan Section and has served AIChE's Separations Division as Area Chair of Extraction (2009–2011), Division Chair (2015), and as Editor of the Separations Division Newsletter (2013–2016). In addition to receiving numerous Dow awards including the President's Environmental Care Award (1992), Dr. Frank is a member of the Dow team recognized by the U.S. Environmental Protection Agency in 1999 with a President's Green Chemistry Challenge Award. He received the AIChE Separations Division Gerhold Award in 2014 and is a Fellow of AIChE.

Bruce Holden retired as Principle Research Scientist at The Dow Chemical Company after 40 years of industrial experience. He received B.S. and M.S. degrees in chemical engineering from Clarkson University (Potsdam, New York) in 1975 and 1977. At Dow, Mr. Holden held a number of technical leadership positions in process R&D, process engineering, and environmental services. Mr. Holden has contributed to several books on separations technology and the proper handling of toxic materials including Section 15, "Liquid-Liquid Extraction and Other Liquid-Liquid Operations and Equipment," in *Perry's Chemical Engineers' Handbook*, 8th and 9th eds. (McGraw-Hill, 2007, 2018); *AIChE Equipment Testing Procedure – Trayed and Packed Columns: A Guide to Performance Evaluation*, 3rd ed. (AIChE/Wiley Press, 2014); and Chapter 7, "Storage of Toxic Materials," in *Handbook of Highly Toxic Materials Handling and Management* (Marcel Dekker, 1995). Mr. Holden has been a strong supporter of the Mid-Michigan AIChE Section, serving as Secretary from 2009 to 2019 and in later years as a Director. He has also served on various AIChE committees on distillation testing methods and as a member of the technical committee at Fractionation Research, Inc. In recognition of his many technical contributions, Mr. Holden has received numerous Dow awards including the Michigan Consultant's Award, The Doug Leng Excellence in Engineering Sciences Award, and 13 Dow Technology Center/Waste Reduction Always Pays (WRAP) awards for development and implementation of commercially successful technology at Dow manufacturing facilities (1990–2013). In 2011, Mr. Holden was named "Chemical Engineer of the Year" by the Mid-Michigan Section of AIChE, and in 2017 he received a Shining Star Award from AIChE for "Volunteer of the Year." He is a Fellow of AIChE.

https://doi.org/10.1515/9783110695052-202

Acknowledgments

A book like ours depends on a great many people who have generously shared their thoughts and expertise in publications, in conference presentations, and perhaps most importantly, in person. We feel fortunate when reflecting on our formal educations long ago in having had many fine teachers, including Lloyd Berg at Montana State University, Richard Nunge at Clarkson University, John Falconer at the University of Colorado, and Joseph Katz at Clarkson University. And during our careers, we benefited from close collaborations with many talented colleagues at The Dow Chemical Company. Early on we were privileged to be members of Dow's Separation & Purification Process R&D Group led by Dr. Lanny Robbins who introduced us to the art and craft of industrial chemical separation. We benefited greatly from Lanny's leadership and keen insights. We also thank George Killat, Pat Au-Yeung, Doug Greminger, Nancy Duty, John Pendergast, Scott Phillips, Ron Leng, Chet Davidson, Tony Avilla, Grant Von Wald, Sumnesh Gupta, Felipe Donate, Joe Downey, Steven Arturo, Jim Ringer, Hank Kohlbrand, Chris Christenson, Richard Helling, and many others at Dow, for sharing their expertise. We also benefited from our interactions with researchers at various universities, including Frank Seibert and Bruce Eldridge (University of Texas), Charles Eckert, Yoshiaki Kawajiri, and Andreas Bommarius (Georgia Institute of Technology), Ed Cussler (University of Minnesota), Phillip Wankat and Rakesh Agrawal (Purdue University), and Carl Lira and the late Kris Berglund (Michigan State University). We are grateful to Mark Rosenzweig and Don Green for their encouragement in our previous publishing efforts, and to Bob Esposito for suggesting that we write a book of this kind and for his guidance and encouragement in getting it started. We also thank the people at De Gruyter, especially Kristin Berber-Nerlinger for supporting our vision for the book and Ria Sengbusch and Dorothea Wunderling for patiently guiding us through the preparation of the manuscript. We also thank David Frank for reviewing early versions and providing helpful feedback. We are also grateful to Tharani Ramachandran and the typesetting team at Integra Software Services for their careful attention to detail in finalizing the look and presentation of the book's text and figures. Last but not least, we thank the staff and industry representatives at Fractionation Research, Inc., members of our local Mid-Michigan AIChE Section, presenters at the Mid-Michigan AIChE Seminar Series, and the many presenters at AIChE Separations Division conference sessions for generously sharing their expertise over the years.

https://doi.org/10.1515/9783110695052-203

Preface

This book deals with chemical separation from an industrial perspective, describing the designer's work process (the engineering method) as well as the inner workings of popular technologies – distillation, extraction, adsorption, crystallization, and the use of membranes. The book is intended as a supplement to the many excellent references and textbooks already available. We have tried to provide a complementary companion book – one that describes key engineering concepts and practical approaches to design with less emphasis on equations. We hope it adds helpful insights for practitioners and students, as well as clear explanations for anyone interested in the subject.

We discuss certain aspects of chemical separation not emphasized in the textbooks, such as how the various technologies have developed over the years (the historical perspective), how specific molecular interactions affect phase equilibrium, and when to use classical theoretical-stage or transfer-unit models. Other aspects we discuss are indeed covered in detail in the textbooks, and for these we offer more practice-oriented explanations, often referring the reader to the textbooks for derivations and solution methods. The result is a conversational-style survey of popular separation methods, setting the stage for more in-depth discussions available elsewhere.

The book has three main parts. Part I introduces chemical separation by discussing general characteristics, by reviewing the history of industrial development, and by describing the engineering method – to show how chemical separation is an integral part of the engineering tradition. Part II then deals with key factors affecting separation performance – such things as phase equilibria, the material balance, mass transport (driving forces and kinetic resistances), effective ways of generating interfacial area, trace chemistry effects, the effect of residence time, and the use of energy. We explain how these factors may be taken into account for practical design purposes (the design fundamentals).

Part III explains key features of various commercial technologies. We refer to the fundamentals introduced in Part II without delving too deeply into specific design calculations. Instead, we describe various process schemes and types of equipment and offer general guidelines regarding overall approaches to design. Many references already cover calculations in detail, and many commercially available software programs may be used for process simulation. But designers need to have a specific process scheme and type of equipment already in mind, and that is the focus of our discussion. We end Part III by discussing how individual operations may be simplified or intensified as needed, and how two or more operations may be advantageously combined into a hybrid process scheme.

In organizing the book's content into these three parts we sometimes repeat ourselves. But we feel this serves to emphasize important points worth repeating in different contexts. And readers can go directly to selected sections of interest, taking something from them without necessarily reading all preceding sections. Of course, much of what we say comes from the published literature, seasoned by our experience working with

https://doi.org/10.1515/9783110695052-204

many talented engineers and chemists. We have listed key background publications and recommendations for further reading in bibliographies following each part of the book. In the text, we cite references when discussing specific material, but not necessarily when discussing well-known material or material already covered by the bibliography publications. Additional information may be obtained by searching the internet (being careful to consider the source). With our book, we hope to pass along useful tips and spread our enthusiasm for this important discipline. Naturally we haven't seen it all, so we encourage readers to consider other approaches to the subject as well and to reflect on their own experiences and insights. We hope our discussion provides helpful background and perspective, inspiring readers to view chemical separation engineering as a valued art and craft based on key fundamentals – as in the art and craft of making music – something we can all appreciate.

July 24, 2023   Tim Frank   Bruce Holden

                           seriousseparationsllc@gmail.com   bsh0@hotmail.com

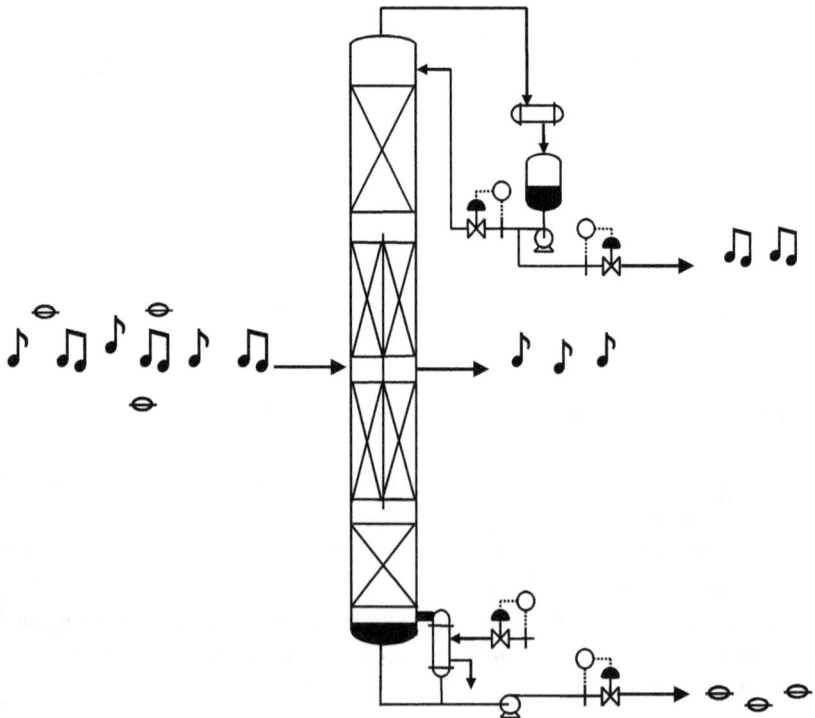

Contents

Part I: What It Is and Why It Matters

Part II: Design Fundamentals

Part III: **Engineering Practice**

List of Symbols

The following are the main symbols used in the text. They often have multiple meanings. The appropriate meaning should be apparent from the context. Specialized symbol notations and units may be found near or directly below the equations that employ them.

Latin Alphabet

a	interfacial area per unit volume
a_i^I	activity of component i in phase
$a_{i,j}$	known reference point, $a_{i,j} = \ln \gamma_{i,j}^{\infty}$ at $T = T_{ref}$
A	$L/(mV)$, solute transfer factor for absorption, called the absorption factor
A_x	column cross-sectional area, m^2
B	bottoms flow rate
B	nucleation or birth rate, crystal nuclei born per unit time per unit volume
B	birth rate for a specific crystal size range, per unit volume, due to nucleation (at small sizes) and agglomeration (at large sizes)
BV	empty bed volumes of feed that have passed through an adsorber vessel
$BVTB$	bed volumes to breakthrough
$BVTB_{50}$	mean bed volumes to breakthrough (mean point of breakthrough curve)
c_i	moles per unit volume
C_{in}	solute concentration at the inlet to the bed
C_i	concentration of component i in the fluid phase
C	solute concentration (moles per unit volume)
C^*	average concentration driving force (deviation from equilibrium)
C_i^{eq}	equilibrium concentration of component i in the fluid phase
C_s	capacity factor, m/s
$(CA)_{in}$	concentration of solute A at the inlet to the adsorber bed
C_{pL}, C_{pV}	specific heat of liquid and vapor, respectively
D_2	recycle stream in a PSD process
D	distillate flow rate
D	adsorber bed diameter or column diameter
D	crystal death rate for a specific crystal size range, per unit volume, due to agglomeration (at small sizes) or breakage (at large sizes)
ε	extraction factor
E	efficiency factor
E_O	fractional overall column or overall stage efficiency
E_{pV}	fractional vapor-phase point efficiency
E_{MV}	fractional Murphree vapor-phase tray efficiency
E_{ML}	fractional Murphree liquid-phase efficiency
f_i^o	vapor pressure (fugacity) of a hypothetical subcooled liquid obtained (fugacity)
F	feed to a PSD process
F	feed-phase flow rate, mass or moles per unit time
F_R	solute reduction factor
F_s	F factor, a simplified capacity factor for the vapor capacity of a distillation column

https://doi.org/10.1515/9783110695052-206

$F_{t,inv}$	inverse t distribution function (statistical function)
$F_{backpurge}$	portion of feed used as backpurge (mole fraction) in PSA process
FOA	fractional open area in tray bubbling area
FPL	tray flow path length, m
g_i	partial molar excess Gibbs energy of mixing, J/mole
G	Gibbs free energy
G	superficial gas mass rate, kg/h-m^2
G	growth rate per unit volume of crystallizer (which may be a function of crystal size)
G	growth rate, crystal length per unit time per unit volume
h_i	partial molar excess enthalpy of mixing, J/mole
h_{weir}	tray outlet weir height, mm
H	enthalpy
H_V	molar enthalpy of vapor
H_L	molar enthalpy of liquid
H_i	Henry's law constant for component i
H'	Henry's law constant, with units of inverse pressure
$\Delta_{fus}H$	enthalpy of fusion for component i, J/mole
HETP	height of packing equivalent to a theoretical plate (or stage)
HETS	height equivalent to a theoretical stage
HTU	height of packing equivalent to a transfer unit
I_i	van't Hoff constant for component i accounting for non-ideal osmotic pressure
j_i	volumetric flux, cm^3 liquid/cm^2 membrane · s
j_i	volumetric flux, cm^3 (STP) of component i/cm^2 polymer · s
k	mass transfer coefficient = first-order rate constant
$k_{second-phase-side}$	individual mass transfer coefficient for film at second-phase side of interface
$k_{feed-side}$	individual mass transfer coefficient for film at feed-phase side of interface
ka	mass transfer coefficient including interfacial area
k_a	linear adsorption rate constant
k_S	Setschenow constant (liter/g-mol)
K_i	partition ratio for solute i distributed between two phases, $K_i = y_i/x_i$
L	bed length
L	length of a particle or a key crystal dimension
L_{total}	total length of adsorber
L_{dist}	length of feed distribution section of adsorber
L_{sat}	length of adsorber saturated with adsorbate from feed (after feed distribution section)
L_{MTZ}	length of adsorber mass transfer zone (following saturated bed)
m_i	slope of equilibrium line, $y_i = m_i x_i + b$
m_{ave}	average slope of equilibrium line (geometric mean)
m_{low}	low value of the slope of the equilibrium line (in the region of interest)
m_{high}	high value of the slope of the equilibrium line (in the region of interest)
M^I	moles or mass of the feed phase (or molar or mass flow rate)
M^{II}	moles or mass of the second phase (or flow rate)
M	mass or mass flow rate
M_{feed}	mass of the initial feed batch
$\dot{M}$	evaporation rate, mass per unit time
$\dot{M}$	mass flow rate
MUC	maximum useable capacity, of distillation column packings or trays
n	number of moles
n_i	moles of component i

n	number density function, volume basis (for a range of crystal sizes)
n	number of components (used in the Underwood equation)
N_{stages}	number of adsorption stages
N	number of theoretical stages
N_{min}	minimum number of theoretical stages
N_T	number of theoretical stages required to achieve the desired separation
N_A	number of actual physical trays required in the column
N_B	fractional theoretical stage required at column bottom
N_B	fractional theoretical stage required at column bottom (the last stage or partial stage)
N_G	number of gas transfer units based on local mass transfer coefficients
N_L	number of liquid transfer units based on local mass transfer coefficients
N_{OL}	number of overall liquid-phase transfer units
N_{OG}	number of overall gas-phase transfer units
NTU	number of transfer units (overall)
NTS	number of theoretical stages
N_{TRAY}/N	number of actual trays per theoretical stage
N_{TRAY}/NTU	number of actual trays per transfer unit
p_{i_0}	partial pressure of component i on the feed side of the membrane
p_{i_l}	partial pressure of component i on the permeate side of the membrane
p_i^{sat}	pure component vapor pressure of component i
P_{total}	total pressure
P_{feed}	operating pressure, feed side, atm
P_{high}	adsorption pressure (absolute)
P_{low}	regeneration pressure (absolute)
$P_{permeate}$	operating pressure, permeate side, atm
P_i^{pervap}	permeance of component i in a pervaporation membrane
P_i^{gas}	effective gas permeance, a lumped transport parameter
ΔP	transmembrane pressure drop (feed side to permeate side)
q_i	adsorption working capacity (grams solute adsorbed per gram of adsorbent)
Δq_i	working adsorption capacity of the adsorbent
q_{ref}	adsorption capacity at $(C_A)_{in}$
q	mole fraction of feed that is saturated liquid, called feed quality, $q = 1$ for a saturated liquid feed
q_{slope}	slope of the feed line (called the q-line) in a McCabe-Thiele graphical calculation
Q	energy flow rate or volumetric fluid flow rate
Q_{reb}	energy input through the reboiler (reboiler duty)
R	universal gas constant, 8.314 J/mole · K
R_m	mass transfer resistance to flow due to membrane material properties and pore structure
R_d	mass transfer resistance due to fouling of the membrane (increasing over time)
R_E	external reflux ratio defined as the ratio of the molar reflux rate to the molar distillate rate
$R_{E,min}$	minimum external reflux ratio
R_{opt}/R_{min}	ratio of optimum to minimum reflux ratio
R_B	boilup ratio
s_i	partial molar excess entropy of mixing, J/mole · K
S	entropy
$\Delta_{fus}S$	entropy of fusion for component i, J/mole/K
S	second-phase flow rate, mass or moles per unit time
S	stripping factor $= mV/L$, also called lambda
t	time or residence time or elapsed time

t_{batch}	batch elapsed time
Δt	total elapsed time or on-line time
t^{**}	characteristic transfer time, the reciprocal of ka
T	absolute temperature, K
T_{ref}	reference temperature
T_D	distillate temperature (near boiling point of light key)
T_B	bottoms temperature (near boiling point of heavy key)
T_{hot}	reboiler heat transfer fluid temperature
T_{cold}	overheads condenser coolant temperature
T_{ref}	reference temperature (Kelvin)
T_m	melting point (or freezing point) temperature
T^{AZ}	azeotropic temperature
T	solute transfer factor: a general term for stripping factor, absorption factor, and desorption factor
u_s	velocity at the inlet to the adsorber bed (superficial velocity or face velocity)
v	average velocity of the feed phase through the column
$v(t)$	bulk fluid velocity
v_S	gas velocity, m/s
V_{vessel}	volume of separation vessel
V	vapor rate leaving the reboiler
V_i	molar volume of pure component i, mL/mole
$\bar{V}$	molar volume of the mixture, mL/mole
V_{actual}	actual volumetric flow rate of gas
V_{ads}	feed volumetric flow rate during adsorption (actual at P_{high})
V_{regen}	backpurge volumetric flow rate during regen (actual at P_{low})
x	concentration of solute in liquid or in a first phase, mole fraction
x_i^{liquid}	mole fraction of component i in the liquid
x_{FA}	mole fraction of the light key component in the feed
$x_{i,F}$	mole fraction component i in feed stream
$x_{i,D}$	mole fraction component i in distillate
x_i^{sat}	solubility of component i in the liquid solution, at saturation
x_i^{AZ}	azeotropic liquid composition of component i in the total liquid including two liquid phases if present
x^*	liquid mole fraction in equilibrium with the vapor
Δx_i	change in composition of component i
X_i^F	mass fraction of component i
y	concentration of solute in vapor or in a second phase, mole fraction
y	concentration of solute in a second phase, mole fraction (or other units)
y_{i_0}	mole fraction of component i on the feed side of the membrane
y_{i_l}	pressure of component i on the permeate side of the membrane
y_i	mole fraction of component i in the vapor phase
y_i^{AZ}	azeotropic vapor composition
y^*	mole fraction of vapor in equilibrium with the liquid
z	total height, or length of a column, or bed length
z	membrane thickness, cm
z	position along the length of the adsorber

Greek Alphabet

α	constant (or the mean) relative volatility of light key to heavy key
$\alpha_{i,j}$	equilibrium-based separation factor for component i with respect to component j
$\alpha_{i,j}\|_{non-equilibrium}$	separation factor given by ratio of mass transfer rates
γ_i	activity coefficient of component i dissolved in a liquid or solid solution
$\gamma_{i,j}$	activity coefficient for component i dissolved in j
γ_i^{sat}	activity coefficient for component i in saturated solution
$\gamma_{i,j}^{\infty}$	activity coefficient of solute i dissolved in solvent j in the limit of infinite dilution
γ_i^{AZ}	activity coefficient of component i at azeotropic conditions of composition and temperature
$\overline{\delta}$	Hansen sol'y parameter, molar volume average, MPa$^{1/2}$
$^i\delta$	Hansen sol'y parameter, for component i, MPa$^{1/2}$
ε_{void}	void fraction (fraction of packed bed volume occupied by empty space)
ε_b	adsorbent bed void fraction
ε_{actual}	actual energy efficiency of an individual separator
ε_{actual}	actual overall energy efficiency of the overall process
ε_{max}	theoretical maximum thermodynamic energy efficiency
$\theta_{i,j}$	empirical constant determined by fitting temperature-dependent data
θ	corrected time (time minus time for liquid to pass through empty bed volume)
θ	Underwood constant
λ	molar latent heat of vaporization
μ_i^I	chemical potential of component i in phase I
μ_G	gas viscosity
μ	liquid viscosity
Π	osmotic pressure
$\Delta\Pi_i$	osmotic pressure difference (permeate side – feed side)
ρ_L	density of liquid
ρ_G	density of gas or vapor
ρ_{bed}	bulk density of packed adsorbent bed, mass per volume of bed
σ	a measure of the supersaturation level
σ	standard deviation for spreading of the curve about the mean value $BVTB_{50}$
τ	average residence time
$\o$	fraction of feed liquid evaporated
φ_{in}	inlet stream number density per unit time, for continuous operation. Equal to zero for batch operation
φ_{out}	outlet stream number density per unit time, for continuous operation. Equal to zero for batch operation
Φ_i	fugacity coefficient of component i in the mixture
Φ_i^{sat}	fugacity coefficient for pure component i
$\hat{\varphi}_i$	fugacity coefficient for component i in the mixture (vapor or liquid)

Part I: **What It Is and Why It Matters**

Numerous books deal with chemical technology and many discuss chemical separation. But few describe the art and craft of this important engineering discipline. The engineering textbooks focus on calculations, which are essential, of course. But there is much more to envisioning an efficient process scheme and devising a balanced design. It is quite a journey that requires knowledge of fundamentals and practical design methods plus a creative vision of an overall process scheme that can meet all the requirements of a given application. It is a journey aided by familiarity with a variety of commercially successful separations, their historical development, and current challenges and opportunities. This book is about that journey and why it matters. It is written especially for engineers and chemists with current or future responsibility for some aspect of chemical separation on a commercial scale – in its development, design, operation, or improvement. The art is in the beauty of an elegant and efficient process. The craft is in the skill with which it is made and used to deliver value.

https://doi.org/10.1515/9783110695052-001

1 Importance and Overall Description

Chemical separations are necessary steps in the production of all kinds of products needed by society – for clean drinking water, food, medicine, clothing, housing, transportation, communications, and more. You name it, and chemical separation is almost guaranteed to be an essential step in the production process. For many years, the chemical and bioprocessing industries have been instrumental not only in the manufacture and purification of specific chemical products (fuels, medicines, fertilizers, pesticides, and so on), but also in the manufacture and purification of all the chemicals that go into essentially all manufactured goods and formulated products. Along the way, tremendous advances have been made not only in the process technology but also in how the various ingredients are formulated, with the development of new solvents and solvent blends and new chemical additives such as emulsifiers, antimicrobials, and antioxidants for better and longer-lasting product performance. And all of this continues today. Those of us involved in separation process design and operation play an important role in this ongoing enterprise.

It is said that society's need for chemical separation derives from the simple fact that "a mixture containing a valuable substance is not necessarily a valuable mixture" (anonymous quotation). In other words, the value is not derived until the desired substance is separated and purified to the extent needed for its intended use. Gaining the value requires application of *cost-effective* separation and purification methods. The descriptor "cost-effective" is emphasized here because the cost of doing the separating is just as important a consideration as the technical capability; that is, the cost of separating and purifying a product, and indeed the total costs involved in producing it (including separation), cannot exceed some measure of affordability with respect to the value of the final product. Total cost includes all costs incurred in manufacturing, marketing, selling, using, and recycling the product or properly disposing of it. In this book we focus on technical capabilities and the design process, but keep in mind that affordability is necessary if a design is to be commercially successful and have a positive impact on society. If it costs too much, it simply won't be used. This does not mean that the least expensive separation technology, in terms of the lowest initial monetary cost, will be the best or most affordable choice in the long run. Many other considerations also must be taken into account when assessing total cost including the safety and environmental impact of the technology. The challenge we face as separation engineers is finding an approach to separation and purification that can manage all of these requirements. Indeed, the chemical and bioprocessing industries and the profession as a whole have developed numerous programs to address these requirements for chemical process development in general. Relevant publications include discussions of the importance of chemical technologies and current challenges [1–8], economic and quality considerations [9–12], safety information and best practices [13–22], and environmental protection

https://doi.org/10.1515/9783110695052-002

[20, 23–25]. Many more can be found by searching the literature and the internet. Also see the publications listed in the Bibliography for Chapter 1.

1.1 Functional Design

In this book we focus on five widely practiced chemical separation technologies that have been employed in one form or another for many years and continue to be a vital part of industrial practice – distillation, extraction, adsorption, crystallization, and membrane-based separations. Society continues to need highly qualified professionals to safely and efficiently operate these technologies in the production plant, to steadily improve them, and to develop new cost-effective applications. In general terms, the book is about mass transfer operations used to separate a mixture of substances into two or more fractions that differ in their chemical composition. These operations involve the selective transfer of some portion of the molecular components present in the feed (normally a liquid or a gas) into another phase (another liquid, gas, or solid). The phases may preexist as in liquid-liquid extraction and liquid-solid adsorption or they may be formed during the process, as in boiling a liquid feed to form a gaseous vapor (distillation) and in cooling a liquid feed to form solid particles (cooling crystallization). Or, separation may involve selective transfer of feed components through a physical barrier placed between the phases (a semi-permeable membrane). Obtaining different compositions in each phase depends on having some difference or combination of differences in the properties of the feed components such as a difference in their physical properties (as in their vapor pressure, density, or molecular size) and/ or a difference in their chemical properties (as in their polarity or hydrophilic/hydrophobic character). A final separation may then be achieved by the physical isolation of each phase.

We are primarily concerned with the functional design of a separation process; that is, devising a process scheme that can meet the requirements of a given application. For a given separation method, this involves identifying suitable operating conditions including temperatures, pressures, choice of material separating agent (if used), and the flow rates of the various streams (for a continuous steady-state process) or the rate of change of a given process variable or variables (for a dynamic non-steady-state process) – plus the required separation capability or separation power as in the required number of theoretical stages or transfer units. The designer's work often involves screening a variety of potential separation methods by assessing physical property differences and potential thermodynamic drivers; and, for a select number of the more attractive ones, running laboratory and miniplant or pilot-plant experiments. This is done to generate data needed to guide further understanding and to quantify (parameterize) process design models, especially with respect to phase equilibrium and mass transfer resistance. Miniplants and pilot plants are also used to test and, if feasible, to demonstrate a desired separation capability before committing additional resources to the design and implementation of a commercial process. Such experiments are most important when the physical/chemical situation is highly complex and difficult to define with sufficient accuracy using theory and calculations alone – a common situation. This fact sets chemical engineering apart from some other engineering disciplines. Edwin R. Gilliland of distillation fame put it this way: "The quantitative approach to (chemical) engineering is rarely purely empirical. However, ultimate relationships are almost always extremely complicated" [26]. This quote is from 1962, and although process simulation has made tremendous progress since that time, especially for distillation, it remains largely true today for the design of many separation processes. Experiments often provide opportunities for discovery and invention, and this can be a particularly exciting and rewarding aspect of the work. (For more discussion of this point, see the introduction to Part III, Engineering Practice).

In addition to selecting a suitable process scheme, specifying a functional design also involves selecting the type of equipment to be used at the commercial scale and the required equipment internals, which may include packings, trays, distributors, nozzles, baffles, and rotary impellers or other types of agitators, among many possibilities. This often is facilitated by the use of miniplants and/or pilot plants. The miniplant concept involves first envisioning what the full-scale equipment will be like and then scaling down this vision to a small-scale operation that allows convenient generation of data that can be used to specify the full-scale version (scaling down to scale up) [27]. This may involve operation of a single tube for a full-scale version involving many similar tubes, or it may involve operation of a small-diameter slice of a larger diameter column, mimicking the same fluid flow patterns and agitation intensity (if agitation is used, as in certain liquid-liquid extraction columns). Or, it may involve operation of a small stirred vessel in the lab, with the understanding that scale-up to a larger stirred vessel will require careful attention to large-scale mixing and heat-transfer capabilities (as in batch cooling crystallization). Larger pilot-plants are sometimes then used to test and refine the scale-up methodology and process control. Although in this book we touch on scale-up in discussing separation equipment, specific scale-up methods are covered elsewhere; for example, in the various handbooks listed at the beginning of the Bibliography for Chapter 1. And although we outline various calculation methods, control strategies, and computer simulation programs, we do not delve too deeply into these methods either. Excellent texts devoted to these aspects of design are already widely available. Instead, in this book we focus our attention on explaining the features and background of popular separation technologies and why we care (the historical perspective), describing fundamental and practical aspects of equipment design and operation (its functional design), and discussing general approaches to envisioning an effective overall process scheme.

It should be noted that the results of such mass transfer operations would be diminished if we were not able to cleanly isolate the resulting phases. That is the job of physical separation methods. Often the ability to separate and discharge individual phases is built into the main separation device, as in many distillation and extraction column designs and in all membrane-based separations. But in other cases it becomes necessary to use stand-alone physical separators. These include liquid-liquid settlers/ decanters and liquid-liquid centrifuges, mist eliminators, hydrocyclones for gas-liquid and solid-liquid separations, cyclones for gas-solid separations, gravity settlers and centrifuges for settling solid particles suspended in liquids, many types of filters for removing solid particles from gases and liquids, and various devices for classifying mixtures of solid particles. For information about these methods, see ref. [28] and other separation textbooks that include discussion of physical separators. We also recommend searching the literature, the internet, and talking with equipment suppliers.

1.2 General Characteristics

Many different chemical separation methods are in use today, each with a great variety of equipment available for their implementation. In this section, we review some general characteristics and list some examples as a way of introducing the subject. The variety of commercial operations is far too great to list all the different versions. But we hope to provide a sense of the many possibilities.

Because mass transfer involves selective transfer of components from one phase to another, the various separation processes often are categorized according to the kinds of phases present. But separation processes may be categorized according to many other differentiating features as well. Box 1.1 lists a number of these. All of this makes for a world of chemical separation that is rich in possibilities. Even so, all separations must follow the same fundamental thermodynamic and kinetic laws of nature, just applied in different ways to address differences in materials and in the goals of specific applications. This common thread is key to understanding and applying separation methods in general and to choosing an approach well-suited to a given application.

Box 1.1: Differentiating Features of Chemical Separation Methods

- **Physical states of matter (phases):** Distillation (transfer of solute from liquid to vapor or vapor to liquid), rectification (vapor → liquid), absorption (gas → liquid), stripping (liquid → vapor or liquid → gas), extraction (liquid → liquid), leaching (solid → liquid), adsorption (gas → solid or liquid → solid), chromatography (liquid → solid → liquid or gas → solid or liquid → gas), crystallization (liquid → solid, gas → solid, or solid → solid), sublimation (solid → vapor), membrane-based separation (gas → gas, liquid → liquid, liquid → vapor, liquid → gas, or gas → liquid, etc.)
- **Physical and chemical properties of molecular components and their mixtures:** These include boiling point, dew point, melting point, critical point, liquid-liquid miscibility, solubility in different types of solvents, molecular weight, molar density, molecular size or volume, liquid-air surface tension, interfacial tension (liquid-liquid, liquid-solid), azeotropic behavior and composition, eutectic behavior and composition, and so on. Chemical characteristics affecting these properties may include electrostatic interactions (dispersion, polarity, hydrogen-bonding, and π-bonds), hydrophilic or hydrophobic (lipophilic) functionality (sometimes both in the same molecule), enthalpic and/or

entropic contributions, ionic strength, pK_a for protonation/deprotonation of an organic solute as a function of pH in aqueous solution, reversible complexation capability, and reactivity. These chemical characteristics are the basis for intermolecular interactions that can significantly affect separation performance.

- **Types of equipment used for phase contacting:** This concerns how the phases are intermixed or contacted to generate large amounts of interfacial area for adequate mass transfer. This involves controlling the velocities and flow patterns of each phase within the equipment. This may involve the use of distributors, trays, and baffles or the use of mechanical agitation, as in the use of rotary impellers, reciprocating plates, or fluid-flow nozzles.
- **Phase dispersion:** This concerns which phase is the continuous or bulk phase and which phase is the dispersed phase. For example, distillation can involve dispersing liquid into a continuous vapor by spreading liquid films onto packing material in a packed column or by spraying droplets into the headspace of a vessel. Or, distillation can involve dispersing vapor into a continuous liquid by bubbling vapor through a liquid layer held on a tray. Liquid-liquid extraction is another example, where one needs to decide which liquid will be dispersed as droplets suspended in the other liquid. Sometimes a given process will be designed to switch between doing it one way and then another. This is the case for the so-called raining-bucket liquid-liquid extractor. Choices also include what form the dispersed phase should take – as smaller or larger gas bubbles, as liquid droplets or films, or as solid layers or solid particles of various sizes and shapes, and so on. Many separation methods have multiple options of this kind that can dramatically impact performance.
- **Process configuration and flow pattern:** Operating options include continuous, batch, fed-batch, countercurrent, co-current, steady state, cyclical, multi-step, with or without recycle, with or without reflux, with or without mechanical agitation, multi-column, multi-vessel, and so on. Equipment often is designed to approach an ideal flow pattern such as completely mixed flow at one extreme and plug flow at the other. Equipment may also be designed with specific zones of fluid-flow for specific purposes within a section of equipment (such as quiescent, phase-settling zones) in otherwise well-mixed or uniform flows.
- **Type of control scheme:** This depends on the type of operation, whether steady-state, dynamic, or cyclical. It may be a feed-back control scheme that relies on measurement of temperature and pressure to infer composition or it may be a more complicated feed forward control scheme using on-line instruments for real-time measurement of composition or some combination of these methods. It may involve use of pre-established (calibrated) control constants or model-based, automatic adjustment of control constants within a specified range (a self-learning algorithm). Or, it may be a pre-determined cycle involving repetition of specific operating steps in a timed sequence.
- **Residence times:** This refers to contact time requirements and limitations for both the feed phase and the product phase. Short contact times may be needed to minimize product degradation during separation, as in the recovery of penicillin from acidified fermentation broth. Long contact times may be needed to give adequate time for solute diffusion as in the devolatilization of polymer melts. The residence time requirement depends on design choices including whether the phase of interest is the continuous phase or the dispersed phase.
- **Feed characteristics:** This concerns the concentrations of desired components and unwanted impurities in the feed, the volume of feed, and whether the thermodynamic behavior of the feed mixture is ideal or non-ideal, etc.
- **Product requirements:** These include the required level of purification (crude cut, sharp cut, mixed product, single-component product, ultra-high purity, etc.), the required recovery of product from the feed (yield), and the required physical form of the final product (liquid, gas, specific type of solid crystal), etc.

–   **Undesirable physical phenomena:** This concerns characteristics that can make mass transfer difficult (high viscosity, presence of fine solids that can foul equipment, tendency to form deposits, stable emulsions, foam, mist), etc.
–   **Type of separating agent:** The separating agent may be heat (thermal energy) and/or a material separating agent such as a solvent, entrainer, adsorbent, membrane, or reagent. Additives and modifiers include co-solvents, wetting agents, emulsifiers, demulsifiers, and defoamers.
–   **Separation difficulty:** This may be measured in terms of the required product enrichment (2×, 10×, 100×, 1,000×, etc.) or the required number of theoretical stages or transfer units or the magnitude of the separation driving force (as in high or low relative volatility in distillation).
–   **Energy utilization:** This concerns energy consumption per unit of product and the energy efficiencies of the individual separator and of the overall process scheme. Energy consumption is determined by how much energy must be expended in doing the separation and by how much of that energy can be recovered from the separator's outlet streams in an overall process scheme that includes integration with other unit operations. Various methods may be used to reduce energy consumption including efforts to approach the maximum thermodynamic energy efficiency of the separator (which varies for different types of separators and different applications) and cross-exchange of energy with other unit operations.
–   **Trace chemistry/reactivity:** This concerns the potential for trace reaction during separation with possible reduction in product purity (by in situ formation of impurities) or detrimental impact on operations through formation of fouling deposits, foam, stable emulsions, plugging of equipment, etc.
–   **Possible combinations of separation methods and unit operations:** A single separation operation often is combined with other operations (sometimes including reaction) in an integrated or hybrid process scheme to improve overall process performance.
–   **High or low cost of separation relative to product value:** This concerns affordability and cost effectiveness. The cost of the process must be affordable relative to the value of the product.

Boxes 1.2 to 1.9 list some examples of commercial separation operations in use today, plus a few with promise for future implementation. More information about many of them is given in subsequent sections of the book. Additional information can be found by searching the internet for specific terms. Some of the process names are likely to be unfamiliar at first, but we recommend revisiting this list from time to time and adding your own examples. Keep in mind that a single separation operation rarely stands alone. Most often it must interact with other operations in a highly integrated process scheme involving other separators or reactors or whatever else may be required. So the design and operation of a given separation unit must take into account its place in the overall process.

Box 1.2: Distillation (Liquid-Vapor, Vapor-Liquid, and Gas-Liquid)

–   General function: Separate feed components by evaporation of liquid (normally) and transfer of components between liquid and vapor (or between liquid and gas).
–   Flash evaporation with partial condensation. Complete vaporization of liquid feed with partial or serial condensation of the resulting vapor, allowing collection of multiple liquid fractions (condensate products).
–   Evaporator types: forced-circulation, rising film (or long tube vertical-tube evaporator), falling film, rising-falling film, rotary agitated thin film, wiped film, multiple effect, natural convection, thermosyphon, jacketed, direct fired, batchwise, continuous, heat pump, vapor recompression, etc.

- Boiler and reboiler types: same as above.
- Alembic distillation or pot distillation. Boiling (partial vaporization) with complete or partial condensation of the resulting vapor and collection of condensate in a separate receiver. Also known as Rayleigh distillation or single-stage distillation without reflux.
- Column distillation. Column still pot. The use of reflux.
- Batch rectification. Multi-stage batch distillation with reflux.
- Batch stripping. Inverted batch distillation. Batch distillation with a stripping column.
- Middle-vessel batch distillation. Batch distillation with both rectification and stripping columns.
- Continuous middle-fed distillation column. Continuous distillation with both rectification and stripping sections. Continuous fractionation.
- Continuous stripping column. Top-fed distillation column. Stripper. Steam stripper. Air stripper.
- Continuous rectification column. Bottom-fed distillation column. Rectifier.
- Ambient-temperature steam stripper (operation under vacuum).
- Gas-liquid absorption column. Scrubber. Solvent-based physical absorption. Reactive absorption. Reaction-enhanced absorption.
- Column internals: liquid distributor, feed distributor, vapor distributor, packing support, tray, chimney tray, etc.
- Mist elimination equipment (removal of suspended liquid drops from vapor).
- Pasteurization (venting of gases or light impurities at the top of a column just above the main overheads removal point).
- Multiple feed points. Feed options: sub-cooled liquid, saturated liquid, vapor.
- Multiple side draws. Dividing-wall design for sharp separation and purification of a side-draw product.
- Representative random packings*: with names such as Raschig ring, Lessing ring, Cross Partition ring, Berl saddle, Intalox saddle, Pall ring, Hy-Pak, IMTP, Nutter ring, Cascade mini-ring, Jaeger Tri-Packs, Intalox Ultra, Raschig Super Ring, Super Ring Plus, Intalox Snowflake, etc.
- Representative structured packings* (wire gauze and corrugated sheets): with names such as Goodlow, Hyperfil, Intalox (wire gauze); Sulzer BX and Mellapak 250Y, Koch-Glitsch Flexipac HC, Montz-Pak A3, and B1 by Julius Montz GmbH, etc.
- Representative tray types* (with and without downcomers): with names such as bubble cap, sieve tray, valve tray (fixed and movable), Thormann tray, tunnel tray, dual-flow, baffle tray, chimney tray, Turbogrid, Koch-Glitsch Ultrafrac, Shell ConSep, UOP MD, etc.
- Molecular distillation. Short-path distillation. Rotary evaporator (under vacuum) with internal condenser.
- Thin-film rotary evaporator with external condenser. Non-contact rotating blade.
- Wiped-film evaporator (with internal or external condenser). Wiper in contact with heat transfer surface. (Thin-film and wiped-film are not the same).
- Thin-film evaporator with ability to handle and discharge solids
- Distillation with internal heat exchange capability, HIDiC column
- Rotating packed-bed distillation. Spinning-band (laboratory-scale) distillation. Higee distillation*.
- Multiple column configurations for multi-component separations. Distillation trains. Combinations of columns in various configurations.
- Thermally coupled distillation configurations (Petlyuk columns). The dividing wall column configuration is an example.
- Enhanced distillation. Extractive distillation using an added solvent. Azeotropic distillation using an added entrainer. Steam distillation (for low-temperature distillation of high-boiling organic oils) and pressure change distillation (taking advantage of change in azeotropic composition with temperature to break an azeotrope).

*No endorsement of any commercial product is intended or suggested. Some names are registered trademarks as indicated in Chapter 7, Section 7.4.

Box 1.3: Extraction (Liquid-Liquid and Solid-Liquid)

- General function: Separate the components of a feed liquid (or a feed composed of solid particles) by contact with a sparingly miscible or partially miscible liquid (the solvent) to induce transfer of a select portion of the feed into the solvent phase.
- Solvent extraction. Liquid-liquid and solid-liquid (leaching).
- Stripping-mode liquid-liquid extraction, also called standard extraction. For high solute recovery. Involves contacting the feed liquid with another solvent or solvent blend (the stripping solvent).
- Fractional liquid-liquid extraction. Combines stripping and washing sections. For high solute recovery and high solute purity (co-solute rejection). Employs a stripping solvent and a wash solvent, one of which may be the feed solvent.
- Fractional extraction with reflux. May employ two added solvents or only one added solvent.
- pH-change extraction. Used to recover and purify weak organic acids or bases by transfer between aqueous and organic solvent phases. Forward and back extraction.
- Single-stage batch extraction (liquid-liquid).
- Continuously fed column extraction (continuous liquid-liquid extraction).
- Spray column extractor. Merichem suspended-fiber contactor*.
- Static column extractors: packed column, sieve-plate column, baffle-tray column.
- Packing types: similar to distillation packing (often available from the same sources).
- Mechanically agitated column extractors*: Scheibel rotating impeller, Karr reciprocating-plate, Kühni rotating impeller, rotating-disc column, pulsed-liquid column, raining-bucket horizontal column, etc.
- Representative mixer-settler box-type equipment for multi-stage countercurrent cascades*: IMI, Kemira, Davy CMS, Lurgi, etc.
- Representative centrifugal extractors*: Podbielniak, CINC, Centrek, Quadronics (Liquid Dynamics), Luwesta (Centriwesta), de Laval, Rousselet-Robatel LX, etc.
- Solid-liquid extraction. Leaching. Multi-vessel countercurrent extraction batteries.
- Solvents: organic solvents including alcohols, esters, ethers, ethylene glycols, propylene glycols, glycol ethers (combining alcohol and ether functionality), aromatics, aliphatics, chlorinated hydrocarbons, etc.
- Aqueous solvents including weak acids, bases, mixtures with polar organic solvents (alcohols, glycol ethers, etc.)
- Mixed solvents (completely miscible solvent blends)
- Supercritical fluids: CO_2, propane, . . . plus additives: methanol, etc.
- Specialty fluids: ionic liquids, deep eutectic solvents, etc.
- Gas-expanded fluids for switchable extraction (between polar and non-polar states).

*No endorsement of any commercial product is intended or suggested. Some names are registered trademarks.

Box 1.4: Adsorption (Liquid-Solid, Gas-Solid, and Vapor-Solid)

- General function: Separate the components of a feed (liquid or gas) by contact with a bed of solid particles to induce selective adsorption of a portion of the feed onto the solid surface area.
- Multi-bed, cyclical adsorption process.
- Temperature swing adsorption (TSA). Adsorption followed by thermal regeneration of the adsorbent.
- Lead-lag TSA adsorber configuration (two columns in series).
- Lead-trim TSA adsorber configuration (three columns in series)
- Pressure swing adsorption (PSA). Adsorption with regeneration using sweep gas at reduced pressure.
- Solvent change desorption/regeneration (liquid phase).
- pH swing adsorption/desorption (liquid phase)
- Countercurrent moving-bed column. Fluidized-bed column or fluidized bed concentrator.

- Continuous countercurrent tray-type adsorber
- Adsorption/desorption/diffusion – with elusion of solute in a mobile phase (chromatography)
- Single-column preparative chromatography.
- Single-column, multiple-pulse-injection, multi-product chromatography, with recycle, with peak-shaving.
- Sequential, multi-column, multi-product chromatography.
- Continuous rotating-annular-bed adsorption/chromatography.
- Simulated moving bed (SMB) chromatography. Binary product chromatography.
- Steady-state recycle chromatography.
- Affinity adsorption/chromatography.
- Adsorbent media: activated carbon, carbon molecular sieve, zeolites, silica, hydrated silica, alumina, activated alumina, clay, $CaCl_2$ beads, porous polymeric beads, ion exchange media (with micro, meso, and macro pores), pyrolyzed polymeric beads (carbon molecular sieve), etc.
- Specialty adsorbents: affinity (molecular recognition) media for highly selective protein adsorption, adsorbents for removing trace metals (zirconium phosphate, thio-functionalized polyacrylonitrile, etc.).
- Chromatography methods: gas adsorption chromatography, two-phase partition chromatography, ion exchange, gel permeation, affinity, thin film, reverse-phase, supercritical-fluid, temperature gradient, mobile-phase composition gradient, hydrophobic interaction, mixed-mode interaction, etc.

Box 1.5: Crystallization (Liquid-Solid and Gas-Solid)

- General function: Separate the components of a feed liquid (or sometimes a feed gas) by altering process conditions to induce a select portion of the feed to come out of solution as a solid phase.
- Batch, fed-batch, and continuous crystallization.
- Cooling crystallization.
- Evaporative crystallization. Solvent vaporization. Evaporative cooling.
- Anti-solvent crystallization. Drowning-out crystallization. Batchwise with continuous anti-solvent addition, and continuously fed.
- Continuous multi-vessel crystallization. Multiple vessels operated in series.
- Countercurrent cascade with partial melting and recrystallization
- Reactive crystallization. Reaction-enhanced crystallization. Adductive crystallization. Reactive precipitation. Batch, fed batch, and continuous.
- Crystallization of eutectic systems versus solid-solution systems. Crystallization of a eutectic system can produce pure component crystals in a single stage of crystallization. Solid solutions require multiple stages to produce pure crystals. Some eutectic mixtures exhibit partial solid-solution behavior which limits crystal purity.
- Continuous, multi-vessel, multiple-effect evaporative crystallization.
- Forced-circulation evaporative or cooling crystallization. With large axial-flow pumps (an impeller in a pipe).
- Continuous crystallization with clear liquor advance.
- Extractive crystallization (used to bypass eutectics)
- Eutectic freeze crystallization of brine. Produces both ice and salt crystals by operating at the eutectic point and taking advantage of differences in crystal density to separate them (by sink/float).
- Deposition crystallization (from the vapor phase). Sublimation of feed solid followed by deposition onto a solid surface.
- Preferential crystallization with dual coupled crystallizers. Dual-vessel eutectic crystallization.
- Melt crystallization. Crystallization from the melt. Solventless crystallization. Ultra-high-purity crystallization.
- Static melt crystallizer. Box-type crystallizer with internal cooling fins. Solventless batch crystallizer.
- Forced-circulation thin-layer melt crystallizer. Forced-circulation, solventless, batch crystallizer.

- Suspension melt crystallizer for continuous, single-stage crystallization.
- Brodie purifier for differential, countercurrent melt crystallization.
- Distillative freezing, combining distillation and crystallization in a single unit.
- Zone melting or zone refining, involving purification of a rod of material by gradual movement of impurities in a moving melt zone.
- Template-assisted crystallization. Addition of template materials that catalyze nucleation of desired crystals.

Box 1.6: Membranes (Selective Permeation through a Semi-Permeable Barrier)

- General function: Separate the components of a feed liquid or gas by permeation of select components through a semi-permeable membrane. Avoids any need to mix and later separate the feed and product phases.
- Microfiltration. Small particle filtration.
- Ultrafiltration. Large molecule filtration.
- Nanofiltration. Small molecule filtration.
- Reverse osmosis involving application of pressure to force permeation of small molecules, usually water out of brine. Desalination membrane.
- Preferential gas permeation. Gas separation membrane.
- Pervaporation. Membrane-based liquid-to-vapor separation. Membrane-enhanced distillation.
- Organic solvent nanofiltration.
- Forward osmosis employing a draw solution.
- Tangential flow membrane configuration. Cross-flow.
- Dead-end flow. Dead-end filtration.
- Tubular membranes. Monolith membrane module. Shell-and-tube module. Hollow-fiber module.
- Spiral-wound membrane module.
- Plate-and-frame membrane module.
- Liquid-liquid extraction membrane. Membrane-based extraction. Supported meniscus membrane.
- Supported liquid membrane.
- Facilitated transport membrane. Reversible reaction membrane.
- Membrane materials: skin-type cellulose acetate, supported thin-film polyamide, ceramic, etc.
- Pressure recovery equipment called pressure-exchangers, to recover pressure from the retentate stream to help drive pumps for other membrane stages.

Box 1.7: Some Other Types of Separations

- Dispersed droplet (liquid-liquid) coalescers: deep-bed coalescer (packed bed) or coalescing filter with porous media, granular media, fiber media. Centrifugal-type coalescers.
- Decantation. Liquid-liquid phase separation. Settling of two sparingly miscible or partially miscible liquids by density difference, under the influence of gravity or centrifugal force.
- Sedimentation, clarification, thickening. Settling of solid particles suspended in a liquid under the influence of gravity or centrifugal force.
- Air-bubble-assisted flotation. Mineral ore cleaning and recovery. Use of small air bubbles to aid flotation of solid particles to form a top layer floating above a feed liquid, which can then be removed by mechanical skimming.
- Electrophoresis. Electrostatics. Induce the selective movement of charged particles or molecules under the influence of an electric field.

- Flocculation. Addition of a chemical coagulant to induce precipitation and settling of fine solid particles (as floc or flake).
- Sublimation (solid to gas). Inducing a portion of a solid feed to vaporize without going through a liquid phase.
- Enantioselective reaction-enhanced purification of chiral compounds (via adsorption, chromatography, liquid-liquid extraction, or crystallization involving use of enantioselective media or chiral reagent).
- Enantioselective crystallization of eutectic enantiomers for purification of chiral compounds.

Box 1.8: Mechanical Solid-Liquid Filtration

- Pressure, gravity, or vacuum filtration. Vertical and horizontal filter configurations. Static and mechanically assisted designs.
- Filter types: plate and frame, filter press, pressure vessel filter, Nutsche filter, leaf filter, drum filter, cartridge filter, belt filter, candle filter, filter-dryer, small-particle membrane filter (microfiltration), etc.
- Cake discharge types. Manual cake removal requiring the opening of equipment. Cake discharge accomplished without opening equipment: pressure-pulse, centrifugal, or gravity-assisted discharge.
- Plate-stack column-filter with rotating (centrifugal) discharge of filter cake.
- Centrifugal filters, basket centrifuge, and pusher centrifuge (optional wash sprays in all types).
- Filter aid added to feed to minimize cake compression/cracking.
- Multi-layer deep-bed sand filter (layers of course, medium, and fine media). With back-pulse regeneration of media (pulsed-bed sand filter).
- Multi-screen particle classification.
- Sedimentation (gravity settling of solids)
- Tangential, spinning flow, and cyclonic separation: cyclone (gas-solid) and hydrocyclone (liquid-solid)

Box 1.9: Select Bioseparations

- Broth extraction. Forward and back extraction. pH-swing extraction. Aqueous to organic solvent, followed by organic to aqueous.
- Aqueous two-phase extraction. Biphasic extraction. Protein separation.
- Micellar extraction/filtration.
- Affinity chromatography. SMB chromatography.
- Anti-solvent precipitation/crystallization.
- Ultrafiltration. Nanofiltration. Dialysis. Electrodialysis.
- Membrane-based washing of proteins (diafiltration).
- Centrifugation
- Bio-designed separation. Inducing cells and plants to produce desired products within specific parts of the organism to simplify recovery and purification.
- Aqueous protein concentration via selective extraction of water
- Precipitation. Sedimentation. Flotation.

References

[1]   National Research Council (USA), *Critical Technologies: The Role of Chemistry and Chemical Engineering*, The National Academies Press, Washington D.C., 1992.

[2]   National Research Council (USA), *Linking Science and Technology to Society's Environmental Goals*, The National Academies Press, Washington D.C., 1996.

[3]   National Research Council (USA), *Separation Technologies for the Industries of the Future*, The National Academies Press, Washington D.C., 1998.

[4]   National Research Council (USA), *Grand Challenges in Environmental Sciences*, The National Academies Press, Washington D.C., 2001.

[5]   M. S. Wrighton (Chair), *The Importance of Chemical Research to the U.S. Economy*, National Academies Press, Washington, D.C., July, 21 2022.

[6]   National Research Council, *America's Climate Choices: Panel on Advancing the Science of Climate Change*, The National Academies Press, Washington D.C., 2010.

[7]   E. Petruzzelli, Editor-in-Chief, "Thinking about Climate," *Chemical Engineering Progress (Bonus Issue)*, December 2020.

[8]   E. Petruzzelli, Editor-in-Chief, "Accelerating Decarbonization," *Chemical Engineering Progress (Special Issue)*, September 2022.

[9]   G. D. Ulrich and P. T. Vasudevan, *Chemical Engineering Process Design and Economics: A Practical Guide*, 2nd ed., Process Publishing, Durham, NH, 2004.

[10] W. M. Vatavuk, "Updating the CE Plant Cost Index," *Chemical Engineering*, vol. 109, no. 1, pp. 62–70, January 2002.

[11] A. Wechsung, C. Buehler, J. James, A. M. Morrison and M. Stern, "Carbonomics: Introduction to Carbon Pricing, Regulations, and Frameworks," *Chemical Engineering Progress*, vol. 118, no. 9, pp. 23–29, 2022.

[12] S. E. Burke and R. T. Silvestrini, *The Certified Quality Engineer Handbook*, 4th ed., ASQ Quality Press, Milwaukee, WI, 2018.

[13] J. E. Olsen, "Conduct an Effective Pre-Startup Safety Review," *Chemical Engineering Progress*, vol. 118, no. 6, pp. 48–56, 2022.

[14] Dow Lab Safety Academy, Dow, Inc., [Online]. Available: https://corporate.dow.com/en-us/science-and-sustainability/innovation/safety-at-dow.html. [Accessed 24 Jun 2022].

[15] N. Hyatt, *Guidelines for Process Hazards Analysis, Hazards Identification, and Risk Analysis*, CRC Press, Boca Raton Florida, 2003.

[16] "Introduction to Hazard Identification and Risk Analysis," Center for Chemical Process Safety (AIChE), [Online]. Available: https://www.aiche.org/ccps/introduction-hazard-identification-and-risk-analysis#:~:text=Hazard%20Identification%20and%20Risk%20Analysis%20(HIRA)%20is%20a%20collective%20term,are%20consistently%20controlled%20within%20the. [Accessed 9 November 2022].

[17] "Globally Harmonized System of Classification and Labelling of Chemicals (GHS Rev. 9)," United Nations, 2021. [Online]. Available: https://unece.org/transport/standards/transport/dangerous-goods/ghs-rev9-2021. [Accessed 9 November 2022].

[18] T. Parker, R. Shen, M. O'Connor, and Q. Wang, "Application of Safety Triad in Preparation for Climate Extremes Affecting the Process Industries," *Process Safety Progress*, vol. 38, no. 3, e12091, Sept 2019.

[19] G. Prpich and R. Unnerstall, "Translating Industrial Lab Safety Practices to Academia," *Chemical Engineering Progress*, vol. 118, no. 5, pp. 29–34, May 2022.

[20] G. Chen and F. Khan, Editor-in-Chief, *Process Safety and Environmental Protection (International Journal Published by the Institution of Chemical Engineers)*, all volumes.

[21] R. J. Willey and J. Murphy, Eds., *Process Safety Progress (International Journal Published by AIChE)*, all volumes.

[22] P. Urben, Ed., *Bretherick's Handbook of Reactive Chemical Hazards*, 7th ed. (Two Volume Set), Academic Press, Cambridge, Massachusetts, 2006.

[23] S. E. Hunter and R. K. Helling, "A Call for Developers to Apply Life Cycle and Market Perspectives When Assessing the Potential Environmental Impacts of Chemical Technology Projects," *Industrial & Engineering Chemistry Research*, vol. 54, no. 16, pp. 4003–4010, 2015.

[24] D. Galafassi, T. M. Daw, L. Munyi, K. Brown, C. Barnaud and I. Fazey, "Learning About Social-Ecological Tradeoffs," *Ecology and Society*, vol. 22, no. 1, 2017, https://doi.org/10.5751/ES-08920-220102.

[25] M. Abraham, Ed., *Environmental Progress & Sustainable Energy (International Journal Published by AIChE)*, all volumes.

[26] E. R. Gilliland, "Eger Vaughan Murphree 1898–1962, A Biographical Memoir," in *Biographical Memoirs*, National Academy of Sciences (USA), Washington, D.C., 1962, pp. 225–238.

[27] L. A. Robbins, "The Miniplant Concept," *Chemical Engineering Progress*, vol. 75, no. 9, pp. 45–47, 1979.

[28] A. B. De Haan, H. B. Eral and B. Schuur, *Industrial Separation Processes: Fundamentals*, De Gruyter, Berlin/Boston, 2020.

2 Scale of Operation

Scale of operation is another distinguishing characteristic. The scale at which a separation is practiced is determined in large part by the size of the market for a given product (in terms of volume) and economic factors that influence the most economical scale of manufacture. The scale of operation greatly impacts the type of equipment and process scheme chosen for a given application. In the following discussion we focus on two categories of chemical production: very large-scale production of commodity chemicals and much smaller-scale production of fine chemicals. An intermediate category of specialty chemicals has characteristics of both. We also discuss mini and micro scales of operation, a special very small-scale category with an increasing number of applications.

2.1 Commodity Chemical Separations

Very large-scale operations may produce several million metric tons of product per year in a single facility (equivalent to billions of pounds per year). The product must meet industry standard specifications as determined by chemical analysis, and often several different grades are established in the marketplace, differing in their purity or composition. Examples include fuels, solvents, basic inorganic chemicals such as acids and bases, organic building-block chemicals used in the synthesis of more complex chemicals, and high-volume chemical additives used in formulated products. At this very large scale of operation, both the cost of equipment (capital cost) and the day-to-day operating costs are critical considerations for commercial success and both drive decision-making in design. Commodity chemicals from different manufacturers generally are not differentiated in terms of their performance properties (or not greatly so), so manufacturers must compete primarily on price and availability in large quantities. In some cases, however, commodity chemicals from different manufacturers have subtle differences in their properties that affect their ultimate use even though all comply with the required industry standard in terms of the overall product assay (wt% purity). This is especially true of commodity polymers which because of differences in polymer structure may offer differentiated performance properties. Also, the impurity composition (or impurity profile) of a commodity chemical may differ from other similar products, especially when different manufacturing processes are used, and such differences, even when small, can cause dramatic differences in how a product performs in an end-user's process or formulated product. If the difference is detrimental, as in contributing an unwanted color or odor, then the product may be eliminated entirely from consideration by certain users, even when offered at a lower price. Even subtle changes of this kind due to an upset in the manufacturing process can quickly become an "all hands on deck" problem for commodity producers.

https://doi.org/10.1515/9783110695052-003

There is also a high barrier to entry into the commodity chemical market due to the very high capital investment that is required, often in the billions of dollars. A key to success is the ability to spread this cost over high production numbers to achieve low cost per unit of product; that is, designing for good economy of scale. Scaling up equipment as a single larger unit tends to reduce cost per unit of production in part because the cost of the equipment per unit volume is lower and various fixed costs related to infrastructure and labor remain about the same. The amount of metal or other material needed to construct the device (per unit volume) generally decreases as volume increases. This is reflected in the famous 0.6 rule for scale-up of commodity-scale equipment (cost generally increases with volume to the 0.6 power). This drives designers to devise larger and larger equipment to enable bigger and bigger world-scale operations. Evolutionary improvements, obtained either through greater production numbers or by improving manufacturing productivity and efficiency, can result in huge savings. For example, saving as little as a single penny per unit of product can yield $10 million in savings at large scale, as a billion units times $0.01 per unit yields $10 million. So, in this context, separation units are designed with goals of maximizing the scale of operation and managing both capital costs and operating costs to minimize the cost per unit of product. This tends to favor design of continuous processing units and simplified process schemes in dedicated manufacturing plants. Equipment is operated to just meet the product specifications and customer requirements for a given commercial grade without exceeding specifications. Any over-purification only adds extra cost that customers are not willing to pay for. Improving process control and management methods to reduce variability and unnecessary over-purification is a major opportunity for cost reduction.

So, the major separation units in a commodity chemical plant must be suited to construction at very large scale, and this favors simpler designs with minimal mechanical complexity. Continuously operated, static designs tend to be favored such as packed and trayed distillation or extraction towers. Rotating or reciprocating equipment generally are not favored because at some point their size becomes limited by sheer weight and high mechanical forces, requiring the use of multiple units operated in parallel which often adds greatly to the cost, and because they typically are more sensitive to fouling and wear, requiring greater maintenance. Complex and sophisticated equipment are not used unless they are needed to achieve a unique function or, in rare cases, they can yield significant savings over and above simpler designs. Some examples of this kind include the use of simulated moving bed chromatography (a type of multi-column adsorption process used to produce fructose-enriched corn syrup and *p*-xylene at large scale), the use of short-contact-time Podbielniak centrifugal extractors in the large-scale production of sensitive organics (such as commodity penicillin which is sensitive to the acidic pH needed for extraction), and the use of high-speed centrifugal gas compressors in conjunction with high-pressure distillation operations in ethylene plants. In each of these special cases, the more complex designs address specific difficulties encountered in the use of simpler designs.

2.2 Fine Chemical Separations

Fine chemicals are complex, high-value products produced in limited quantities. Examples include active compounds and specialty monomers used in pharmaceuticals, in agricultural pesticides, and in high-performance coatings. In this case, manufacturers often compete in the market according to their product's unique performance properties and cost-effectiveness in use, with less sensitivity to price per unit of product. Process design goals often are set by the need to carry out multi-step complex chemistry or microbiology, operations that require isolation of intermediate compounds as well as purification of one or more final products, often using the same equipment.

Fine chemical manufacturing plants most often are multi-purpose plants designed with processing flexibility in mind. They are used to manufacture a specific product with the idea that at some future date they may be modified (relatively easily) to manufacture another product. Or, sometimes they are used to manufacture several different products in different campaigns throughout the year. Another factor that can come into play is the need to minimize capital risk with the introduction of a new product. This tends to favor use of an existing multi-purpose, highly flexible, and relatively small manufacturing plant. Larger-scale, dedicated plants are implemented only after the product has proved sufficiently successful in the marketplace to support larger scale production.

For these reasons, flexibility is highly prized in the design of separation units, and this often favors use of batch processing. Furthermore, complex and expensive separation units are not uncommon such as short-path evaporation/distillation equipment (high-speed rotating evaporators) and low-throughput chromatography units containing expensive, specialty adsorbents. Investment decision-making tends to focus on the multi-use capability of processes and equipment and is not as sensitive to operating costs per unit of product (compared to commodity scale operations). This does not mean that this kind of facility can afford to be wasteful of resources. Efficiency in manufacturing is still a critical design goal as well as flexibility.

Sometimes the suitability of a given separation method is determined through performance testing of the finished product, not by whether the product composition is within a required purity specification as determined by chemical analysis. This kind of operation produces a *product by process*, meaning that the product's desirable performance attributes are acquired during manufacture using a specific processing scheme. Note that this is a practical definition of the term which may differ from the legal definition used in writing patent claims and determining patentability. Box 2.1 lists some examples. Alternative schemes are not used unless it can be proved that they yield a product with the same or better attributes. This situation can be frustrating to process designers because attempts to implement newer, more cost-effective methods may be thwarted simply because the resulting product will need to be qualified in testing by customers. Such tests can be time-consuming and expensive, so the historical technology may effectively be locked in place. For example, the process used to make photoresist chemicals for microlithography in the manufacture of electronic chips (including separation and purification operations) must be qualified in extensive tests of actual chip performance, a very expensive proposition that takes months to complete. This is also the situation in the manufacture of most active compounds for pharmaceuticals and agricultural pesticides, where regulatory rules require use of a specific process scheme because that scheme was used to produce samples for end-use testing (clinical trials or field tests) in the course of registering the product. These rules avoid inadvertent introduction of new impurities with unknown effects. So, in the beginning when registering a new product, the range of process options and operating conditions evaluated in the registration process (including those for separation and purification) should be as wide-ranging as practical considerations will allow in order to enhance manufacturing flexibility.

Box 2.1: Example Products by Process

- Specialty foods or drinks such as a distilled spirits with unique flavors and aroma
- Active ingredients with specific types and concentrations of trace contaminants (purity profile or matrix), such as active compounds used in pharmaceuticals or agricultural pesticide formulations
- Chemicals with unique performance properties used in the manufacture of electronic microchips (microlithography)
- Polymers of specific structure (crystallinity and crosslinking) yielding unique physical properties

---

2.3 Mini- and Micro-Separations

As discussed above, in the manufacture of many chemical products, achieving good economy per unit of production is advanced by operating at larger and larger scales. But some technologies are advanced by designing them to operate at smaller and smaller scales. This trend will not replace large-scale mass production of commodity chemicals, which we believe will continue to benefit from very large-scale operation for good economy of scale. Rather, very small-scale operation is reserved for special purposes where it offers an advantage or may even be a requirement. For example, small-scale separations are used in personal medical equipment such as membrane-based blood dialysis machines (hemodialysis) and portable pressure swing adsorption (PSA) machines used to concentrate oxygen from air for those needing assistance breathing. Another example of small-scale separation technology is the development and fabrication of hand-held chemical sensors and analytical chemistry devices. In some cases, the scale of operation is so small that some of the same techniques used in fabricating electronic devices may also be applied in fabricating an analytical method involving chemical separation. Such small-scale separations include thin-film or micro-channel liquid-liquid extraction and small-scale membrane devices for degassing liquids.

Somewhat larger mini-scale operation can also be beneficial for certain niche chemical processing applications where it is important to minimize the volume of chemicals used in the process. An example involves efforts to minimize inventories of toxic or explosive intermediate chemicals such as methyl isocyanate used in the production of certain carbamate pesticides. Use of continuous, small-scale reactors and separators offers advantages over traditional batch processing methods because the amount of intermediate chemical held within the equipment at any one time is kept small so that any accidental release can be contained within the facility.

Small-scale processing may also be advantageous in applications where space is severely limited such as for chemical processing on an offshore oil rig or onboard a ship. The purpose of a ship-based chemical process is to provide a mobile platform that can be taken from one source of raw material to another, rather than the other way around, and may allow for some processing during shipment. A downsized chemical process may also be advantageous for use in remote locations. An example is the mini methanol reformer which involves catalytic reaction of methanol and water at high pressure (perhaps 20 bar) to form hydrogen and byproduct CO_2 ($CH_3OH + H_2O \rightarrow 3H_2 + CO_2$). Technologies of this type may use a semi-permeable membrane or a mini PSA unit for hydrogen recovery and purification. The resulting hydrogen may then be used in a fuel cell to generate electricity at a remote location.

Small-scale processing also finds application in the manufacture of certain very high-value, very low-volume active compounds. For example, separations used to isolate and purify certain very low-volume biopharmaceuticals are performed using small, disposable containers to ensure sterile conditions. Small-scale operation is also used for high-throughput chemical testing (combinatorial experimentation) to identify most effective chemical formulations, catalysts, and operating conditions. Here, small-scale chemical separation is needed to sort through and identify the most promising candidates. Techniques include head-space gas chromatography and optical (photo or video) detection of crystal formation. Small-scale chemical processing methods may also find use in quickly testing and developing new chemical processes in the laboratory, replacing traditional batch laboratory methods. More information about such technologies may be found by searching the internet.

3 Part of a Valued Tradition

Those in our society who earn a living by contributing to the improvement of society's infrastructure and standard of living are part of an engineering tradition that has long been recognized and celebrated. In 1907, this tradition was honored by Rudyard Kipling in his poem, *The Sons of Martha*, which describes the role of workers (engineers and others) as "simple service simply given to his own kind in their common need," a sentiment inspired by the biblical story of Martha (Luke 10:38–42). Beginning in the 1920s, Kipling's poem became popular within engineering societies for its inspirational message, particularly in Canada. And today, engineers and scientists continue this tradition by playing lead roles in furthering technology and industrial practices. These efforts often go unseen by the general public, but whenever something disrupts the smooth operation of the technologies we depend upon, the importance of maintaining and further improving them quickly becomes apparent.

In the next several sections, we briefly review some of the history of technological development and industrial-scale manufacturing to show how chemical manufacturing and chemical separation fit into the bigger picture. This is an ongoing story of technological change that affects everyone. This is how engineers address society's changing wants and needs. With this brief introduction we hope to stimulate interest and provide some perspective. We encourage you, the reader, to take it from here by pursuing the literature that interests you and by discussing your thoughts and impressions with others to form your own view of how this all works and your particular role in it. Reflecting on these things can be enlightening, thought-provoking, and we think, encouraging.

Sources of information about this history are listed in the Bibliography for Chapter 3. Much information is also available on the internet. These sources include Wikipedia and the various websites of the American Chemical Society (ACS), the American Institute of Chemical Engineers (AIChE), the European Federation of Chemical Engineering (EFCE), the Institution of Chemical Engineers (IChemE), the Royal Society of Chemistry (RSC), and the European Route of Industrial Heritage. Other helpful sources of information include articles in *Chemistry World* published by the RSC, articles in *Distillations* magazine published by the Science History Institute and the Chemical Heritage Foundation (Philadelphia, Pennsylvania), and articles in *The Bridge: Linking Engineering and Society* published by the National Academy of Engineering (USA). And Phillip Wankat has written a short history with comments about the future of separations technology [1].

https://doi.org/10.1515/9783110695052-004

3.1 Historical Perspective

The engineering tradition began long ago with antecedents from many parts of the ancient world. Consider Roman aqueducts built to deliver clean water to distant cities, Greek screw-type pumps known as Archimedes' screws used to irrigate crops, Chinese canals built to enable efficient shipping of grain from farming regions to cities, and Greek waterwheels and gears used to power grain mills. Regarding chemical separation, consider ancient Egyptian distillation vessels used to obtain drinking water from salty water and perfumes from plant extracts. To be successful, all of these practices required some knowledge of engineering principles in such areas as hydraulics, mechanical levers and gears, and methods of heating and cooling.

These practices generated significant value for the communities that gave rise to them. And other communities in other places also found them to be useful as information about how they worked spread from place to place. Over time, as ideas concerning these and other useful practices spread, many of them were adapted to meet the needs of local communities, often leading to further improvements. This seems to be the case with the invention of the movable-type printing press in Europe in the fifteenth century, a technology benefiting from earlier practices in paper-making and woodblock printing in Asia. And with the printing press in place, the cost of producing a book was greatly reduced, so many more books could be produced and many more people could afford them. Best practices of all kinds could be communicated faster than ever. Over the next several centuries, this fostered the sharing of information and ideas about religion, logic, reason, mathematics, mechanics, materials, medicine, worldwide geography, astronomy, philosophy, government, individual rights, craft methods, storytelling (plays and novels), and more – a period of time known as the Age of Enlightenment. Over the years this helped stimulate new ways of thinking about human progress and daily life; or, more to our point here, how to make the tasks required of daily life easier and more productive – as in how best to make cloth for clothes or how best to plow land for growing crops. A new approach to work began to flourish: seeking changes that increase the productivity of individuals and organizations, a trend that has driven a continual increase in the general standard of living. The engineering tradition, with roots in ancient times, became a major force in society. This began a transition from a mostly agrarian and artisan lifestyle to a more highly specialized, far-reaching, and interconnected (industrial) way of life.

By the mid-eighteenth century, this movement toward industrialization accelerated into what is now commonly called the Industrial Revolution. This changed how many products were made from the making of made-to-order, hand-made items in homes and local shops, to larger-scale production of standardized products at regional factories with sales and distribution over a much wider area. Larger-scale production addressed the severe scarcity of finished goods common at that time, making useful products such as cloth and labor-saving tools more readily available and affordable, and this improved the standard of living for many. It also contributed to more productive agricultural practices and changed how many people could earn a living – away from the hard work but flexible schedule of farming toward a more rigid schedule working specialized jobs in mines, factories, and offices. This new way of working involved difficult and sometimes dangerous work, but many people found that they could have a better lifestyle by working in industry, notwithstanding the difficulties. And in the beginning, the Earth's environment was viewed by many as a resource to be used for the benefit of people, with little attention paid to controlling emissions. At that time, making a basic living was a struggle for many and protecting the environment was not often a priority. But with advances in technology over many years, the general standard of living has improved dramatically, and now working conditions and environmental protection are indeed priorities for many people in industry and in society at large. Jobs held by those responsible for the technology and its operation have developed into specific occupations that require specialized training and credentials. Simply stated, the Industrial Revolution introduced a way of working that encourages inventiveness and rewards operational efficiency, and over the years this has enabled continual improvement in productivity and in the care taken to address the needs of all stakeholders. Like all human endeavors, how the work is done and for what purpose are choices to be made according to available resources, perceived needs and priorities, continually changing standards, and value judgments. For more discussion of this point, see Chapter 4: Addressing Universal and Changing Needs.

The Industrial Revolution began in Britain and spread from there. As it happened, a number of key factors all came together in one place at the same time: the presence of skilled workers, talented inventors, material resources, and supportive institutions. Institutional support came from a developing system of commercial and contract law that assigned specific rights to persons engaged in business, often through partnerships or joint-stock companies (forerunners of modern corporations). Support also came from a developing banking system that fostered investment in new business ventures. All helped enable the financing of large industrial projects beyond what an individual inventor or business person could possibly finance on their own. Another factor was the development of patent law, which gave specific rights to patent owners for a limited period of time in exchange for sharing information about the workings of their inventions, thereby stimulating further innovation by others. Also critical to success were key material resources such as local deposits of iron ore and coal (other materials could be imported by ship) plus substantial sources of energy, mainly water power and steam made from burning coal. Having a large and dependable supply of energy was critical because operating at large scale necessarily required large amounts of energy to do the work of producing many items (or large volumes of material) with the potential to minimize the energy consumed per item.

This new, more-highly integrated system of work benefited from numerous innovations in mining technology, in metal smelters and foundries, in metallurgy, and in the use of interchangeable parts made to precise specifications using new kinds of measuring instruments and machine tools. These included micrometers, the lathe, milling machines, and large presses, punches, and forges. Large-scale operation was also enabled by the development of specialized process equipment such as pumps and compressors, reliable pressure vessels, and mixing vessels. The transition from small- to large-scale production gained additional momentum in the nineteenth century with the development of long-range transportation networks that allowed for more economical shipping of raw materials to factories and finished goods to consumers. In addition to building more and better roads, this included the implementation of new canals, locks, and dams to expand inland waterways and the introduction of the train locomotive and the building of extensive railways. The necessary steel rails were made possible by the new Bessemer process (a kind of reactive iron-purification process) for large-scale production of inexpensive steel. That same steel made possible all kinds of new and better machines, ships, and buildings. Later in the nineteenth century, electric power generation and distribution networks were introduced along with many new devices powered by electricity: early electric motors, the rechargeable lead-acid battery, the telegraph, the telephone, and electric lighting. The nineteenth century also saw the rise of the oil and gas industry, supplying lubricants for the new machines and fuels for cooking, lighting, and heating. These same fuels powered early internal combustion engines which began to replace the steam engine. This drove the initial development of oil refining technology to address the need for more highly refined fuels and lubricants. Vulcanization of natural rubber was invented and used to make waterproof clothes, inflatable rafts, and eventually, pneumatic tires. Food processing innovations included large-scale production of cereals and extruded and dried pasta. These and other developments stimulated inventors and manufacturers to continually offer consumers new and better products for the home and workplace. As a result, more people were able to buy the things that made life easier and more comfortable such as ready-to-wear clothing and shoes, furniture, timepieces, domestic appliances, tools for easier and faster construction, implements for more productive farming, and a greater variety of processed foods.

The pace of technological change accelerated even further into the twentieth century – not only in the number of new inventions but also in the steady pace of improvement in the efficiency, precision, and control of manufacturing in general, yielding continual improvement in product consistency, quality, and affordability. The early twentieth century saw great improvement in the systems of weights and measures needed for precise manufacturing including widespread application of C. E. Johansson's standard gauge blocks, establishment of exact conversions between different systems of measurement (e.g., 1 inch was defined as 25.4 mm exactly), and development of sophisticated automatic machine controls (electromechanical and electropneumatic). The first proportional-integral control schemes were developed to improve the steering mechanisms of navy ships, spreading from there to many other industries, ultimately becoming the proportional-integral-derivative (PID) control now universally applied in industry. The introduction of computers and computer controls beginning in the mid-twentieth century took this to an even higher degree of sophistication and capability. Today, almost every machine and chemical process imaginable depends in some way on electronic controls (hardware and software). With these continual advancements in technology, finished products could be designed and produced with greater consistency and precision (less variability) and offered in various standard grades at various price points, yielding far greater options for consumers to choose from. Scientific discoveries and the advancement of scientific understanding also enabled development of new products such as new medicines, new materials, and new electronic devices. Just think of the many new inventions and products that exploded onto the scene in the twentieth century, the collective result of vigorous activities by a multitude of individuals and companies, all working to compete in the marketplace (Box 3.1). This stimulated the chemical and bioprocessing industries to produce all kinds of new chemical products, also in various grades and at various price points – for manufacturers of finished goods to use in making and improving their products (Box 3.2). All of this stimulated the development of new and improved chemical separation technologies.

In summary, the Industrial Revolution set into motion an age of continual technological change toward greater productivity and efficiency enabled by:
- The engineering tradition
- Widespread communication of ideas and information
- Educational institutions
- Access to capital for financing new innovations and large projects
- Commercial and contract law establishing legal rights and responsibilities
- Patent protection in exchange for disclosure of technical innovations
- Availability of key raw materials and large sources of energy
- Introduction of new and improved materials and chemicals
- Standardization of parts and manufacturing practices
- Large-scale production for good economy of scale
- Long-range, efficient transportation and shipping networks
- Advances in precision manufacturing
- Development of automated controls
- Scientific understanding of underlying mechanisms
- Discovery of new scientific phenomena leading to new and better products.

These enabling factors continue today with the expansion of the internet and continual change in technologies of all kinds including chemical separation technologies.

Box 3.1: Finished Products Introduced in the Twentieth Century

Advanced internal combustion engines, the automobile, the mobile traction engine (tractors), machine-powered boats and ships, mechanized planting and harvesting implements for more productive agriculture, new plant varieties and cultivars, new animal breeds, air flight, skyscrapers, electric power plants, the power grid, efficient electric motors, incandescent light bulbs, air conditioning, the telephone network, radio, radar, microwave cooking, movies, high fidelity audio recordings, television, computer hardware and software, the jet engine, interstate highways, electronic vacuum tubes, solid-state transistors, analog computers, fluorescent lighting, integrated circuits, digital computers and software, microelectronic devices, fiber optics, light emitting diode lighting, synthetic polymers, reverse-osmosis membranes, advanced medicines and healthcare, genetically modified bacteria, plants, and other organisms with desired properties, high-capacity batteries, rocket engines, space flight, wireless communications, satellite communications, the internet and the world-wide web, personal computers and smart phones, high-speed information networks and search engines, social media, the global positioning system (GPS), etc.

Box 3.2: Chemical Products Introduced in the Twentieth Century

New adhesives, synthetic organic dyes, flavors, fragrances, lubricants, refrigerants, refined solvents, heat-transfer fluids, aluminum and steel alloys, composite materials, solid foams, insulation, metals, magnets, battery materials, semiconductors, bio-derived chemicals including alcohols and acetone; highly refined fuels, synthetic solvents, and petrochemicals such as isopropanol, chlorinated solvents, ethylene glycol, glycol ethers, sulfonamides, poly α-olefin and polyalkylene glycol lubricants, and alkyl benzene sulfonate detergents; organic polymers including nitrocellulose, methylcellulose, and other cellulosics, phenol-formaldehyde polymers, alkyd polyesters, dimethyl terephthalate polyesters, polyamides (nylon), polyvinylchloride, polystyrene and styrene-divinylbenzene, epoxies, polycarbonates, acrylics, polyethylene, polypropylene, polyethylene-propylene, synthetic rubber (styrene-butadiene and other chemistries), sulfonated styrene-divinylbenzene-based ion-exchange resins, cellulose acetate membranes, polysulfone membranes, composite polymeric-ceramic membranes, and more; silicone polymers, antibiotics, pharmaceuticals, nitrogen fertilizer based on N_2 fixation chemistry, agricultural chemicals, antimicrobials, biopharmaceuticals, chemically modified bio-diesel (fatty acid alkyl esters), polylactic acid biopolymer, synthetic sunscreens such as aromatic ketones and esters, plus specialty metals such as silicon and lithium.

The development of the chemical and bioprocessing industries and the chemical separations that they require in many ways resembles the history of industrial development as described above. In fact, it is an integral part of the story. Up until the eighteenth century, chemical processing methods were artisan techniques used mostly for processing water, food, drinks, leather, wood, and metal products such as iron and copper-based metals. These techniques were also used in the laboratory – in the alchemist tradition and in newer scientific studies of the properties of gases, liquids, and solids. Techniques for the mass production of chemicals emerged in the eighteenth and nineteenth centuries and their development accelerated throughout the twentieth century. This trend continues to the present day. The result is a kind of chemical assembly line in which chemical separation plays a key role, along with raw material acquisition, chemical reaction or biological action, and product formulation and distribution. The success of today's chemical manufacturing enterprise depends on many factors including access to education, economies of scale, efficient logistics, availability of reliable and affordable energy, control of emissions, and more. These factors are further described in Box 3.3.

Box 3.3: Characteristics of Modern Industrial Chemical Manufacturing and Separation

- **Education.** Industry relies on dedicated teachers in grade schools, high schools, colleges, and universities to prepare future employees, inventors, and entrepreneurs.
- **A strong economy.** Economic and market forces reward companies and individuals for devising useful products that address society's wants and needs. These include finished chemical products such as fuels, medicines, fertilizers, and pesticides as well as chemicals and materials that enable the production of finished goods of all kinds. The productivity of industrial operations often contributes to greater economic activity, so industry both benefits from and contributes to a strong economy.
- **Economies of scale.** Operation at large-scale delivers low cost per unit of production. It is normally most effective, with some exceptions as discussed in Chapter 2: Scale of Operation.
- **Efficient logistics.** A manufacturing plant's location is chosen to provide cost-effective access to key raw materials, utilities, and end-user facilities. This requires access to electricity, fuel, and water, plus access to shipping terminals and distribution networks which may include ocean shipping ports, inland waterways, railways, highways, pipelines for transporting liquids and gases, and raw material storage (such as underground salt-dome storage of oil and gas). These logistic considerations have grown to include supply chain efficiencies involving the coordination of supplier production with end-user production – optimizing the amount of raw materials and products stored in warehouses to manage both inventory costs and the ability to adjust to market changes. Managing inventories has become a major part of chemical production planning, including chemicals manufactured in limited campaigns for seasonal demand such as agricultural pesticides. Another important consideration in the supply chain is having more than one source of supply for key raw materials and intermediates for leverage in negotiating prices and in case of disruptions.
- **Reliable and economical sources of large amounts of energy.** For many years energy supplies have come mainly from coal, petroleum, natural gas, hydroelectric power, or nuclear energy, depending on local availability. Alternative sources such as wind, solar, geothermal, and biofuels are available in limited quantities and are undergoing further development. Natural gas is likely to remain a major source of energy for industrial operations for the next several decades, replacing coal and oil as a cleaner burning fuel (as well as a raw material for chemical production).
- **Efficient use of energy.** With commodity-scale production and cost-conscious market forces, the industry is driven to reduce production costs. This provides incentives to improve production efficiencies and reduce the amount of energy used per unit of production.
- **Continual innovation.** After 1920, many chemical companies set up internal R&D departments to support process design and manufacturing in the pursuit of new and better ways to compete in the marketplace. This has resulted in the development of larger and larger equipment for "world-scale" operations, plus ready exploitation of new insights into phase equilibrium, solute diffusion, molecular interactions, fluid dynamics, organic and inorganic chemistry, microbial functioning, and so on, for greater equipment efficiency and productivity, new routes to chemicals, and later on, for new ways of formulating products and catalyzing reactions.
- **Patent protection.** Advancement of chemical technology requires significant investment of resources. The patent law system has fostered such developments by giving inventors the right to exclusive use of their invention for a limited time in exchange for full disclosure. This provides time for inventors and their investors to try and make back their investment and hopefully to profit from it to allow for further investment in innovation. The first statutory patent system was introduced by the Republic of Venice in 1474, and similar patent laws were established in many countries by the late nineteenth century. Refinements introduced since then have led to the patent systems in force in many regions of the world. A current issue concerns the potential for a patent owner to avoid or delay commercial implementation while at the same time preventing others from doing so. This may in some cases serve as an impediment to progress.

- **Generation of chemical and physical property data.** Since the late nineteenth century, the generation of chemical data and the establishment of databases containing all kinds of chemical and physical property information, for pure compounds and for mixtures, has greatly improved the precision and reliability of chemical plant design. Much of this work was initially done in Europe but now is done by many institutions around the world. In 1901, for example, The National Bureau of Standards (now The National Institute of Standards and Technology) was established in the U.S. to promote standards for weights and measures and for generation of physical and chemical data needed by industry. Data generation was especially active in the period 1930 to 1970. The need to improve and expand these databases, as well as the need to accurately correlate and predict data, continues today. This need also exists for biological and bioprocessing data.
- **Highly engineered processing methods.** This involves the development of specialized equipment or "chemical processing machines" beyond laboratory methods. Chemical manufacturing is not simply the scale-up of laboratory methods and devices. Rather, engineering concepts not found in the laboratory provide much higher efficiency and productivity. Examples include continuously fed, semi-batch, and cyclical modes of operation, which allow use of smaller equipment for a given production rate. Other engineering concepts include the use of reflux and countercurrent processing for greater efficiencies in separations, selective use of centrifugal force, and operation at conditions outside of normal laboratory temperatures and pressures. The development of automated controls is another example of highly engineered processing. For chemicals with seasonal demand, another engineering approach involves design of flexible processes and equipment capable of manufacturing multiple products in separate campaigns carried out within a single facility. The development of bioprocessing technology is another example. Many of these engineering innovations have come from industrial R&D programs.
- **Continual improvement in the individual steps of mass production.** Now called unit operations, these are the individual steps in the chemical assembly line or block flow diagram. This includes optimization of fluid flow patterns within a specific piece of equipment and improvement in the distribution and intermixing (contacting) of multiple phases.
- **Integration of multiple unit operations.** For optimization of the overall process. Examples: Recycling of unreacted raw materials from a reactor through separation, recovery, and recycle back to the reactor for higher overall conversion, and development of interconnected separation operations such as distillation for primary recovery followed by adsorption for final purification, with a portion of the adsorber's reject stream recycled back to the upstream distillation operation and the remainder sent to waste treatment. Integration also enhances opportunities for cross-exchanging energy between units for greater heating and cooling efficiencies. The development of multi-functional hybrid processes, as in combining reaction with separation in a single operation, is another example.
- **Integration of multiple chemical plants.** At a single, large industrial site. This allows the use of central power plants and other shared utilities for low utility costs, central facilities for improved waste handling and management, and shared shipping facilities. It also allows for added efficiency in cross-exchanging energy and in using products or byproducts from one industry as raw material for another, as in using naphtha, ethane, and propane obtained from petroleum refining, or ethane and propane obtained from natural gas processing, as raw materials for petrochemicals.
- **Multifaceted supply chain networks.** Raw materials needed for chemical manufacturing come from many places around the world. Often a plant is located near a source of the major materials needed for production or the major source of energy such as hydroelectric or nuclear power, but normally many other raw materials and intermediate chemicals must be brought in from other places often far away. Raw material costs, transportation logistics, and the need to avoid disruptions drive decisions concerning how to ensure affordable and dependable sources of supply, and this often leads to agreements with multiple suppliers for the same materials. Having more than one

source of supply for key raw materials and intermediates allows for leverage in negotiating prices and provides alternative sourcing options in case of disruptions due to manufacturing problems or changes in business relationships.

- **Specialized job functions and training.** This provides in-depth expertise in specific core areas for greater organization and productivity of the overall enterprise. Examples: process R&D and equipment design (including separations expertise), process monitoring and control, scheduling and production management, production planning, raw material testing and qualification, product quality control, safety, reliability, equipment maintenance and repair, waste treatment and emissions control, specialized trades (operators, pipe fitters, welders, instrument technicians, laboratory technicians, electricians, machinists, and other skilled trades), supply chain management, purchasing, commercial sales and marketing (to generate orders needed to keep a facility running at full capacity), and more. No single person can possess all the necessary skills and experience (or the time needed to do it all).

- **Safe and environmentally responsible operations.** Safe working conditions, the safety of the surrounding community, and minimizing the impact of operations on the environment are major priorities for today's chemical and bioprocessing industries. Programs have been developed to focus on safety awareness and training, cooperation with fire and rescue organizations, and advancement in environmental protection and waste minimization. Safety is also a key consideration in the design of chemical products and in their application and disposal.

The characteristics of industrial chemical manufacturing and separation that we recognize today began to emerge in the eighteenth century with the production of certain basic inorganic chemicals:

- Sulfuric acid for treating metals and processing other materials (1730s).
- Lime and slaked lime (calcium oxide and calcium hydroxide) made by burning limestone and absorbing it in water for processing other materials and making building products (1750s).
- Soda ash (sodium carbonate) made from sodium chloride salt, lime, sulfuric acid, and coal for softening water and for making glass (1790s).
- Hydrochloric acid, made by absorbing HCl off-gases from soda ash manufacture, for neutralizing bases and processing other materials (ca. 1800).
- Chlorine gas (first obtained from aqueous HCl obtained from soda ash manufacture) for bleaching textiles.
- Bleaching powder (made from chlorine gas and lime) for safer bleaching of textiles (1799).

Note that one innovation seems to have led to another. The introduction of new inorganic products continued into the nineteenth century with:

- Black powder (potassium nitrate *carefully* ground together with charcoal and sulfur) as an early gun powder and explosive (ca. 1800).
- Portland cement (made with calcium silicates and aluminates and other water-incorporating minerals) for making concrete (1824).
- Calcium phosphate salts (such as dicalcium phosphate) as fertilizer for more productive farming and as a supplementary ingredient in animal feed (ca. 1860s).

–   Gold extraction and refining via cyanidation (MacArthur-Forrest process, ca. 1887).
–   Alumina from bauxite ore for use in textile manufacturing (Carl Bayer process, ca. 1890).

The nineteenth century also saw the development of electrochemical batteries and this led to electroplating of dissolved metals discovered by Luigi Brugnatelli around 1805. This, combined with the Carl Bayer process for extracting aluminum hydroxide (precursor to alumina) and the MacArthur-Forrest process for extracting gold, marked the beginning of the hydrometallurgy industry. Soon after came the development of electrolysis for production of chlorine and bromine from brine for use in chemical processing and the manufacture of specialty chemicals (around 1890). Electrochemical production of chlorine (and co-product sodium hydroxide) in particular has grown to be a major foundational industry known as the chlor-alkali industry. Film photography is another industry with its beginnings in the nineteenth century (1837). It grew with the invention of photo-sensitive emulsions containing silver halide crystals deposited on glass (ca. 1850) and later on celluloid film (ca. 1885).

The chemical industry continued its development with the production of organic compounds, employing new innovations in synthetic organic chemistry made in England, in France, in German speaking countries, in Northern and Eastern Europe, and in Russia with later contributions coming from America. New organic products included nitroglycerine for ammunition and explosives (1850s), synthetic organic dyes for better textiles (1856), and refined mineral wax (paraffin wax) for smokeless candles and semi-solid lubricants (1860s). A new synthetic medicine offering better pain and fever relief was introduced in 1899 (aspirin from the German company Bayer), marking the beginning of the modern pharmaceutical industry. These products were derived from a variety of precursor materials including animal and vegetable-derived fats and oils (used to make glycerol for nitroglycerin), coal (used to make synthetic dye), petroleum (used to make paraffin wax), and various plant materials (such as willow tree bark initially used to make acetylsalicylic acid, the active ingredient in aspirin). Coal became the source for many more organic chemicals, eventually including a wide variety of aromatics (benzene and others), anilines, pyridines, phenolics, polycyclic aromatics, heterocyclic compounds, and creosote, plus various liquid products (such as methanol and acetic acid) and fuels (coal-tar naphtha and coal gas). The coal tar industry was the major source of organic chemicals and their precursor compounds for many years into the early twentieth century. Other sources at that time, in lower volumes, included materials derived from plants such as pine tar and plant extracts, and various products derived from petroleum including caulking and adhesive materials, lubricants, wax for candles and for preserving food in jars, and fuel for lamps and heaters. The making of products from petroleum was just getting started.

The microbial bioprocessing industry also emerged in the later years of the nineteenth century with a growing awareness and understanding of microbes. The existence of microbes was known as early as the seventeenth century as reported by Robert Hooke and by Antonie van Leeuwenhoek, and the art of fermentation and the making of craft liquor was described by George Smith in 1729 [2]. But it was Louis Pasteur who, in the latter half of the nineteenth century, described the principles of microbial growth. This fostered improvements in the traditional ways of making alcohol and vinegar by understanding them as microbial fermentations of yeast, a single-celled fungus. And in the 1890s and early twentieth century, the use of microbes was expanded to improve the treatment of sewage for better sanitation in cities – exploiting the metabolism of naturally occurring bacteria. Scientists and engineers began to better understand the role that bacteria play in breaking down sewage, introducing the concept of biochemical oxygen demand in the aerobic digestion of organic waste [3]. Later, specialty microbes and bioprocesses were introduced for commercial production of certain industrial chemicals, notably acetone, butanol, and ethanol produced via bacterial fermentation beginning around 1916. This process is known as the Chaim Weizmann ABE process which utilizes *Clostridium acetobutylicum* bacteria. Around the same time, industrial production of glycerol began via a yeast fermentation process known as the Carl Neuberg process which promotes production of glycerol instead of ethanol but utilizes the same yeast used in brewing. The Weizmann and Neuberg processes marked the beginning of large-scale microbial bioprocessing for chemical manufacture. Traditional food preparation methods involving the action of yeast and bacteria and the various enzymes produced by microbes and plants were also modified and improved for large-scale processing of foods such as breads, cheeses, pickled foods, alcoholic beverages, and more. Otto Röhm of Röhm and Haas fame (known for acrylic polymers) is also known for isolating enzymes for use in detergents (1914) and in food processing (ca. 1920s). Interestingly, Eger Murphree, who would become famous for quantifying distillation tray efficiency and co-inventing fluid catalytic cracking of petroleum, began his career by studying enzymes, earning a Master's degree in 1921 titled: A Study of Enzyme Action and the Effect of Certain Antiseptics upon Malt Diastase (at the University of Kentucky).

In the 1920s, petroleum began to surpass coal and bio-feedstocks as the preferred raw material for chemical manufacture. Petrochemical processes often proved to be simpler and offered greater economy because pumping liquid crude oil out of the ground and processing it generally is simpler and less expensive than mining and processing coal or growing microbes. As a result, petrochemicals replaced many coal-based chemicals as well as bio-based ABE (but not entirely). Since the 1940s, petrochemical manufacturing has dominated the organic chemical market, providing all kinds of organic chemicals in large volumes. Many are synthesized with one or more atoms of oxygen, chlorine, bromine, nitrogen, sulfur, or other substituent atoms and functional groups. Although called the petrochemical industry because of the historical use of petroleum feedstocks, this industry depends on many other sources of raw materials as well. Natural gas is a fundamental source of carbon-based materials in addition to petroleum. Other raw materials include:

– Oxygen and nitrogen. Originally supplied as raw air, today these gases are available in pure form, produced in large volumes using cryogenic distillation of air (among other methods).
– Hydrogen from steam reforming of natural gas and subsequent water-gas shift reaction, as well as co-product hydrogen from chemical processing such as in cracking ethane to ethylene.
– Sulfuric acid and nitric acid from sulfur and nitrogen combustion products (SO_3 and NO_2).
– Phosphoric acid from treating phosphate rock (calcium phosphate and hydroxy-apatite minerals) with sulfuric acid.
– NaOH and Cl_2 made from electrolysis of sodium chloride brine mined from underground salt domes (the chlor-alkali industry).

These raw materials are used to produce the building blocks and reagents of organic chemical synthesis, which include:
– Alkanes (methane, ethane, propane, and butane)
– Alkenes (ethylene, propylene, and butadiene)
– Aromatics (benzene, toluene, and xylenes)
– Inorganic oxides (carbon monoxide and hydrogen peroxide)
– Organic oxides (ethylene oxide, propylene oxide, and phosgene ($COCl_2$))
– Methanol
– Organic phosphates (esters of phosphoric acid)
– Chlorinated organics (ethylene dichloride (EDC), vinyl chloride, chloromethane, and allyl chloride)
– Sulfur compounds (sulfonic acid and sulfonyl chloride)
– Nitrogen compounds (ammonia and nitrobenzene)
– Plus amines (trimethylamine and trimethylammonium hydroxide).

These materials provide the foundation for making all kinds of organic chemicals, from relatively simple compounds such as alcohols, acetic acid, cumene (isopropyl benzene), chlorinated hydrocarbons, and aniline, to highly complex organic molecules such as synthetic pharmaceuticals, agricultural pesticides, and polymers of various kinds. Box 3.4 lists some examples.

Box 3.4: Example Petrochemical Processes

– The first petrochemical product is reported to be isopropanol made from propylene in 1920 by Standard Oil Company of New Jersey (later Esso and now ExxonMobil) at its Linden, New Jersey manufacturing site. The alcohol was synthesized by hydrating propylene by reacting propylene with sulfuric acid to form isopropyl sulfates followed by hydrolysis to form the alcohol
– Steam-modulated thermal cracking of ethane to ethylene for polyethylene
– Propylene on purpose (catalytic cracking of propane to propylene)

- Pygas (pyrolysis gasoline) from steam cracking of ethane to ethylene which forms byproduct aromatics (BTX)
- Partial oxidation of ethylene to ethylene oxide for manufacture of ethylene glycol, polyglycols, and glycol ethers
- Partial oxidation of propylene to propylene oxide for analogous products
- Benzene to cumene (alkylation with propylene) to phenol and acetone (partial oxidation)
- Benzene to cumene (alkylation with propylene) to bisphenol-A (via condensation of acetone and phenol) to epoxy resin (via reaction with epichlorohydrin made from allyl chloride made by chlorination of propylene) and to polycarbonate resin (via reaction with phosgene)
- Benzene to cyclohexane to caprolactam and adipic acid to adiponitrile to polyamide resin (nylon)
- Catalytic partial oxidation of propylene to acrolein and then to acrylic acid for acrylic polymers
- Toluene to dinitrotoluene (nitration) to toluene diamine (hydrogenation) to toluene diisocyanate (TDI) (phosgenation), followed by reaction with polyol to make polyurethane
- Benzene to nitrobenzene (nitration) to aniline (hydrogenation) to methylenedianiline (MDA) (via reaction with formaldehyde) to methylene diphenyl diisocyanate (MDI) (phosgenation), followed by reaction with polyol to make polyurethane
- p-Xylene to terephthalic acid to polyester resin
- Chlorination of ethylene to ethylene dichloride (EDC), followed by dehydrochlorination of EDC to vinyl chloride, followed by polymerization of vinyl chloride to polyvinyl chloride (PVC)
- Production of hydrogen cyanide (HCN) via reaction of ammonia, natural gas, and air over a platinum catalyst. HCN is a precursor to adiponitrile for nylon production and methyl methacrylate for polymer manufacturing, among other products

Coal, which served as the major source of industrial organic chemicals in the eighteenth and nineteenth centuries, continues to be a major feedstock for certain commodity chemicals such as carbon monoxide, benzene, toluene, naphthalene, anthracene, phenols, biphenyl, and derivatives, which may be used as precursor chemicals in the manufacture of methanol, formaldehyde, acetic anhydride, methyl acetate, synthetic dyes, polyester resin and fiber, antimicrobials, and much more. Coal is also a major source of activated carbon adsorbents, carbon fillers and binders, pavement sealants, and roofing materials. Although declining as a source of fuel for power generation due to high CO_2 emissions relative to natural gas, coal reserves are plentiful and likely will continue to be a major source of chemicals and materials [4].

Biomanufacturing, the industry that supplied ABE at the beginning of the twentieth century, has since become a competitive source of a wide variety of commodity and specialty chemicals, taking advantage of the natural ability of a microbe or plant to produce chemicals more efficiently than certain multi-step petrochemical routes (in special cases). This includes large-scale production of commodity chemicals such as citric acid and specialty chemicals such as antibiotics (penicillin and others) and certain agricultural pesticides such as spinosad manufactured by Corteva (formerly Dow AgroSciences). This also includes production of recombinant proteins via bacterial fermentation or via animal cell cultures for use as therapeutic proteins (biopharmaceuticals). Therapeutic proteins may also be produced in modified plant crops, as in virus-based production of various therapeutics involving the Tobacco mosaic virus. Some of these bio-derived compounds are later chemically modified to improve their properties, yielding semi-synthetic biomolecules. Biopharmaceuticals (or biologics) are used in treating cancer, in gene therapies, and in preparing vaccines. In another form of bioprocessing, industrial and specialty enzymes are used for conversion of starches to sugars (amylases), breakdown of proteins in food processing (proteases), and scission of stain molecules in washing clothes with detergents (lipases). The current bioprocessing industry also includes production of industrial alcohol and acetic acid from fermentation of sugars (often from corn starch) and the production of biodiesel from transesterification of soybean oil triglycerides to make fatty acid alkyl esters as well as other plant-based bioprocessing possibilities. The production of chemicals from corn and other grains has grown to become a large industry with large facilities often referred to as biorefineries (analogous to oil refineries). The biomanufacturing industry now utilizes all kinds of organisms to produce a wide range of products including beer and wine, vinegar, biofuels, carboxylic acids, fragrance and flavor compounds, antibiotics, agricultural pesticides, amino acids, peptides, industrial and specialty enzymes, various kinds of therapeutic proteins, and specialty polymers such as polylactic acid (polylactide) (see Section 13.1: Bioseparations and Section 13.2: Bio-designed Separations).

The polymer industry began around 1860 with the commercialization of Parkesine, a resin made by treating cellulose with nitric acid and then with a solvent and an oil such as chloroform and castor oil. Improved nitrocellulose and other cellulosic resins followed. In 1910 the industry grew larger with the introduction of Bakelite, a thermoset polymer made from phenol and formaldehyde in an autoclave. This was followed by a highly productive period of new polymer development in the 1920s and 1930s. Emulsion polymerization for production of synthetic polyisoprene and styrene-butadiene rubber was introduced in the 1920s. Acrylic polymers were first commercialized in the 1920s based on acrylic acid and then in the 1930s based on methyl methacrylate. And polyamide polymer (nylon) was invented around 1935. Epoxy resins, known since the 1890s, were developed in the 1930s and 1940s and commercialized at large scale in the 1950s. Similarly, polyethylene was first manufactured in the 1930s, and new catalytic processes for improved ethylene and propylene polymerization were introduced in the 1950s. Other polymers introduced in the twentieth century include a variety of phenolic resins, styrenic polymers, polyols, alkyd resins and other polyesters, polyvinyl chloride, polyvinyl esters, polyurethanes, fluorinated polymers, and more [5]. And new co-polymers and specialty polymers continue to be introduced to this day. In many cases, rigorous purification of the monomer reactants is required for successful polymerization.

In addition to all the industries producing chemicals based on carbon, the twentieth century also saw a great expansion in products based on silicon from silica sand (silicon dioxide mineral). Silicon and aluminum oxides have long been used in making concrete, bricks, glass, and ceramic products, but beginning in the 1940s, the silicone polymer industry began producing polysiloxane polymers from silica. A polysiloxane or *silicone* polymer is a kind of synthetic polymeric silica with methyl groups (or other organic groups) attached to the –Si–O–Si–O– backbone, so it differs from natural silica in that it contains carbon and hydrogen in addition to silicon and oxygen. Elemental silicon is also a major ingredient in specialty steel and aluminum alloys, its largest use by volume (called metallurgical grade silicon). In the 1960s and 1970s, processes for making ultrapure grades of polycrystalline silicon were devised to provide feedstock for making silicon single-crystal wafers used in making microelectronic devices (requiring 99.9999999% pure Si) and for photovoltaic solar cells (99.9999 +% pure Si). The need for microelectronic chips with more and more electronic capabilities (with tens of billions of transistors per chip) drove development of specialized photo-sensitive chemicals needed for precision microlithography used in fabricating complex microcircuits onto each silicon wafer. Fabrication of structures at or even below 10 nm in size is currently possible. Single-crystal silicon carbide is also being used to make microchips, ones designed to operate reliably in extreme environments.

Products based on other metal oxide raw materials also saw strong growth including the refining of aluminum metal from hydrated alumina (obtained from bauxite using the Carl Bayer Process) via chemical dissolution and electrolysis. The basic process was discovered in the late nineteenth century, but it developed into a major industry in the early twentieth century, greatly increasing availability and reducing the cost of aluminum metal. Other products derived from metal oxides include ceramic armor and refractory materials introduced around 1910 and molecular sieve adsorbents with well-defined internal pore sizes (synthetic zeolites) made from alumina and silica (aluminosilicates) introduced at large scale in the 1950s. Hydrometallurgical methods for isolating metals such as uranium, zinc, and copper from acidic ore-leachate solutions also became more prevalent including the development of reaction-enhanced liquid-liquid extraction in the 1940s and the development of ion exchange processes for hydrometallurgy in the 1950s and 1960s [6].

3.2 Pioneers

A great number of people have contributed to the advancement of chemical manufacturing and chemical separation. The history of chemical separation is inseparable from the history of artisan processing and industrial manufacturing in general because these activities drove (and continue to drive) the need for inventing and advancing chemical separations. Various sources of information are listed in the Bibliography for Chapters 3. The internet is also a good source of information. Excellent descriptions of distillation and other methods, including their history, may be found in Wikipedia and other web-based references. And a number of engineering companies and equipment suppliers provide webpages describing the history of their particular technology.

In Boxes 3.5 to 3.9 we list some of the pioneers of separation technologies. This is only a sampling of the many people who have made contributions in one way or another. We encourage everyone to look into the histories of the separation technologies of particular interest to them to learn more about the people behind them – their motivations, the challenges they faced, and how they overcame them. Knowing something of this background can be both informative and inspiring.

Box 3.5: Pioneers of Distillation

- In ancient times, sea water was boiled to obtain salt-free overheads for drinking water and to distill wine (noted by Aristotle, ca. 350 B.C.), although it appears the practice was not common.
- Egyptians practiced some form of pot distillation from around 200 B.C. or perhaps earlier.
- **Mary the Jewess** (Alexandria, Egypt), early alchemist, is purported to be the inventor of the alembic batch distillation apparatus (or its precursor), sometime between the first and third centuries.
- Primitive versions of sieve trays were used by the Greeks in the early first millennium.
- **Jabir ibn Hayyan** improved alembic distillation methods in the ninth century.
- **Abu Ali Al-Husssein Ibn Abdullah Sina** (980–1057), also known as Avicenna, described a steam distillation method used to recover essential oils in the tenth century.
- **Taddeo Alderotti** (ca. 1250) described a fractional batch distillation method.
- **Robert Boyle** (1627–1691) studied distillation of alcohol obtained from wood and fermented grain or fruit (ca. 1650).
- **John French** (1616–1657) authored a book titled *The Art of Distillation: Or, a Treatise of the Choicest Spagyrical Preparations Performed by Way of Distillation.* The word spagyrical refers to the alchemist's art. The book describes distillation methods and applications known at the time, with illustrations of various alembic distillation apparatus including vessels and heaters (1651, printed by Richard Cotes, London).
- **George Smith** authored books describing distillation practices. The first book, titled *The Practical Distiller: Or, a Brief Treatise of Practical Distillation*, is often attributed to George Smith. It was published in 1718 (printed by Bernard Lintot, London). An expanded version, titled "A Compleat Body of Distilling, Explaining the Mysteries of that Science, in a Most Easy and Familiar Manner," was published in 1725, also printed by Bernard Lintot. The books describe distillation methods for making alcoholic liquors.
- **Jean-Édouard Adam** (1768–1807) developed a batch distillation method incorporating controlled reflux (ca. 1800).
- **William Henry** (1774–1836) studied gas solubility and introduced Henry's law (1803).
- **John Dalton** (1766–1844) introduced the law of partial pressures, called Dalton's law, and described the absorption of gases by water and other liquids (1805).
- **François Marie Raoult** (1830–1901) introduced concepts involving the behavior of liquid solutions, especially concerning freezing point depression and vapor pressure (ca. 1870s and 1880s).
- **Jean-Baptiste Cellier-Blumenthal** (1768–1840) developed an early continuous distillation tower with a vertical arrangement (or cascade) of trays similar to modern-era bubble cap trays and used to distill alcohol from fermented sugar beets and fruit (ca. 1813).
- **Jean-Jacques Saint Marc** patented a batch still with a rectifying head employing reflux (1824).
- **Anthony Perrier** (in 1822), **Robert Stein** (in 1826), and **Aeneas Coffey** (in 1830) developed improvements for continuous fractional distillation.
- **Aeneas Coffey** (1780–1839) is most famous for developing the Coffey still (1830), a two-column distillation process using trays similar to modern sieve trays to produce alcohol at greater than 90% concentrations (still in use today).
- **William Gossage** (1799–1877) developed the scrubber column for absorption of HCl into water (1836). Used for control of HCl emissions from production of soda ash.

- **Justus Baron von Liebig** (1803–1873) promoted the use of distillation methods in the organic chemistry laboratory (especially during the 1830s and 1840s) including the use of the Liebig condenser, a countercurrent water-cooled heat exchanger based on earlier designs, and the use of reflux.
- **Norbert Rillieux** (1806–1894) invented the multiple-effect evaporator, which may be applied in distillation and crystallization processing (ca. 1840s).
- **Ernest Solvay** (1838–1922) patented a tray column design (1877).
- **Adolf Eugen Fick** (1829–1901), **James C. Maxwell** (1831–1879), and **Josef Stefan** (1835–1893) introduced theories of diffusion (ca. 1855–1871). Fick described his first law of diffusion in 1855. Maxwell and Stefan independently introduced expressions for multicomponent diffusion (1871).
- **Ernest Sorel** (1850–1904) developed a theoretical analysis of distillation. Introduced the theoretical stage concept (1889, 1893).
- **George E. Davis** (1850–1906) wrote *A Handbook of Chemical Engineering*, a two-volume work describing distillation methods and other chemical processing methods (1901, 1904). Considered a founder of chemical engineering.
- **Franciscus Antonius Hubertus Schreinemakers** (1864–1945) and **Friedrich Wilhelm Ostwald** (1853–1932) described fundamental features of phase diagrams and residue curve mapping (1901–1903). Around the same time, Ostwald also helped to pioneer the concept of the meta-stable zone in studies of crystallization (meta-stable zone mapping).
- **Lord Rayleigh (John William Strutt)** (1842–1919) introduced a theoretical analysis of single-stage distillation (1902).
- **Warren K. Lewis** (1882–1975) developed a framework for mathematical analysis of fractional distillation (1909). Considered a founder of modern chemical engineering.
- **Clark Shove Robinson** (1888–1947) and **Edwin Richard Gilliland** (1909–1973) wrote *Elements of Fractional Distillation*, 1st ed. (1920) by C. S. Robinson, through to the 4th ed. (1950) updated by E. R. Gilliland.
- **Walter G. Whitman** (1895–1974) developed mass transfer coefficients for distillation design (ca. 1920s).
- **Allan P. Colburn** (1904–1955) developed the concept of mass transfer units (ca. 1930s).
- **Eger Vaughan Murphree** (1898–1962) described factors affecting distillation tray performance and efficiency (ca. 1924).
- **Gilbert N. Lewis** (1875–1946) and **Merle Randall** (1888–1950) defined and quantified the liquid activity coefficient and fugacity in the study of phase equilibria (ca. 1920s).
- **Wendell Mitchell Latimer** (1893–1955), **Worth Huff Rodebush** (1887–1959), and **Maurice L. Huggins** (1897–1981) proposed the concept of hydrogen bonding (1920).
- **Worth Huff Rodebush** (1887–1959) introduced an early graphical method of distillation design (1922), a forerunner of the famous method of McCabe and Thiele.
- **M. Ponchon** (in 1921) and **R. Savarit** (in 1922) independently developed a graphical distillation design method including thermal effects.
- **Warren L. McCabe** (1899–1982) and **Ernest W. Thiele** (1895–1993) developed their famous graphical theoretical-stage calculation method (1925).
- **Merrell R. Fenske** (1904–1971) introduced the concept of what is now called separation power in his famous equation relating separation power to separation factor and theoretical stages at total reflux (1932).
- **Edwin R. Gilliland** (1909–1973) introduced a short-cut correlation for the number of theoretical stages versus reflux ratio (1940).
- **Arthur Joseph Victor Underwood** (1897–1972) introduced a method for estimating the minimum reflux ratio for multicomponent mixtures (1946, 1948).
- **Sidney Young** (1857–1937) and **Emily C. Fortey** (1866–1946) introduced the physical chemistry principles of azeotropic distillation (1902).

- **Konrad Kubierschky** developed a continuous azeotropic distillation process for drying alcohols (1915).
- **Ernest W. Dean** (1888–1959) and **David D. Stark** (1893–1979) applied azeotropic distillation in the laboratory for petroleum analysis (ca. 1920).
- **Aylmer H. Maude** introduced an azeotropic distillation process for drying acetic acid using benzene (1929).
- **Donald B. Keyes** (b. 1891) further developed azeotropic distillation for drying alcohols (1928).
- **Warren K. Lewis** (1882–1975) patented pressure-change distillation for drying alcohols (1928).
- **Robert Roger Bottoms** invented acid gas absorption employing aqueous amines at The Girdler Corp. (1930).
- **Joel Hildebrand** (1881–1983) and **George Scatchard** (1892–1973) are known for solubility parameter and regular solution theory.
- **Władisław (Walter) Joseph Podbielniak** (1899–1978) developed low-temperature (cryogenic) distillation of close-boiling components (1928) and a centrifugal distillation apparatus (1932). Also known for developing the Podbielniak centrifugal extractor (ca. 1945).
- **Clarence L. Dunn** and **Robert B. McConaughy** of Shell Dev. Co. introduced extractive distillation (ca. 1940).
- **Richard O. Wright** (born ca. 1917) patented the placement of a dividing-wall within a distillation column, citing prior art involving placement of vertical partitions to form subsidiary fractionation zones, at Standard Oil Dev. Company (1949).
- **Felix B. Petlyuk** (b. 1934) introduced the theory of thermally coupled distillation relating to dividing-wall distillation and other distillation configurations (ca. 1965).
- **Gerd Kaibel** (1944–2017) led development and implementation of commercial applications of dividing-wall distillation, at BASF (ca. 1980s).

Box 3.6: Pioneers of Extraction

- Solvent extraction has been used since ancient times to process natural materials using water or acidic or basic solutions made from natural materials (such as tannic acid, citric acid, and even urine). Used to wash and chemically process natural materials to obtain specific properties or to remove unwanted colors and odors or to extract desired components such as fragrances.
- **Carl Wilhelm Scheele** (1742–1786) discovered that gold would dissolve in aqueous solutions of potassium cyanide (1783).
- **John Stewart MacArthur** (1856–1920), **Robert W. Forrest**, and **William Forrest** invented the MacArthur-Forrest cyanidation process for large-scale extraction and refining of gold (1887). The process exploits the ability of cyanide to complex with gold, first observed by Scheele. It is an early example of hydrometallurgy.
- **Carl Josef Bayer** (1847–1904) developed a process for extraction of aluminum hydroxide from bauxite ore by using hot caustic soda (1887–1892). The resulting hydroxide is later crystallized and calcined to form alumina. The process initially produced alumina for the textile industry and later for production of aluminum metal. The Carl Bayer process is another early example of hydrometallurgy and an early industrial application of extraction + crystallization.
- **James Shank** developed a multi-bed countercurrent liquid-solid extraction process (a countercurrent cascade or battery of multiple leaching beds called an extraction battery) (ca. 1841).
- **Pierre Eugène Marcellin Berthelot** (1827–1907) and **Emil Jungfleisch** (1839–1916) introduced the partition ratio (1872).
- **Th. Göring** patented a countercurrent extraction process for recovery of acetic acid, German patent 28064 (1883).

- **Louis Charles Reese** patented a stirred extraction column, U.S. patent 679,575, assigned to Paul Pfleiderer (1901).
- **Karl Gustaf Patrik de Laval** (1845–1913) developed centrifugal milk-cream separators and steam turbines (ca. 1900).
- **Lazăr Edeleanu** (1861–1941) developed an oil refining process involving extraction of aromatics from kerosene using liquid SO_2 (1908). This is the first commercial use of a solvent to extract aromatics from aliphatics. Liquid SO_2 was later replaced by phenol.
- **Clarence L. Dunn** and **Robert B. McConaughy** of Shell Dev. Co. combined extraction and distillation in a single fractionation column using alkyl phenol as a heavy polar solvent to extract aromatics away from aliphatics and into the distillation bottoms. They called it vapor phase extraction in the presence of a relatively high-boiling selective solvent or extractive distillation (ca. 1940).
- **Kühni** rotating-impeller column extractor (developed by the Kühni Company, Switzerland, founded in 1932).
- **Manhattan Project** (USA) ca. 1940 developed a process for the recovery of uranium from acidic ore leachate solution, followed by nitric acid treatment and liquid-liquid extraction, followed by precipitation steps and roasting to form purified uranium oxide. The uranium oxide product was converted to uranium halides, and isotopes of uranium hexafluoride were ultimately separated using gas-phase membrane diffusion. The initial liquid-liquid extraction has become a general approach in hydrometallurgy. It has been modified and improved with different extractants for different purposes.
- **Władisław Joseph Podbielniak** (1899–1978) is most famous for his centrifugal extractor used for penicillin recovery (developed in the 1940s). Also known for laboratory apparatus: cryogenic fractional distillation (employing a packed column) and early gas chromatography for sample analysis (1920s–1960s).
- **Lee H. Horsley** (1911–1995), **Victor S. Morello** (1915–2005), and **Noland I. Poffenberger** (1906–1995) developed fractional extraction with extract reflux for isolation of benzene, toluene, and xylenes from mixed hydrocarbons using a high-boiling polar solvent (diethylene glycol in water) (ca. late 1940s at The Dow Chemical Company). Lee Horsley is also known for pioneering work compiling azeotropic data in the 1970s.
- **Edward G. Scheibel** developed various versions of a rotating-impeller column extractor (1950s and 1960s).
- **Andrew E. Karr** developed a reciprocating-plate column extractor (1960s).

Box 3.7: Pioneers of Adsorption

- **Axel Fredrik Cronstedt** (1722–1765) coined the term zeolite for minerals that gave off steam when heated (ca. 1756).
- **Carl Wilhelm Scheele** (1742–1786) known for discovering gaseous elements and organic acids. Studied the uptake of gases by charcoal (ca. 1773). Also known for discovering dissolution/extraction of gold into aqueous potassium cyanide.
- **Johann Tobias Lowitz** (1757–1804) studied decolorization of aqueous solutions using charcoal (ca. 1785).
- Anonymous, at English sugar refinery, introduced industrial use of charcoal to remove color from aqueous sugar solutions (ca. 1794).
- **Alfred Eaton**, distillery owner, introduced the Lincoln County process for making Tennessee whiskey which employs charcoal filtration/steeping to remove grainy and harsh flavors, probably based on prior practices in the making of craft spirits (ca. 1825). Filtration/steeping is now recognized as an adsorption step.

-   **John Hunter** (Queen's College, Belfast) studied the uptake of gases by charcoal made from coconut shell using the term *absorption* to refer to the phenomenon. He published numerous papers in *J. Chem. Soc.* concerning various gases and the effect of temperature and pressure (1865–1872).
-   **Heinrich Kayser** (1853–1940) studied the uptake of gases by charcoal, this time using the term adsorption as suggested to him by Emil du Bois-Reymond (1881).
-   **Emil du Bois-Reymond** (1818–1896) coined the term *adsorption* in 1881 to distinguish between the adherence of gases at a solid surface versus diffusion into a bulk material, in communication with Heinrich Kayser who then used the term in his writing. Emil du Bois-Reymond studied the biological action of nerves and muscles and proposed a mechanism of muscle action involving *electric molecules* (now called ions). Presumably, his use of the term adsorption specifically referred to the adherence of molecules at a solid surface (molecular adsorption). The molecule had already been introduced by Amedeo Avogadro in 1811, and molecular diffusion was described by Thomas Graham in 1831.
-   **Mikhail Tsvet** (1872–1919) invented liquid adsorption chromatography (ca. 1900) and coined the term chromatography (1906).
-   **Robert Gans** invented the use of ion exchange for water softening. He synthesized sodium aluminosilicate cation exchange zeolite (later marketed as Permutit) to exchange calcium and magnesium for sodium, using aqueous NaCl brine for regeneration (ca. 1905). The same process concept is used today for residential water softening (using styrene-divinylbenzene-based ion exchange resins).
-   **Walter A. Patrick** (1888–1969) invented synthetic silica gel (1918).
-   **Archer Martin** (1910–2002) and **Richard Synge** (1914–1994) invented/developed two-phase partition chromatography, both liquid-liquid and gas-liquid (ca. 1940s).
-   **Erika Cremer** (1900–1996) independently invented gas adsorption chromatography (ca. 1944).
-   **Gaetano F. D'Alelio** (1909–1980) invented styrene-divinylbenzene-based ion exchange resin (ca. 1944).
-   **Richard Maling Barrer** (1910–1996) was the first to synthesize a zeolite material with no natural counterpart (1948). He also advanced the science of gas permeability in membranes.
-   **Robert M. Milton** (1920–2000) and **Donald W. Breck** (1921–1980) developed synthetic zeolites A, X, and Y (late 1940s to 1950s).
-   **Heinrich Kahle** (at Linde in Germany), **Pierre Guerin de Montgareuil** and **Daniel Domine** (at Air Liquide SA in France), and **Charles W. Skarstrom** (at Standard Oil Co. in the U.S.) introduced pressure change concepts underlying pressure swing adsorption (PSA) processing (ca. 1955).
-   **Charles W. Skarstrom** (1906–1984) developed the heatless dryer for removing moisture from air, a PSA process scheme with a four-step cycle involving use of a portion of the product gas as backpurge, a kind of reflux (1960).
-   **Donald B. Broughton** (1917–1984) and **Clarence G. Gerhold** (1908–2001) invented simulated moving bed chromatography (ca. 1960).
-   **J. J. Collins** introduced the length of unused bed (LUB) method of fixed-bed adsorber design (1967).
-   **Frederick Wells Leavitt** (1928–2019) developed advanced PSA process technology for gas separations (ca. 1970s and 1980s).

Box 3.8: Pioneers of Crystallization

-   In ancient times, sea salt was obtained by evaporation of water from brine in sun-lit ponds (early batch evaporative crystallization).
-   **Jabir ibn Hayyan** described crystallization methods in the ninth century.
-   Cooks and bakers over many generations developed empirical cooking and baking methods for preparing foods with desired properties (controlling crystallization in bread, candy, ice cream, chocolate, etc.)

- **Norbert Rillieux** (1806–1894) invented a multiple-effect evaporator for crystallizing sugar (continuous evaporative crystallization) (ca. 1843).
- Various industrial processes involving crystallization were introduced for making alumina, ammonium sulfate byproduct from coal gasification, paraffin wax from petroleum, aspirin powder, etc. (ca. 1850 to 1900).
- **Carl Josef Bayer** (1847–1904) (not the Bayer of aspirin fame) developed a process for production of alumina involving extraction of aluminum hydroxide from bauxite ore by using hot caustic soda (1887–1892). The extraction is followed by crystallization of purified aluminum hydroxide and calcination to form alumina, an early industrial use of extraction + crystallization.
- **Frederick Guthrie** (1833–1886) coined the term *eutectic* for mixtures of pure single-component co-crystals that exhibit a minimum melting temperature (1884).
- **Friedrich Wilhelm Ostwald** (1853–1932), **Henry Alexander Miers** (1858–1942), and **Florence Isaac** introduced the concepts of Ostwald ripening and mapping of the metastable zone (ca. 1900–1910). Ostwald also contributed to the concept of residue curve mapping in distillation.
- **Max Theodor Felix von Laue** (1879–1960), **Walter Friedrich** (1883–1968), and **Paul Knipping** (1883–1935) discovered the diffraction of X-rays by crystals (ca. 1912), the basis for determining crystal lattice structure.
- **William Lawrence Bragg** (1890–1971) introduced the theory of crystallography and Bragg's law (1913), while working with his father, William Henry Bragg (1862–1942).
- ProABT static melt crystallization process introduced in France (ca. 1950).
- **Frederick Charles Frank** (1911–1998) introduced theories of crystal dislocations and other types of defects (ca. 1950).
- **Alan D. Randolph** (1934–1999) and **Maurice A. Larson** (1927–1999) introduced the crystallization population balance (ca. 1960–1970).
- **John Alfred Brodie** (1908–1996) invented a countercurrent melt crystallizer, Union Carbide Australia (ca. 1970).
- **Kurt Saxer** invented a falling-film melt crystallizer (ca. 1960–1970) in Switzerland.
- **TNO** developed a suspension melt crystallizer hydraulic wash column (ca. 1970–1980) in The Netherlands.
- Oscillatory-flow reactor/crystallizer. A new process concept introduced in the 2000s based on fundamental work on pulsed flow in tubes by Prof. M. R. Mackley and others (ca. 1990).

Box 3.9: Pioneers of Membrane-Based Separations

- **Jean-Antoine Nollet** (1700–1770), also known as Abbé Nollet, described osmosis phenomena in natural membranes after observing transfer of water through a pig's bladder into alcohol (1748).
- **René Joachim Henri Dutrochet** (1776–1847) studied osmosis and membrane equilibrium in detail. Showed that osmosis is purely a physical phenomenon (ca. 1820–1830).
- **John Kearsley Mitchell** (1793–1858) studied gas permeation through natural rubber (1831).
- **Thomas Graham** (1805–1869) used a membrane apparatus in the laboratory to remove excess urea from blood, the first demonstration of the concept (1854). He named the process dialysis and suggested its use for treating patients with kidney failure. He also introduced the term diffusion (1831) and described gas separation via selective permeation through natural rubber (1866).
- **Adolf Eugen Fick** (1829–1901) introduced Fick's law of diffusion (1855) and studied diffusion through porous collodion (cellulose nitrate) (1857).
- **John Jacob Abel** demonstrated dialysis on animals (1913).
- **Philip Adolf Kober** coined the term pervaporation. Introduced pervaporation process concepts with experimental demonstration (1917).
- **Richard Zsigmondy** and **Wilhelm Bachmann** invented microfiltration using a membrane filter to remove colloid-size droplets from liquids (1922).

- **Georg Haas** performed the first human hemodialysis for blood purification (1924). Commercial development came later.
- **Willem Johan Kolff** (1911–2009) and **Hendrik Berk** developed the high-surface area, rotating drum artificial kidney for hemodialysis (ca. 1940).
- **Sartorius AG**, introduced commercial membranes for microfiltration (ca. 1930).
- **U.S. Army**. Membranes were used for water filtration (ca. 1940).
- **Manhattan Project** (USA) ca. 1940 developed enrichment of uranium hexafluoride isotopes via gas-phase diffusion through microporous inorganic materials (an early type of ceramic membrane) using approx. 1,000 membrane stages. A highly specialized and unique application not duplicated in normal chemical separation practice, but it introduced many membrane process concepts.
- **Richard Maling Barrer** (1910–1996) advanced the science of gas permeability in membranes. The Barrer unit of gas permeability is named for him. He also was the first to synthesize a zeolite material with no natural counterpart (1948).
- **Sol W. Weller** (1918–2008) and **Waldo A. Steiner** introduced membrane-based gas separation theory and proposed use of non-porous thin films supported on a porous support layer (ca. 1950).
- **Robert C. Binning** and **Robert J. Lee** published research into pervaporation membranes and processes (ca. 1950).
- **Charles Edward Reid** (1917–2000) introduced the concept of desalination by reverse osmosis membranes (1953).
- **Sidney Loeb** (1917–2008) and **Srinivasa Sourirajan** (1923–2022) developed the first commercial reverse osmosis membrane, "skin-type cellulose acetate" (ca. 1960).
- **Monsanto Company** introduced the first commercial membranes for gas separation: Permea polysulfone hollow fiber membranes used to separate H_2/N_2 (1979).
- **John E. Cadotte** (1925–2005) invented a thin-film composite membrane for reverse osmosis (polyamide barrier film on polysulfone support) (ca. 1977) and co-founded FilmTec (1979).
- **Gesellschaft für Trenntechnik** (GFT at Hameln, Germany, now part of Sulzer) commercialized polyvinylalcohol pervaporation membranes and process for ethanol + water separation (ca. 1982).
- **Louis Cot** and others at Université de Montpellier (France) published work on ceramic membranes (ca. 1980-present).
- **Andrew G. Livingston** led development of organic solvent nanofiltration membranes (made with cross-linked polyamide-imide polymer) at Imperial College London (2002).

We end this section by highlighting four distinguished pioneers of chemical separation, each with their own unique story and contribution: Norbert Rillieux (1806–1894), Władisław J. Podbielniak (1899–1978), Charles W. Skarstrom (1906–1984), and Donald B. Broughton (1917–1984). Their stories show how people can advance separations technology with great impact on the chemical and bioprocessing industries and society. Each person identified an opportunity for improvement through technological innovation showing how:

- Large-scale processing methods can differ greatly from established laboratory methods and yield significantly better results (Rillieux's multiple-effect evaporator)
- A process may be intensified for better performance and minimal product degradation (Podbielniak's centrifugal extractor)
- A process may be simplified and reduced in size for a given production rate [Skarstrom's pressure swing adsorption (PSA) process, the Heatless Dryer]

–   A process may be configured to achieve both practical operation and efficient operation [Broughton's simulated moving-bed (SMB) chromatography process]

We discuss some of the technical aspects of these technologies in later sections of the book. Here we focus on their individual stories and impact.

Norbert Rillieux invented the multiple-effect evaporator for efficient concentration of sugar solutions and crystallization of refined sugar, a technology widely practiced today not only for sugar but for a wide variety of products [7]. Rillieux's approach to process development – thinking beyond scale-up of laboratory methods – has had a great impact on chemical process development in general. Here we refer to the introduction of large-scale process concepts that significantly improve process performance but are not easily or conveniently demonstrated in the laboratory. In our eyes, the impact of Rillieux's work on chemical manufacturing certainly is comparable to that of Eli Whitney (1765–1825) and his famous cotton gin with its impact on mechanized agriculture (ca. 1800).

Born in 1806 Louisiana in a former French settlement, Norbert Rillieux was a free black man in pre-civil war America. Sent to France for his education, Rillieux studied at the prestigious École Centrale Paris des Arts et Manufactures (Central School of Paris for the Manufacturing Arts). He must have impressed his teachers, as he later became an instructor there in applied mechanics. Around 1830 while working in France, he introduced the idea that several evaporators could be economically operated in series, by operating them at successively lower pressures using the steam vapors emanating from the first (direct-fired) evaporator to heat the second evaporator (at reduced pressure and thus a lower boiling-point temperature in order to generate the ΔT needed for heat transfer), and using the steam from the second evaporator to heat a third one operated at a further reduced pressure. This allowed repeated use of the fuel energy input to the first evaporator to operate the second and third evaporators. It also made for operation at successively lower temperatures, which helped minimize thermal degradation of the product. Rillieux introduced his invention into commercial operation in Louisiana in 1843. He received U.S. patent No. 4879, Improvement in Sugar-Making, dated December 10, 1846. The Rillieux process greatly reduced the consumption of fuel (usually wood at that time) and minimized exposure to high temperatures that over time would generate unwanted color and burnt flavors. The technology greatly improved the processing of sugar in terms of both greater efficiency and better product quality. Rillieux later returned to France to pursue other interests and died there in 1894. Today, Norbert Rillieux is remembered as one of the very first industrial chemical engineers and a true pioneer, having invented a key unit operation long before the terms chemical engineer or unit operation came into being. Rillieux's story is an inspiration to process designers everywhere because it shows that improvements in efficiency, affordability, and product quality are not necessarily mutually exclusive.

Władisław Podbielniak (known as Walter or Pod to some) is most famous for the Podbielniak centrifugal liquid-liquid extractor used in petrochemical processing and for recovery of penicillin from fermentation broth [8, 9]. Introduced around 1945, the Podbielniak centrifugal extractor is still being produced today, in Saginaw, Michigan by the B&P Littleford Company. Walter Podbielniak is also known for developing a variety of laboratory-scale chemical separation apparatus for sample analysis: low-temperature/cryogenic fractional distillation units for hydrocarbon analysis and early gas chromatography instruments. Related to the use of cryogenics, he also pioneered the use of mixed refrigerants for efficient liquefaction of multicomponent gases. Pod-bielniak was actively engaged in separations technology development from the 1920s into the 1960s, much of that time as an inventor and entrepreneur. He was born in 1899 in Buffalo, New York to Polish-born parents. He studied analytical chemistry at Buffalo University and chemical engineering at the University of Michigan, where he completed a Ph.D. in 1928. He maintained a strong interest in his Polish heritage. He played the piano and appreciated the music of Chopin in addition to pursuing his interests in chemistry and engineering. Dr. Podbielniak commercialized the centrifugal extractor in partnership with his first wife, Władzia (Gajda) Podbielniak, also an accomplished chemist and engineer. They enjoyed great success from sales of their Pod-bielniak extractor, mostly related to penicillin production but also used in a variety of petrochemical applications. Podbielniak and his wife became fabulously wealthy, and for a time they were rock stars of chemical engineering. They owned glamorous cars, a large yacht on Lake Michigan, and several mansions. One of these was later sold to Hugh Hefner to become the Chicago Playboy Mansion.

Podbielniak's centrifugal extractor enabled rapid extraction of antibiotics out of acidified fermentation broth into an organic solvent, typically amyl acetate (called banana oil due to its characteristic scent and use as a non-toxic flavoring agent). Here, the pH must be lowered below the pK_a of the product compound to yield the protonated form for efficient extraction, but doing so can catalyze hydrolysis and degradation of the product. Short contact time is needed to minimize this side effect. Podbielniak's centrifugal extractor takes advantage of centrifugal force to achieve very rapid liquid-liquid contacting and phase separation with short diffusion path length and very little holdup of liquid within the extractor. In this way, beginning around 1945, the Podbielniak extractor helped enable large-scale production and availability of penicillin, the first truly effective treatment for bacterial infection and a life saver.

Charles W. Skarstrom is known for inventing a variety of pressure swing adsorption (PSA) processes for purifying gases [10]. He was born in 1906 in New York City and raised in the Boston area. His father, William Skarstrom, was born in Sweden, held an M.D. degree from Harvard University, taught at Columbia Teachers College, and later became a prominent professor of physical education at Wellesley College. The elder Skarstrom's work focused on teaching and further developing the Swedish gymnastics tradition known as gymnastic kinesiology, emphasizing the mechanics of muscle action

[11]. No doubt influenced by his father but following his own path, the younger Skarstrom studied physics at Columbia University and completed a Ph.D. at the University of Virginia in 1939. During the 1940s, he joined the Manhattan Project as a member of the group working to isolate fissile uranium-235 using gas-phase centrifugation. This approach was promising but ultimately was set aside in favor of a method involving gaseous diffusion through porous membranes. After the war, Charles Skarstrom began working for the Standard Oil Company of New Jersey (later Esso and now ExxonMobil). He is most famous for clearly describing a four-step, PSA process scheme for separating gases, a process later commercialized as the Heatless Dryer for removing moisture from air [12]. Because of this, Skarstrom is often cited as the originator of the PSA process scheme although similar process concepts were reported around the same time. In 1953, Heinrich Kahle, working at Linde in Germany, was granted a patent in which he describes a different process scheme involving elevated adsorption pressure [13, 14]. Pierre Guerin de Montgareuil and Daniel Domine of the French company Air Liquide SA described pressure change concepts similar to Skarstrom's in a patent application filed in 1957 and granted in 1964 [15]. This kind of near-simultaneous, independent introduction of similar concepts by different inventors in different parts of the world is not uncommon in a competitive industry. All of these early inventors of PSA concepts deserve recognition for their contributions to what has become a highly successful process scheme. Charles Skarstrom realized that by regenerating the adsorbent at reduced pressure with an appropriate amount of clean sweep gas, no heat input would be needed at all, thereby greatly simplifying the process and allowing use of shorter adsorption cycles and therefore much smaller fixed-bed adsorbent columns (compared to processes requiring heating and cooling of the columns). Over the course of his career, Dr. Skarstrom further developed PSA and was granted patents for processes involving concentration of oxygen and purification of hydrogen as well as a related process for recovering gasoline vapors from tank vents [16]. The latter technology has influenced the design of carbon-canister vapor recovery systems used to control gasoline-tank vent emissions on most gasoline-fueled cars and trucks built since the 1970s.

The basic PSA concepts introduced by Skarstrom in the U.S., by Kahle in Germany, and by Guerin de Montgareuil and Domine in France, have since been further developed and widely applied. Commercial applications include generation of concentrated oxygen in a convenient, portable package for people needing help breathing, plus many other beneficial applications in industry including the production of industrial nitrogen, oxygen, and hydrogen, breaking of alcohol + water azeotropes, and recovery of organic vapors from air vents for emissions control.

Donald B. Broughton is known for separation technologies first introduced into the petroleum and petrochemical industries and now practiced in many other industries as well [17, 18]. He was born in Rugby, England in 1917, but his family soon moved to the U.S., first to Altoona, Pennsylvania and later to Philadelphia where Donald completed high school. He was a talented student, earning a merit scholarship to Pennsylvania

State University to study chemical engineering. From there he completed a Master's degree at the Massachusetts Institute of Technology (MIT) in 1940, gained some industrial experience working for the Rohm and Haas Chemical Company as an assistant engineer, and then returned to MIT to complete a D.Sc. degree in 1943. Dr. Broughton served in the U.S. Navy in Washington, D.C. before teaching and doing research full time at MIT. He then joined Universal Oil Products (UOP, now Honeywell UOP) in the early 1950s. At UOP, Broughton devised sophisticated chemical separations for the petroleum refining and petrochemical industries in collaboration with many talented engineers and process chemists including Clarence G. Gerhold, another separations pioneer. Broughton was a thoughtful man, well-liked by his colleagues and known for his ability to explain complex concepts in ways easy to understand. Dr. Broughton's accomplishments include key contributions to commercial processes involving fractional liquid-liquid extraction as well as the invention of SMB chromatography (ca. 1960, with Clarence Gerhold) [19]. His experience developing extraction processes producing raffinate and extract streams surely contributed to his ideas for SMB chromatography, a process that mimics the functions of fractional extraction. In 1976, he was elected to the National Academy of Engineering (USA) in recognition of his pioneering work [17].

Donald Broughton showed us how to efficiently utilize adsorbents to separate the components of a liquid feed – obtaining process efficiencies approaching those of a true countercurrent adsorption process without actually moving the adsorbent particles. Compared to standard fixed-bed adsorption processes, SMB chromatography requires much less elution solvent for the same production rate. SMB chromatography has greatly impacted the growth of the corn refining industry by helping reduce the cost of high fructose corn syrup and related products. It has also been successfully applied in the separation of certain petrochemicals, most famously for separating xylene isomers (its first commercial application), as well as the separation of certain pharmaceuticals and other specialty chemicals.

3.3 The Engineering Method

As we have seen in the previous section, the engineering tradition can be traced back to ancient times, well before the scientific method was first defined. David Rowe (1925–2021) at the University of York put it this way: *"It is often supposed that pure science is a necessary precursor of technological development but a study of history reveals many cases in which scientific understanding of technology lags behind the technology, sometimes by a long way"* [20]. A prime example in the field of chemical separation is the use of naturally occurring solids to remove unwanted colors, odors, or flavors from liquids, a practice long preceding the scientific concept of adsorption.

So what is the difference between the engineering method and the scientific method? As we all know, the scientific method is used to observe nature, to formulate hypotheses and construct models of nature, and to generate data needed to test and improve these ideas – all activities aimed at better understanding the principles and mechanisms of nature. On the other hand, engineers employ the engineering method and science to change how work is done and how products are made. The methods are certainly related, as engineering is fundamentally grounded in scientific principles, logic, and the testing of new designs and prototypes.

According to Billy Vaughn Koen, the engineering method involves formulation of heuristic rules for the purpose of creating new, state-of-the-art designs that enable useful, balanced improvements or *best changes* in how society operates to meet its wants and needs [21–23]. These are changes that produce something that is valued by society while dealing with real-world constraints of limited resources, incomplete knowledge (uncertainties), and the need to manage risks and tradeoffs within acceptable bounds with respect to performance targets, cost, safety, and environmental concerns. And, we believe, the concept of best change also involves managing the transition in ways that mitigate the impact of disruptions.

In practicing the engineering method, important skills include curiosity, problem solving ability, aptitude for mathematics and accounting, economic and social awareness, sensitivity to priorities and balance, and the application of the scientific method to better understand underlying mechanisms. Many examples of how these skills may be used to make changes that generate value are apparent in descriptions of the various chemical and bioprocessing industries [20, 24–29]. Consider the pioneering contributions highlighted earlier:

- Norbert Rillieux (1806–1894). Multiple-effect evaporator used to concentrate sugar solutions and crystallize sugar. This process reduces energy consumption by utilizing the steam emanating from a first evaporator to heat other evaporators in series. It also improves sugar quality by reducing operating temperatures, avoiding generation of unwanted color and burnt flavor.
- Władisław J. Podbielniak (1899–1978). Podbielniak centrifugal extractor for isolating penicillin produced by fermentation. This is an engineering innovation unrelated to the discovery of penicillin that has proved important for its economical manufacture, helping enable penicillin's implementation and benefit to society.
- Charles W. Skarstrom (1906–1984). Four-step PSA process for separating gases. This involves switching valves in sequence to change or "swing" the operating pressure from a high pressure for adsorption to a lower pressure for regeneration, using a small amount of clean product gas as reflux to aid desorption. This avoids any need to heat the adsorbent which allows for a reduction in the size of the adsorbent vessels, even allowing for design of small personal oxygen generators to aid breathing.
- Donald B. Broughton (1917–1984). SMB chromatography process. This involves switching the flow of liquid through beds of solid particles in such a way as to approach the performance of true countercurrent processing without actually having to move the particles (which would cause significant attrition over time), making for an efficient and robust process widely used across industries.

In general, the work of the engineer involves devising safe and workable systems and equipment by maximizing desired functions, integrating multiple operations, devising efficient controls, and in a balanced way, dealing with practical requirements. These include dealing with deadlines (often sooner rather than later), sourcing supplies, managing wastes, and reducing costs. The work may involve such activities as careful measurement, performance and durability testing, problem-solving diagnostics, continual refinement in operational efficiency, calibration and adjustment, and learning from failure analysis – in addition to theory and mathematical modeling. Furthermore, high standards of ethical behavior and professionalism are critical to the engineer's role including the need for fair and honest behavior, prudent use of resources, safety in design, and environmental protection. These are important responsibilities. Change has consequences, and not all change is best change. For discussions of professional responsibilities and ethics, see refs. [30–33].

References

[1] P. C. Wankat, "Separations: A Short History and A Cloudy Crystal Ball," *Chemical Engineering Education*, vol. 43, no. 4, pp. 286–295, Fall 2009.

[2] G. Smith, *The Nature of Fermentation Explain'd: With the Method of Opening the Body of Any Grain or Vegetable Subject, so as to Obtain from It a Spirituous Liquor*, B. Lintot, London, 1729.

[3] J. A. Tarr, "Industrial Wastes and Public Health: Some Histroical Notes, Part I, 1876–1932," *American Journal of Public Health*, vol. 75, no. 9, pp. 1059–1067, 1985.

[4] H. H. Schobert and C. Song, "Chemicals and Materials from Coal in the 21st Century," *Fuel*, vol. 81, no. 1, pp. 15–32, 2002.

[5] D. Feldman, "Polymer History," *Designed Monomers and Polymers*, vol. 11, pp. 1–15, 2008.

[6] M. Streat, Section 2.13, "Ion Exchange Processes in Hydrometallurgy," in *Ion Exchangers*, K. Dorfner, Ed., De Gruyter, Berlin, 1991, pp. 1061–1072.

[7] "Norbert Rillieaux and the Multiple Effect Evaporator," [Online]. Available: https://www.acs.org/content/acs/en/education/whatischemistry/landmarks/norbertrillieux.html. [Accessed 12 Feb 2022].

[8] A. Sella, "Podbielniak's Contactor. How a new spin on separation produced petroleum, penicillin and much more," 3 Sep 2021. [Online]. Available: https://www.chemistryworld.com/opinion/podbielniaks-contactor/4014234.article. [Accessed 20 Oct 2021].

[9] J. O. Wilkes, *A Century of Chemical Engineering at the University of Michigan*, Univ. of Michigan, Ann Arbor, 2002, pp. 61–65, 191.

[10] A. Sella, "Skarstrom's Separator. Producing Oxygen and Other Gases from the Atmosphere," 4 Jun 2021. [Online]. Available: https://www.chemistryworld.com/opinion/skarstroms-separator/4013737.article. [Accessed 12 Feb 2022].

[11] R. Renson, "Kinesiologists: Raiders of the Lost Paradigm?," *Kinesiology*, vol. 34, no. 2, pp. 210–221, 2002.

[12] C. W. Skarstrom, "Method and Apparatus for Fractionating Gaseous Mixtures by Adsorption," U.S. Patent 2,944,627, 12 July 1960.

[13] H. Kahle, "Process for the Purification and Separation of Gas Mixtures," U.S. Patent 2,661,808, 8 Dec 1953.

[14] C. N. Kenney and N. F. Kirkby, "Pressure Swing Adsorption," in *Zeolites: Science and Technology*, Vol. 80, F. R. Ribeiro, A. E. Rodrigues, L. D. Rollmann and C. Naccache, Eds., *NATO ASI Series*, Springer, Heidelberg, 1984, pp. 657–694.

[15] P. Guerin de Montgareuil and D. Domine, "Process for Separating a Binary Gaseous Mixture by Adsorption," U.S. Patent 3,155,468, 3 Nov 1964.

[16] W. F. Biller and C. W. Skarstrom, "Process and Device for Preventing Evaporation Loss," US Patent 3,352,294, 14 Nov 1967.

[17] C. G. Gerhold, "Donald B. Broughton, 1917–1984," [Online]. Available: https://www.nae.edu/29641/Dr-Donald-B-Broughton. [Accessed 12 Feb 2022].

[18] D. B. Broughton, R. W. Neuzil, J. M. Pharis, and C. S. Brearley, "The Parex Process for Recovering Paraxylene," *Chemical Engineering Progress*, vol. 66, no. 9, pp. 70–82, Sept 1970.

[19] D. B. Broughton and C. G. Gerhold, "Continuous Sorption Process Employing Fixed Beds of Sorbent and Moving Inlets and Outlets," US Patent 2,985,589, 23 May 1961.

[20] D. J. M. Rowe, "History of the Chemical Industry 1750 – 1930 – An Outline," 1998. [Online]. Available: edu.rsc.org. [Accessed 13 Dec 2021].

[21] B. V. Koen, *Discussion of the Method: Conducting the Engineer's Approach to Problem Solving*, Oxford University Press, Oxford, 2003.

[22] B. V. Koen, "Toward a Definition of the Engineering Method," in *Proceedings of the ASEE-IEEE Frontiers in Education Conference*, Philadelphia, Pennsylvania, 1984.

[23] B. V. Koen, "Engineering Method," in *Ethics, Science, Technology, and Engineering: A Global Resource*, 2nd ed., J. Britt Holbrook and C. Metcham, Eds., Macmillan Reference USA, Farmington Hills, Michigan, 2015, pp. 90–92.

[24] G. T. Austin, *Shreve's Chemical Process Industries*, 5th ed., McGraw-Hill, New York, 1984.

[25] M. F. Ali, B. M. E. Ali and J. G. Speight, *Handbook of Industrial Chemistry: Organic Chemicals*, McGraw-Hill, New York, 2005.

[26] G. Bugliarello, Eds., "Special Issue on the Biotechnology Revolution," *The Bridge: Linking Engineering and Society, Fall issue, National Academy of Engineering (USA)*, vol. 34, no. 3, 2004 (entire issue).

[27] A. K. Frolkova and V. M. Raeva, "Bioethanol Dehydration: State of the Art," *Theoretical Foundations of Chemical Engineering*, vol. 44, no. 4, pp. 545–556, 2010.

[28] K. A. Berglund, "Artisan Distilling: A Guide for Small Distillers," Department of Chemical Engineering and Materials Science, Michigan State University, East Lansing, Michigan, 2004.

[29] M. Strickland, *Batch Distillation: Science and Practice*, White Mule Press, Hayward, California, 2021.

[30] "Code of Ethics," National Society of Professional Engineers (USA), [Online]. Available: https://www.nspe.org/resources/ethics/code-ethics. [Accessed 28 December 2022].

[31] S. Sheppard, A. Colby, K. Makatangay and W. Sullivan, "What Is Engineering Practice," *International Journal of Engineering Education*, vol. 22, no. 3, pp. 429–438, 2006.

[32] C. B. Fleddermann, *Engineering Ethics*, 4th ed., Prentice Hall, Upper Saddle River, New Jersey, 2012.

[33] D. L. Grubbe, "Ethics – Examining Your Engineering Responsibility," *Chemical Engineering Progress*, vol. 111, no. 2, pp. 21–29, Feb 2015.

4 Addressing Universal and Changing Needs

Separation and purification operations continue to be developed and improved for numerous purposes aimed at generating value. These include:

- Product recovery. Recover desired components from a crude feed or reaction mixture (often termed primary recovery). Examples include the use of isoamyl acetate solvent to extract penicillin antibiotic out of microbial fermentation broth and the use of distillation to recover aniline from nitrobenzene after reaction with hydrogen.
- Product fractionation (of different classes of compounds). This involves selective isolation of a specific class of compounds from other compounds present in the same feed. An example is the extraction of aromatic compounds (benzene, toluene, and xylenes) from aliphatic hydrocarbons present in crude hydrocarbon feeds such as pyrolysis gasoline (pygas) produced as a byproduct of steam cracking ethane to ethylene.
- Product purification. Purifying a crude feed. This can involve purification of raw materials, intermediates, or products after primary recovery. Examples include purification of natural gas, purification of penicillin after its recovery from fermentation broth (via crystallization), and purification of aniline after recovery from nitrobenzene (via distillation) prior to its use in making other products such as isocyanates for polyurethanes.
- Ultra-purification. Removal of trace impurities to ultra-low levels as needed to ensure satisfactory product performance in use. This may involve removing impurities present at only 10 or 100 parts per million (ppm) down to 1 ppm or even lower. Examples include removal of trace aldehydes from acrylic latex for low-odor paint, removal of trace limonin and other bitterness compounds from orange juice to improve the taste, and removal of musty-smelling trichloroanisole impurity from cork for fresher smelling wine-bottle corks (to avoid what is known as cork taint). Another example involves removal of ultra-low trace metals from isopropanol (IPA) solvent for use in the manufacture of microelectronic devices to avoid defects caused by trace metals. And in certain cases, the production of monomers for polymers also requires a high level of purification.
- Delivering a desired form of product. Sometimes a product must be delivered in a specific form needed for its intended use. This is especially a concern for crystallization processes, where it is often important to deliver a crystalline product with a specific crystal structure, shape, and size distribution.
- Concentrating a feed. Reducing the overall volume of feed or product by selectively removing solvent or inert components. Concentration of feed may be needed to minimize the size and cost of downstream processing equipment. Examples include aqueous feed concentration by using an evaporator or a membrane module to remove a

https://doi.org/10.1515/9783110695052-005

portion of the water content or the use of a partially-miscible, hydrophilic, oxygenated solvent to extract the water.

- Solvent recovery. Recover and purify a solvent for reuse and recycle to minimize the need for make-up solvent and for solvent disposal. This is a standard requirement of most liquid-liquid extraction solvents and many crystallization solvents. An example is the use of distillation to recycle amyl acetate back to the fermentation-broth extractor (used for primary recovery of penicillin).
- Solvent exchange. Exchange one solvent for another, one that is better suited to a downstream operation or product formulation. Examples include solvent exchange in multi-step batch manufacture of pharmaceutical active compounds prior to crystallization from solution (as in switching from methylene chloride to isopropyl alcohol, e.g.), and back extraction of penicillin from amyl acetate into water and away from impurities.
- Reaction. Remove reaction products from a reaction mixture to minimize unwanted series reactions (avoid adding additional substituents to the product molecule) or to further drive conversion. Examples include control of nitration reactions by adjustment of liquid-liquid contacting conditions and enhancement of the transesterification reaction used to make fatty acid methyl esters (FAME for biodiesel fuel) by reaction of methanol with triglycerides in vegetable oil, yielding second-phase glycerol as a byproduct. Removal of glycerol helps drive conversion.
- Waste treatment and environmental protection. Treating a waste stream to remove pollutants prior to discharge to air, water, or landfill is a key aspect of environmental protection. Waste may be treated to meet or exceed regulatory requirements or as a pretreatment process to remove constituents that would interfere with a final treatment process or make it impractical. Examples include removal of volatile organic compounds from air vented from process equipment, removal of salt from waste brine prior to biotreatment and disposal in a river, and the adsorption of toxic components (such as phenol) out of wastewater prior to biotreatment to avoid injury to the health and activity of the bacteria in the treatment plant.
- Recycling and upcycling. Upgrade previously used materials or waste materials so they can be used as raw material for new products (recycling and upcycling). Upcycling refers to recycling previously used materials for higher-value uses. These kinds of operations rely heavily on separation methods.

These purposes are universal in the sense that they drive the development of separation applications across many industries. And they continue to change in the sense that the specific needs and goals change over time as the standards of the day change, presenting new challenges and opportunities. Although separation technologies such as distillation and extraction may be considered mature after many years of continual development, many new opportunities to generate significant value through continued improvement are driven by society's need to produce better materials and better products while reducing the consumption of scarce resources, improving safety, and reducing environmental impact. Box 4.1 lists a number of challenges and opportunities of great importance today.

Box 4.1: Current Challenges and Opportunities in Advancing Chemical Separations

- Continual attention to improving safety in design and operation. This includes better understanding of safety hazards associated with each new application and implementation of safety measures designed to provide safe working conditions for operators and other workers, to avoid incidents, and to prevent injury in the event of an accident.
- The need for reduction in energy consumption and improvements in the energy efficiency of separation (for resource conservation). This involves replacing older processes with newer, more efficient designs that allow a reduction in energy consumption (both for individual separators and for the overall process scheme). This can involve updating a given separation method with a newer process scheme or replacing an older method with an entirely different one.
- Greater use of alternative feedstocks and energy sources (especially renewable ones) and therefore a need to separate and purify new feedstock materials and to power the required processes in new ways.
- Better equipment designs that provide higher capacity, higher efficiency, lower capital cost, and more robust operation (less fouling or corrosion, better mechanical reliability),
- Better materials for use as separating agents. For example, more selective, higher capacity, and less expensive specialty adsorbents and membrane materials,
- Reconditioning or retrofitting of older installed equipment to improve throughput or separation performance; for example, by installing improved agitators, packing, trays, adsorbents, etc., avoiding the need to install a more costly brand-new operation,
- Better design methods (faster, more reliable, and more aware of overall system optimization),
- Better chemical property estimation methods (faster, more reliable),
- Process intensification (enabling smaller, more productive equipment),
- Process simplification (fewer processing steps, lower capital cost),
- Continuous mindfulness of potential hazards and the never-ending need to strengthen attitudes, procedures, and designs to ensure safe operations,
- Mindfulness of the need for sustainable operation over the long term.
- Removal of trace impurities including residual solvents, monomers, and reactants to lower and lower levels in the final product. This includes reduction of trace impurities in paints and coatings and in food packaging materials down to ppm levels or even lower.
- Development and use of new solvent materials with desirable properties (low toxicity, environmentally friendly, easy to recycle, low cost, and available in large industrial quantities).
- Reduction in the production of waste products (higher efficiency in the use of raw materials and reduction in the volume of waste material that is produced in the production of desired products),
- Finding economically-viable uses for left-over or used materials or for waste streams (recycling and upcycling) to reduce or avoid sending waste material to landfill.
- Improving technologies for pollution prevention or reduction of emissions to air, water, and land. New developments in separation technology will be key to meeting society's goals toward further reducing emissions to the environment.
- Developing new technologies for treating groundwater and land contamination resulting from certain past industrial practices. Current efforts include groundwater remediation at heritage manufacturing sites in the U.S. and elsewhere.
- Development of strategies and technologies to address society's goal of reducing CO_2, methane, and other greenhouse gas emissions.

Protecting the environment has been a concern of many involved in the chemical industry for many years, and this concern has grown to be of great importance to the industry overall. Some of the first industrial-scale pollution control technologies were introduced in the late 1800s and early 1900s. The use of condensers and absorbers/scrubbers on vent streams and the implementation of industrial waste-water treatment plants are good examples. The Alkali Act of 1863 in Great Britain required that soda ash factories reduce HCl acid gas emissions [1], and this was accomplished in part by absorbing HCl into water using a scrubber column. In the 1930s, biotreatment processes first developed for treatment of municipal wastewater [2] were being used at large industrial sites such as the Dow facility in Midland, Michigan specifically for treatment of industrial wastes [3]. The concept of biochemical oxygen demand was used to quantify the requirement for aerobic digestion and waste destruction [4]. However, even with these advancements, up until the 1960s many chemical processes were only half a process in the sense that their waste streams were not highly processed to remove or destroy pollutants. Instead, after minimal processing, some wastes were sent directly to the atmosphere or to rivers or to the ocean. The focus was on dilution and dispersion into the environment rather than elimination or destruction at the source. Historical examples include the construction in 1919 of one of the world's tallest smoke stacks at the copper smelter near Anaconda, Montana (USA) [5]. This stack was designed to disperse pollutants high into the atmosphere. The 178 m (585 ft) tall brick chimney structure, with a diameter of 23 m (75 ft) at the base, remains standing as a reminder of this past practice. The site was in operation until 1980 and is now a Montana State Park. The odor of industrial emissions was tolerated at the time as the smell of progress and an acceptable tradeoff. But this is no longer considered acceptable. Industrial practices have greatly improved, driven by greater awareness on the part of the chemical industry and society along with the passage of laws and emissions regulations beginning in the 1960s and 1970s. For an example of current requirements, see ref. [6]. In addition, cap and trade policies were developed in the 1980s and enacted in the United States in 1990 [7]. These policies allowed the trading of power plant emission credits to provide market-based economic incentives to invest in emission-reduction technologies – with the ultimate goal of reducing total emissions of pollutants that cause acid rain. With changes of this kind, chemical and power plant facilities have made significant progress since the 1960s in controlling emissions [8], but more work is needed. Today, many in industry are actively involved in tracking emissions, assessing toxicological risks (e.g., see ref. [9]), and setting goals for further reductions through continued technological advancement and investment. The reduction of CO_2 emissions is a current focus.

Publications explaining the issue of atmospheric CO_2 and climate change include refs. [10–15]. Some studies have proposed that the level of CO_2 in the atmosphere needs to be kept to a level no higher than about 450 ppm through 2050 and beyond [16]. It is said that achieving such a goal will require concerted efforts in two areas: (1) significant reduction in CO_2 emissions from all the various sources including transportation, housing, power plants, and industrial facilities and (2) recovery and sequestering of CO_2. This will require tremendous change in how society generates and uses energy and how the cost of CO_2 emissions is taken into account. How this issue is addressed will depend on many factors including the time and investment required for development and large-scale implementation of various options, the kinds of tradeoffs that must be made, and the real possibility of yet-to-be discovered innovations. A variety of initiatives already are underway. Box 4.2 lists some of these, many relying on the development of advanced separation technologies. Additional information may be obtained by searching the literature and the internet. Life Cycle Assessment is a technique that may be used to evaluate and compare the various options for their impact on reducing CO_2 in the atmosphere [17].

Box 4.2: Current Initiatives for Reduction of CO_2 in the Atmosphere

- **Improving energy utilization and reducing energy demand**. Improving efficiency in the use of energy and practicing conservation to reduce the demand for energy are good ideas whether CO_2 reduction is the goal or not, as such activities can pay for themselves over time through reduced utility costs. Opportunities for reducing energy consumption by improving the efficiency of specific separation operations are discussed in later chapters. Examples include retrofitting older distillation columns with newer, more efficient internals (distributors, trays, and packings), cross-exchange of energy between unit operations (pinch technology), improving process control to reduce variability in operation and avoid unnecessary over-purification [18, 19], and combining processing steps into hybrid process schemes such as reaction with in-situ separation (see Chapter 14). Other approaches include replacing older distillation configurations with newer, more efficient ones, such as the use of dividing wall columns and more-efficient multi-column distillation trains [20] and the use of various heat pump cycles for more efficient evaporators and dryers [21, 22].
- **Transitioning to cleaner-burning hydrogen-rich fuels for on-site steam boilers and furnaces that drive reactions and separations**. Substituting natural gas for coal can reduce CO_2 emissions by 40%. Replacing oil with natural gas can reduce emissions by 25% [23]. Because of this and its natural abundance, natural gas will continue to be a major source of energy for the foreseeable future.
- **Advancement in the production, purification, and distribution of carbon-neutral hydrogen, for direct combustion, for fuel cells, and as an environmentally friendly reducing agent**. One example involves development of so-called green steel manufacturing that uses hydrogen instead of fossil fuels [24]. However, the challenge of supplying carbon-neutral hydrogen at the required scale (perhaps tens of millions of tons per year) is great. At present, most industrial hydrogen is generated from steam reforming of hydrocarbons, which co-produces CO plus some CO_2. Use of this hydrogen will require processes for purifying the hydrogen, dealing with the CO as a chemical reagent or as a fuel gas burned to yield CO_2 which will then need to be captured and sequestered in some way. Or, other non-CO_2-generating sources of hydrogen will be needed at very large scale (much larger than exist today) such as electrolysis of water powered by electricity from solar, wind, hydroelectric, or nuclear power [25].
- **Greater use of combined heat and power (cogeneration) at electric power plants**. This involves generating steam and tempering heat-transfer fluids for nearby industries using elevated temperatures generated by the power plant such as the exhaust-gases from gas-fired turbines [26]. Such a power plant may be of any type that generates heat (nuclear, solar, biomass-fired, natural-gas-fired, coal-fired, etc.) and may include energy-efficient thermally driven refrigeration (such as the ammonia absorption/thermal stripping refrigeration cycle).
- **Implementation of advanced electric power generation technologies such as natural-gas-fired combined-cycle technology**. This involves generating electricity using gas-fired turbines along with downstream secondary turbines driven by steam generated by using the hot exhaust-gases. Future

potential advances include the natural-gas-fired Allam power cycle which promises more efficient CO_2 capture. The natural-gas-fired Allam power cycle utilizes enriched oxygen for combustion and CO_2 as diluent to control flame temperature (instead of using nitrogen in air). It generates a stream of highly concentrated CO_2 that is ready for sequestration [27].

– **Further development of CO_2-free electric power generation – nuclear, hydroelectric, deep-well geothermal, wind, solar, and even ocean tidal power.** An example of how CO_2-free energy can replace hydrocarbon-based energy in powering the chemical industry is the development of carbon-free electric furnaces to replace natural-gas fired furnaces at hydrocarbon cracking plants (steam-moderated thermal cracking of alkanes to olefins) [28]. Heat from the furnaces can power subsequent distillations and reduce their CO_2 emissions.

– **Development of more efficient technologies for recovery and purification of materials needed to enable CO_2-free energy generation, storage, and use** such as lithium, nickel, copper, and rare earth metals used in batteries, electronics, electric generators, and electric motors [29, 30].

– **Expanded use of recycled and upcycled materials to avoid the higher energy costs of resourcing and manufacturing from original raw materials.** Well-known examples include repurposing and recycling of steel, copper, plastics, motor oils, rubber products, concrete, and batteries of all types including lead-acid and lithium ion batteries. This can involve either direct recycling of old materials or dissolution or decomposition of used materials followed by separation and purification to yield as-new raw materials. For example, see ref. [31].

– **Use of wastes as feedstock for new products (a form of upcycling).** Well-established examples include the use of heavy residues at oil refineries to make asphalt for roads as an alternative to more-energy-intensive cement for concrete, the use of bio-mass fiber wastes in making compressed fiber board for residential and commercial construction, and the use of left-over animal fat and tallow from rendering plants as oleochemical products such as lubricants used in steel rolling operations and emulsion control agents (emulsifiers and de-emulsifiers).

– **Development of new alternative materials requiring less energy to produce and use.** Examples include Nexiite® and Glasscrete®, alternatives to concrete with claims of up to 30% less energy required in their manufacture.

– **Near-carbon-neutral production of bio-derived chemicals using feedstocks obtained from growing grain crops, vegetable oils, and cellulosic plants, as precursor chemicals and final products** [32]. Current research includes genetic modification to enhance desired plant properties. This approach may be used for the production of alcohols, esters, diols, carboxylic acids, oleochemicals, and furans as well as complex organics such as polymers, antibiotics, agricultural pesticides, and biopharmaceuticals. For example, a well-known approach to reducing net CO_2 emissions involves replacing petroleum-derived fuels with bio-derived diesel and jet fuel. Bio-derived chemicals come from growing crops that consume and sequester CO_2 which offsets later emissions and reduces the net emission of CO_2. Potential bioprocessing operations include microbial fermentation, fungi and algal cultivation, and enzyme catalyzed conversion steps. Not all potential bioprocessing schemes are sufficiently carbon neutral, and some are burdened by costly separations required to isolate desired compounds from dilute mixtures loaded with solids and impurities. Processes that produce deleterious off-gases such as CO_2 need additional steps for emissions control.

– **Carbon sequestration and resource regeneration in agriculture and forestry.** Chemical companies in the agricultural market are working directly with farmers to increase the use of agricultural CO_2 sequestration and soil regeneration practices. These include large-scale planting of cover crops between cash crops and reduced tillage of crop lands to increase the carbon content of the soil. Forestry initiatives involve planting specific types of trees in specific patterns with carefully managed harvesting of timber. See ref. [33].

– **Capture of CO_2 out of industrial streams such as those at hydrocarbon processing operations for acid gas removal (mostly CO_2).** This involves use of aqueous amine absorption fluids already used for hydrocarbon sweetening or other similar-acting absorption fluids or adsorbents. This approach is also being considered for removal of CO_2 at much lower concentrations from flue-gas at coal-fired power plants. See refs. [34, 35].

– **Development of synthetic fuels as carbon neutral, near-net-zero fuels.** Various bio-derived fuels and synthetic fuels derived from sequestered CO_2 are being developed as alternatives to petroleum and natural gas-based fuels in order to reduce the net emission of CO_2 from combustion. Bio-derived fuels such as bio-ethanol, bio-butanol, and soy-based biodiesel have already been commercialized. Another approach involves production of synthetic fuels from sequestered CO_2 and green hydrogen. The high energy consumption required to manufacture such fuels presents a significant challenge. See ref. [36].

References

[1] P. Reed, *Acid Rain and the Rise of the Environmental Chemist in Nineteenth-Century Britain*, Routledge, Abingdon-on-Thames, Oxfordshire, 2014.

[2] G. Lofrano and J. Brown, "Wastewater Management through the Ages: A History of Mankind," *Science of the Total Environment*, vol. 408, pp. 5254–5264, 2010.

[3] I. F. Harlow, T. J. Powers and R. B. Ehlers, "Industrial Wastes: The Phenolic Waste Treatment Plant of the Dow Chemical Company," *Sewage Works Journal*, vol. 10, no. 6, pp. 1043–1059, 1938.

[4] J. A. Tarr, "Industrial Wastes and Public Health: Some Historical Notes, Part I, 1876–1932," *American Journal of Public Health*, vol. 75, no. 9, pp. 1059–1067, 1985.

[5] "Anaconda Smoke Stack State Park," Montana State Parks, [Online]. Available: https://fwp.mt.gov/stateparks/anaconda-smoke-stack#:~:text=The%20old%20Anaconda%20Copper%20Company,Monument%20is%20555%20feet%20tall. [Accessed 27 December 2022].

[6] "Hazardous Air Pollutants," United States Environmental Protection Agency, [Online]. Available: https://www.epa.gov/haps. [Accessed 28 December 2022].

[7] R. Conniff, "The Political History of Cap and Trade," *Smithsonian Magazine*, August 2009.

[8] "Progress Cleaning the Air and Improving People's Health," United States Environmental Protection Agency, [Online]. Available: https://www.epa.gov/clean-air-act-overview/progress-cleaning-air-and-improving-peoples-health#:~:text=Between%201970%20and%202020%2C%20the,the%20air%20that%20we%20breathe. [Accessed 27 December 2022].

[9] D. G. Kovacs, R. R. Rediske, S. Marty, P. J. Spencer, D. Wilson and B. Landenberger, "Contemporary Concepts in Toxicology: A Novel Multi-Instructor General Education Course to Enhance Green Chemistry and Biomedical Curricula," *Green Chemistry Letters*, vol. 12, no. 2, pp. 136–146, 2019.

[10] W. D. National Research Council, *America's Climate Choices: Panel on Advancing the Science of Climate Change*, The National Academies Press, Washington D.C., 2010.

[11] E. Petruzzelli, Editor-in-Chief, "Thinking about Climate," *Chemical Engineering Progress (Bonus Issue)*, December 2020.

[12] E. Petruzzelli, Editor-in-Chief, "Accelerating Decarbonization," *Chemical Engineering Progress (Special Issue)*, September 2022.

[13] F. J. Dyson, "The Science and Politics of Climate," *APS News*, vol. 8, no. 5, pp. 12, 1999.

[14] S. Roussanaly, N. Berghout, T. Fout, M. Garcia, S. Gardarsdottir, S. M. Nazir, A. Ramirez and E. S. Rubin, "Towards Improved Cost Evaluation of Carbon Capture and Storage from Industry," *International Journal of Greenhouse Gas Control*, 106, March 2021 (Open Access) 103263, ISSN 1750-5836.

[15] "IPCC Reports," Intergovernmental Panel on Climate Change (founded by the World Meteorological Organization and the United Nations Environment Programme in 1988), [Online]. Available: https://www.ipcc.ch/reports/. [Accessed 28 December 2022].

[16] L. Walter, "Zero by 2050: Understanding the Challenge before Us," Third Way, [Online]. Available: https://www.thirdway.org/memo/zero-by-2050-understanding-the-challenge-before-us#:~:text=This% 20would%20put%20us%20on,pre%2Dindustrial%20levels%20by%202100. [Accessed 27 December 2022].

[17] S. E. Hunter and R. K. Helling, "A Call for Developers to Apply Life Cycle and Market Perspectives When Assessing the Potential Environmental Impacts of Chemical Technology Projects," *Industrial & Engineering Chemistry Research*, vol. 54, no. 16, pp. 4003–4010, 2015.

[18] L. A. Robbins, *Distillation Control, Optimization, and Tuning: Fundamentals and Strategies*, CRC Press, Boca Raton, Florida, 2011.

[19] D. C. White, "Optimize Energy Use in Distillation," *Chemical Engineering Progress*, vol. 108, no. 3, pp. 35–41, 2012.

[20] R. T. Gooty, P. Mobed, M. Tawarmalani and R. Agrawal, "Optimal Multicomponent Distillation Column Sequencing: Software and Case Studies," *Computer Aided Chemical Engineering*, vol. 44, pp. 223–228, 2018.

[21] V. Minea, Eds., *Advances in Heat Pump-Assisted Drying Technology*, CRC Press, Boca Raton, Florida, 2016.

[22] B. W. Hackett, "The Essentials of Continuous Evaporation," *Chemical Engineering Progress*, vol. 114, no. 5, pp. 24–28, 2018.

[23] "Green House Gas Equivalencies Calculator – Calculations and References," United States Environmental Protection Agency, [Online]. Available: https://www.epa.gov/energy/greenhouse-gases-equivalencies-calculator-calculations-and-references. [Accessed 26 December 2022].

[24] D. Kindy, "Fossil Fuel-Free 'Green' Steel Produced for the First Time," *Smithsonian Magazine*, pp. On-line at https://www.smithsonianmag.com/smart-news/green-steel-produced-first-time-180978550/, 31 August 2021.

[25] M. Tao, J. A. Azzolini, E. B. Stechel, K. E. Ayers and T. I. Valdez, "Review – Engineering Challenges in Green Hydrogen Production Systems," *Journal of Electrochemical Society*, vol. 169, no. 054503, 2022, DOI 10.1149/1945-7111/ac6983.

[26] "Combined Heat and Power: A Sleeping Giant May Be Waking," *Power Magazine*, pp. On-line at https://www.powermag.com/combined-heat-and-power-a-sleeping-giant-may-be-waking/, 1 March 2019.

[27] E. W. Kennedy, S. Metew, R. Chaturvedi, W. D. Seider, L. A. Fabiano and B. M. Vrana, "Carbon Capture by Allam Power Cycle: A Comparative Analysis," *Chemical Engineering Progress*, vol. 117, no. 12, pp. 29–34, 2021.

[28] A. H. Tullo, "Firms Explore Greener Ethylene," *Chemical and Engineering News*, 22 June 2020.

[29] D. Castelvecchi, "Electric Cars and Batteries: How Will the World Produce Enough?," *Nature*, vol. 596, pp. 336–339, August 2021.

[30] H. Royen and U. Fortkamp, "Rare Earth Elements – Purification, Separation, and Recycling (Report No. C211)," Swedish Environmental Research Institute, Stockholm, 2016.

[31] S. C. Kosloski-Oh, Z. A. Wood, Y. Manjarrez, J. P. De Los Rios and M. E. Fieser, "Catalytic Methods for Chemical Recycling or Upcycling of Commercial Polymers," *Materials Horizons*, vol. 8, pp. 1084–1129, 2021.

[32] R. Ahorsu, F. Medina and M. Constantí, "Significance and Challenges of Biomass as A Suitable Feedstock for Bioenergy and Biochemical Production: A Review," *Energies*, vol. 11, no. 3366, 2018. doi:10.3390/en11123366.

[33] J. J. Hutchinson, C. A. Campbell and R. L. Desjardins, "Some Perspectives on Carbon Sequestration in Agriculture," *Agricultural and Forest Meteorology*, vol. 142, no. 2–4, pp. 288–302, 2007.

[34] P. Bains, P. Psarras and J. Wilcox, "CO_2 Capture from the Industry Sector," *Progress in Energy and Combustion Science*, vol. 63, pp. 146–172, 2017.

[35] C. N. Schubert and T. C. Frank, "Regeneration of Acid Gas Containing Treatment Fluids". U.S. Patent 8,303,685, 6 Nov 2012.

[36] L. Snowden-Swan, S. Li, J. Jenks, J. Phillips, J. Askander, J. Holladay, L. T. Knighton and D. S. Wendt, "Synthetic Hydrocarbon Fuels from H_2 and Captured CO_2," in *Encyclopedia of Nuclear Energy*, Ehud Greenspan, Editor-in-Chief, Elsevier, Amsterdam, 2021, pp. 94–112.

Bibliography

Chapter 1

D. W. Green and M. Z. Southard, Eds., *Perry's Chemical Engineers' Handbook*, 9th ed., McGraw-Hill, 2018.

R. W. Rousseau, Ed., *Handbook of Separation Process Technology*, Wiley, Hoboken, New Jersey, 1987.

P. A. Schweitzer, Ed., *Handbook of Separation Techniques for Chemical Engineers*, 3rd ed., McGraw-Hill, New York, 1997.

J. L. Humphrey and G. E. Keller II, *Separation Process Technology*, McGraw-Hill, New York, 1997.

P. C. Wankat, *Separation Process Engineering: Includes Mass Transfer Analysis*, 4th ed., Prentice Hall, Upper Saddle River, New Jersey, 2016.

A. B. De Haan, *Process Technology: An Introduction*, 2nd ed., De Gruyter, Berlin/Boston, 2015.

A. B. De Haan, H. Burak Eral and B. Schuur, *Industrial Separation Processes: Fundamentals*, De Gruyter, Berlin/Boston, 2020.

T. Kevin Swift, M. Gilchrist Moore, H. R. Rose-Glowacki, E. Sanchez and Z. Saifi, *2019 Guide to the Business of Chemistry*, American Chemistry Council, Washington, D.C., 2019.

Anonymous, *Critical Technologies: The Role of Chemistry and Chemical Engineering, National Research Council (USA)*, National Academies Press, Washington D.C., 1992.

D. A. Crowl and J. F. Louvar, *Chemical Process Safety: Fundamentals and Applications*, 4th ed., International Series in the Physical and Chemical Engineering Sciences, Prentice-Hall, Upper Saddle River, New Jersey, 2019.

Center for Chemical Process Safety (AIChE), https://www.aiche.org/ccps.

Mary Kay O'Connor Process Safety Center, Texas A&M University, https://psc.tamu.edu.

M. Sam Mannan, Ed., *Lees' Loss Prevention in the Process Industries*, 4th ed., Butterworth-Heinemann, Oxford, England, 2012.

Responsible Care Program. International Council of Chemical Associations, https://icca-chem.org/focus/responsible-care. Accessed Nov. 22, 2022.

T. Sowell, *Basic Economics*, 5th ed., Basic Books, New York, 2014.

G. P. Towler and R. K. Sinnott, *Chemical Engineering Design: Principles, Practice, and Economics of Plant and Process Design*, 2nd ed., Butterworth-Heinemann, Oxford, England, 2013.

G. D. Ulrich and P. T. Vasudevan, *Chemical Engineering Process Design and Economics: A Practical Guide*, 2nd ed., Process Publishing, Durham, New Hampshire, 2004.

J. E. Anderson, "Determining Manufacturing Costs," *Chemical Engineering Progress*, vol. 105, no. 1, pp. 27–31, 2009, and references cited therein.

J. E. Anderson and A. Fennell," Calculate Financial Indicators to Guide Investments, "*Chemical Engineering Progress*, vol. 109, no. 9, pp. 3440, 2013, and references cited therein.

T. Brown, *Engineering Economics and Economic Design for Process Engineers*, CRC Press, Boca Raton, Florida, 2007.

E. Petruzzelli (Editor-in-Chief), "Accelerating Decarbonization," *Chemical Engineering Progress*, (special Issue), September 2022.

Chapter 2

C. E. Thomas, *Introduction to Process Technology*, 4th ed., Cengage Learning, Boston, 2015.

J. Harmsen, *Industrial Process Scale-up: A Practical Innovation Guide from Idea to Commercial Implementation*, 2nd ed., Elsevier, Amsterdam, 2019.

https://doi.org/10.1515/9783110695052-006

A. Cybulski, J. A. Moulijn, M. M. Sharma and R. A. Sheldon, *Fine Chemicals Manufacture: Technology and Engineering*, Elsevier, Amsterdam, 2001.
G. Gilleskie, C. Rutter and B. McCuen, *Biopharmaceutical Manufacturing: Principles, Processes, and Practices*, De Gruyter, Berlin/Boston, 2021.
V. Hessel, H. Löwe, A. Müller and G. Kolg, *Chemical Micro Process Engineering: Processing and Plants*, Wiley-VCH, Hoboken, New Jersey, 2005.

Chapter 3

Chemical and engineering society websites: American Chemical Society (ACS), American Institute of Chemical Engineers (AIChE), European Federation of Chemical Engineering (EFCE), Royal Society of Chemistry (RSC), and the European Route of Industrial Heritage, *Chemistry World* published by the Royal Society of Chemistry.
Distillations magazine published by the Science History Institute (Philadelphia, Pennsylvania).
The Bridge: Linking Engineering and Society published by the National Academy of Engineering (Washington, D.C.).
T. Sowell, *Conquests and Cultures: An International History*, Basic Books, New York, 1998.
J. Appleby, *The Relentless Revolution: A History of Capitalism*, Norton, New York, 2010.
J. Mokyr, *The Enlightened Economy: An Economic History of Britain 1700–1850*, Yale University Press, New Haven, Connecticut, 2012.
R. C. Allen, *The Industrial Revolution: A Very Short Introduction*, Oxford University Press, Oxford, England, 2017.
William S. Hammack, *The Things We Make: The Unknown History of Invention from Cathedrals to Soda Cans*, Sourcebooks, Naperville, Illinois, 2023.
S. Pinker, *Enlightenment Now: The Case for Reason, Science, Humanism, and Progress*, Viking, New York, 2018.
D. J. M. Rowe, "History of the Chemical Industry 1750–1930 – An Outline," 1998. [Online]. Available: https://edu.rsc.org [Accessed 13 Dec 2021]. Search the worldwide web for author and title.
D. A. Hounshell, *From the American System to Mass Production 1800 – 1932: The Development of Manufacturing Technology in the United States*, Studies in Industry and Society, Johns Hopkins University Press, Baltimore, 1984.
S. Matar and L. F. Hatch, *Chemistry of Petrochemical Processes*, 2nd ed., Gulf Professional Publishing, Houston, 2001.
F. Aftalion, *History of the International Chemical Industry*, 2nd ed., Chemical Heritage Foundation, Philadelphia, 2005.
G. Lofrano and J. Brown, "Wastewater Management through the Ages: A History of Mankind," *Science of the Total Environment*, vol. 408, pp. 5254–5264, 2010.
J. A. Tarr, "Industrial Wastes and Public Health: Some Histroical Notes, Part I, 1876–1932," *American Journal of Public Health*, vol. 75, no. 9, pp. 1059–1067, 1985.
H. A. Wittcoff, B. G. Reuben and J. S. Plotkin, *Industrial Organic Chemicals*, 3rd ed, Wiley, Hoboken, New Jersey, 2013.
J. G. Speight, *Handbook of Industrial Hydrocarbon Processes*, 2nd ed., Gulf Professional Publishing, Houston, 2019.
M. Farhat Ali, B. M. El Ali and J. G. Speight, *Handbook of Industrial Chemistry: Organic Chemicals*, McGraw-Hill, New York, 2005.
J. G. Speight, *The Chemistry and Technology of Coal*, 3rd ed., CRC Press, Boca Raton, Florida, 2012.
C. Alan Heaton, Ed., *The Chemical Industry*, 2nd ed., Springer, Heidelberg, Germany, 1994.

J. A. Kent and V. Tilak, "Bommaraju," in *Handbook of Industrial Chemistry and Biotechnology*, D. B. Scott, Ed., 13th ed., Springer, Heidelberg, Germany, 2017, formerly Riegel's Handbook of Industrial Chemistry.

G. Jagschies, E. Lindskog, K. Lacki and P. Galliher, *Biopharmaceutical Processing: Development, Design, and Implementation of Manufacturing Processes*, Elsevier, Amsterdam, 2017.

W. Noll, *Chemistry and Technology of Silicones*, Academic Press, Cambridge, Massachusetts, 1968.

E. G. Rochow, *An Introduction. Chemistry of the Silicones*, Wiley, New York, 1946, Macnutt Press, 2008.

H. Geng, *Semiconductor Manufacturing Handbook*, 2nd ed., McGraw-Hill, New York, 2017.

N. Kockmann, "History of Distillation," in *Chapter 1, 1–43 in Distillation: Fundamentals and Principles*, A. Górak and E. Sorensen, Eds. Academic Press, Cambridge, Massachusetts, 2014.

N. Kockmann, "200 Years in Innovation of Continuous Distillation", *Chemical BioEngineering Reviews*, vol. 1, pp. 40–49, 2003.

Dąbrowski, "Adsorption – From Theory to Practice," *Advances in Colloid and Interfacial Science*, vol. 93, no. 1–3, pp. 135–224, 2001.

E. Robens, "Some Intriguing Items in the History of Adsorption," in *Characterization of Porous Solids III, Studies in Surface Science and Catalysis*, Vol. 87, J. Rouquerol, F. Rodríguez-Reinoso, K. S. W. Sing and K. K. Unger, Eds., Elsevier, Amsterdam, pp. 109–177, 1994.

H. M. Schoen, C. S. Grove Jr. and J. A. Palermo, "The Early History of Crystallization," *Journal of Chemical Education*, vol. 33, no. 8, pp. 373–375, 1956.

H. S. Muralidhara, "History Overview," Section 2.1 In Chapter 2, "Challenges of Membrane Technology in the XXI Century," in *Membrane Technology, Butterworth-Heinemann*, Z. F. Cui and H. S. Muralidhara, Eds., 2010, pp. 19–32.

Chapter 4

Anonymous, "The Global Chemical Industry: Catalyzing Growth and Addressing Our World's Sustainability Challenges," International Council of Chemical Associations (ICCA), Arlington, Virginia, March 11, 2019.

D. T. Allen and D. R. Shonnard, *Green Engineering: Environmentally Conscious Design of Chemical Processes*, 2nd ed., Prentice-Hall, Upper Saddle River, New Jersey, 2015.

R. Turton, R. Bailie, W. Whiting, J. A. Shaeiwitz and D. Bhattacharyya, *Analysis, Synthesis, and Design of Chemical Processes*, 4th ed., Prentice Hall, Upper Saddle River, New Jersey, 2013.

D. T. Allen and D. R. Shonnard, *Sustainable Engineering: Concepts, Design, and Case Studies*, Prentice-Hall, Upper Saddle River, New Jersey, 2011.

P. T. Anastas and J. C. Warner, *Green Chemistry: Theory and Practice*, Oxford University Press, Oxford, England, 2000, Describes the 12 Principles of Green Chemistry.

S. E. Hunter and R. K. Helling, "A Call for Developers to Apply Life Cycle and Market Perspectives When Assessing the Potential Environmental Impacts of Chemical Technology Projects," *Industrial & Engineering Chemistry Research.*, vol. 54, no. 16, pp. 4003–4010, 2015.

M. Bidstrup, "Life Cycle Thinking in Impact Assessment – Current Practice and LCA Gains", *Environmental Impact Assessment Review*, vol. 54, 72–79, 2015.

Part II: **Design Fundamentals**

By becoming familiar with the underlying principles that govern chemical separation, a designer gains valuable insight into what is needed for good process performance. This is not unlike the music profession, where a working musician gains in proficiency, versatility, and self-confidence through familiarity with musical scales, rhythms, chord structures, and chord progressions for various styles of music (the fundamentals). As working engineers and chemists responsible for separation performance, we also gain in proficiency, versatility, and self-confidence through familiarity with the fundamentals. In our case, these include:

- The material balance and its implications,
- Molecular interactions underlying phase equilibrium,
- Phase diagrams as the map or landscape on which a process operates,
- Thermodynamic drivers and kinetic resistances controlling mass transfer rates,
- Favorable multi-phase fluid-flow patterns and process schemes,
- Ways of generating sufficient interfacial area needed for industrial-scale mass transfer,
- Ways of minimizing energy consumption for a given operation,
- The wide variety of methods or "styles" of separation.

Just as musicians focus on the fundamentals to become better at performing their craft, so do process engineers and chemists.

In reviewing process fundamentals, like many other authors we have chosen to divide the discussion into two main categories: phase equilibria (in Chapter 5: Thermodynamics) and factors affecting the rate of mass transfer (in Chapter 6: Mass Transport). An analysis of process performance often requires making an assessment as to whether phase equilibrium is the dominant factor controlling performance or whether performance is limited by the rate at which the system moves toward equilibrium. With this understanding, a design may be improved by focusing on those process variables that address the main issue. If the issue is thermodynamic in nature, then a different temperature or a different separating agent such as a different solvent or a different adsorbent may be needed. If the issue is a rate limitation, then devising ways of increasing interfacial area, reducing diffusion path length, or reducing liquid viscosity (perhaps by increasing temperature) may be in order. Of course, such changes may affect both equilibrium and mass transport to some degree, but one often dominates the other. With this in mind, a number of rules have been devised by practitioners to highlight the main factors controlling performance for a variety of separation methods. For example:

- Distillation begins to be mass transfer rate limited when the transfer factor (the stripping factor or absorption factor) for key components is greater than $\mathcal{T} = 2$ or so. Below $\mathcal{T} = 2$, the process operates close enough to equilibrium that mass transfer rate limitations become less important. This determines the type of process model needed for accurate design. When vapor and liquid molar flow rates are comparable, this general rule translates into the rough guideline: Distillation

https://doi.org/10.1515/9783110695052-007

begins to be mass transfer rate limited when the relative volatility of key components is greater than α = 2. In many distillation designs with α < 2, vapor and liquid rates are indeed comparable and the process operates close to equilibrium.

– As with distillation, extraction tends to be mass transfer rate limited when the extraction factor for the key component is greater than about ε = 2.

– A fixed-bed adsorption process should be operated with face velocities at the entrance to the bed of less than 1 ft/sec for gases (about 0.3 m/sec) or less than 1 or perhaps 2 gal/min per ft^2 for liquids (40 to 80 liters/min per m^2). At higher flow rates the process begins to be limited by the rate of solute diffusion into and out of the adsorbent's internal pore structure. Of course, this depends on fluid viscosity and the size and tortuosity of the pores.

– A continuously operated cooling crystallization process should be operated with no more than about 5 Celsius degrees of sub-cooling below the solution's saturation temperature to avoid formation of too many small difficult-to-filter crystals. With greater sub-cooling (greater supersaturation), the crystal nucleation rate begins to dominate the crystal growth rate, resulting in the formation of too many crystals that cannot grow to be very large.

We will explain the bases for these rules in our discussions of fundamentals in Chapters 5 and 6. They are only rough guidelines that vary depending on the specific application. The point we make here is that thinking about process fundamentals in order to make an assessment as to whether phase equilibrium or a rate limitation is the main factor controlling performance often is a key first step toward understanding and improving separation process performance. Sometimes the assessment indicates that both need to be addressed.

In starting a discussion of fundamentals, we also want to point out a traditional distinction in terminology that once was emphasized but now seems less popularly recognized: The outcomes of experimental work and the direct measurement of real things are *data*, and the outcomes of model calculations and simulation programs are *results* that need to be confirmed by comparison with data. The distinction between data and results emphasizes the need to experimentally validate model calculations which necessarily involve assumptions and simplifications. This distinction simply reflects an inability to anticipate all possibilities when dealing with complex physical/chemical systems. Of course, theory and calculations are critical to good design, and simulation programs are getting better all the time. But well-planned experiments allow us to check our thinking, they allow for discovery, and they provide direct feedback from nature to use in making design improvements – so we believe making a careful distinction between data and results is helpful. This does not mean that every situation will require generation of new data. A model or simulator may already be validated for applications similar to those of current interest, and useful data (exact or very similar) may already be available in a database. This often is the case for distillation design.

Readers may notice another common theme in the discussion of fundamentals: In many aspects of chemical engineering design, simplified ideal models are used as a framework for process data and as a guide for design, with model parameters being adjusted or modified to match data for real systems and real processes. Examples (some to be discussed later) include the following:

- Phase equilibrium data fitted by modifying ideal mixture equations using the concepts of activity and fugacity,
- Solubility model descriptors (such as Hansen solubility parameters and MOSCED parameters) determined by fitting solubility data,
- Representation of dispersed liquid droplets as ideal spheres in calculating interfacial area in liquid-liquid dispersions, with application of shape factors to correct for non-ideal shapes,
- Fluid flow models that frame real flow behavior between that for ideal plug flow and ideal completely mixed flow,
- Theoretical stage models with application of stage efficiencies to match data,
- The mass transfer unit model used with the assumption of constant transfer unit height in packed columns,
- Mass transfer coefficient correlations developed by using idealized models as a framework (modifications of two-film theory, penetration theory, and surface renewal theory),
- Simplified short-cut calculations used to examine and troubleshoot an existing separation unit operating in the manufacturing plant, such as calculation of the overall number of theoretical stages or transfer units achieved by the process. The results, although approximate, can highlight whether the unit is operating as intended or if there is a problem (an opportunity for improvement).

These simplified models are useful in framing our understanding of real processes, in terms of how well they describe general trends, but also in terms of how actual performance may deviate from the ideal model because the inherent assumptions are not always valid. In other words, thinking about why the simplified models sometimes fit actual data (or not) can be a useful mental exercise in thinking about how real equipment may deviate from ideal assumptions – in identifying potential rate-limiting steps, in optimizing a new design, or in troubleshooting poor performance of an existing unit. Is performance limited by deviation from ideal flow patterns or by inadequate interfacial area, by inadequate residence time or, on the other hand, by excessive residence time in the separator which may allow time for partial decomposition of the product and generation of impurities? Is performance limited by fouling and plugging of the equipment internals; or, as sometimes happens, by damage of the internals or loss of the internals altogether due to corrosion? And so the thinking goes. This line of thought can help focus the effort and clear the way to finding a good solution for the separation problem at hand.

Of course, the development and use of more complex models for greater under-standing of fundamental mechanisms also is important. For example, computational fluid dynamics (CFD) can be used to better understand fluid flow in the design of equip-ment internals, and molecular simulations can be used to better understand phase equi-libria. Computing power is no longer a major limitation. In principle, calculations can be carried out to simulate almost any molecular mixture and any process scheme. But chemical systems are very complex, and in many cases there will be significant un-certainties in physical properties, fluid flow, phase equilibria, trace reaction, or mass transfer resistance – uncertainties that require some experimentation and careful measurement to address. It is a bit like trying to forecast the weather a week ahead of time. Model calculations or simulations can be relied upon to highlight the most likely possibilities, but there are too many uncertainties to allow for a precise prediction. Calcu-lation methods such as CFD often are used to explore various geometries and flow pat-terns in order to zero in on several options for testing in the laboratory or miniplant. The idea is to minimize the amount of experimentation needed to finalize a design. So in prac-tice, both experiments (or data from a database) and calculations (models or simulations) are needed for commercial success. The degrees to which experimentation and calcula-tion are emphasized will vary depending on the particular separation method, but also on the experiences, areas of expertise, and preferences of the design team (and available resources). In some cases, as in the specification of certain distillation and extraction col-umns, straight-forward design calculations and process simulation using well-established databases can quickly yield all the information needed for design. On the other hand, designing adsorption and crystallization processes often relies heavily on experimental work. In many cases, a new separation application is first developed using laboratory and miniplant experiments (or existing data) along with sufficient process simulation to support design – with the goal of implementing a reliable commercial process in a timely manner and within budget. Once a technology has proved to be valuable and commer-cially successful, then additional resources may be used for additional studies of key fac-tors and underlying mechanisms and to develop more detailed simulations – with the goal of further refining the technology and improving its performance as needed. This is a practical approach that strives to balance upfront investment of time and resources needed to gain confidence in a new design (for design success), with the need to enter the market in a timely manner (for business success). Both involve risk, and the appropriate balance is a judgment call.

5 Thermodynamics: Characterizing Dynamic Equilibrium

Thermodynamics is the branch of science that deals with energy transformation and the state of equilibrium in macroscopic systems. The concept of dynamic phase equilibrium is central to separation process design. The following is a brief overview highlighting some of the key concepts for designers to keep in mind. Specific references and textbooks such as those listed in the bibliography at the end of Part II should be consulted for more detailed discussions. We also strongly recommend consulting thermodynamics specialists when developing phase equilibrium models, especially for complex, multi-component systems.

5.1 Ideal and Real Mixtures

The concept of an ideal mixture of components provides a basis for characterizing real mixtures – in terms of the magnitude of the difference between real and ideal behavior [1]. The components are the different chemical substances that constitute the mixture, which may be a gaseous mixture or a liquid or solid solution. Normally the component present in highest concentration is called the solvent and the other components are the solutes. But when considering the thermodynamic properties of mixtures, the solvent for a given solute is actually the mixture of all components that surround that solute. That is, the solvent composition includes the main solvent plus all dissolved solutes.

Several definitions of ideal behavior have been developed for different purposes. In one definition, the contribution of each component to the overall properties of the mixture is assigned according to their actual molecular concentration in the mixture. Normally this is expressed as mole fraction for liquid and solid solutions and as partial pressure for gaseous mixtures (equal to mole fraction for an ideal gas). With this definition of an ideal mixture, each component in the mixture behaves as if it is surrounded by its own kind. There are no synergistic or contrary effects due to mixing that would require correcting molecular concentrations in determining the mixture's properties. The behavior of real mixtures is then conveniently expressed by applying correction factors to mole fraction and to partial pressure to obtain corrected or effective concentrations of each component, using the equations for the ideal mixture as a framework. *Activity* is the name given to a corrected mole fraction for a liquid or solid solution. *Fugacity* is the name for corrected partial pressure for a gaseous mixture. The activity coefficient, normally designated by the Greek letter γ, is the multiplier applied to mole fraction, and the fugacity coefficient, designated by Φ, is the multiplier applied to partial pressure. Strictly speaking, fugacity may also be used to describe the liquid state. But for many practical purposes, it is often convenient to

https://doi.org/10.1515/9783110695052-008

focus on activity for the liquid and fugacity for the vapor. (There are exceptions, of course, as when using an equation of state, as we discuss later.) With this framework, the only requirements for the mixture (gas, liquid, or solid) to be considered ideal (such that activity and fugacity coefficients are unity) are that the energy change associated with taking a component molecule out of its pure-component reference state and placing it in the mixture is zero, and there is no change in the volume or density of the mixture. Fortunately, when considering vapor-liquid equilibria it is often reasonable to assume fugacity coefficients are unity and then focus solely on quantifying activity coefficients – because at pressures and temperatures well below the critical point, the non-ideal behavior of the system is mostly due to non-ideal molecular interactions in the liquid (the much denser phase). This is a practical way of looking at solution behavior. It should be noted that activity and fugacity coefficients are not simply correction factors but true thermodynamic properties of the components in solution related to fundamental energy relationships, as we discuss below.

Another point to mention here is the fact that an ideal vapor mixture defined in terms of a fugacity coefficient of unity does not necessarily behave like an ideal gas defined by the ideal gas law, PV = nRT. When the fugacity coefficient is unity, fugacity is equal to partial pressure, but that pressure may not follow the ideal gas law exactly even at standard reference conditions. Although the pure-component reference gas at standard temperature and pressure normally follows the ideal gas law very closely, there can be exceptions due to association of same molecules in the gas phase which forms some amount of dimer, and this can cause the pure-component reference and the mixture to deviate from the ideal gas law to some extent.

Another type of behavior is defined by Henry's law which describes the linear behavior of solute in a dilute liquid solution. Activity and Henry's law are sometimes viewed as two different ways of defining an ideal limit: solute acting as if it is surrounded by its own kind in one case (when the activity coefficient is unity), and solute surrounded by solvent molecules in the other (the linear Henry's law regime). To avoid confusion, in this book we use the terms ideal and non-ideal as they are defined by activity and fugacity coefficients being equal to or deviating from unity. The linear Henry's law regime is not an ideal limit in that sense. Instead, Henry's law describes the real behavior of both the liquid and the vapor (or gas) in the limit of dilute liquid solution; that is, in the linear regime. So measurement of the Henry's law constant includes any non-idealities in both liquid and gas phases.

A concept related to Henry's law involves the activity coefficient of a given solute in the limit as its concentration in solution decreases to very low levels and it becomes completely surrounded by solvent molecules (called infinite dilution). The activity coefficient at infinite dilution normally differs significantly from unity. And, at low to moderate pressures, the Henry's law constant and the solute's infinite-dilution activity coefficient are directly related. The value of one can be determined from the value of the other. However, at higher pressures when the behavior of the vapor or gas phase is non-ideal, unlike the activity coefficient at infinite dilution, the Henry's law constant reflects any non-ideal effects in the vapor as well as the liquid. And Henry's law is better able to deal with dissolved gases (as we will see).

For solids, there is another possibility called *eutectic* behavior, and this may be viewed as another kind of ideal behavior. In a eutectic system, a solid phase can form out of a multi-component liquid solution in the form of a crystal lattice structure containing only components of its own kind at equilibrium. That is, in a eutectic system an individual component may crystallize out of a liquid solution by forming a pure crystal lattice, rejecting all other components as impurities that do not fit or belong in the structure of the lattice. Not all systems behave this way. Other systems form complete or partial solid solutions instead of pure crystals. The story is even more complex for molecules that form solid compounds, some with variable composition, leading to highly complex phase diagrams. But eutectic behavior is an ideal that is fairly common and highly desirable for ease of chemical separation, as we discuss later in describing crystallization methods.

5.2 Dynamic Equilibrium and Gibbs Free Energy

The components of a chemical system are always in motion, with continuous change of position in space and continuous exchange of energy between components. Even a molecular or ionic solid exhibits some motion or vibration within the solid structure, although it is highly constrained. The total energy content of the system includes all chemical bonds and electrostatic interactions. At dynamic equilibrium, all of the forces that drive motion and energy exchange within the system are balanced, and all the macroscopic thermodynamic properties of the system are constant.

A chemical process spontaneously moves toward a state of dynamic equilibrium corresponding to minimum Gibbs free energy (G_{min}). The change in Gibbs free energy in going from some initial state to the final equilibrium state (ΔG) is given by:

$$\Delta G = \Delta H - T\Delta S \tag{5.1}$$

where ΔG = change in Gibbs free energy
ΔH = change in enthalpy
T = absolute temperature
ΔS = change in entropy

This fundamental relationship was introduced by the great engineer and physicist J. Willard Gibbs in the 1870s, building on the work of Rudolf Clausius and James C. Maxwell [2]. Gibbs's work laid the theoretical foundation for the entire field of chemical thermodynamics. These concepts were later developed by Johannes Diderik van der Waals, Gilbert N. Lewis, Edward A. Guggenheim, and many others to quantify phase equilibria and chemical reaction equilibria.

Enthalpy is defined as the internal energy of a closed system plus the product of its pressure and volume. Most chemical processes are conducted at constant pressure, including chemical separation. In this case, the change in enthalpy is equal to the amount of heat or thermal energy transferred between the chemical system and its surrounding environment. Furthermore, the change in enthalpy is indicative of the relative strength of molecular interactions (or component interactions) formed within the system during the process. For example, when cooling a system to freeze liquid to solid (cooling crystallization) or to condense vapor to liquid (condensation), the thermal energy that transfers from the system to the surroundings is equal to the change in the enthalpy of the system (equal to the heat of solidification or the heat of condensation). And this energy change corresponds to a strengthening of the interactions or bonds between components as the more condensed phase forms (solid forming out of liquid or liquid forming out of vapor). On the other hand, melting a solid or vaporizing a liquid requires heating the system (input of energy) to weaken or disrupt those bonds. The same concepts apply in thinking about formation of more or less attractive molecular interactions in chemical mixtures in general, with the formation of stronger, more attractive interactions corresponding to a decrease in enthalpy. A process or a molecular interaction is exothermic when it results in a decrease in enthalpy from removal or release of energy ($\Delta H < 0$), and it is endothermic when it results in an increase in enthalpy from input of energy ($\Delta H > 0$).

Entropy is defined in terms of the reversible flow of energy within a system:

$$\Delta S = \frac{\Delta Q}{T} \tag{5.2}$$

where ΔQ = the flow of energy within the system as the process proceeds.

So, the change in Gibbs free energy (ΔG) is a special energy function that represents nature's balance between change in the energy content of a system (ΔH) and the flow of energy within the system (ΔS). An increase in entropy ($\Delta S > 0$) represents the tendency for energy to flow or spread out and become dispersed in space. This corresponds to the natural tendency for heat to spontaneously flow from a hot object to a colder one until the temperature of the system (comprised of both objects) becomes uniform. The concept of entropy is useful in describing other kinds of spontaneous processes as well, including those involving chemical mixtures and mass transfer. In this case, entropy is often described in terms of a tendency of nature to move toward a state of greater disorder ($\Delta S > 0$); or, in going from a state of well-defined structure to a more chaotic state. More precisely, $\Delta S > 0$ refers to the spreading of energy in space which results in the dispersal of energy and matter, and this yields less ordered regions of energy (including bond energies) and a reduction in the segregation of molecules. The field of statistical mechanics defines ΔS in terms of the statistical behavior of molecules (or the particles that comprise a system), by considering probabilities associated with potential microstates of particle energy and position in space and how these relate to the macroscopic thermodynamic properties of the system. The more microstates available to a system, the more chaotic (stochastic) it can be and the greater the entropy.

We all know from the second law of thermodynamics that the entropy of the universe must increase. But in analyzing a chemical process we are concerned with a material system (separate from the surrounding environment) that allows for exchange of thermal energy with the surroundings (not exchange of matter). Such a system may undergo a change in entropy that is either positive or negative. This is made possible by the exchange of thermal energy with the surroundings. Consider a homogenous liquid mixture that spontaneously splits into two partially miscible liquids on standing. This hypothetical mixture might be formed by condensing at room temperature a vapor-phase mixture consisting of oil + water, the vapor having been formed by boiling a mixture of oil + water at a pressure near 25 mm Hg. Such a liquid is unstable. As the liquid forms out of the homogeneous vapor, oil molecules will begin to cluster together to minimize contact with water molecules – in order to avoid higher-energy (relatively repulsive) hydrophobic interactions. The savings in energy that results from this clustering or ordering is sufficient to offset the corresponding reduction in the entropy of the system. The general concepts to keep in mind, for our purposes here, are that dynamic equilibrium is determined by both enthalpic and entropic contributions, as expressed in the Gibbs free energy equation, and that spontaneous or favorable processes can involve either $\Delta S > 0$ or $\Delta S < 0$ depending on the enthalpic contribution. A process is favored whenever it involves moving to a lower free energy state ($\Delta G < 0$), and dynamic equilibrium corresponds to minimum Gibbs free energy for the system (G_{min}).

Getting back to crystallization, the formation of solid out of liquid may appear to defy thermodynamics with regard to entropy, because it involves going from a relatively disordered fluid state to a highly ordered rigid structure. Yet, this occurs spontaneously because crystallization processes are driven by nature's constant move toward minimum Gibbs free energy. As discussed above, in the process of forming solid out of liquid by cooling a system, the enthalpy of the system decreases due to formation of stronger interactions between components. At practical temperatures and pressures, the enthalpy of solidification is always exothermic ($\Delta H < 0$), with energy released to the environment, and this energy savings is sufficient to overcome the corresponding decrease in entropy. The reverse process (melting a solid) is always endothermic, requiring input of thermal energy, with a corresponding increase in entropy. Analogous concepts apply to liquid condensation and vaporization.

5.3 Chemical Potential and Mass Transfer

The chemical potential of a solute in solution is a measure of its chemical energy level. More precisely, the chemical potential of component i in a chemical mixture is the incremental change in the Gibbs free energy of the mixture that results from an incremental change in the number of molecules of component i present in the mixture, all other component amounts remaining unchanged. This is written:

$$\mu_i \equiv \left.\frac{\partial G}{\partial n_i}\right)_{n_{j \neq i}} \tag{5.3}$$

Given two phases in contact with one another, solute will naturally transfer from a state of high chemical energy in one phase (high chemical potential) to a state of lower energy (lower chemical potential) in the other – in the course to seeking to minimize the total Gibbs free energy for the system.

Consider a multi-component, two-phase system with phases I and II. A given component i will transfer from one phase to the other due to a chemical potential difference $\Delta\mu_i$. As a simple example, at constant temperature and pressure, the expression for transfer between two immiscible liquids can be written:

$$\Delta\mu_i = RT\ln\frac{a_i^I}{a_i^{II}} \tag{5.4}$$

The resulting transfer of solute from one phase to the other alters the compositions of each phase, which alters the solute's chemical potential in each phase, and mass transfer continues in this way until phase equilibrium is attained when chemical potential becomes balanced; that is, when the solute's chemical potential is the same in each phase. At this point equilibrium is attained, and

$$\mu_i^I = \mu_i^{II}, \ a_i^I = a_i^{II}, \ \gamma_i^I x_i^I = \gamma_i^{II} x_i^{II} \text{ at equilibrium} \tag{5.5}$$

where μ_i^I = chemical potential of component i in the indicated phase (I or II)
a_i^I = activity of component i in the indicated phase
γ_i^I = activity coefficient
x_i^I = mole fraction

Other expressions may be written in terms of fugacity. The point we make here is that the chemical potential of component i in a given phase is fundamentally related to the activity of component i in the mixture comprising that phase (for a liquid or solid) or the fugacity of component i (for a vapor). In practice, we normally do not deal directly with chemical potential but rather with the related concepts of activity, fugacity, or an equation of state in determining phase equilibrium. Then, for the design of mass transfer equipment, mass transfer is analyzed in terms of the deviation from equilibrium; that is, in terms of the difference between the current concentration of solute and the concentration at equilibrium. Performance can then be quantified in terms of achieving a theoretical stage worth of mass transfer or in terms of the mass transfer rate.

So, a deviation from equilibrium is the key driver for separation processes such as distillation, extraction, adsorption, and crystallization. These processes all involve solute diffusion within phases and across interfaces between phases. An exception is the use of membranes for separation, where the application of pressure normally (but not always) is the primary driving force. In Chapter 6: Mass Transport, we discuss driving forces along with other key factors affecting mass transfer, such as favorable fluid flow patterns and generation of high interfacial area.

5.4 Specific Molecular Interactions

In this section we describe common kinds of relatively attractive or repulsive interactions that can take place between molecules, with a focus on how they affect phase equilibrium. Key questions are: 1) Is a given mixture likely to behave as an ideal mixture, in which case our analysis can be greatly simplified; and 2) If the mixture is likely to deviate appreciably from ideal behavior, then why? That is, what specific types of intermolecular interactions are likely to cause non-ideal behavior and how are those interactions best handled or modeled? Specific molecular interactions are due to intermolecular forces of relative attraction or repulsion caused by separation of electric charges within and between molecules. In simple terms, opposite charges attract and like charges repel. Intermolecular forces are weak compared to the intramolecular forces that bind atoms together, as in covalent and ionic bonding. Intermolecular interactions result from variations in electron densities within a molecule's structure, phenomena that are beyond the scope of our discussion. However, a qualitative understanding of molecular interactions can serve as a useful guide in sorting through potential separation options and trends.

Note that here we discuss molecular interactions primarily in terms of single interactions occurring between two specific atoms or individual groups of atoms, one in each of two interacting molecules. But real mixtures can have multiple interactions that occur simultaneously, with non-linear proximity effects. The complex nature of these multiple interactions is a topic of computational chemistry and molecular modeling and a primary reason why experimental studies are required to validate phase equilibrium models.

5.4.1 Non-ideal intermolecular interactions

J. Willard Gibbs, working in the 1870s, was concerned with visualizing geometrical representations of macroscopic thermodynamic properties in devising his famous free energy expression, even building a 3D model that he shared with James C. Maxwell, the great Scottish scientist. In later years he also thought about individual particles and their possible microstates in devising the science of statistical mechanics, building on the work of Maxwell and Ludwig Boltzmann [2]. But in doing so, Gibbs did not speculate or theorize about the nature of the specific molecular interactions responsible for macroscopic behavior. This was left for others to develop beginning in the late nineteenth century and early to mid-twentieth century, in particular by Johannes Diderik van der Waals, Gilbert N. Lewis, and Edward A. Guggenheim, among others. In 1920, Wendell M. Latimer, Worth H. Rodebush, and Maurice L. Huggins proposed the concept that has become known as hydrogen bonding as an outgrowth of Lewis theory [3, 4].

Many of the phase equilibrium models that are now routinely used in industry were developed with a focus on specific molecular interactions and serve as a framework for correlating phase equilibria. Such models often are quantified by regression of select data to ensure sufficient accuracy needed for design. Extrapolation beyond the data can be useful in exploring potential process conditions and identifying candidate solvents, but keep in mind that extrapolated or predicted results generally cannot be relied upon for a final design.

As discussed earlier, a first step in thinking about molecular interactions is to think about whether a given interaction is likely to be enthalpic or entropic in nature (or both):

– Enthalpic (energetic) interactions: These are weak intermolecular interactions (relative to stronger intramolecular chemical bonding forces) due to non-uniform distribution of electrons within molecules (variation in electron density). They result in relatively attractive or repulsive forces acting between molecules called van der Waals forces.
– Entropic effects: These are structural effects influenced by energetic interactions. They are due to change in molar density on mixing molecules of different sizes or structures and to clustering together of like molecules within a mixture to avoid relatively repulsive interactions with unlike molecules (as in oil + water mixtures that form two liquid phases on mixing).

The forces acting between molecules are due to electrostatic interactions within and between molecules. These interactions are mainly due to:

– Dispersion. Caused by resonance in the separation of charges within a molecule. This refers to the potential for the distribution of electrons within a non-polar molecule (its electron density) to fluctuate, causing the molecule to electrostatically interact with other molecules. This effect is called London dispersion or London resonance. It produces an induced dipole moment with potential for relatively weak electronic interactions with other molecules.
– Polarity. Due to permanent separation of charges within a molecule (a permanent dipole moment). Depending on molecular structure, polarity varies from weakly polar to moderately polar to highly polar extremes. The difference in the electronegativities of interacting atoms largely determines how polar the molecule will be. In general terms, electronegativity is the ability of an atom or group of atoms to attract electrons to itself. Table 5.1 lists relative electronegativities of common atoms.

– Hydrogen bonding. This is a special kind of polar interaction between molecules involving hydrogen. A relatively electropositive hydrogen in one molecule (the proton donor) interacts with a relatively electronegative atom in another molecule (the proton acceptor). The order of electronegativity for atoms commonly involved in hydrogen bonding is F > O > Cl > N > S > C. Another type of hydrogen bonding involves hydrogen that is made particularly active by its close proximity to an electronegative atom within the same molecule. This makes the hydrogen more electropositive and therefore more able to interact with an electronegative atom in another molecule. Examples include halogenated hydrocarbons with two or three halogens on the same carbon as the hydrogen atom, such as dichloromethane, trichloromethane, and 1,1-dichloroethane. Common examples of hydrogen bonding include:

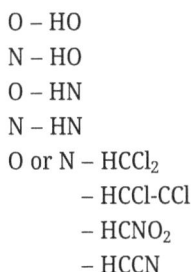

$$O - HO$$
$$N - HO$$
$$O - HN$$
$$N - HN$$
$$O \text{ or } N - HCCl_2$$
$$- HCCl\text{-}CCl$$
$$- HCNO_2$$
$$- HCCN$$

– π bond interactions. This involves sharing of electrons with a π bonded aromatic ring or organic complex. These are polar interactions that tend to be weaker than hydrogen bond interactions because the electronegativity of the aromatic ring or complex is weaker and the distance of interaction is greater [5]. Examples of π bond interactions include benzene + methanol and benzene + phenol (the OH hydrogen interacting with the benzene ring) and complexation of metals with π-bonding complexes of CO carbonyl or CN cyanide groups.

Table 5.1: Relative Electronegativities of Select Atoms [6].

K	0.82
Na	0.93
Si	1.90
P	2.19
H	2.20
C	2.55
S	2.58
N	3.04
Cl	3.16
O	3.50
F	3.98

The factors that determine the phase equilibrium properties of molecular mixtures may then be classified as follows:
- Ideal interactions between the same or very similar molecules
- Dipole interactions (induced dipole and permanent dipole)
  - non-polar + non-polar (induced-dipole + induced dipole = weak dispersion forces)
  - non-polar + polar (induced-dipole + dipole)
  - polar + polar (dipole + dipole)
  - association of molecules of the same kind (as in the formation of dimer and trimer)
- Hydrogen bonding (a special kind of dipole + dipole interaction)
  - electropositive hydrogen on polar molecule + electronegative atom on another polar molecule
  - active hydrogen on carbon next to electronegative atom on the same carbon (as in dichloromethane) + electronegative atom on polar molecule
  - active hydrogen on carbon next to electronegative atom on the same carbon along with one or more electronegative atoms on an adjacent carbon atom (as in 1,2-dichloroethane) + electronegative atom on polar molecule
- π bond interactions
  - polar molecule + aromatic molecule
  - metal ion + aromatic molecule
  - metal ion + π-bonding organic complex
- Entropic effects due to minimization of energy or change in molar densities. Examples include:
  - polar molecule + non-polar molecule (segregation resulting from minimizing energetically unfavorable interactions)
  - association of molecules to form dimer and trimer
  - eutectic behavior of solids (a kind of self-association or self-assembly)
  - polymer + solvent or polymer + monomer (change in molar densities on mixing)

Various chemical descriptors have been devised to represent the potential a given compound has for participating in these kinds of interactions. These include Kamlet-Taft descriptors, LSER (linear solvation energy relationship) and QSAR (Quantitative Structure-Activity Relationship) models, Hansen solubility parameters, and MOSCED (Modified Separation of Cohesive Energy Density) parameters. Descriptors related to dispersion, polarity, hydrogen bonding, and sometimes aromatic π bonding are quantified with reference to specific kinds of chemical behavior such as vapor-liquid equilibria, liquid-liquid miscibility, or solubility in a variety of solvent types. Lanny Robbins has further classified these into 12 sub-categories for specific classes of chemicals [7]. A modified version of Robbins' Chart is also available [8]. We highly recommend consulting these charts which are readily available elsewhere.

Adsorption phenomena may also be understood in terms of the interactions described above. These are the interactions that occur between molecules in a gas or a liquid and the molecular groups that comprise a solid surface. For example, an activated alumina surface containing surface hydroxyl groups can form relatively strong interactions with oxygenated organics and with water. An activated carbon surface containing surface methylene groups can form relatively strong interactions with organic compounds but only weaker interactions with water.

5.4.2 Classification of chemical mixtures

In thinking about specific molecular interactions, a first step is to consider whether such interactions are likely to be favorable:

Favorable	Unfavorable
$\Delta H < 0$ exothermic	$\Delta H > 0$ endothermic
$\Delta S > 0$ spreading	$\Delta S < 0$ segregating

With this in mind, we have classified binary chemical mixtures into 7 types that differ in terms of enthalpic and entropic effects [8, 9], as described below. To keep it simple, we have focused our attention on normal process temperatures, which we define as 0 to 100 °C. Classifying binary mixtures in this way shows how enthalpic and entropic effects can manifest themselves in common types of phase behavior (away from the critical point). Keep in mind that this is simply a way of classifying mixtures and analyzing behavior, not a way of predicting behavior. In the following discussion, the notations $g_{i,j}$, $h_{i,j}$ and $s_{i,j}$ refer to the change in enthalpy and entropy on mixing (called partial molar excess properties).

Type I. Chemically Similar, Small Molecules (Near Zero Enthalpy and Entropy Change on Mixing)
Type I mixtures are completely miscible and nearly ideal across a wide temperature range. These are mixtures comprised of components of the same chemical type or class, all of similar size, such as all small-molecule linear hydrocarbons or all similarly sized alcohols. Mixtures with significant non-idealities beyond Type I normally contain components from more than one chemical class (of similar size). Type I mixtures include propane + butane, benzene + toluene, and ethanol + methanol. Many hydrocarbon mixtures in the oil refining industry are Type I mixtures. As size differences increase, the entropic effect becomes significant and the mixture becomes more non-ideal in character (approaching the behavior of Type VII).

Type II. Regular-Solution-Like (Endothermic with Near-Zero or Slightly Positive Excess Entropy)
A regular solution is a classic type of mixture for which molecular interactions are weakly repulsive (compared to pure component interactions), with little hydrophobic

interaction, so $h_{i,j}$ is positive (endothermic) and $s_{i,j}$ is near zero or slightly positive. Examples include benzene + n-heptane, ethanol + 2-propanone, and methanol + n-heptane. Such mixtures become more ideal as temperature increases. They may be completely miscible or partially miscible but not sparingly miscible. If partially miscible, mutual solubility, increases with increasing temperature.

Type III. Endothermic with Negative Excess Entropy
For Type III binaries, $h_{i,j}$ is positive (or near zero) and $s_{i,j}$ is negative, and mixture behavior is sensitive to temperature. Type III mixtures may be completely miscible, partially miscible, or sparingly miscible depending on the magnitude of hydrophobic interactions. Examples include cyclohexane + 1,2-ethanediol, toluene + n-butanol, n-pentane + diethylene glycol, and C_4 to C_7 alcohols dissolved in water.

Type IV. Exothermic with Negative Excess Entropy
In this case, $h_{i,j}$ is negative and $s_{i,j}$ is negative. Type IV binaries typically involve water plus a surfactant-like molecule that has a hydrophilic (hydrogen bonding) group at one end of the molecule and a hydrophobic group at the other end. This allows for attractive interactions at one end with a high degree of segregation due to relatively repulsive interactions at the other end. Such mixtures exhibit partial miscibility and a lower critical solution temperature such that mutual solubility decreases with increasing temperature – so-called inverse solubility. This may be due to the disruption of hydrogen bonding with increase in temperature, allowing hydrophobic effects to become more significant. Examples include water mixed with a glycol ether such as propylene glycol *n*-butyl ether or with a non-ionic surfactant such as polyoxyethylene (polyethers) or with polymeric methylcellulose ethers.

Type V. Net Attractive Interactions with Negative Excess Entropy
Similar to Type IV mixtures, $h_{i,j}$ is negative and $s_{i,j}$ is negative. But unlike Type IV mixtures, Type V mixtures exhibit particularly strong attractive interactions such that activity coefficients are less than unity. Type V includes binaries that form new solute-solvent interactions not available to solute in its pure-component state. For example, a solute with hydrogen bond-accepting capability but lacking an active hydrogen may form strong attractions with a solvent that possesses an active hydrogen. This is the case for 2-propanone (proton acceptor) dissolved in an active-hydrogen compound such as chloroform (proton donor). Other examples include chloroform + 2-propanone and chloroform + ethyl acetate.

Type VI. Net Attractive Interactions with Positive Excess Entropy
Here, $h_{i,j}$ is negative and $s_{i,j}$ is positive. These are completely miscible mixtures of small and large molecules involving attractive interactions. Examples include alcohols dissolved in N-methylpyrrolidone polar aprotic solvent.

Type VII. Chemically Similar, Wide Molecular Size Distribution
Here, $h_{i,j}$ is slightly positive and $s_{i,j}$ is positive. Type VII binaries generally are completely miscible mixtures of small and large molecules without significant attractive interactions, so the entropic effect dominates. Examples include mixtures of C_4 to C_{10} hydrocarbons dissolved in C_{12} to C_{36} hydrocarbons. The mixture tends to become more sensitive to change in temperature as the difference is molecular size increases, due to the $T\Delta s$ term in the Gibbs energy equation.

In the following sections, we review popular phase equilibrium expressions commonly used to model these kinds of mixtures for design purposes, models that are commonly available in popular process simulation software. Of course, many other models are described in the literature and may also be used.

5.4.3 Activity coefficients

Activity coefficients have long been used to represent non-ideal molecular interactions in liquid solution. The Lewis-Randall activity coefficient [10] is a thermodynamic property that is directly related to the change in Gibbs free energy that occurs when pure compounds are mixed together.

$$RT\ln\gamma_i = g_i = h_i - Ts_i \tag{5.6}$$

where γ_i = activity coefficient of solute i dissolved in the mixture (the solvent)
 g_i = partial molar excess Gibbs energy of mixing, J/mole
 h_i = partial molar excess enthalpy of mixing, J/mole
 s_i = partial molar excess entropy of mixing, J/mole · K
 T = absolute temperature, K
 R = Universal gas constant, 8.314 J/mole · K

Here, the solvent is defined as the liquid comprised of all the components in the mixture including all solutes. As we discussed earlier, activity can be thought of as a "corrected mole fraction" for a component dissolved in the liquid (given by $a_i = \gamma_i x_i$), with the activity coefficient serving as the correction term. The same concept may be applied to solid solutions, as well. The activity coefficient is equal to unity for an ideal solution. For other mixtures where net solute-solvent interactions are repulsive relative to solvent-solvent interactions or lead to segregation of the mixture (negative entropic effects), the solute's activity coefficient is greater than unity and the solute behaves as if there is more of it present in the mixture than its mole fraction would indicate ($a_i > x_i$, a positive deviation from ideality). For mixtures dominated by positive entropic effects or by attractive solute-solvent interactions relative to solvent-solvent interactions, the solute's activity coefficient is less than unity and its effective mole fraction is reduced ($a_i < x_i$, a negative deviation from ideality). Activity coefficient data compilations are available in various literature references and commercial databases. For example, see ref. [11] and the Dortmund Data Bank [12] available from DDBST GmbH [13].

The activity of solute dissolved in solvent often is characterized by the value of γ for a binary pair at infinite dilution, denoted by $\gamma_{i,j}^{\infty}$. This refers to the limiting value of γ as the concentration of solute decreases, to the point where solute i becomes completely surrounded by solvent j. The value of $\gamma_{i,j}^{\infty}$ generally is the largest value for that binary pair interaction, and so it serves to characterize the behavior of the pair. The value of $\gamma_{j,i}^{\infty}$ for component j dissolved in component i does not necessarily have the same value. If it does, it is called a symmetric mixture. In the absence of data, a mixture is sometimes assumed to be symmetric for estimation purposes. The behavior of multi-component mixtures normally can be estimated reasonably well by using activity coefficient correlation equations parameterized with values for all binary pairs at infinite dilution (but not always).

Many activity coefficient correlation equations or models have been developed over the years. Several of the commonly used ones are listed in Box 5.1, illustrated here for binary mixtures. Their derivations are discussed in standard textbooks on solution thermodynamics [14–17]. These equations model the composition and temperature dependence of the activity coefficient for non-ionic solutes in solution. They are commonly used in modeling phase equilibrium involving liquid solutions (see Section 5.5.1: Vapor-Liquid Equilibria and Section 5.5.2: Liquid-Liquid Equilibria), but they may also be used to model thermodynamic properties of any kind involving liquid solutions, such as interfacial tension. Current research includes efforts to incorporate association terms into the standard activity coefficient models to better model associating systems such as those containing alcohols or carboxylic acids [18].

Typical equations are of two types: those based only on enthalpic considerations (partial molar energy of solvation) and those that take into account both enthalpic and entropic interactions (partial molar excess Gibbs energy), as summarized in Box 5.1. The resulting equations are widely used to correlate experimental vapor-liquid and liquid-liquid phase equilibrium data. Model constants are adjusted to fit the data and the best-fit equations are then used to interpolate values between data points. In practice, these equations often are developed to represent data within a narrow composition range of interest needed for design and not across the entire composition range – such as for very dilute solutions (near infinite dilution, where solute is completely surrounded by solvent) or for concentrated solutions (at moderate to high concentrations), or anywhere within a specific composition and temperature range of interest. These equations are not generally reliable for extrapolation beyond the range of data used to develop them, so extrapolation as a function of composition or temperature normally is not recommended, and if it is done, it must be done with care. The equations outlined in Box 5.1 are for binary mixtures. Multi-component mixtures are modeled using similar relationships plus certain mixing rules [1, 15, 17]. Other versions of these same models may be used for prediction purposes. Box 5.2 lists various ones commonly used in industry, but many others can be found in the literature.

Box 5.1: Popular Activity Coefficient Correlation Equations (shown for a binary mixture, $i + j$)*.

	Description	Ref.
Enthalpic (Solvation Models)		
Margules	$RT\ln\gamma_i = A(1 - x_i)^2$	[1] [15]
van Laar (a regular solution, such that $RT\ln\gamma_i$= constant at constant composition)	$RT\ln\gamma_i = A\left(1 + \dfrac{A\,x_i}{B\,x_j}\right)^{-2}$ = constant at constant x_i, x_j	[1] [15] [19]
Skatchard-Hildebrand Sol'y Parameter (a regular solution, such that $RT\ln\gamma_i$ = constant at constant composition)	$RT\ln\gamma_i = V_i^L \phi_j^2 \left(\delta_i - \delta_j\right)^2,\ \ \delta \equiv \left(\dfrac{\Delta U_i}{V_i^L}\right)^{\frac{1}{2}}$ where $\dfrac{\Delta U_i}{V_i^L}$ is the cohesive energy density of the liquid	[1] [15] [19]

Enthalpic + Entropic Excess Gibbs Energy (Local Composition Models)		
Wilson (for completely miscible mixtures)	$\ln\gamma_i = -\ln\left(x_i + \Lambda_{ij}x_j\right) + x_j\left(\dfrac{\Lambda_{ij}}{x_i + \Lambda_{ij}x_j} - \dfrac{\Lambda_{ji}}{x_j + \Lambda_{ji}x_i}\right)$ $\ln\gamma_i^{\infty} = -\ln\Lambda_{ij} - \Lambda_{ji} + 1$	[1] [15]
Non Random Two Liquid (NRTL)	$\ln\gamma_i = x_j^2\left[\tau_{ji}\left(\dfrac{G_{ji}}{x_i + x_j G_{ji}}\right)^2 + \dfrac{\tau_{ij}G_{ij}}{\left(x_j + x_i G_{ij}\right)^2}\right]$ $\ln\gamma_i^{\infty} = \tau_{ij}\exp\left(-a_{ij}\tau_{ij}\right) + \tau_{ji}$	[1] [15]
Universal Quasi Chemical (UNIQUAC)	$\ln\gamma_i = \ln\dfrac{\Phi_i}{x_i} + \dfrac{z}{2}q_i\ln\dfrac{\theta_i}{\Phi_i} + \Phi_j\left(l_i - \dfrac{r_i}{r_j}l_j\right)$ $- q_i\ln\left(\theta_i + \theta_j\tau_{ji}\right) + \theta_j q_i\left(\dfrac{\tau_{ji}}{\theta_i + \theta_j\tau_{ji}} - \dfrac{\tau_{ij}}{\theta_j + \theta_i\tau_{ij}}\right)$	[1] [15]

*This Table illustrates the general form of these equations with the usual notation and symbols. Consult the references for symbol definitions.

Box 5.2: Popular Activity Coefficient Estimation Models.

Model	Description	Ref.
UNIQUAC Functional Group Act'y Coef. (UNIFAC)	Chemical group contribution method	[20] [21]
Modified Cohesive Energy Density (MOSCED)	Modified regular soln. = f(chem. properties). Models infinite-dilution activity coefficients. Employs multiple parameters including two for hydrogen bonding.	[11] [22] [23]
Hansen Solubility Parameters	Modified regular soln. = f(chem. properties). Employs three parameters for dispersion, polarity, and hydrogen bonding.	[24] [25]
NRTL-SAC	Modified NRTL = f(chem. properties). Employs molecule segment descriptors to account for molecular interactions.	[26] [27]
COSMOtherm	Calculates activity = f(electron density profile)	[28] [29] [30]

With UNIFAC and other group contribution methods, values for specific groups are determined by regression of data for mixtures of compounds containing those groups. These models can then be used to estimate properties for other mixtures containing compounds comprised of the same groups. The same general approach may be used to determine values of chemical descriptors for the models that employ them. This is the approach often taken with the Hansen, MOSCED, and NRTL-SAC models. Molecular modeling methods may also be used to estimate model parameters; for example, for MOSCED [23]. And COSMOtherm methods may be used to estimate values in the absence of data (see Section 5.4.6: COSMOtherm).

Activity coefficients typically are strong functions of temperature as well as composition, and most activity correlation equations include some temperature-dependent terms. However, in most cases these equations can only approximate the effect of temperature. Many industrial activity coefficient models are modified to include empirical adjustment of the model constants as a function of temperature – by careful regression of temperature-dependent data. This normally involves parameterizing a linear expression of a model parameter as a function of temperature, but more extensive series expansions may also be used. However, for screening purposes and initial studies of potential process options, the temperature dependence over a temperature range of 50 Celsius degrees or so may be represented by another empirical expression of the following form [9]:

$$\ln \gamma_{i,j}^{\infty} \ (at\ T) = a_{i,j} \left(\frac{T_{ref}}{T}\right)^{\theta_{i,j}}$$

(5.7)

where $a_{i,j}$ = known reference point, $a_{i,j} = \ln\gamma_{i,j}^{\infty}$ at $T = T_{ref}$

$\quad T_{ref}$ = reference temperature (Kelvin)

$\quad T$ = temperature of interest (Kelvin)

$\quad \theta_{i,j}$ = empirical constant determined by fitting data

Equation 5.7 is derived assuming that the ratio of entropic to enthalpic effects $(Ts_{i,j}/h_{i,j})$ is constant over the temperature range of interest such that $\theta_{i,j} = 1/\left[1 - \left(Ts_{i,j}/h_{i,j}\right)\right]$. Specific values of $\theta_{i,j}$ for representative binary pairs are given elsewhere [9]. Typical values for real mixtures are in the range of −5 to +5 depending on the type of mixture. A value of $\theta_{i,j} = 0$ indicates ideal behavior, and $\theta_{i,j} = 1$ indicates regular solution behavior $(RT\ln\gamma_{i,j}$ = constant). Negative values indicate inverse-solubility behavior. Table 5.2 lists a summary of typical values of $\gamma_{i,j}^{\infty}$ and $\theta_{i,j}$ for the various classes of compounds described earlier in Section 5.4.2: Classification of Chemical Mixtures.

Table 5.2: Typical Values of Activity Coefficients and Temperature Dependence. Taken from ref. [9]. Reprinted with permission from *AIChE J.*, copyright 2014, American Institute of Chemical Engineers.

Mixture type	Typical characteristics	Typical $\gamma_{i,j}^{\infty}$ values and change with temperature		Typical $\theta_{i,j}$ values
I	Chemically similar, small molecules (Nearly Ideal)	$0.8 < \gamma_{i,j}^{\infty} < 1.2$,	$\frac{\partial\gamma_{i,j}^{\infty}}{\partial T} \approx 0$	$\theta_{i,j} \approx 0$
II	Regular-solution-like	$1.2 < \gamma_{i,j}^{\infty} < 1000$,	$\frac{\partial\gamma_{i,j}^{\infty}}{\partial T} < 0$	$1 < \theta_{i,j} < 5$
III	Endothermic with negative excess entropy	$\gamma_{i,j}^{\infty} > 2$,	$\frac{\partial\gamma_{i,j}^{\infty}}{\partial T} < 0$	$0.15 < \theta_{i,j} < 1$
IV	Exothermic with negative excess entropy (inverse temperature dependence)	$\gamma_{i,j}^{\infty} > 2$,	$\frac{\partial\gamma_{i,j}^{\infty}}{\partial T} > 0$	$-3 < \theta_{i,j} < -0.3$
V	Net attractive interactions with negative excess entropy	$0.2 < \gamma_{i,j}^{\infty} < 1$,	$\frac{\partial\gamma_{i,j}^{\infty}}{\partial T} > 0$	$1 < \theta_{i,j} < 5$
VI	Net attractive interactions with positive excess entropy	$0.2 < \gamma_{i,j}^{\infty} < 1$,	$\frac{\partial\gamma_{i,j}^{\infty}}{\partial T} > 0$	$0.3 < \theta_{i,j} < 1$
VII	Chemically similar, wide molecular size distribution	$0.6 < \gamma_{i,j}^{\infty} < 1$,	$\frac{\partial\gamma_{i,j}^{\infty}}{\partial T} < 0$	$-5 < \theta_{i,j} < -0.5$

5.4.4 Henry's law

Henry's law is used to model the equilibrium behavior of solute at dilute conditions where equilibrium properties are linear functions of solute concentration in the liquid. This is the regime in which solute dissolved in the liquid is mostly surrounded by solvent molecules.

The application of Henry's law is an alternative to the use of infinite-dilution activity coefficients. For example, Henry's law may be used to relate vapor phase composition to liquid phase composition for a dilute liquid solution:

$$y_i = H_i x_i \tag{5.8}$$

where x_i and y_i are in units of mole fraction, and H_i is the dimensionless proportionality constant. With this model, The Henry's law constant represents the real behavior of liquid and vapor (or gas) in this dilute regime. That is, the non-idealities of both the liquid phase and the vapor phase (or gas phase) are lumped into the value of the Henry's law constant. (Here, when we refer to non-idealities, we mean those defined in terms of activity and fugacity coefficients differing from unity.) Advantages of using this model include its simplicity plus the fact that Henry's law constants have been measured for numerous binary pairs and their values are readily available in various literature and commercial compilations for standard conditions [8]. Much of the data are for low to moderate pressures, so the vapor phase is nearly ideal. In that case, the Henry's law constant is directly related to the infinite-dilution activity coefficient:

$$H_{i,j} = \gamma_{i,j}^{\infty} \frac{p_i^{sat}}{P_{total}} \quad \text{at low to moderate pressures} \tag{5.9}$$

The temperature dependence of this type of Henry's law constant depends on that of both the activity coefficient and vapor pressure. So the temperature dependence of $H_{i,j}$ may be correlated in a manner similar to that described above for activity coefficients, multiplied by the temperature dependence of vapor pressure as determined from vapor pressure data or estimated from vapor pressure correlation equations.

Another application of Henry's law involves modeling the solubility of gases in liquids as a function of gas partial pressure:

$$x_i^{sat} = H' p_i = \left(\frac{1}{H_i P_{total}} \right) \times p_i \tag{5.10}$$

where H' has the units of inverse pressure. Here, the conventions used to define H' and H_i are different (they are reciprocal relationships), but they are related as indicated above whenever the vapor is nearly ideal. There are many other ways of expressing Henry's law, which vary depending on the choice of units and whether the expression is used to calculate a property of the liquid or of the vapor. So, in using Henry's law constants it is important to determine how they are defined. For more discussion of Henry's law versus activity coefficient models, see Section 5.5.1: Vapor-Liquid Equilibria.

5.4.5 Equations of state

An equation of state (EOS) represents the physical properties of a chemical substance at thermodynamic equilibrium, especially its pressure and density as a function of temperature. The substance may be a pure component or a mixture of components. The classic example is the ideal gas law, PV = nRT. Many equations of state may be viewed as modifications of the ideal gas law, modified to better model the behavior of real vapors, gases, and liquids. Well-known EOS models of this type include those of van der Waals, Redlich-Kwong, Redlich-Kwong-Soave, and Peng-Robinson, as well as the Virial equation of state (an expansion in density terms). Much of the current research into improved equations of state focuses on improving accuracy in modeling a wide composition range below and above the critical point of the mixture.

For example, the Peng-Robinson EOS may be used to model hydrocarbon mixtures at high pressures where non-ideal effects in the vapor phase are more significant than those in the liquid phase [31]. It has the form:

$$P = \frac{RT}{V-b} - \frac{a}{V(V-b)+b(V-b)} \tag{5.11}$$

where a and b are best-fit model parameters.

Another type of EOS is the Statistical Association Fluid Theory (SAFT) model introduced by Walter Chapman and others [32] and the subsequent perturbed-chain model, PC-SAFT. Many models of this type are based on the perturbation association theory of Michael S. Wertheim (1931–2021) [33]. Such models describe fluid density as a function of temperature and pressure with reference to a hard-chain fluid, modeling the various components in the mixture in terms of the number of spheres or segments per chain and the number of bonds between chains (associating bonds). These models are often used to correlate data for polymer + solvent solutions.

5.4.6 COSMOtherm

An important method of another kind is the conductor-like screening model (COSMO or COSMOtherm) introduced by Andreas Klamt and colleagues [28, 29, 34–36]. This method utilizes computational chemistry to calculate a two-dimensional electron density profile to characterize a given molecule. This profile is then used to estimate phase equilibrium using Gibbs energy expressions and statistical mechanics. The Klamt model is called COSMO-RS (for realistic solvation). A similar model is COSMO-SAC (segment activity coefficient) published by Lin and Sandler [37]. Databases of two-dimensional electron density profiles (sigma profiles) are available in the literature and in commercial software. For example, a sigma-profile database of more than a thousand molecules is available from the Virginia Polytechnic Institute and State

University [38]. Once determined, these profiles allow convenient calculation of phase equilibria using available software.

Such methods are used for estimation purposes when lacking experimental data. In essence, they are similar to group contribution models such as UNIFAC because they relate the contribution of specific atoms or groups of atoms, as characterized by specific parts of the sigma profile, to the overall phase behavior of the molecule. COS-MOtherm models have an advantage in that they do not require prior analysis of a representative phase equilibrium data set to determine group-contribution values. These are determined by calculation of the sigma profile. As with UNIFAC, the accuracy of such calculations is often reasonably good depending on the chemical system; however, the results should be validated with actual phase equilibrium data for process design purposes.

5.5 Key Expressions of Phase Equilibria

Most phase equilibrium correlation equations used in industry are based an excess Gibbs energy expression, a Henry's law expression, or an equation of state [39]. For an excess Gibbs energy expression, think of summing the partial molar properties of all components in the mixture, beginning with pure component properties and then making adjustments to account for non-ideal interactions. In using a Henry's law expression, think of linear effects at dilute liquid conditions. For an equation of state, think of a modified ideal gas law that is adjusted to take into account non-ideal interactions between molecules including liquid-phase interactions.

Many separation design calculations and process simulation programs require the user to specify the phase equilibrium model and suitable values for the model parameters, often by regression of phase equilibrium data. One of the main sources of error in separation design involves an inappropriate choice of model or its use outside of its validated range. For best results, the designer should consult with a thermodynamics specialist, especially in cases involving highly non-ideal, multi-component systems. Here we describe well-known standard models used for design, but much work continues in developing improvements using computational chemistry methods and statistical mechanics. These include molecular dynamics simulations and free energy calculations, electron density functional theory (DFT) calculations, use of Monte Carlo simulations to identify minimum energy states, and chemical perturbation theory to understand chemical self-association.

5.5.1 Vapor-liquid equilibria

Vapor-liquid equilibrium for design of distillation processes often is expressed in terms of a modified Raoult's law:

$$\gamma_i x_i p_i^{sat} = y_i P_{total} \frac{\Phi_i}{\Phi_i^{sat}} \tag{5.12}$$

where γ_i = activity coefficient of component i in the mixture
       p_i^{sat} = pure component vapor pressure of component i
       x_i = liquid phase mole fraction of component i in the mixture
       y_i = vapor phase mole fraction of component i in the mixture
       P_{total} = total pressure
       Φ_i = fugacity coefficient of component i in the mixture
       Φ_i^{sat} = fugacity coefficient for pure component i

This expression comes from combining the concepts of Raoult's law (for ideal liquid solution behavior) with Dalton's law of partial pressures (for ideal vapor mixtures) plus corrections for mole fraction and vapor pressure (activity coefficients and fugacity coefficients). For more discussion, see Section 5.1: Ideal and Real Mixtures. At low to moderate pressures, the non-idealities of the system are mainly present in the condensed phase (the liquid). In that case:

$$\gamma_i x_i p_i^{sat} \approx y_i P_{total} \tag{5.13}$$

In distillation, the relative volatility of key components is represented by equilibrium ratios called partition ratios or K values:

$$K_i = \frac{y_i}{x_i} \tag{5.14}$$

Relative volatility $\alpha_{i,j}$ is defined as the ratio of K values for two key components i and j:

$$\alpha_{i,j} = \frac{K_i}{K_j} \tag{5.15}$$

It follows from eqs. (5.13) to (5.15) that relative volatility may also be expressed as:

$$\alpha_{i,j} = \frac{\gamma_i\, p_i^{sat}}{\gamma_j\, p_j^{sat}} \tag{5.16}$$

where p_i^{sat} is pure-component vapor pressure. For a binary distillation (or distillation of light key from heavy key components), the relationship between key components at equilibrium is given by:

$$y_i = \frac{\alpha_{i,j} x_i}{1 + (\alpha_{i,j} + 1) x_i} \tag{5.17}$$

Alternatively, for dilute solutions vapor-liquid equilibrium may be represented in terms of a Henry's law constant:

$$y_i = H_i x_i \tag{5.18}$$

where H_i is a dimensionless proportionality constant. Although not often stated as such, the Henry's law constant (in mole fraction units) is simply the K value of a given solute at dilute liquid conditions. Here, the non-idealities of both liquid and vapor (or gas) are lumped into the value of the constant, which normally is determined in laboratory experiments. The result is a simple and convenient model for applications involving dilute solutions such as stripping-mode distillation.

In applications involving condensable vapors (no permanent gases), the relationship between the Henry's law constant and the activity coefficient at infinite dilution is determined by relating eqs. (5.12, 5.13, and 5.18). For solute i dissolved in solvent j, this gives:

$$H_{i,j} = \gamma_{i,j}^{\infty} \frac{p_i^{sat}}{P_{total}} \frac{\Phi_i^{sat}}{\Phi_i} \approx \gamma_{i,j}^{\infty} \frac{p_i^{sat}}{P_{total}} = a_{i,j}^{\infty} \tag{5.19}$$

Henry's law is also used to model the solubility of gases in liquids. The model relates liquid phase composition to the partial pressure of the gas in the vapor phase:

$$x_i^{sat} = H' p_i = \left(\frac{1}{H_i P_{total}} \right) \times p_i \tag{5.20}$$

where H' has the units of inverse pressure. Note that H' and H_i are defined using different conventions and have different units. This is sometimes the source of considerable confusion, but the two conventions are related as indicated above when the vapor is nearly ideal. Otherwise their relationship may only be approximate. Henry's law can be expressed in a variety of ways using different units and measured at different pressures, so in applying Henry's law it is important to clearly understand how the Henry's law constants to be used have been defined and measured.

In modeling the solubility of gases in liquids, Henry's law has a clear advantage over the use of an activity coefficient model which requires invoking a hypothetical pure-component vapor pressure in representing the partial pressure of the gaseous component. And unlike the activity coefficient, the Henry's law constant represents non-idealities in the vapor (or gas phase) as well as in the liquid. For this reason, some process simulators that use activity coefficient models to calculate vapor-liquid equilibria will provide the option of representing non-condensable components (permanent gases) as separate Henry's law components. This avoids any need to enter a pseudo vapor pressure to represent the concentration of gas dissolved in the liquid, which often is very low, on the order of 100 ppm or less. Trying to represent gas solubilities using an activity coefficient correlation equation often leads to calculation of unreasonably high gas solubilities. Some process simulators also include an extended

Henry's law model which includes terms for the temperature dependence of the Henry's law constant.

Vapor-liquid equilibrium may also be expressed in terms of an equation of state (EOS). For example, the Peng-Robinson EOS often is used to represent VLE for hydrocarbon mixtures at high pressures where non-ideal effects in the vapor phase are more significant than those in the liquid phase. The Peng-Robinson EOS is given by:

$$P = \frac{RT}{V-b} - \frac{a}{V(V-b)+b(V-b)} \tag{5.21}$$

where a and b are best fit model parameters. An EOS may be used to calculate fugacity coefficients for the liquid phase as well as for the vapor phase:

$$\ln \hat{\varphi}_i = \frac{b_i}{b(Z-1)} - \ln(Z-B) - \frac{A}{2\sqrt{2}B}\left(\frac{2\sum_j z_j a_{ij}}{a} - \frac{b_i}{b}\right)\ln\left[\frac{Z+(1+\sqrt{2})B}{Z+(1-\sqrt{2})B}\right] \tag{5.22}$$

$$B = \frac{bP}{RT}, \ A = \frac{bP}{(RT)^2}, \ Z = \frac{PV}{RT} \tag{5.23}$$

where liquid and vapor phases have different binary interaction parameters. Vapor-liquid equilibrium is then given by:

$$y_i = \frac{\hat{\varphi}_i^{liquid}}{\hat{\varphi}_i^{vapor}} x_i \tag{5.24}$$

More detailed discussions of these relationships and their derivations are given elsewhere [31].

5.5.2 Liquid-liquid equilibria

In liquid-liquid extraction, solute transfers between two liquid phases until equilibrium is attained. At that point, the driving force for solute transfer goes to zero, and a given solute has the same activity in each phase. So, the equilibrium state (without chemical reaction) is given by:

$$a_i^I = a_i^{II} \tag{5.25}$$

$$(x_i \gamma_i)^I = (x_i \gamma_i)^{II} \tag{5.26}$$

where the superscripts I and II denote the two liquid phases. An equation of state such as PC-SAFT also may be used to represent liquid-liquid equilibrium, but this approach is not commonly used for extractor design. The equilibrium ratio or partition ratio, K, is then defined by:

$$K = \frac{(x_i)^{II}}{(x_i)^{I}} = \frac{(\gamma_i)^{I}}{(\gamma_i)^{II}} \tag{5.27}$$

Note that liquid-liquid equilibrium is determined solely by non-ideal molecular interactions with no contribution from differences in physical properties such as pure component vapor pressure.

But limited liquid-liquid miscibility does have an effect on solute volatility. For example, consider a sparingly miscible liquid-liquid system. At equilibrium, each liquid is saturated with the other. Within a given liquid phase, the relative volatility of the dissolved component (the solute) is determined in part by its solubility limit in that phase (the solvent), such that:

$$a_{i,j} \approx \frac{1}{x_i^{sat}} \frac{P_i^{sat}}{P_{total}} \tag{5.28}$$

where I denotes the dissolved solute, j denotes the solvent liquid, and x_i^{sat} is the saturation amount of solute in that liquid phase. These relationships are discussed with regard to the phase diagram in Section 5.6: Phase Diagrams: Equilibrium Curves, Azeotropes, and Eutectics and in Section 7.3.3: Continuous Stripping and Absorption.

At equilibrium, each liquid phase becomes saturated with solute from the other, and the saturation amount (the solubility of solute in the liquid) varies as a function of temperature. An equation derived from the Margules activity coefficient equation and commonly used to correlate the temperature dependence for a narrow temperature range has the form:

$$x_i^{sat} \approx \exp\left(-\frac{A}{RT}\right) \tag{5.29}$$

where A often is determined by fitting mutual solubility data. Additional activity coefficient correlation equations such as UNIQUAC and NRTL may be used to model liquid-liquid phase splitting and mutual solubility.

5.5.3 Solid-liquid equilibria

Solid-liquid equilibrium is, of course, a critical factor in crystallization. Measuring solid solubility and the solubility curve for the operating region of interest is a necessary first step in evaluating any crystallization process. In many cases, solid-liquid equilibrium may be modeled as follows [19]:

$$\ln x_i^{sat} = \frac{\Delta_{fus}S}{R}\left(1 - \frac{T_m}{T}\right) - \ln y_i^{sat} \quad \text{for } T \leq T_m \tag{5.30}$$

$$\Delta_{fus}S = \frac{\Delta_{fus}H}{T} \tag{5.31}$$

$$x_i^{sat}\big|_{ideal} \approx \exp\left[6.8\left(1 - \frac{T_m}{T}\right)\right] \tag{5.32}$$

where x_i^{sat} = solubility of solid component i in the liquid solution, mole fraction
   y_i^{sat} = activity coefficient for component i in saturated solution
   $\Delta_{fus}S$ = entropy of fusion for component i, J/mole/K
   $\Delta_{fus}H$ = enthalpy of fusion for component i, J/mole
   T_m = melting point (or freezing point) temperature, K

One result of these expressions is their prediction that very high-melting solids are only sparingly soluble in most solvents at moderate temperatures, which is the general behavior observed in practice. Furthermore, at temperatures above the melting point, the system becomes a liquid-liquid system which may be modeled as described in the previous section.

Equations (5.30) to (5.32) are written for the common situation involving eutectic solids. These are solid-liquid systems that do not form solid solutions. That is, the solid crystal lattice rejects foreign components as it forms out of solution. This amounts to assuming that the solid-phase activity coefficient for component i is equal to unity for the pure solid i. If a solid solution does form, then the equations must be altered. A simplified example is given by:

$$x_i^{sat}y_i^{sat}\big|_{liquid} \approx y_i^{sat}\big|_{solid}\exp\frac{\Delta_{fus}S}{R}\left(1 - \frac{T_m}{T}\right) \tag{5.33}$$

where

$$y_i^{sat}\big|_{solid} < 1 \tag{5.34}$$

Other possible complications include formation of solid intermolecular compounds or solvate compounds and solid phase transitions due to the existence of more than one crystalline form (polymorphism). Normally it is safe to assume a eutectic system whenever solute, solvent, and impurity compounds have very different molecular structures.

For a sparingly soluble solid dissolved in liquid solution, the relative volatility of the dissolved solute may be expressed as:

$$a_{i,j} \approx \frac{1}{x_i^{sat}} \exp\left[\frac{\Delta_{fus}S}{R}\left(1 - \frac{T_m}{T}\right)\right] \times \frac{f_i^o}{P_{total}} \qquad (5.35)$$

where f_i^o = vapor pressure (fugacity) of a hypothetical sub-cooled liquid obtained by extrapolating liquid vapor pressure below the melting point (the freezing point).

This is similar to the situation for a sparingly miscible liquid-liquid system discussed earlier, but now the melting point and entropy of fusion are additional factors. For more discussion, see Section 5.6: Phase Diagrams: Equilibrium Curves, Azeotropes, and Eutectics and Section 7.3.3: Continuous Stripping and Absorption.

Finally, it should be noted that in the laboratory it is often convenient to fit solubility data as a function of temperature by using a semi-empirical equation. Typical examples have the following form:

$$\ln x_i^{sat} = A\left(1 - \frac{T_m}{T}\right) - \frac{B}{T} \quad \text{for } T \leq T_m \qquad (5.36)$$

$$\ln x_i^{sat} = A - \frac{B}{T} \quad \text{for } T > T_m \qquad (5.37)$$

where A and B are best-fit constants determined from the slope and intercept obtained by plotting $\ln x_i^{sat}$ versus $1/T$.

5.5.4 Adsorption isotherms

For liquid-solid adsorption, the capacity of a solid adsorbent to adsorb solute out of a liquid solution depends on the concentration of solute in the feed and two competing solute affinities: the affinity of solute for the solid surface and the affinity of solute to remain solvated in the liquid. So, the adsorption capacity depends not only on the properties of the solid adsorbent, but also on the properties of the liquid solution. For significant adsorption to occur, the affinity a solute has for the liquid phase must be overcome by a stronger affinity for the solid phase. As an example, consider adsorption of a weak organic acid out of aqueous solution using an activated carbon adsorbent. The solid adsorbent's capacity is much higher for the fully protonated (non-dissociated, non-ionized) organic acid in aqueous solution compared to the dissociated (ionized) acid – because the fully protonated organic has much less affinity for interacting with water in solution. Adjusting the pH of the solution above or below the pKa of the acid to control the extent of solute ionization greatly affects the adsorption capacity. As another example, adding a mineral salt to the liquid phase will tend to reduce the affinity of an organic solute for the liquid and increase the adsorption capacity, an effect called the salting-out effect. In general, liquid feed composition and temperature matter greatly. This is also true for gas-solid adsorption, where adsorption capacity can be very sensitive to competitive adsorption by co-solute as well as any change in temperature. In addition, change in pressure matters greatly. Because of these many factors, experimental measurement using representative feed solutions is the most reliable way of quantifying adsorption capacity for design purposes as opposed to predictive methods. The results, normally obtained at constant temperature (isotherm data) can then be fit to various empirical or semi-empirical expressions as a function of solute concentration in the liquid or gas phase.

In practice, adsorption equilibrium often is expressed as the grams of solute adsorbed per gram of adsorbent as a function of solute concentration in the feed-phase at equilibrium. For liquid-solid adsorption, generating equilibrium adsorption data involves preparing vials of solution containing the solute of interest, adding known weights of adsorbent, and placing the vials on a shaker table over night. Measurements can be made at room temperature or at a controlled temperature using a heated or cooled sample holder. After equilibration, the liquid phase is analyzed and equilibrium adsorption capacity is calculated from:

$$q_i = \frac{M_{feed}}{M_{ads}} \left(x_{i,feed} - x_{i,equil} \right) \tag{5.38}$$

where q_i = equilibrium adsorption capacity, g solute adsorbed/g adsorbent

$x_{i,feed}$ = concentration of solute in the feed solution, wt. fraction

$x_{i,equil}$ = concentration of solute in the liquid at equilibrium, wt. fraction

M_{feed} = weight of feed solution, grams

M_{ads} = weight of dry adsorbent added to the vial, grams

Similar weight-change methods may be used to measure gas-liquid adsorption equilibrium.

There are many types of adsorption behavior [40]. In this book we focus on self-sharpening adsorption known as type I behavior, a common type of physical adsorption. Adsorption capacity initially increases linearly with increasing solute concentration and then plateaus to a constant high level. This type of behavior often is modeled using the Langmuir isotherm equation:

$$q_i = \frac{a x_{i,equil}}{1 + b x_{i,equil}} \tag{5.39}$$

where a and b are empirical constants determined by fitting available data. For many applications the Langmuir isotherm is quite satisfactory and even preferred over some other equations such as the Freundlich equation, because it resolves to a Henry's law (linear) isotherm at low concentrations and reaches a constant loading capacity at high concentrations. A useful approach to fitting the Langmuir isotherm is to fit the equation using two or more appropriately chosen data points. This may conveniently be done using a spreadsheet programmed to fit the data by adjusting the Langmuir model parameters.

Sometimes the adsorption behavior cannot be modeled adequately using a simple Langmuir expression. An example involves the adsorption of water out of organic solvents using zeolite molecular sieves. In this case, adsorption involves filling and saturation of micropores at low concentrations, followed by filling and saturation of larger channels within the adsorbent. This bimodal behavior may be modeled using a dual-mode Langmuir equation with four adjustable parameters:

$$q_i = \frac{ax_{i,\,equil}}{1 + bx_{i,\,equil}} + \frac{cx_{i,\,equil}}{1 + dx_{i,\,equil}} \tag{5.40}$$

Sometimes a simpler dual-mode equation is satisfactory:

$$q_i = \frac{ax_{i,\,equil}}{1 + bx_{i,\,equil}} + cx_{i,\,equil} \tag{5.41}$$

This equation resolves to a Henry's law isotherm at low concentrations, but cannot model a constant saturation loading capacity at high concentrations.

Another popular adsorption isotherm is the Freundlich isotherm:

$$q_i = ax_i^{1/n} \quad \text{for liquids} \tag{5.42}$$

$$q_i = aP^{1/n} \quad \text{for gases} \tag{5.43}$$

where P is the system pressure or partial pressure of component i.

These expressions can be useful for representing gas adsorption at moderate pressures and liquid adsorption at low liquid concentrations when the adsorption behavior does not follow Henry's law. However, a Freundlich expression must be used with caution because it can predict unrealistically high adsorption capacity at high solute concentrations outside the range of the correlated data.

Note that these equations do not capture the temperature dependence. Normally, the temperature dependence is determined simply by fitting a suitable isotherm expression at each temperature – using the same model framework for all temperatures. The best-fit model values for the model constants may then be fit to an empirical expression as a function of temperature.

No matter how adsorption capacity is modelled, it is important to deal with the working capacity of the regenerated material; that is, the adsorbent that has been repeatedly regenerated using the method envisioned for the overall process. The capacity of fresh, unused adsorbent can be significantly higher than material that has undergone numerous cycles of adsorption and regeneration. This is because adsorption into small micropores can be difficult to regenerate, so this micro-pore capacity may not be utilized in practice. This is relevant for adsorbents used in a pressure swing adsorption process (regenerated by reduction in pressure and use of a sweep gas) and for adsorbents used in a liquid-solid adsorption process (regenerated by using steam, for example). This means that while isotherms provided by adsorbent suppliers are useful for screening materials, the actual adsorption capacity of the re-generated material will need to be measured for use in modeling and process design.

Furthermore, for feeds containing multiple adsorbing solutes, the prediction of multi-solute adsorption capacity from knowledge of single-solute isotherms can be quite unreliable. This is because the adsorption of a given solute can alter the surface and change how other solutes adsorb in ways that are difficult to predict. A more strongly adsorbing component can reduce adsorption capacity for a less-strongly ad-sorbing component. It can even cause desorption of the less-strongly adsorbed compo-nent. It is therefore not valid to add capacities from single-solute isotherms to estimate the capacity for mixed-solute feeds, or to assume that all solutes behave the same as a given solute of the same class [41]. For this reason, it is normal practice to measure iso-therms using representative mixed-solute feeds. As a practical example, it is well known that high levels of humidity in the air significantly alter the adsorption capacity of activated carbon used to adsorb volatile organic compounds from air vents, so hu-midity must be controlled when measuring the adsorption isotherms used in designing carbon adsorbers used for air emissions control [42].

5.6 Phase Diagrams: Equilibrium Curves, Azeotropes, and Eutectics

Having discussed phase equilibrium relationships in previous sections, in this section we focus on how they all come together in a phase diagram. The phase diagram is the equilibrium roadmap or landscape on which a separation process operates. It shows regions of temperature, pressure, and composition that correspond to the various phases or states of matter at equilibrium: vapor or gas, one or more liquid phases, one or more solid phases, and the supercritical region. Various kinds of diagrams may be constructed to focus on specific phases and specific conditions of interest.

A separation process normally is driven by a deviation from equilibrium, so the actual path of the process deviates somewhat from equilibrium, and many factors can affect the rate of mass transfer (see Section 6.1: Driving Forces and Key Descriptors). But no matter how fast or slow a process proceeds, the phase diagram can be a reliable map showing where it can go. At least that is mostly true. Exceptions involve formation of non-equilibrium, meta-stable states of matter that may persist for extended periods of time, such as stable liquid-liquid emulsions and solids that contain defects. Further-more, phase diagrams for some systems can be quite complex, showing regions with many different kinds of solid phases. Detailed discussions are available elsewhere [43, 44]. Our aim here is to describe some typical phase diagrams that relate to popular sep-aration methods.

5.6.1 Equilibrium curves

A simple kind of phase diagram may be constructed simply by plotting an equilibrium line or curve showing the relationship between equilibrium concentrations for different phases. For example, in designing a binary distillation process, we plot the vapor-liquid equilibrium curve for the binary system and then position the operating curve next to it, as in a McCabe-Thiele diagram (see Section 7.3.4: Continuous Fractional Distillation). The equilibrium curve is the phase diagram or landscape on which the process operates. The operating curve is the chosen path. The phase diagram shows where the process can go (or not go).

To illustrate, consider the representative vapor-liquid phase diagrams shown in Figures 5.1 to 5.4 for binary mixtures containing benzene: benzene + p-xylene, benzene + acetone, and benzene + methanol. The system benzene + p-xylene (Figure 5.1) may be considered nearly ideal. It exhibits nearly symmetric equilibrium curves, and standard distillation may easily be used to obtain two pure products. On the other hand, benzene + acetone (Figure 5.2) is moderately non-ideal, and this non-ideality can be seen in the shape of its equilibrium curve. The equilibrium curve begins to hug the x-y diagonal line at high concentrations of acetone. This type of behavior is extreme in the case of the more highly non-ideal benzene + methanol binary. This binary exhibits a homogeneous azeotrope such that the equilibrium curve actually crosses the x-y diagonal, and this severely limits the purities that can be achieved by using standard distillation (Figure 5.3). But the composition of the azeotrope changes with change in temperature (Figure 5.4), and this suggests a work-around for obtaining pure products. We discuss this option in Section 8.3: Pressure Change Distillation.

The phenol + water system is another interesting non-ideal system, shown in Figure 5.5. This system is unusual because it forms a homogeneous azeotrope at atmospheric pressure (nbp = 99.2 °C, x_{H2O}^{AZ} = 0.9793 mole fraction equal to 90.1 wt% water). But on condensing the azeotropic vapor and cooling the condensate to ~30 °C, the condensate phase-splits into two liquid layers. This is behavior that emulates a heterogeneous azeotrope. This system also exhibits a eutectic point at ~0 °C and pure solid phases [45]. So the separation options for this system include distillation at temperatures near 70 °C with sub-cooling of the overheads condensate to 30 °C to form two liquid layers, and crystallization of pure or nearly pure crystals at temperatures below 0 °C. The general features of phase diagrams involving azeotropes and eutectics like these are described below.

A Note of Caution. The chemical systems discussed above are meant to provide informative examples, but keep in mind that handling these and other chemicals can be dangerous. In particular, acetone and methanol vapors can be harmful if inhaled in sufficient amounts, benzene is considered a carcinogen, *and phenol can be lethal on skin contact with only very small amounts*. So appropriate personal protective equipment and safe operating procedures *must* be carefully determined and always used whenever handling these and other toxic chemicals. (See Chapter 1 references [13–22].)

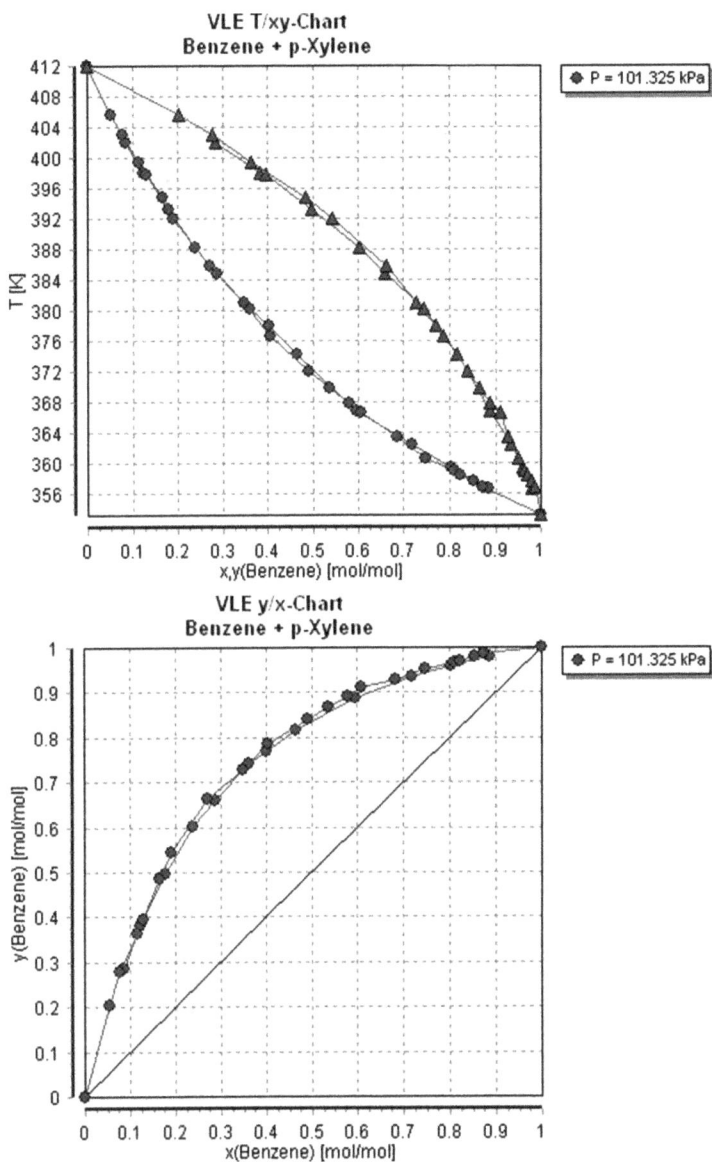

Figure 5.1: Benzene + p-Xylene Vapor-Liquid Equilibrium. Taken from the Dortmund Data Bank. Reprinted with permission, courtesy of DDBST GmbH [13].

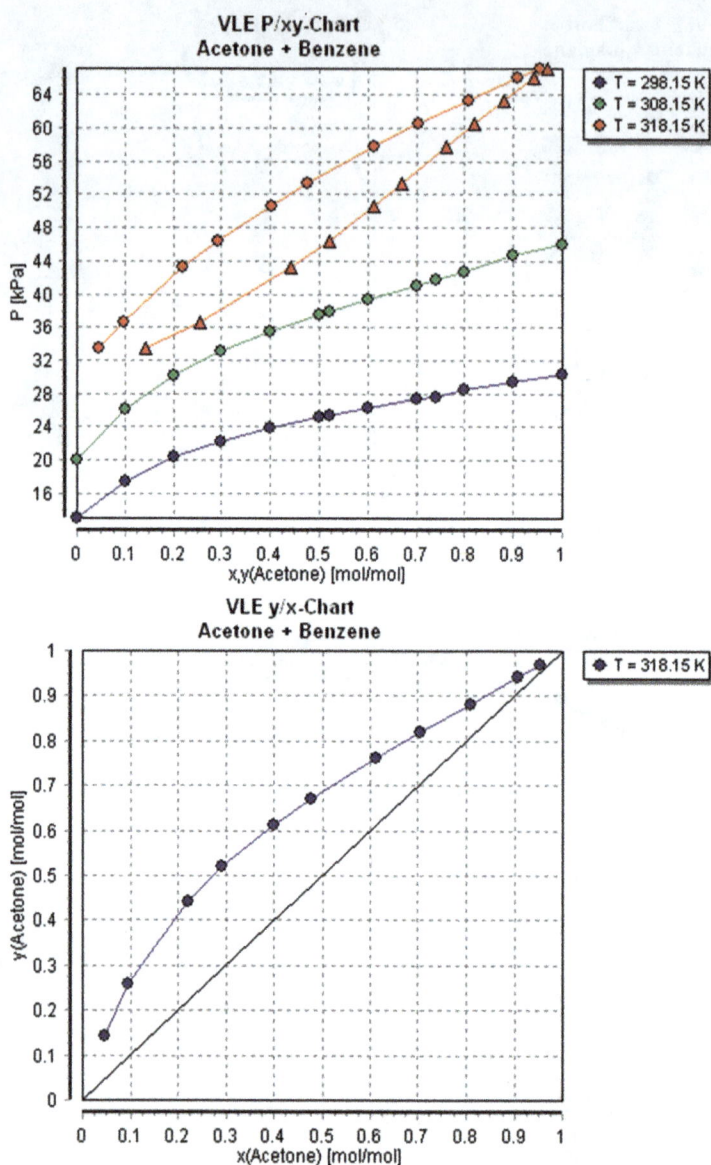

Figure 5.2: Acetone + Benzene Vapor-Liquid Equilibrium. Taken from the Dortmund Data Bank. Reprinted with permission, courtesy of DDBST GmbH [13].

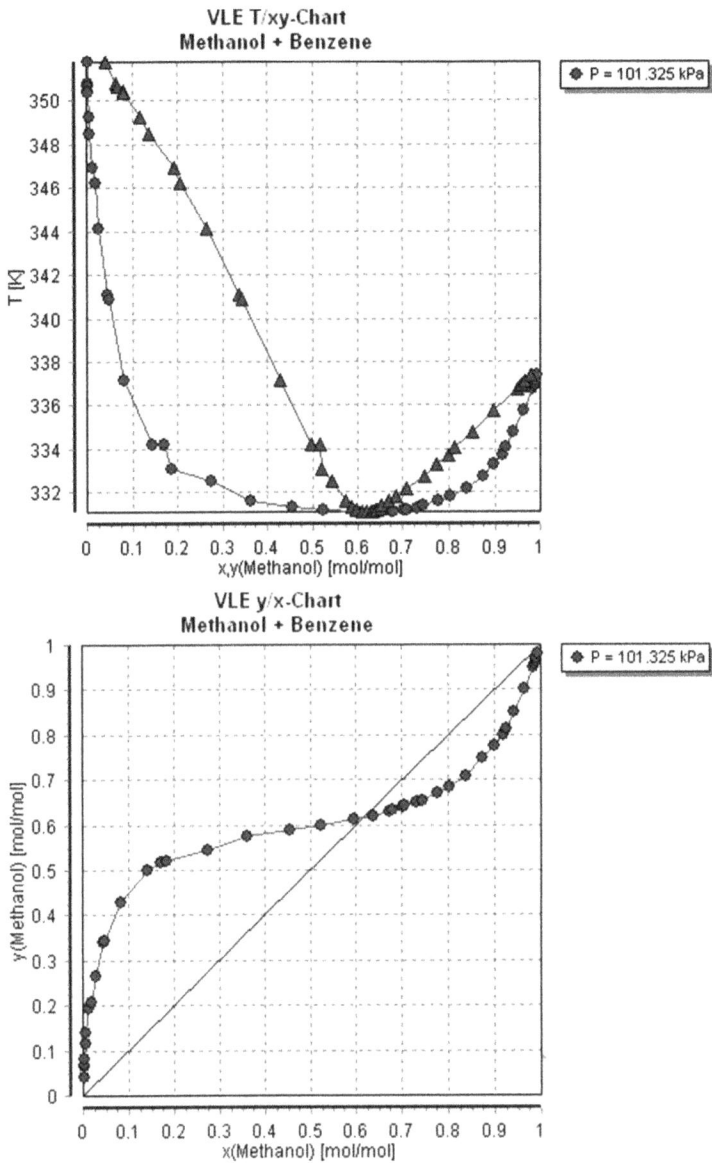

Figure 5.3: Methanol + Benzene Vapor-Liquid Equilibrium. Taken from the Dortmund Data Bank, Reprinted with permission, courtesy of DDBST GmbH [13].

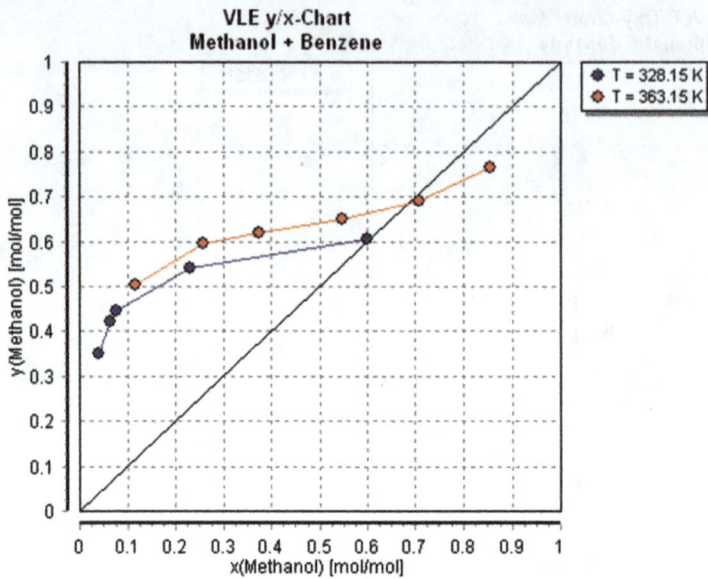

Figure 5.4: Methanol + Benzene VLE as a Function of Temperature. Taken from the Dortmund Data Bank. Reprinted with permission, courtesy of DDBST GmbH [13].

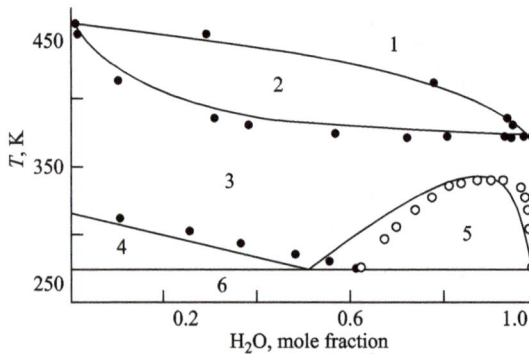

Figure 5.5: Phase Diagram for Phenol + Water Showing Vapor-Liquid, Liquid-Liquid, and Solid-Liquid Equilibria: 1) vapor phase; 2) region of vapor-liquid equilibrium; 3) liquid phase; 4) region of coexistence of solid phenol and solution; 5) region of liquid-liquid equilibrium; 6) region of coexistence of phenol crystals and ice. Taken from ref. [45]. Reprinted with permission from *Russian Journal of Applied Chemistry*, copyright 2007, Springer Nature.

5.6.2 Homogeneous azeotropes

The term homogeneous azeotrope refers to liquid and vapor phases of equal composition in equilibrium, as illustrated in Figures 5.3 and 5.4. The liquid is a single liquid phase. If two liquid phases are present at equilibrium, it is called a heterogeneous azeotrope. A good deal of information is available about azeotropes. The classic compilation of azeotropic data is by Lee Horsley [46]. Today, large data compilations are available in the Dortmund Data Bank [13] and other commercial sources. It is best to have real data, of course, but lacking data, azeotropic composition may be estimated by using various activity coefficient correlations or predictive models [15, 47].

The presence of an azeotrope indicates significant non-ideal molecular interactions in solution. In most cases, solute + solvent interactions are somewhat repulsive relative to solvent + solvent interactions, and as a result the activity coefficients at infinite dilution for the binary pair are significantly greater than unity. In these cases, the vapor pressure of the solution is somewhat elevated compared to an ideal mixture, so the boiling point of the mixture is less than that of either pure component. For this reason, most azeotropes, either homogeneous or heterogeneous, are classified as minimum-boiling azeotropes. In the extreme case (generally for activity coefficients greater than $\gamma = 7$), this shows itself not only through azeotropic behavior (minimum-boiling) but also through partial miscibility [15]. On the other hand, a few strongly interacting binaries form a maximum-boiling azeotrope with an elevated boiling point. This can happen when activity coefficients at infinite dilution are less than unity indicating strong attractive interactions in solution that suppress the vapor pressure of the mixture. An example of a maximum-boiling azeotrope is HCl + water.

For the reasons just mentioned, most azeotropes are at least somewhat sensitive to change in temperature. The effect of temperature on the composition of a homogeneous minimum-boiling binary azeotrope, like that shown in Figure 5.4 for methanol + benzene, may be estimated by extrapolation using Joffe's method. This method is based on the van Laar activity coefficient correlation and often gives good results [48]. First, the activity coefficients corresponding to the azeotropic data are calculated by setting $y_i^{AZ} = x_i^{AZ}$ in eq. (5.13). This gives:

$$\gamma_i^{AZ} = \frac{P_T}{P_i^{sat}} \text{ at the given value of } T^{AZ} \text{ and } x_i^{AZ} \text{ at } P_T \tag{5.44}$$

Then, the van Laar constants are calculated from the known azeotropic data point:

$$A = RT^{AZ} \ln \gamma_i^{AZ} \left(1 + \frac{x_j^{AZ} \ln \gamma_j^{AZ}}{x_i^{AZ} \ln \gamma_i^{AZ}} \right)^2 \tag{5.45}$$

$$B = RT^{AZ} \ln \gamma_j^{AZ} \left(1 + \frac{x_i^{AZ} \ln \gamma_i^{AZ}}{x_j^{AZ} \ln \gamma_j^{AZ}} \right)^2 \tag{5.46}$$

Next, the quantity C is calculated for a new value of T^{AZ}:

$$C = \sqrt{\frac{AB}{1 - \left(\frac{A-B}{AB} RT^{AZ}\ln\left(\frac{p_i^{sat}}{p_j^{sat}}\right)\right)}} \tag{5.47}$$

The corresponding azeotropic concentrations at the new temperature are then given by:

$$x_i^{AZ} = \frac{C - B}{A - B} \tag{5.48}$$

$$x_j^{AZ} = \frac{A - C}{A - B} = 1 - x_i^{AZ} \tag{5.49}$$

and the corresponding pressure at the new temperature is given by:

$$P_T = \exp\left(\frac{BA^2 \left(x_i^{AZ}\right)^2}{RT^{AZ} C^2}\right) p_2^{sat} \tag{5.50}$$

5.6.3 Heterogeneous azeotropes

The term heterogeneous azeotrope refers to vapor in equilibrium with two liquid phases, as illustrated in Figure 5.6. For sparingly soluble liquids, the composition and temperature dependence of the heterogeneous azeotrope may be calculated from pure component vapor pressures. In this case, each liquid phase independently exerts nearly its entire vapor pressure independent of the amount of each liquid present; that is, the total system pressure is close to the sum of pure component vapor pressures. For a binary mixture, azeotropic composition may then be calculated as follows:

$$y_i^{AZ} = x_i^{AZ} = \frac{p_i^{SAT}}{p_i^{SAT} + p_j^{SAT}}, T = T^{AZ}, p_i^{SAT} + p_j^{SAT} = P_{total} \tag{5.51}$$

where x_i = mole fraction of component i
x_j = mole fraction of component j
p_i^{SAT} = vapor pressure of pure component i at the azeotropic boiling point
p_j^{SAT} = vapor pressure of pure component j at the azeotropic boiling point
P_{total} = total pressure

Equation (5.51) provides a good estimate of y_1^{AZ} for systems with sparingly miscible liquids. It is normally a fair estimate for partially miscible liquids, as well, especially when both pure-component vapor pressures are comparable. Here, x_i^{AZ} is the concentration of component i in all of the liquid made up of both liquid phases. That is, x_i^{AZ} does not equal x_i^{SAT}, which is the concentration of component i dissolved in a single liquid phase at saturation.

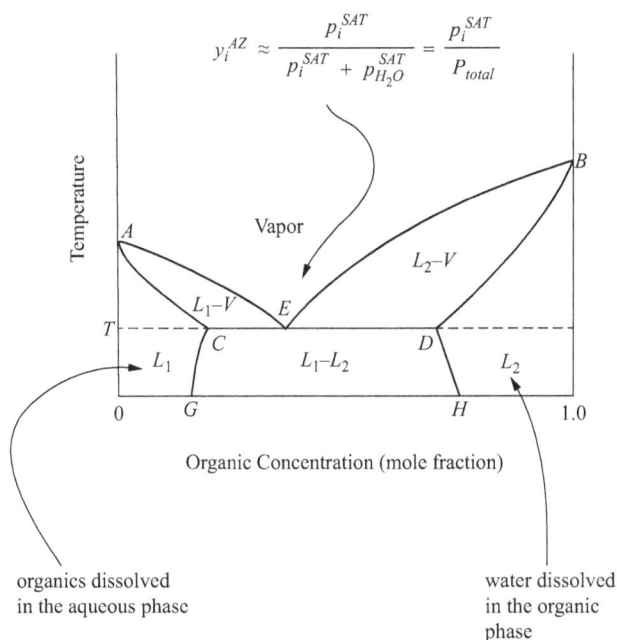

Figure 5.6: Representative Vapor + Liquid + Liquid Phase Diagram for an Organic + Water System Showing the Presence of a Heterogeneous Azeotrope. The solubility limits at the left and right-hand sides of the diagram represent the behavior of a partially miscible system.

All heterogeneous azeotropes exhibit relatively repulsive solute + solvent interactions responsible for liquid-liquid phase instabilities (phase splitting), and so all such azeotropes are minimum-boiling azeotropes. The total vapor pressure of the mixture is elevated compared to what it would be if the mixture were ideal. Values of y_i^{AZ} and T^{AZ} corresponding to a particular total pressure may be estimated simply by evaluating vapor pressures in eq. (5.51) as a function of temperature. A simple, iterative procedure starts with an estimate or first guess of the azeotropic boiling point. If the sum of the pure component vapor pressures at that temperature is equal to the total pressure of interest, then the calculation is finished. If not, the temperature is adjusted unit the calculation converges on the desired total pressure. The azeotrope's boiling point will be somewhat below the boiling point of the lightest component in the azeotrope – because all heterogeneous azeotropes are minimum-boiling azeotropes. So, as a first guess, simply choose the light component's boiling point, and adjust that temperature downward until the calculation converges on the desired pressure. For example, partially miscible water + organic systems normally will exhibit a heterogeneous azeotrope that boils only slightly below the pure-component boiling point of either water or the organic depending on which is lightest as a pure component. This method is easily used to estimate heterogeneous azeotropic composition at atmospheric pressure from knowledge of normal boiling points and vapor pressures of pure components at several temperatures. A similar approach is described by Horsley [46].

Of course, heterogeneous azeotropic composition can also be obtained directly from a vapor-liquid-liquid phase diagram corresponding to some constant total pressure, as in Figure 5.6. This provides a convenient way of obtaining values for the relative volatility of dissolved solutes in each of the two liquid phases [49]. Consider relative volatility for an organic component dissolved in the aqueous phase of an organic + water binary system. Values of y_i^{AZ}, x_i^{SAT}, and thus $\alpha_{i, H_2O}^{\infty}$ can be taken directly from the phase diagram (Figure 5.6):

$$y_i^{AZ} \approx \frac{p_i^{SAT}}{p_i^{SAT} + p_{H_2O}^{SAT}} = \frac{p_i^{SAT}}{p_{total}} = \text{composition at pt. } E \tag{5.52}$$

$$\alpha_{i, H_2O}^{\infty} \approx \frac{1}{x_i^{SAT}} \frac{p_i^{SAT}}{p_{total}} = \frac{y_i^{AZ}}{x_i^{SAT}} = \frac{\text{composition at pt. } E}{\text{composition at pt. } C} \tag{5.53}$$

Thus, the value of y_i^{AZ} is taken from point E on the diagram, and x_i^{SAT} is taken from point C which represents the composition of the aqueous liquid in equilibrium with the vapor. Turning this around, when such information is available from other sources, it may be used in constructing the appropriate phase diagram.

Although azeotropes have a bad reputation as a limiting factor in distillation, sometimes an azeotrope may actually enhance distillation. See Chapter 8: Enhanced Distillation: Taking Advantage of Specific Molecular Interactions (for discussion of azeotropic distillation and steam distillation) and Section 7.3.3: Continuous Stripping and Absorption (for discussion of steam stripping).

5.6.4 Liquid-liquid phase behavior

There are various ways to present liquid-liquid equilibrium data on a phase diagram. These include triangular composition diagrams for ternary systems as in Figures 5.7 and 5.8, and temperature-composition diagrams for binary systems as in Figures 5.9 to 5.11. Triangular diagrams are often used to show how a dissolved solute distributes between feed and solvent phases at equilibrium. They are used to show equilibrium tie lines for liquid-liquid extraction (LLE) operations. Other liquid-liquid diagrams are used to show the effect of a change in temperature on liquid-liquid miscibility and the effect of pressure on boiling behavior, as indicated in Figures 5.9 to 5.11. This temperature sensitivity can be important in optimizing LLE and other processes involving liquids (as in whether the overheads from a distillation operation will phase split and to what degree). In many cases, the two liquids become more soluble in each other with increasing temperature. That is, partial miscibility (mutual solubility) between the two liquid phases increases with increasing temperature until a temperature is reached at which only one liquid phase exists, as in Figures 5.9 and the upper part of

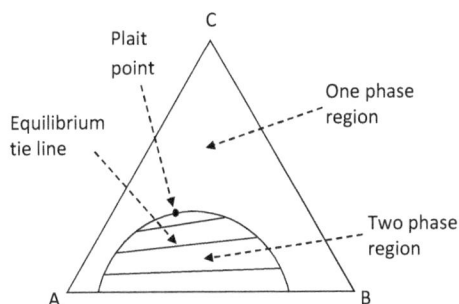

Figure 5.7: Liquid-Liquid Ternary Diagram in which Components A and B are Partially Miscible. The plait point is the point at which the two-phase tie-line compositions merge, becoming a single-liquid phase.

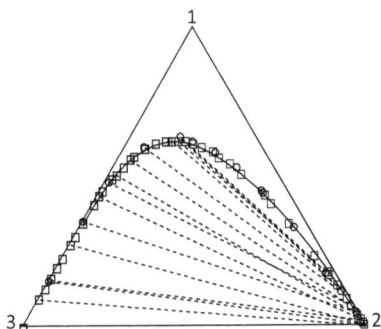

Figure 5.8: Liquid-Liquid Ternary Diagram for Methanol (1) + Benzene (2) + Water (3) at 293 K. Taken from ref. [50]. Reprinted with permission from *J. Phys. Chem. Ref. Data*, copyright 1999, AIP Publishing.

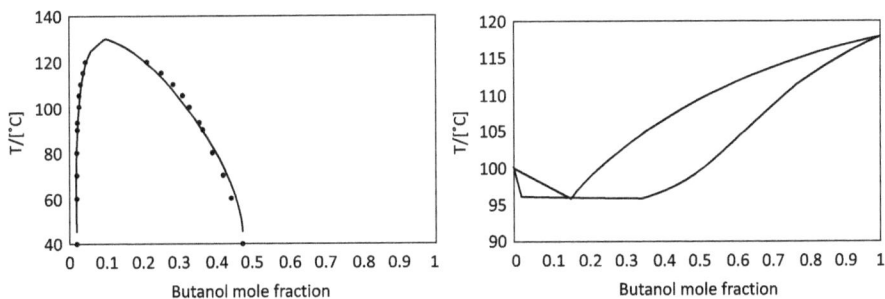

Figure 5.9: Temperature Composition Phase Diagrams for n-Butanol + Water. Left-hand side) under sufficient pressure to prevent formation of vapor (suppressed boiling); right-hand side) at 1 atmosphere with vapor formation (boiling conditions). Taken from ref. [51]. Reprinted with permission from *Computer Aided Chemical Engineering*, copyright 2016, Elsevier.

Figure 5.10. This is the upper critical solution temperature (also called the upper conso-lute temperature). As illustrated in Figure 5.9, whether the liquid boils to vapor in this region of the phase diagram depends on pressure.

Some systems exhibit a lower critical solution temperature (lower consolute temper-ature). In this region of the diagram, as temperature decreases, partial miscibility actually increases (the components become more soluble in each other), until a temperature is reached at which the system is completely miscible (only one liquid phase exists), as shown in Figures 5.10 and 5.11. This behavior is sometimes called inverse solubility. It is typical of surfactant + water systems and glycol ether + water systems. See the discussion of these systems (which we label Type IV) in Section 5.4.2: Classification of Chemical Mix-tures. A few systems exhibit both an upper and a lower critical solution temperature, as shown in Figure 5.10. Boiling of liquid to vapor may hide the presence of the upper critical solution temperature if pressure is not high enough to suppress boiling (as in Figure 5.9).

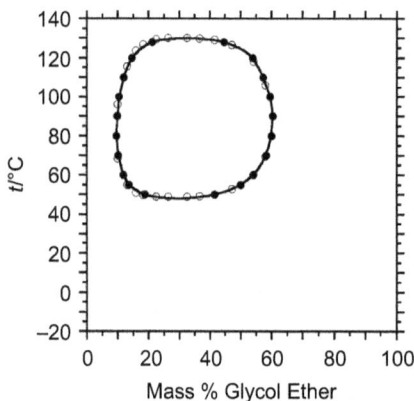

Figure 5.10: Temperature Composition Phase Diagram for Ethylene Glycol n-Butyl Ether + Water. P = 1 atm. Taken from ref. [52]. Reprinted with permission from *Journal of Chemical and Engineering Data*, copyright 2005, American Chemical Society.

5.6.5 Solid-liquid eutectic behavior

A eutectic in a binary solid-liquid phase diagram (Figure 5.12) is analogous to a hetero-geneous azeotrope in a binary vapor-liquid phase diagram (Figure 5.6). For a system comprised of two components, a simple eutectic is the low-melting point in the liquid + solid + solid equilibrium diagram at which two different solid phases (pure or nearly pure) can exist at equilibrium along with liquid containing both components. The components of the system are not miscible in either solid phase. Similarly, a het-erogeneous azeotrope is the low-boiling point in the vapor + liquid + liquid diagram

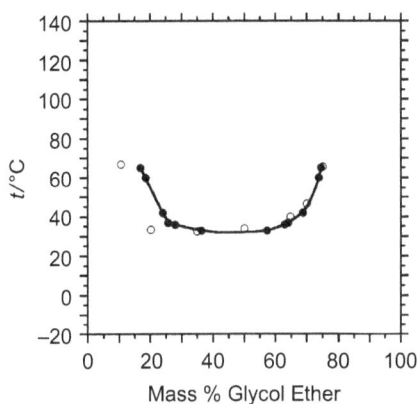

Figure 5.11: Temperature Composition Phase Diagram for Propylene Glycol n-Propyl Ether + Water. P = 1 atm. Taken from ref. [52]. Reprinted with permission from *Journal of Chemical and Engineering Data*, copyright 2005, American Chemical Society.

at which two different liquid phases (pure or nearly pure) can exist at equilibrium along with vapor containing both components. In this case, the components of the system are not miscible in either liquid phase. Such diagrams can become much more complex, of course, especially when adding more components to the mixture. Eutectic systems in particular can become highly complex because more than just two different equilibrium solid phases may form depending on conditions. Furthermore, molecular mobility is greatly restricted in a solid, so non-equilibrium or meta-stable solid phases may also form and persist. Non-equilibrium solids may result from rapid cooling (or some other kinetic driver) and then persist even though they are not the lowest energy form of solid. As a result, kinetic factors must be controlled to obtain the desired pure solid in the desired form. Nevertheless, a eutectic system is of particular importance in crystallization operations because such a system has the potential to form highly pure crystals out of a multi-component mother liquor in a single stage of crystallization. On the other hand, a non-eutectic system may form when two or more components of the system are very similar in molecular structure so they form solid solutions of variable composition and require multiple stages of crystallization for purification [53] (more on this later).

For the simple binary eutectic system shown in Figure 5.12, the individual solid-liquid equilibrium lines converge at the eutectic point, and both types of crystals may form out of the solution at that point (pure crystals of each type). Most discussions of eutectic systems seem to focus on the low-melting behavior at the eutectic point. But for chemical separation, the fact that a eutectic system can produce pure crystals is what we take advantage of. How much of a desired solid can be produced in a single stage of crystallization depends on the position of the liquid feed composition relative to the eutectic point. In Figure 5.12, if the feed composition is on the right side of the

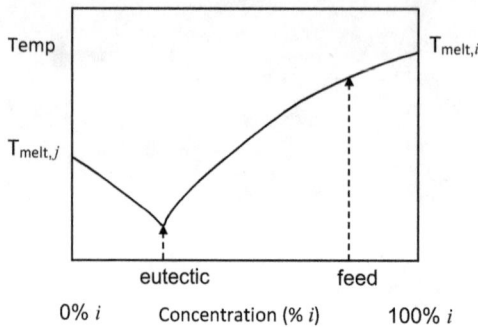

Figure 5.12: Representative Liquid + Solid + Solid Phase Diagram for a Binary System ($i + j$) with Immiscible Solid Phases Showing the Presence of a Eutectic.

eutectic point, pure i crystals form on cooling. On the other side of the eutectic point, pure j crystals form on cooling. At the eutectic point, both types of crystals, pure i and pure j can form and exist together at equilibrium. In a typical crystallization process, the feed to the crystallizer will be concentrated in the desired component (say it is component i), at perhaps 90% i. If the eutectic composition is 55% i, then as the crystallization proceeds and pure crystals of i come out of solution, the liquid phase will gradually be enriched in component j until the eutectic composition is reached. At this point, both pure i and pure j crystals will come out of liquid solution. So, for a process designed to produce crystals of pure i (and none of j), the maximum yield; that is, the maximum recovery of the desired crystalline product, is limited by the feed composition and the eutectic composition. For a feed containing 90% i relative to a eutectic point at 55% i, the maximum recovery of component i is 86%, as given by the material balance:

$$\text{Maximum Recovery } (\% \, i) = \frac{100}{1 - x_{i,\text{eutectic}}^{SAT}} \left[1 - \frac{x_{i,\text{eutectic}}^{SAT}}{x_{i,\text{feed}}} \right] \quad (5.54)$$

This is one reason why crystallization processes typically are used to process feeds that are already fairly concentrated in the desired component. Crystallization is often preceded by a primary recovery step such as distillation or liquid-liquid extraction, and crystallization is used for final purification.

The eutectic point is the point on the phase diagram where the liquid becomes saturated in both components. At this point, the material balance gives:

$$x_i^{sat} + x_j^{sat} = \frac{1}{\gamma_i^{sat}} \exp\left[\frac{\Delta_{fus} S_i}{R} \left(1 - \frac{T_{m,i}}{T} \right) \right] + \frac{1}{\gamma_j^{sat}} \exp\left[\frac{\Delta_{fus} S_j}{R} \left(1 - \frac{T_{m,j}}{T} \right) \right] = 1.0 \quad (5.55)$$

Note that separation of components i and j is possible even when their melting points are identical, although a significant difference in melting points can be important in obtaining high recovery, as this impacts the location of the eutectic point relative to the

feed point. The addition of a solvent does not necessarily change this fundamental eutectic behavior (involving formation of pure crystals), but it can reduce the temperatures
at which pure crystals form. For a multi-component system containing two solutes and a
solvent, the presence of the solvent must be included in the material balance. At the eutectic point:

$$x_i^{sat} + x_j^{sat} + x_{solvent} = 1.0 \qquad (5.56)$$

$$= \frac{1}{\gamma_i^{sat}} \exp\left[\frac{\Delta_{fus}S_i}{R}\left(1 - \frac{T_{m,i}}{T}\right)\right] + \frac{1}{\gamma_j^{sat}} \exp\left[\frac{\Delta_{fus}S_j}{R}\left(1 - \frac{T_{m,j}}{T}\right)\right] + x_{solvent}$$

$$(5.57)$$

So addition of solvent may very well retain eutectic behavior, and the temperature
and composition of the eutectic point for the system including the solvent can be determined by solving this equation. In this case, the eutectic point is called a double
saturation point.

The phase diagrams and relationships described here are relatively simple ones.
Actual phase diagrams may be more complex [54]. Potential complications include:

- Formation of solid solutions with complete miscibility in the solid phase (as mentioned above). With such systems, the composition of the solid phase varies as a
function of liquid-phase composition. The melting point of the crystals is always
between that of the pure solids from which they are made. In this case, obtaining
pure crystals requires multi-stage crystallization.
- Formation of partially miscible solid solutions. These systems exhibit a eutectic
point, but partial solid miscibility occurs at the dilute ends of the phase diagram
(nearly pure i and nearly pure j). This limits the purity of the crystals that can be
made in a single stage of crystallization. A phase diagram of this type is analogous
to that shown in Figure 5.6 for two-liquid systems.
- The presence of numerous impurities or intentional additives (which adds complexity and can affect solid-phase purity).
- The existence of more than one crystalline form (lattice structure) of a given component
(polymorphism).
- Formation of hydrate or solvate compounds on addition of solvent, by reaction of
solute with solvent.
- Reaction of different solutes in solution to form less-soluble compounds (reactive
crystallization).
- Dissociation of one or more solutes in solution and reaction of ions to form less-
soluble compounds.
- Formation of compounds with somewhat variable composition that crystallize within
a specific liquid composition range. Formation of so-called non-stoichiometric hydroxyapatite minerals (hydroxy calcium phosphate solids) is a classic example.

An example of a common eutectic system is the mixture of water and a common salt such as NaCl or CaCl$_2$ (Figure 5.13). Spreading salt on icy roads in winter will cause the ice to melt due to the low-melting behavior of the mixture (−21 °C for the eutectic mixture containing 23 wt% NaCl, and −51 °C for the eutectic mixture containing 29% CaCl$_2$) [55]. This behavior often is described in terms of the colligative property of freezing-point depression. This refers to the same phenomenon as eutectic behavior, just described in a different way. Freezing point depression focuses on the change in the freezing point of water (the melting point of ice) on addition of a non-volatile salt to the liquid solution. The melting point depends on the amount of salt in solution; that is, on the position of the liquid composition along the liquid-solid equilibrium line, as shown in Figure 5.13. The relationship to eutectic behavior becomes clearer when we look at this phenomenon in terms of the composition of the solids instead of the liquid. Crystallizing solids out of the salt solution will produce crystals of either pure salt or pure ice (or both at the eutectic point) but no crystals containing both components (except for defects). This phenomenon is the basis for a crystallization process called eutectic freeze crystallization [56]. Glycerol is another common chemical that forms a eutectic system with water [57]. In this case, the eutectic point is at −46 °C with 66 wt% glycerol.

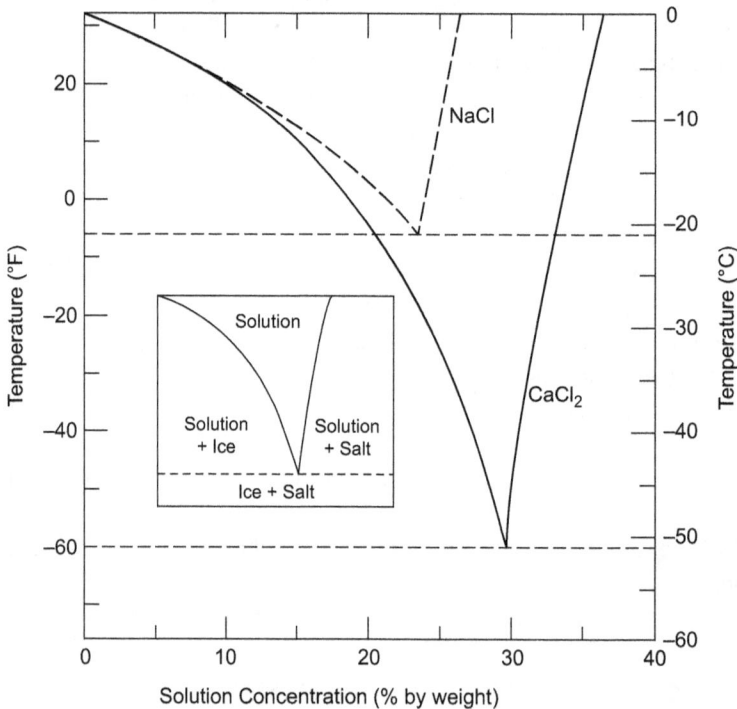

Figure 5.13: Solid-Liquid Phase Diagram for Common Salt + Water. Taken from ref. [55], a U.S. federal government publication in the public domain.

In the field of metallurgy, metallurgists focus on the low-melting behavior of a eutectic mixture in formulating low-melting metal solders. Well-known examples include tin + lead (63/37) and tin + silver (96.5/3.5) in their eutectic proportions (wt% indicated in parentheses). But it is not uncommon for metal alloys to form solid solutions (called substitutional alloys) instead of eutectic alloys. Brass is a good example, made from copper + zinc in varying proportions. In this case, the composition of the solid-solution (alloy) crystals varies with the composition of the melt, and the melting point of an alloy crystal is in between those of the pure metals. A metallurgist may also want to form a particular meta-stable solid with desired properties. Techniques include sudden cooling (quenching), cold-work hardening (hammer or shot peening), and tempering (reheating followed by slow cooling), all intended to form a meta-stable (but long-lasting) solid structure with specific properties, and all falling short of complete annealing which yields the equilibrium solid structure.

Polymorphism occurs when a given atom or compound can form more than one lattice structure. A well-known example is carbon which has diamond and graphite polymorphs. Water also can form various polymorphs depending on temperature and pressure. Some complex organic compounds are reported to have numerous polymorphs, and oftentimes the exact number is unknown. The structure that actually forms depends on crystallization conditions and the composition of the mother liquor. Some polymorphs are stable crystals and others may be meta-stable crystals that transition to a more stable form if conditions allow. Nucleation and growth of meta-stable crystals is a kinetic phenomenon that locks in a crystal structure that is not easily altered – so meta-stable crystals may appear to be stable for a time, but then transition to another form given the appropriate conditions and sufficient time. The presence of polymorphs is of particular interest in the pharmaceutical industry because different polymorphs can have differing biological activity due to differences in solubility, dissolution rates, chemical stability, and bioavailability.

Hydrates and solvates are solid compounds that form between solute and solvent in solution. In industrial practice, a desired solute may be conveniently crystallized in the form of a solvate crystal and then released in a subsequent step. Hydrate or solvate crystals sometimes can be dehydrated (or desolvated) simply by heating to elevated temperatures for a time to release water or solvent vapor. In that case, they are sometimes called adducts, a term used to indicate a molecular compound crystal that is easily decomposed to release constituent molecules (usually by applying heat). The phenol-Bisphenol-A adduct that forms on crystallizing Bisphenol-A in phenol is a well-known example. Many other examples involve formation of hydrates when crystallizing a product solute from aqueous solution. Hydrates and solvates may also form as unwanted solid deposits in processing equipment of any kind, causing fouling or other processing problems. For example, methane hydrates can foul natural gas distribution pipelines if water leaks into the pipes or the feed gas is not sufficiently dry.

Fortunately, even with so many possible complications, many compounds behave as simple eutectic compounds and may be crystallized as pure (or nearly pure) crystals in a single stage of crystallization. This is because in many cases the molecular structures of the product solute, the various impurities, and the solvent are significantly different, so the product molecule is able to reject most foreign molecules from its crystal lattice as it forms. But there are exceptions, of course, especially when impurities have nearly the same structure as the desired product [53]. In such cases, complete or partial miscibility in the solid phase may limit the ultimate thermodynamic purity that can be achieved in a single stage of crystallization. Differential scanning calorimetry may be used to determine whether a given system exhibits true eutectic behavior, solid solution behavior, or partial solid solution behavior [54, 58]. Purity may also be limited by formation of crystal defects such as occlusions that trap mother liquor within the crystal lattice (a kind of kinetic impurity). But defects may be reduced through careful design and operation of the crystallization process. So, whenever efforts to increase the purity of crystals obtained in a single stage of crystallization are no longer effective, the cause may be due to partial miscibility in the solid phase (a true thermodynamic limit) or the formation of defects (a kinetic phenomenon). Knowing which is the cause in a particular case is important in understanding whether an additional stage of crystallization is truly needed or may be avoided by changing the crystallization process. For discussion of process options, see Chapter 11: Crystallization from Solution and from the Melt: Taking Advantage of Eutectic Behavior.

5.7 Solvent Selection Methods

Solvent selection methods help a designer sort through the many hundreds of commercially available solvents (not to mention solvent blends) by quickly generating a short and much more manageable list of candidates. One of the earliest and most successful solvent selection methods (originally used to identify solvents for resins and polymers used in paint) was introduced by Charles M. Hansen in the 1960s – the use of Hansen solubility parameters. This model is a modified version of the Hildebrand-Scatchard solubility parameter model. The Hansen model divides the overall solubility parameter into three parameters that characterize a molecule's ability to interact with solvents through 1) dispersion, 2) polar interactions, and 3) hydrogen bonding interactions. (For further discussion of these and other interactions, see Section 5.4.1: Non-ideal Intermolecular Interactions.) Hansen showed that the solubility of a solid (or solute) in a given solvent was greatest when the solubility parameters for solute and solvent were close, as in like dissolves like. More precisely, he introduced the concept of a distance measure on a 3D map of a solubility sphere with axes for each parameter, and showed that solubility was likely to be greatest when the distance between points on the sphere for solute and solvent was shortest. Hansen parameters and the Hansen sphere model are widely used in the coatings industry to optimize paint formulations, and they have been used to optimize many other product + solvent formulations as well [24, 25]. A major advantage of Hansen's method is the very large database of Hansen parameters that have been determined since its introduction in the 1960s.

Hansen's general approach has since been leveraged for use with other phase equilibrium models. In general, a mathematical expression is derived from a given thermodynamic model and used as a framework for phase equilibrium data. In the case of solid solubility, the model parameter values are determined by fitting solubility data. For example, suitable parameter values may be determined for a solid of interest by first measuring a given solid's solubility in a variety of pure solvents, each a member of a different class of solvents, and then regressing the data to determine suitable model parameter values for the given solid. The mathematical expression can then be used to calculate the solid's solubility (or relative solubility) in other solvents not included in the original data set. This requires determining a database of known model parameters for a variety of solvents. This is the general approach that Hansen introduced in devising his sphere model.

There are many ways to implement this approach – as many ways as there are models used to represent solid-liquid equilibrium and solubility (and phase equilibrium in general). For example, at Dow we worked with Sumnesh Gupta and Joe Downey on an approach that uses a Hansen solubility parameter model expressed in terms of activity coefficients, called the extended Hansen model [19]. With this model, solubility is given by:

$$\ln\gamma_i^{sat} \approx \frac{V_i}{RT}\left\{\left(^i\delta_d - \overline{\delta_d}\right)^2 + 0.25\left[\left(^i\delta_p - \overline{\delta_p}\right)^2 + \left(^i\delta_h - \overline{\delta_h}\right)^2\right]\right\} + \ln\frac{V_i}{\overline{V}} + 1 - \frac{V_i}{\overline{V}} \qquad (5.58)$$

$$x_i \approx \frac{1}{\gamma_i^{sat}}\exp\left[\frac{\Delta_{fus}S_i}{R}\left(1 - \frac{T_m}{T}\right)\right] \approx \frac{1}{\gamma_i^{sat}}\exp\left[6.8\left(1 - \frac{T_m}{T}\right)\right] \qquad (5.59)$$

where $\overline{\delta_d}$ = Hansen sol'y parameter (dispersion), molar volume average, MPa$^{1/2}$
$^i\delta_d$ = Hansen sol'y parameter (dispersion), for component i, MPa$^{1/2}$
$\overline{\delta_p}$ = Hansen sol'y parameter (polarity), molar volume average, MPa$^{1/2}$
$^i\delta_p$ = Hansen sol'y parameter (polarity), for component i, MPa$^{1/2}$
$\overline{\delta_h}$ = Hansen sol'y parameter (H bonding), molar volume average, MPa$^{1/2}$
$^i\delta_h$ = Hansen sol'y parameter (H bonding), for component i, MPa$^{1/2}$
V_i = molar volume of pure component i, mL/mole
$\overline{V}$ = molar volume of the mixture, mL/mole

The other symbols are defined as before. The solubility parameters used in this expression are molar-volume weighted average values for the mixture including all solvent components and the solute [19]. Figure 5.14 summarizes how the model may be used within a spreadsheet to identify useful solvents for a given solid of interest. The solid may be any solid material the user needs to dissolve, such as a sample of deposits taken from process equipment needing to be cleaned, or it may be a pharmaceutical compound or other specialty chemical that needs to be formulated in solution. The program may also be used to identify potential solvents for liquid-liquid extraction or crystallization from solution or any process that involves liquid solutions of solids.

To use the spreadsheet, the first step is to determine Hansen parameter values for the solid of interest. This is done by measuring its solubility in six different solvents, each a member of a different class of solvents – such as a paraffinic hydrocarbon, aromatic hydrocarbon, alcohol, ether, ketone, and a chlorinated solvent. The solubility data are then regressed within the spreadsheet to determine best-fit parameter values for the solid. Normally, this involves use of a least squares regression routine to minimize the difference between measured and calculated solubility values, by adjustment of Hansen parameter values for the solid. Once the regression converges on a solution, the best-fit values are the values used to represent the solubility characteristics of that solid.

The solubility of the solid can now be calculated for all the solvents in the database and listed in order of highest to lowest solubility. The program can be programmed to include a database with a large number of commercially available solvents. In effect, the program sorts through all the many available solvents to generate a short and much more manageable list of solvent candidates. The calculated solubilities are not sufficiently accurate for design, but the value of the program comes from its ability to suggest potential solvents to evaluate in the lab. In particular, the spreadsheet may suggest several solvents the user is not aware of. The program may be used by anyone with a need to dissolve a given solid. Those candidates that are obviously unsuitable due to toxicity or reactivity can quickly be eliminated, focusing only on those that appear to be practical.

Other similar methods have been developed based on other activity coefficient models. These include the MOSCED model [11, 22] and NRTL-SAC [26]. The use of MOSCED has an advantage over the use of the Hansen model in that the ability to participate in hydrogen bonding is represented in terms of two parameters instead of just one: one for proton donor ability and one for proton acceptor ability. It also can represent π bond interactions and the effect of temperature. But the database of solvents is smaller than the Hansen database. The use of NRTL-SAC has been shown to be particularly effective in representing the solubility of pharmaceutical compounds. Each model has its advantages depending on the availability of parameter values for a variety of solvents and its ability to model specific kinds of molecular interactions.

In general, a great variety of phase equilibrium models may be used to estimate phase equilibrium properties of any kind for a variety of purposes – distillation, extraction, crystallization, and so on. These include various group-contribution methods such as UNIFAC, as well as various computational chemistry methods such as COSMOtherm [29, 34]. COSMOtherm may also be used to estimate MOSCED parameters to fill out a MOSCED database of parameter values [59]. And a group contribution method may be used to calculate MOSCED parameters, as well [60]. Of course, models such as UNIFAC and COSMOtherm may be used to model solubility and other phase equilibria

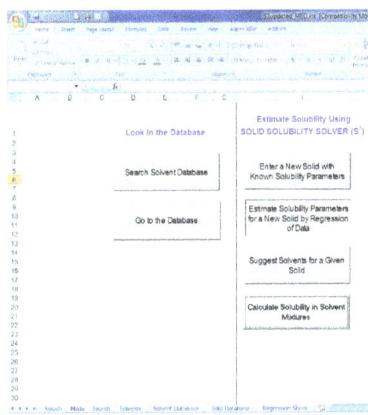

- Large database of solvents.
- Automated data regression →
  solute parameters.
- Solubility calculated for all solvents
  in the database.
- Handles mixtures/blends.

Quickly generates a list of candidate
solvents and solvent blends.[a]

Similar strategies with

- NRTL-SAC (Chen and Song, 2004)[b]
- MOSCED (Lazzaroni, *et al.*, 2005)[c]

- Hansen's solubility parameter model
  expressed as a modified regular solution equation.

- Unknown parameters determined by regressing solubility data.

  - Measure solubility in 6 solvents, each a member of a different class.

  - After regression, calculate solubility in all solvents.

- Quickly generates a list of candidate solvents.

- May suggest solvents the user would not have considered otherwise.

a Frank, Gupta, Downey, *Chemical Engineering Progress* **1999**, *95*(12), 41-61.
b Chen, Song, *Ind. Eng. Chem. Res.* **2004**, *43*(26), 8354-8362.
c Lazzaroni, Bush, Eckert, Frank, Gupta, Olson, *Ind. Eng. Chem. Res.* **2005**, *44*(11), 4075-4083.

Figure 5.14: Example Solvent Selection Spreadsheet.

directly. See Section 5.4: Specific Molecular Interactions, for more discussion. Future
work in this area no doubt will continue to offer improved methods.

5.8 Energy Consumption and Energy Efficiency

Energy consumption and energy efficiency are related terms, but they refer to different aspects of energy utilization. And their meanings depend on the context of the
discussion, whether we are talking about an individual separator or the overall process scheme including all the other unit operations that interact with that separator.
Energy consumption simply refers to how much energy is consumed in separating
and purifying a unit of product. This depends on how much energy is expended in

doing the work of separating and on how much of that energy may be recovered through integration with other operations. On the other hand, the thermodynamic energy efficiency of an individual separator refers to how much of the energy expended in doing the separating goes into the outlet streams; that is, how much may be recovered from those streams versus how much is lost work or lost heat that cannot be recovered (as when using a high reflux ratio). Whether the energy in the outlets from the separator is recovered in a useful form depends on the overall process scheme. Note that here we are discussing efficiency in terms of energy use, not efficiency for mass transfer as in the efficiency of a single contacting stage in approaching equilibrium. But improving mass transfer efficiency to improve separation capability normally allows for a reduction in the work required to do the separation (as in the use of a lower reflux ratio), and this improves energy efficiency.

So, in talking about the energy utilization of a separation process, it is important to consider the energy consumption of the unit itself, how much of that energy is contained in the outlet streams with the opportunity to recover it, and whether the overall process is designed to actually recover that energy for useful purposes. This means that for an individual separator that internally consumes a large amount of energy, it may be possible to recover a good portion of that energy for other uses if the separator itself is energy efficient (with relatively little lost work and lost heat). In that case, the opportunity exists to reduce the overall consumption of energy. This may be accomplished by cross-exchange of thermal energy with other unit operations, by using the pressure of a separator outlet stream to drive a pump or turbine, or by other means.

Understanding maximum thermodynamic efficiency in the use of energy allows the process designer to gauge how close a given separator design is to the theoretical limit, and from that knowledge, assess the opportunity for reduction in energy consumption, both from an internal perspective and an overall perspective. The designer can then determine if the potential for energy savings is worth the added complexity and capital cost that may be required. Such an analysis also allows comparison of competing design options, allowing the designer to quickly evaluate options according to their potential for reducing overall energy consumption (among many other factors to be considered) [61].

The actual efficiency of a given separator design may be calculated from the enthalpy contained in the outlet streams versus the total amount of energy input to the separator. On the other hand, the theoretical maximum energy efficiency of the separator may be calculated by analyzing the exergy cycle for the process. Exergy is the name given the energy expended by a process that results in useful, recoverable work. An exergy analysis considers the total energy used by the process and accounts for losses in useful work due to an increase in the entropy of the system or due to loss of heat to the surroundings. The term *exergy* was coined by Zoran Rant in 1956, combining the Greek terms ex and ergon to mean *from work* [62].

An exergy analysis envisions the process as a type of Carnot heat engine coupled with its surroundings, which in our case are hot and cold heat transfer fluids and the surrounding ambient environment (considering heat lost through insulation, for example). In general, the loss of useful work may be reduced by minimizing temperature extremes, by better insulating the separator from the environment, and in some cases, by increasing the concentration of the desired product in the feed, minimizing the change in concentration that is required in carrying out the separation. The latter may coincide with reducing the volume of feed that the process must handle, which results in reduced energy consumed by the separator even if the specific energy efficiency of the separator is unaffected. And as mentioned above, improving mass transfer capability is another way to reduce the required separation work and improve energy efficiency. In this regard, minimizing reflux is especially important for distillation, and minimizing compression work is especially important for membrane-based separations. The latter may be accomplished by reducing the required number of membrane contacting stages through improvement in membrane selectivity.

We touch on the use of exergy analysis and the effect of feed composition in our discussion of distillation operations in Section 7.6: Energy Utilization. A more detailed discussion of energy efficiency in general is outside the scope of our book. A review of energy efficiency for a variety of separation processes is available elsewhere [63]. Here, our focus is on general rules for reducing energy consumption in process design that derive from exergy and thermal analysis:

- Avoid extreme temperatures.
- Minimize the volume of the feed stream.
- Maximize the concentration of product in the feed.
- Minimize dilution of the product during processing.
- Minimize pressure drop through the equipment.
- Match concentration profiles whenever mixing recycle or multiple feed streams within the overall separation process.
- Match pressures whenever mixing recycle or multiple feed streams.
- Increase mass transfer efficiency and separation capability. For more discussion of separation capability, see Section 6.5.1: Material Balance Split and Separation Power.

For a given application, some of these rules may not be practical, but they are important rules to keep in mind and take advantage of when feasible. In addition, general strategies for more efficient overall use of energy include:

- Process the outlet streams from the separator to recover a portion of their energy content in some useful form.
- Combine two or more separation operations into a hybrid process scheme, each chosen to operate within a specific composition range such that energy consumption is reduced within that range.

- Locate plants near power generation facilities to minimize energy-transmission losses and to tap into heat expelled from power generation plants as an economical heating source (called combined heat and power).

A number of energy-saving measures follow from these rules. Perhaps the quickest and easiest ones involve minimizing thermal losses and unwanted heat transfer from process systems; for example, by installing better insulation between the process and the environment, by replacing older boilers and heaters with newer, more energy-efficient designs, by using steam traps to discharge condensate from steam distribution lines with only minimal loss of steam, and by using more efficient refrigeration systems. Another way of reducing energy consumption is to replace older, outdated separation equipment with newer designs. For example, replacing older distillation column internals (distributors, trays, or packings) with newer, more mass transfer efficient ones that achieve many more stages or transfer units may allow operation with a lower reflux ratio for the same column height. And, as stated earlier, using newer membranes with improved selectivity may reduce the required number of membrane contacting stages, thereby reducing the need for compression of fluids.

The need for extreme process temperatures may be reduced somewhat by increasing residence time in a separator (by using larger equipment) to compensate for any reduction in mass transfer rates at more moderate temperatures. Or, in the case of distillation, the operating pressure may be adjusted, either lower to reduce the required reboiler temperature and avoid any need for a high-temperature furnace, or higher to increase the required overheads condensing temperature and minimize or eliminate the need for refrigeration. And pre-concentrating the feed stream will reduce the volume of feed that must be processed. This will reduce the total amount of energy that is required by the main separator regardless of the temperatures used, whether extreme or not. A pre-concentrator such as a membrane module or even an efficient evaporator may allow for reduction in the total energy consumed by the overall process, even though a pre-concentration step will require some amount of energy itself. Another example involves the use of liquid-liquid extraction to pre-concentrate a feed, by transfer of desired components into a smaller-volume solvent stream that is then fed to another separation unit, usually a distillation column. Here, the desired product is first recovered and concentrated using extraction to generate a new feed phase from which the product can more easily be isolated and purified by using distillation. In general, an analysis of a proposed process scheme with and without pre-concentration may indicate an opportunity for significant reduction in energy consumption.

Energy integration approaches emphasize the sharing and reusing of energy between various processing streams, either internally by cross-exchanging energy between streams generated by the separation process itself, or externally, by cross-exchanging energy between other processes operating at the same overall site. A typical example involves cross-exchanging energy between the bottoms stream from a distillation column and the incoming feed stream. Another approach involves cross-exchange of energy between reactors and separators. Well-known examples include the use of heat generated in an HCN reactor to operate the downstream ammonia recovery column [64], and the use of heat generated in a steam-modulated ethane to ethylene cracking furnace to operate downstream processing steps [65]. Even in cases where the ΔT driving force is low, there may be options that allow transfer of sufficient energy by specifying large heat-transfer area. This may open up opportunities to utilize heating or cooling sources that would otherwise be wasted. Also, whenever evaporation or distillation methods are used, energy consumption may be reduced by using internal integration methods such as multiple-effect evaporation and mechanical vapor recompression (a kind of heat-pump cycle). These options are described elsewhere [66].

The use of hybrid processing to reduce energy consumption involves combining a main separator with another kind of separation unit to reduce the load on the main separator. The added separator is chosen because it doesn't have to work as hard as the main separator within a specific region of the overall product concentration range. There are many examples of this general approach, which we discuss in Chapter 14: Inventive Engineering: Process Simplification, Intensification, and Hybrid Processing. A straightforward example is the use of a pre-concentrator as mentioned earlier, as in using a membrane module to pre-concentrate the feed to a distillation column or the use of liquid-liquid extraction to do so. A related approach involves going so far as to combine multiple operations within a single vessel, as in the use of extractive distillation which is then followed by product isolation and solvent recycle in another unit operation, usually another distillation. Another example is dividing-wall distillation, which combines multiple distillation functions within a single column. This can reduce reboiler and condenser duties by up to 30%. Yet another approach involves the use of various multi-column thermally coupled column configurations [67]. Of course, when considering any of these energy-saving options, the cost of implementation must be taken into account in a cost/benefit analysis.

A final point that should be emphasized no matter what energy-saving options are used is the need to avoid re-processing of product or over-purification of product. Obviously, if the product has to be re-processed or over-purified, extra energy will be required and the benefits of any energy-saving options will be diminished. So, a separation process should be carefully designed and operated to avoid production of off-spec product which then must be discarded, recycled for reprocessing, or blended with over-purified product to meet specifications. This may be done by improving process control to minimize variability in the process operation to the point where any need for re-processing or blending is eliminated. For more discussion in the context of distillation process control, see Section 7.5: Process Control Considerations.

References

[1]   B. E. Poling, J. M. Prausnitz and J. P. O'Connell, *The Properties of Gases and Liquids*, vol. 8, 5th ed., McGraw-Hill, New York, 2001, pp. 111–118.

[2]   H. A. Bumstead and R. G. V. Name, *The Scientific Papers of J. Willard Gibbs*, Vol. 1 and 2, Ox Bow Press, Woodbridge, Connecticut, 1993 reprint of 1906 original.

[3]   D. Quane, "The Reception of Hydrogen Bonding by the Chemical Community", *Bulletin for the History of Chemistry*, vol. 7, pp. 3–13, 1990.

[4]   W. M. Latimer and W. H. Rodebush, "Polarity and Ionization from the Standpoint of the Lewis Theory of Valence", *The Journal of the American Chemical Society*, vol. 42, no. 7, pp. 1419–1433, 1920.

[5]   M. Solá, "Aromaticity Rules", *Nature Chemistry*, vol. 14, pp. 585–590, 2022.

[6]   J. R. Rumble, Ed., *CRC Handbook of Chemistry and Physics*, 103rd ed., CRC Press, Boca Raton, Florida, 2022.

[7]   L. A. Robbins, "Liquid-Liquid Extraction: A Pretreatment Process for Wastewater," *Chemical Engineering Progress*, vol. 76, no. 10, pp. 58–61, 1980.

[8]   J. R. Elliott, C. T. Lira, T. C. Frank and P. M. Mathias, "Section 4, Thermodynamics," in *Perry's Chemical Engineers' Handbook*, 9th ed., D. W. Green and M. Z. Southard, Eds., McGraw-Hill, New York, 2018.

[9]   T. C. Frank, S. G. Arturo and B. S. Holden, "Framework for Correlating the Effect of Temperature on Nonelectrolyte and Ionic Liquid Activity Coefficients," *AIChE Journal*, vol. 60, no. 10, pp. 3675–3690, 2014.

[10] G. N. Lewis and M. Randall, *Thermodynamics and the Free Energies of Chemical Substances*, McGraw-Hill, New York, 1923.

[11] M. J. Lazzaroni, D. Bush, C. A. Eckert, T. C. Frank, S. K. Gupta and J. D. Olson, "Revision of MOSCED Parameters and Extension to Solid Solubility Calculations (Plus Supplemental Information. Database of Published Limiting Activity Coefficients)," *Industrial & Engineering Chemistry Research*, vol. 44, no. 11, pp. 4075–4083, 2005.

[12] U. Onken, J. Rarey-Nies and J. Gmehling, "The Dortmund Data Bank: A Computerized System for Retrieval, Correlation, and Prediction of Thermodynamic Properties of Mixtures," *International Journal of Thermophysics*, vol. 10, no. 3, pp. 739–747, 1989.

[13] Anonymous, "Dortmund Data Bank," DDBST GmbH [Online]., Available: http://www.ddbst.com/. [Accessed 7 October 2022].

[14] J. M. Smith, H. C. V. Ness and M. M. Abbott, *Introduction to Chemical Engineering Thermodynamics*, 7th ed., McGraw-Hill, New York, 2005.

[15] J. M. Prausnitz, R. N. Lichtenthaler and E. G. D. Azevedo, *Molecular Thermodynamics of Fluid-Phase Equilibria*, 3rd ed., Prentice Hall PTR, Upper Saddle River, New Jersey, 1999.

[16] J. R. Elliot and C. T. Lira, *Introductory Chemical Engineering Thermodynamics*, 2nd ed., Prentice Hall, 2012.

[17] S. M. Walas, *Phase Equilibria in Chemical Engineering*, Butterworth-Heinemann, Oxford, 1985.

[18] S. Gupta, J. R. Elliott, A. Anderko, J. Crosthwaite, W. G. Chapman, and C. T. Lira, "Current Practices and Continuing Needs in Thermophysical Properties for the Chemical Industry," *Industrial & Engineering Chemistry Research*, vol. 62, no. 8, pp. 3394–3427, 2023.

[19] T. C. Frank, J. R. Downey and S. K. Gupta, "Quickly Screen Solvents for Organic Solids," *Chemical Engineering Progress*, vol. 95, no. 12, pp. 41–61, 1999.

[20] A. Fredenslund, R. L. Jones and J. M. Prausnitz, "Group-Contribution Estimation of Activity Coefficients in Nonideal Liquid Mixtures," *AIChE Journal*, vol. 21, no. 6, pp. 1086–1099, 1975.

[21] J. Gmehling, J. Li and M. Schiller, "A Modified UNIFAC Model. 2. Present Parameter Matrix and Results for Different Thermodynamic Properties," *Industrial & Engineering Chemistry Research*, vol. 32, no. 1, pp. 178–193, 1993.

[22] E. R. Thomas and C. A. Eckert, "Prediction of Limiting Activity Coefficients by a Modified Separation of Cohesive Energy Density Model and UNIFAC," *Industrial & Engineering Chemistry Process Design and Development*, vol. 23, no. 2, pp. 194–209, 1984.

[23] R. T. Ley, G. B. Fuerst, B. N. Redeker and A. S. Paluch, "Developing a Predictive Form of MOSCED for Nonelectrolyte Solids Using Molecular Simulation," *Industrial & Engineering Chemistry Research*, vol. 55, no. 18, pp. 5415–5430, 2016.

[24] C. M. Hansen, *Hansen Solubility Parameters: A User's Guide*, 2nd ed., CRC Press, Boca Raton, Florida, 2007.

[25] A. F. M. Barton, *CRC Handbook of Solubility Parameters and Other Cohesive Parameters*, 2nd ed., Routledge, Abingdon-on-Thames, Oxfordshire, 1991.

[26] -C.-C. Chen and Y. Song, "Solubility Modeling with a Non-Random Two-Liquid Segment Activity Coefficient Model," *Industrial & Engineering Chemistry Research*, vol. 43, no. 26, pp. 8354–8362, 2004.

[27] -C.-C. Chen and Y. Song, "Extension of Nonrandom Two-Liquid Segment Activity Coefficient Model for Electrolytes," *Industrial & Engineering Chemistry Research*, vol. 44, no. 23, pp. 8909–8921, 2005.

[28] A. Klamt, "Conductor-like Screening Model for Real Solvents: A New Approach to the Quantitative Calculation of Solvation Phenomena," *The Journal of Physical Chemistry*, vol. 99, no. 7, pp. 2224–2235, 1995.

[29] A. Klamt, *From Quantum Chemistry to Fluid Phase Thermodynamics and Drug Design*, Elsevier, 2005.

[30] R. Xiong, S. I. Sandler and R. I. Burnett, "An Improvement to COSMO-SAC for Predicting Thermodynamic Properties," *Industrial & Engineering Chemistry Research*, vol. 53, no. 19, pp. 8265–8278, 2014.

[31] S.-E. K. Fateen, M. M. Khalil and A. O. Elnabawy, "Semi-Empirical Correlation for Binary Interaction Parameters of the Peng-Robinson Equation of State with the van der Waals Mixing Rules for the Prediction of High-Pressure Vapor-Liquid Equilibrium," *Journal of Advanced Research*, vol. 4, pp. 137–145, 2013.

[32] W. G. Chapman, K. E. Gubbins, G. Jackson and M. Radosz, "New Reference Equation of State for Associating Liquids," *Industrial & Engineering Chemistry Research*, vol. 29, no. 8, pp. 1709–1721, 1990.

[33] M. S. Wertheim, "Exact Solution of the Percus-Yevick Integral Equation for Hard Spheres," *Physical Review Letters*, vol. 10, no. 8, pp. 321–323, 1963.

[34] A. Klamt and F. Eckert, "COSMO-RS: A Novel and Efficient Method for the a Priori Prediction of Thermophysical Data of Liquids," *Fluid Phase Equilibria*, vol. 172, no. 1, pp. 43–72, 2000.

[35] F. Eckert and A. Klamt, "Fast Solvent Screening via Quantum Chemistry: COSMO-RS Approach," *AIChE Journal*, vol. 48, no. 2, pp. 369–385, 2002.

[36] H. Grensemann and J. Gmehling, "Performance of a Conductor-Like Screening Model for Real Solvents Model in Comparison to Classical Group Contribution Methods," *Industrial & Engineering Chemistry Research*, vol. 44, no. 5, pp. 1610–1624, 2005.

[37] S.-T. Lin and S. I. Sandler, "A Priori Phase Equilibrium Prediction from A Segment Contribution Solvation Model," *Industrial & Engineering Chemistry Research*, vol. 41, no. 5, pp. 899–913, 2002.

[38] E. Mullins, R. Oldland, Y. A. Liu, S. Wang, S. I. Sandler, -C.-C. Chen, M. Zwolak and K. C. Seavey, "Sigma-Profile Database for Using COSMO-Based Thermodynamic Methods," *Industrial & Engineering Chemistry Research*, vol. 45, no. 12, pp. 4389–4415, 2006.

[39] J. Gmehling, M. Kleiber, B. Kolbe and J. Rarey, *Chemical Thermodynamics for Process Simulation*, Wiley-VCH, Hoboken, New Jersey, 2012.

[40] C. Tien, *Introduction to Adsorption: Basics, Analysis, and Applications*, Elsevier, Amsterdam, 2019.

[41] M. Ilić, D. Flockerzi and A. Seidel-Morgenstern, "A Thermodynamically Consistent Explicit Competitive Adsorption Isotherm Model Based on Second-order Single Component Behaviour," *Journal of Chromatography A*, vol. 1217, p. 2132–2137, 2010.

[42] J. L. Sorrels, A. Baynham, D. D. Randall and K. S. Schaffner, *Chapter 1, Carbon Adsorbers*, 7th ed., U.S. Environmental Protection Agency, Research Triangle Park, North Carolina (USA), 2018.

[43] B. Predel, M. Hoch and M. J. Pool, *Phase Diagrams and Heterogeneous Equilibria: A Practical Introduction*, Springer, Heidelberg, 2004.

[44] S. M. Lai, M. Y. Yuen, L. K. S. Siu, K. M. Ng and C. Wibowo, "Experimental Determination of Solid-Liquid-Liquid Equilibrium Phase Diagrams," *AIChE Journal*, vol. 53, no. 6, pp. 1608–1619, 2007.

[45] I. V. Prikhod'ko, F. Tumakaka and G. Sadowski, "Application of the PC-SAFT Equation of State to Modelling of Solid-Liquid Equilibria in Systems with Organic Components Forming Chemical Compounds," *Russian Journal of Applied Chemistry*, vol. 80, no. 4, pp. 542–548, 2007.

[46] L. H. Horsley, "Azeotropic Data – III," in *Advances in Chemistry Series*, vol. 116, R. F. Gould, Ed., American Chemical Society, Washington, D.C., 1973.

[47] A.-H. Li, A. S. Paluch and Z.-Y. Liu, "Prediction of Azeotrope Formation in Binary Mixtures with Pure Component Properties and Limiting Activity Coefficients," *Fluid Phase Equilibria*, vol. 565 no. 113664, 2023.

[48] T. C. Frank, "Break Azeotropes with Pressure-Sensitive Distillation," *Chemical Engineering Progress*, vol. 93, no. 4, pp. 52–63, 1997.

[49] L. A. Robbins, "Method of Removing Contaminants from Water". US Patent 4236973, 1980.

[50] A. Skrzecz, D. Shaw and A. Maczynski, "IUPAC-NIST Solubility Data Series 69. Ternary Alcohol – Hydrocarbon – Water Systems," *Journal of Physical & Chemical Reference Data*, vol. 28, no. 4, pp. 983–1235, 1999.

[51] C. Bildia, I. Patraşcu, J. G. Hernandez and A. Kiss, "Enhanced Downstream Processing of Biobutanol in the ABE Fermentation Process," *Computer Aided Chemical Engineering*, vol. 38, pp. 979–984, 2016.

[52] S. P. Christensen, F. A. Donate, T. C. Frank, R. J. LaTulip and L. C. Wilson, "Mutual Solubility and Lower Critical Solution Temperature for Water + Glycol Ether Systems," *Journal of Chemical & Engineering Data*, vol. 50, no. 3, pp. 869–877, 2005.

[53] M. Lusi, "A Rough Guide to Molecular Solid Solutions: Design, Synthesis and Characterization of Mixed Crystals," *Crystal Engineering Commerce*, vol. 20, no. 44, pp. 7042–7052, 2018.

[54] A. Newman, "Specialized Solid Form Screening Techniques," *Organic Process Research & Development*, vol. 17, no. 3, pp. 457–471, 2013.

[55] Anonymous, *Manual of Practice for an Effective Anti-Icing Program. Appendix B. Freezing Point of Chemical Solutions*. Publ No. FHWA-RD-95-202, U. S. Department of Transportation, Federal Highway Administration Research and Technology, Washington, D.C., 1996.

[56] A. J. Barduhn and A. Manudhane, "Temperatures Required for Eutectic Freeze Crystallization of Natural Waters," *Desalination*, vol. 28, no. 3, pp. 233–241, 1979.

[57] Anonymous, *Glycerine: An Overview*, The Soap and Detergent Association (industry trade group), New York, 1990.

[58] H. Takiyama, H. Suzuki, H. Uchida and M. Matsuoka, "Determination of Solid–Liquid Phase Equilibria by Using Measured DSC Curves," *Fluid Phase Equilibria*, vol. 194–197, pp. 1107–1117, 2002

[59] J. R. Elliott and M. Gnap, "Estimation of MOSCED Parameters from the COSMO-SAC Database," *Fluid Phase Equilibria*, vol. 470, pp. 241–248, 2018.

[60] P. Dhakal, S. N. Roese, E. M. Stalcup and S. P. Andrew, "GC-MOSCED: A Group Contribution Method for Predicting MOSCED Parameters with Application to Limiting Activity Coefficients in Water and Octanol/Water Partition Coefficients," *Fluid Phase Equilibria*, vol. 470, pp. 232–240, 2018.

[61] N. Asprion, B. Rumpf and A. Gritsch, "Work Flow in Process Development for Energy Efficient Processes," *Applied Thermal Engineering*, vol. 31, no. 13, pp. 2067–2072, 2011.

[62] Z. Rant, "Exergie, Ein Neues Wort Für "Technische Arbeitsfähigkeit"," *Forschung Auf Dem Gebiete Des Ingenieurwesens*, vol. 22, pp. 36–37, 1956.

[63] J. L. Humphrey and G. E. Keller II, "Chapter 6, Energy Considerations," in *Separation Process Technology*, McGraw-Hill, New York, 1997.

[64] J. M. Pirie, "The Manufacture of Hydrocyanic Acid by the Andrussow Process," *Platinum Metals Review*, vol. 2, no. 1, pp. 7–11, 1958.

[65] H. Chang and J.-J. Guo, "Heat Exchange Network Design for an Ethylene Process Using Dual Temperature Approach," *Tamkang Journal of Science and Engineering*, vol. 8, no. 4, pp. 283–290, 2005.

[66] B. W. Hackett, "The Essentials of Continuous Evaporation," *Chemical Engineering Progress*, vol. 114, no. 5, pp. 24–28, 2018.

[67] R. T. Gooty, P. Mobed, M. Tawarmalani and R. Agrawal, "Optimal Multicomponent Distillation Column Sequencing: Software and Case Studies," *Computer Aided Chemical Engineering*, vol. 44, pp. 223–228, 2018.

6 Mass Transport: Characterizing Mass Transfer Drivers and Resistances

The term mass transport refers to all the mechanisms responsible for the relative movement and separation of molecules, primarily molecular diffusion and bulk flow. Molecular diffusion is the main mechanism for movement of solute out of one phase through an interface and into another phase. It is primarily due to thermal motion, which may or may not involve phase change. It is a function of temperature, viscosity of the fluid, and solute molecular weight. The rate of diffusion also depends on the amount of interfacial area that exists between the phases. Bulk flow refers to the relative flow of the phases, often with a wide distribution of phase velocities. It includes convection (movement of a fluid due to density gradients) and advection (movement of a fluid due to velocity and momentum gradients). These are the basics. However, instead of dealing directly with diffusion, interfacial area, and bulk flow, in this chapter we focus on various design terms that have been devised to characterize mass transfer performance specifically for design purposes. These include such terms as separation factor, transfer factor, material balance split, separation power, theoretical stage, mass transfer unit, and the related mass transfer coefficient. Of these, the latter term is the only one that deals directly with the mass transfer rate and interfacial area per unit volume (sometimes at least). These terms are central to the language of separation process design, especially for distillation, absorption, stripping, and extraction. They reflect all the factors that affect process performance – such things as separation driving force, multiphase fluid flow patterns, interfacial area, and the material balance – in succinct and easy-to-comprehend ways. And they are directly related to process design choices. Even when rigorous process simulators are used to model distillation or extraction, it is often useful to express the results in standard design terms. For other separation processes, we focus on other design terms, such as bed volumes to breakthrough and face velocity (adsorption), the meta-stable zone (crystallization), and flux and permeability (membrane-based separations).

So even though the mechanisms involved in mass transport can be highly complex and difficult to elucidate for a given application, designers have devised clever ways of characterizing performance that can manage this complexity – by focusing on key performance measures. As is often said, the more you learn about how something works, the more you can appreciate its intricacies and how much more there is to learn. But there are ways of dealing with this complexity to accomplish what is needed.

https://doi.org/10.1515/9783110695052-009

6.1 Driving Forces and Key Descriptors

We begin by discussing separations driven by deviation from equilibrium. We also touch on certain aspects of membrane-based separations to point out key similarities and differences.

6.1.1 Separations driven by deviation from equilibrium

Separation processes such as distillation and extraction must operate away from equilibrium if they are to transfer mass from the feed phase to a second phase. This is true of adsorption and crystallization, as well. By operating at conditions that deviate from equilibrium, these processes harness nature's tendency to move toward the equilibrium state, and it is this tendency that drives the selective transfer of components between phases. In Section 5.3: Chemical Potential and Mass Transfer, this tendency is described in terms of a difference in the chemical potential of a given solute in different phases. For mass transfer to occur, the difference in the chemical potential of that solute, when comparing the feed phase to the second phase, must be positive in going from a higher energy state to a lower energy state. For practical design purposes, it is standard practice to express this driving force in terms of the difference between the current concentration of solute in a given phase and the concentration at equilibrium.

A process driven by deviation from equilibrium may operate at steady-state conditions (with constant flow rates, etc.), but this only means that process conditions are constant over time, not at equilibrium. With steady-state operation, the compositions of feed and effluent phases change with position in the equipment (not with time) as the phases naturally move toward equilibrium on their path through the equipment. On the other hand, for a non-steady-state process, both process conditions and phase compositions change with position and with time. In any case, for the above mentioned processes a deviation from phase equilibrium is required, and this is why up to this point we have focused so much on what equilibrium is and how to model it. An exception involves the use of pressure to drive selective transfer of components through a semipermeable membrane placed between two phases.

6.1.2 Drivers for membrane-based separations

Membrane-based separations differ from other separation methods, because in many cases mass transfer across the membrane is not driven by a deviation from equilibrium. Instead, the membrane serves as a physical barrier and a kind of sieve to effectively filter out molecules of specific sizes or with differing chemical properties. Normally, elevated feed pressure is needed to overcome mass transfer resistance in the form of significant pressure drop across the membrane. Pressure does the work of de-mixing the feed, and in so doing, the process actually works against the natural forces that would tend to move the system back toward a mixed equilibrium state if the membrane were removed. The de-mixing work is accomplished by the use of mechanical or thermal compression energy. The non-membrane-based methods (distillation and others) also need to overcome some form of pressure drop for adequate flow through their specific equipment, but for these methods, pressure is not the force that drives the separation.

This is not to say that phase equilibrium does not or cannot play a role in membrane-based separations. Although mass transfer flux in many commercial membrane-based processes is not driven by a deviation from equilibrium, the chemical composition of the membrane can significantly affect the selectivity of the process through specific thermodynamic interactions between feed components and the membrane material. These can yield differences in the adsorption, dissolution, and diffusion of feed components within the membrane structure, occurring within pores or within the bulk material, and this can significantly affect the molecular discrimination and flux that takes place. This may involve discrimination according to the relative polarity of the molecules as well as relative size, yielding different rates of transfer across the membrane barrier for different types of molecules. But normally this still requires use of elevated feed pressure to overcome significant pressure drop to achieve commercially viable flux and product recovery. Examples include separations involving reverse osmosis as well as many gas separations. Various mechanisms of membrane-based mass transfer are discussed elsewhere; for example, see refs. [1–3].

In certain special kinds of membrane-based separations, elevated feed pressure (or pressure drop) may be reduced or even eliminated by harnessing a specific and strong form of thermodynamic (equilibrium seeking) driver. These special types of membrane-based processes include:

- Pervaporation. This involves selective permeation of liquid through a membrane, with evaporation on the permeate side of the membrane. Such a process takes advantage of relative rates of solute permeation driven by evaporation. These are influenced by thermodynamic interactions between solute and the membrane material. Pervaporation is used commercially to dry water-wet alcohols. See Section 12.5: Pervaporation.
- Forward osmosis. This process takes advantage of naturally driven osmosis to drive solute out of a liquid feed, across a membrane barrier, and into a more thermodynamically favored draw solution [4, 5]. For example, this may be used to dry a water + organic liquid solution by selective transfer of water out of the feed, through the membrane, and into concentrated brine.
- Facilitated transport. This process involves reaction-enhanced transfer of solute through a membrane by reversible reaction with a carrier agent contained within the membrane [6]. The membrane may be a supported liquid or a fixed polymer containing a mobile or fixed carrier. Solute interacts with the carrier and is carried through the membrane via diffusion with the carrier or by hopping from one carrier site to another [7]. The reaction is reversed on the other side of the membrane

to release solute and regenerate the carrier. In principle, such a process may involve mass transfer from liquid to liquid, gas to liquid, liquid to gas, or gas to gas. The process is analogous to cellular transport processes in biology and to reaction-enhanced liquid-liquid extraction driven by reversible chemical reaction between solute and an extractant species dissolved in the extraction solvent. Potential applications include removal of CO_2 from nitrogen rich flue gas [7, 8].

– Membrane distillation. This is a membrane-based thermal process driven by a difference in the vapor pressures (or solute partial pressures) exerted by the liquids on each side of a membrane [9, 10]. This may involve hot liquid on the feed side of the membrane and cold liquid on the permeate side, such that transfer of solute involves permeation of solute vapor within the membrane, followed by condensation of vapor within the cold liquid on the permeate side. Membrane distillation concepts are still under development. Compared to standard distillation, the membrane-based process may allow for higher interfacial area per unit volume and a reduction in energy consumption by allowing for localized heat transfer within the contacting device. Other configurations have also been proposed. Pervaporation is a kind of membrane distillation.

– Liquid-liquid extraction membranes. This process involves placing a membrane barrier between a feed liquid and an extraction solvent. Mass transfer occurs at micro-scale meniscus held within the pores of the membrane. The process is driven by deviation from liquid-liquid equilibrium at the liquid-liquid interface, just as in standard liquid-liquid extraction. The membrane serves to form and control interfacial area, avoiding any need for droplet formation during extraction and droplet coalescence after extraction followed by liquid-liquid phase separation.

Pervaporation is the most commercially successful of these specialized thermodynamically driven processes. It is used to permeate water away from alcohols and other organics. However, at present, pressure-driven membrane-based processes constitute the great majority of large-scale industrial applications, mainly for aqueous-liquid separations such as desalination and ultrafiltration and for gas separations. For more discussion, see Chapter 12: Membrane-Based Separations: Adding a Semipermeable Barrier between Phases.

It should be noted that when elevated feed pressure is required, it can sometimes be significantly reduced if the designer is willing to sacrifice some product recovery and reduce volumetric efficiency. Here, we are talking about the ΔP required to achieve a desired volumetric flow rate per unit area of membrane (see Section 6.7.5: Membrane Permeance and Separation Factor). A well known example is the use of reverse osmosis (RO) in a point-of-use application in homes (such as under the kitchen sink) to produce soft water from a hard water supply [11]. Here, the concentration of calcium salts on the feed side of the membrane is not allowed to build up to be very high as water permeates the membrane, so the osmotic pressure that develops is never very high and the pressure drop across the membrane is kept to a low level. This allows the use of normal residential water pressures (about 60 psig or about 400 kPa differential pressure) to produce satisfactory flow of soft water for drinking. But doing so requires sending at least 3 liters of feed water to the drain for every liter of soft drinking water produced (often much more), depending on feed water composition and the particular RO system design. Reducing the amount of water sent to the drain requires operating at a pressure significantly above the typical municipal or well-water supply pressure, which then requires use of a pressure-boosting pump at significantly greater cost. (We discuss osmotic pressure in Section 6.7.5: Membrane Permeance and Separation Factor and reverse osmosis in Section 12.2: Water Treatment.) A similar tradeoff is found with pressure-driven gas separations. The separation of air into enriched oxygen and enriched nitrogen streams is a well-known example. Commercial units employ membranes that preferentially permeate oxygen, and they are rated according to the air-to-product ratio that they require; that is, according to how much air must be processed to generate the required product flow rate (with the product being either enriched oxygen or enriched nitrogen). The air-to-product ratio is a measure of product recovery, which is important even though air is free because it affects the volumetric efficiency of the process and how large the feed compressor must be for a given production rate. A typical design involves a tradeoff whereby a reduction in the compressor operating pressure is obtained at the expense of handling a higher volumetric flow rate. A balance is needed to minimize the cost of the compressor [12]. See Sections 6.7.5 and 12.4.

6.1.3 Partition ratio, separation factor, and transfer factor

In analyzing a separation process involving transfer of solute from a feed phase to a second phase and driven by deviation from equilibrium, it is helpful to focus on three key descriptors in analyzing performance: the equilibrium partition ratio, the separation factor, and the transfer factor.

Partition Ratio. The equilibrium partition ratio is defined as:

$$K_i = \frac{y_i}{x_i} \tag{6.1}$$

where x_i = concentration of component i in the feed phase
$\qquad y_i$ = concentration of component i in the second phase, in equilibrium with x_i.

The quantity K_i is also called the distribution coefficient, the equilibrium constant, or the K-value. It represents the equilibrium partitioning of solute between two phases. Its value depends on the choice of concentration units, so it is important to know the specific definition used by authors. Typical units are mole fraction, mass fraction, mass per unit volume, or moles per unit volume.

Separation Factor. The separation factor is a relative partition ratio, which reflects the ability to separate one solute from another solute or to separate solute from solvent. It measures the relative enrichment of a given solute in the second phase after one theoretical stage of contacting. Enrichment may be further increased with additional stages. The overall enrichment achieved by an entire process scheme including all stages is called separation power. Separation power is a function of the single-stage separation factor and other factors that contribute to separation performance. Think of separation power as a magnified version of the separation factor, magnified by the process (see Section 6.5.1: Material Balance Split and Separation Power). The separation factor is used to characterize the selectivity of a separation process for a given solute. Its value is independent of the choice of concentration units, as long as they are consistent. For example, the separation factor for distillation is relative volatility. It is given by:

$$\alpha_{i,j} = \frac{y_i/y_j}{x_i/x_j} = \frac{y_i/x_i}{y_j/x_j} = \frac{K_i}{K_j} \tag{6.2}$$

For binary distillation, this gives:

$$\alpha_{i,j} = \frac{y_i/x_i}{y_j/x_j} = \frac{y_i x_j}{y_j x_i} = \frac{y_i(1-x_i)}{y_j(1-x_j)} \tag{6.3}$$

which upon rearranged gives

$$y_i = \frac{\alpha_{i,j} x_i}{1 + (\alpha_{i,j} - 1)x_i} \tag{6.4}$$

The separation factor can also be expressed in terms of equilibrium relationships and physical properties. In the case of distillation, substituting the equilibrium relationships for distillation, discussed in Section 5.5: Key Expressions of Phase Equilibrium, yields the following:

$$\alpha_{i,j} = \frac{p_i^{sat}}{p_j^{sat}} \quad \text{for an ideal liquid mixture} \tag{6.5}$$

$$\alpha_{i,j} = \frac{\gamma_i \, p_i^{sat}}{\gamma_j \, p_j^{sat}} \quad \text{for a real liquid mixture} \tag{6.6}$$

$$\alpha_{i,j} \approx \frac{1}{x_i^{sat}} \frac{p_i^{sat}}{P_{total}} \quad \text{for a sparingly soluble liquid solute} \tag{6.7}$$

$$\alpha_{i,j} \approx \frac{1}{x_i^{sat}} \exp\left[\frac{\Delta_{fus}S}{R}\left(1 - \frac{T_m}{T}\right)\right] \times \frac{f_i^o}{P_{total}} \quad \text{for a sparingly soluble solid} \tag{6.8}$$

where f_i^o = vapor pressure (fugacity) of a hypothetical subcooled liquid obtained by extrapolating the liquid's vapor pressure below the melting (or freezing) point.

The separation factor as defined above characterizes separation ability strictly in terms of phase equilibria. But for processes that operate far from equilibrium, the actual separation such a process can deliver depends on the relative rates of component mass transfer. The process might then be better characterized by a non-equilibrium separation factor based on relative rates of mass transfer:

$$\alpha_{i,j}|_{non-equilibrium} \equiv \frac{R_i}{R_j} \tag{6.9}$$

where R_i is the overall rate of mass transfer from the feed phase to the second phase for an individual component (in the presence of all other components). The separation power of the entire process is then a function of $\alpha_{i,j}|_{non-equilibrium}$. We discuss the non-equilibrium separation factor for membrane-based separations in Section 6.7.5: Membrane Permeance and Separation Factor. The concept of a non-equilibrium separation factor may be viewed as an application of non-equilibrium thermodynamics, an area of study that considers processes that are not in equilibrium but may be described in analogous thermodynamic terms [13–15].

However, in general practice a non-equilibrium separation factor is not dealt with directly in the design of separation process equipment. Instead, process performance is characterized in terms of phase equilibrium (expressed by the equilibrium separation factor) and the mass transfer capability (expressed by the number of actual stages or by the number of transfer units or another rate-based model). Phase equilibrium is a major factor even when a process operates far from equilibrium, because it affects both the separation driving force and the relative capacities of the various phases to carry solute. These things are reflected in the magnitude of the solute transfer factor.

Transfer Factor. The transfer factor for a given solute is defined as the slope of the equilibrium line for that solute multiplied by the relative amounts of second phase to feed phase:

$$\text{Transfer Factor } \mathcal{T}_i \equiv \frac{dy_i}{dx_i} \frac{M^{II}}{M^I} = m_i \frac{M^{II}}{M^I} \tag{6.10}$$

$$\mathcal{T}_i \approx K_i \frac{M^{II}}{M^I} \text{ at low concentrations of } x_i \tag{6.11}$$

where M^I = moles or mass of the feed phase (or molar or mass flow rate)
M^{II} = moles or mass of the second phase (or flow rate)
m_i = slope of equilibrium line, $y_i = m_i x_i + b$
K_i = partition ratio for component i, $K_i = y_i/x_i$

In principle, any concentration units may be used as long as they are consistent. Of course, when using phase equilibria correlations expressed in mole fractions (the normal practice in distillation design), mole fractions should be used for design calculations and later converted to mass units. The transfer factor $\mathcal{T}_i$ is a measure of the carrying capacity of the second phase relative to that of the feed phase. We have chosen to call it the transfer factor in general because of its relevance to a wide range of separation processes and the fact that that its value greatly affects the ability of a process to transfer solute between phases. It goes by many specific names depending on the particular type of separation process being discussed, such as stripping factor (also called lambda), extraction factor, and desorption factor. In order to transfer a high percentage of desired solute from a feed phase into a second phase, the transfer factor must be greater than unity. And the greater its value, the lesser is the number of theoretical stages or mass transfer units needed for high recovery. On a McCabe-Thiele diagram, the transfer factor is the slope of the equilibrium line divided by the slope of the operating line. Figure 6.1 illustrates the transfer factor for stripping-mode distillation involving transfer of solute from liquid to vapor. In this case, the transfer factor is called the stripping factor (or lambda), given by:

$$S = m\frac{V}{L} \approx K\frac{V}{L} \text{ (for dilute solutions)} \tag{6.12}$$

As illustrated in the figure, the number of stages needed for a given separation decreases as the value of S increases. At S equal to unity, the two lines become parallel, so an infinite number of theoretical stages (or mass transfer units) are needed to accomplish any finite separation. The same analysis applies to standard (stripping mode) liquid-liquid extraction. In that case, the analogous transfer factor is called the extraction factor. In the rectification section of a distillation process, the transfer factor is called the absorption factor. It is written as the reciprocal of the stripping factor, because solute transfers from vapor to liquid instead of from liquid to vapor. In pressure swing adsorption (PSA), a related kind of transfer factor is called the desorption factor. In this cyclical type of process, the transfer factor may be defined in terms of the ratio of backpurge gas to feed gas on a volumetric basis – to represent relative carrying capacity during regeneration (see Section 10.7: Pressure Swing Adsorption). And so it goes for other types of separation processes driven by deviation from equilibrium.

Transfer units can be thought of as stages that never reach equilibrium, so the number of transfer units is always greater than the number of theoretical equilibrium stages, as illustrated in Figure 6.1 and labeled N_{OL} for the number of overall liquid phase mass transfer units. We discuss all of this in more detail in later sections. Our purpose here is to introduce the basic concepts. Revisiting this page after reading the later discussions should help clarify their meaning.

The partition ratio, separation factor, and transfer factor are not necessarily constant throughout a given process. This depends on whether solute concentrations are sufficiently dilute as to be within a linear y_i versus x_i concentration range, or more concentrated and outside this range. Dilute conditions are common for stripping-mode distillation and extraction operations, as well as for many gas absorption operations. But for fractionation processes, solute concentrations may be much more concentrated. And the definition and value of the key descriptors will vary from one end

$$S = m\, V/L$$

Small S

slope = m_1

y

x_{out} x_{in}

Large S

equilibrium line
slope = $m_2 > m_1$

operating line
slope = L/V

y

x_{out} x_{in}

Transfer Unit Concept

slope = m_1

N_{OL} > No. theoretical stages

slope = L/V

y

x_{out} x_{in}

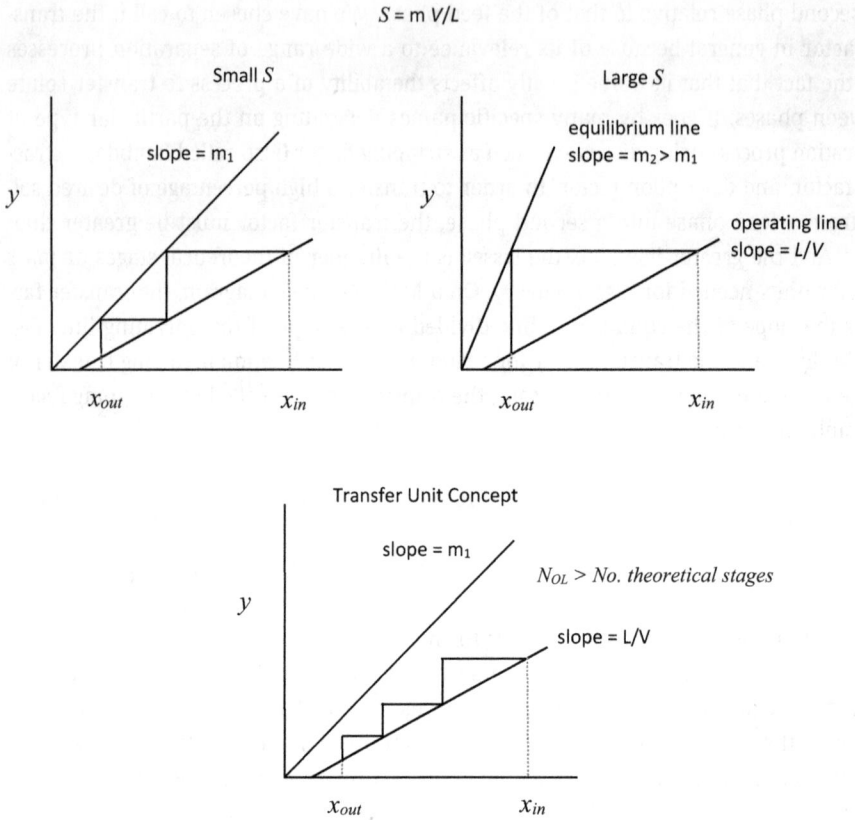

Figure 6.1: The Impact of the Stripping Factor on the Number of Theoretical Stages and the Number of Transfer Units.

of the process to the other as the concentration profile changes. In fractional distillation, for example, we pay attention to the magnitude of the absorption factor within the rectification section, because in this section the focus is on the transfer of solute from vapor to liquid. On the other hand, we pay attention to the magnitude of the stripping factor in the stripping section, because in this section the focus is on transfer of solute from liquid to vapor. And furthermore, the transfer factor often is a useful concept within a specific concentration range, as when analyzing a given distillation tray, because the equilibrium and operating lines that relate to operation within that range are fairly constant in slope. See Section 6.6: Theoretical Stage Models and Section 6.7: Rate-Based Models for more discussion.

6.1.4 Assessing rate limitations and thermodynamic consistency

In developing a phase equilibrium expression to quantify the partition ratio, separation factor, and transfer factor, the model expression may fit the available data reasonably well, but without being thermodynamically correct. For example, in determining relative volatility for use in distillation design, both pure component vapor pressures and activity coefficients contribute to the calculation of relative volatility (as in eq. (6.6)). If the available vapor pressure data are incorrect, the activity coefficient model may be adjusted to fit available vapor-liquid equilibrium (VLE) data, but the resulting values of the activity coefficients will not be correct in terms of the underlying fundamentals. This can easily lead to poor designs, especially when using the expression outside the range of the original data fit.

So how accurate or correct does a phase equilibrium model need to be when designing a separation process for a given application, and how do we know if it is correct? The answers vary widely depending on the application. For some distillation applications, an approximate phase equilibrium model will suffice simply because the process operates far from equilibrium, as when steam stripping sparingly soluble organics with high relative volatilities with respect to water and correspondingly high stripping factors. In that case, the design is limited by mass transfer rates related to solute diffusion and transport properties, not by phase equilibrium. So when relative volatility is above 50, for example, whether it is exactly 50 or 60 or even 100 does not really matter. But the designer must recognize this fact – so that the appropriate design methods based on mass transfer resistance are used instead of assuming that equilibrium is attained (for more discussion, see Section 7.3.3: Continuous Stripping and Absorption). On the other hand, in many other cases, relative volatility is not extreme, so the distillation process operates much closer to equilibrium. Then, having an accurate phase equilibrium model is critical to achieving a successful design. An example of this kind of process is the use of azeotropic or extractive distillation to distill a highly non-ideal feed mixture (see Chapter 8: Enhanced Distillation). In general, the required accuracy may be gauged by studying the sensitivity of calculated separation performance to variation in the value of relative volatility or whatever the thermodynamic driving force may be for the particular separation process of interest. The general rule is that rate limitations become significant whenever the transfer factor is greater than about 2.0. In many cases, this indicates a separation factor greater than 2.0.

In evaluating the separation factor for distillation (relative volatility), one technique used to evaluate the validity of the chosen phase equilibrium model is to plot, for isothermal vapor-liquid equilibrium data, the quantity $ln\gamma_i/\gamma_j$ versus mole fraction of the more volatile component in a binary mixture, from $x_i = 0$ to $x_i = 1.0$. The areas above and below the line $ln\gamma_i/\gamma_j = 1.0$ should be equal. If not, the phase equilibrium model used to calculate relative volatility is not thermodynamically consistent with the Gibbs-Duhem equation and should not be relied upon to accurately model relative volatility in all scenarios [16, 17]. Note that one can also plot $ln\alpha_i$ versus mole fraction for isothermal VLE data, no matter how α_i is determined (whether via modeling or by experiment), and look for equal areas above and below the line $ln(\alpha_i)_{ideal} = lnp_i^{sat}/p_j^{sat}$. If the areas differ significantly, further investigation into how vapor pressures and activity coefficients (or other model parameters) are determined is warranted. This kind of consistency test is not always possible for other types of phase equilibria. For example, these tests are not valid with isobaric (non-isothermal) VLE data because activity coefficients and vapor pressures vary with temperature. But in any case, calculated values may be compared with experimental data to at least gauge how well the model represents the data in terms of statistical measures.

For liquid-liquid extraction (and other separation processes as well), the data should be evaluated for accountability; that is, whether the material balance closes for individual components. This requires measuring solute concentration in both phases. Sometimes, designers will use data developed by measuring solute concentration in only one phase and then calculate values for the other phase by material balance, assuming 100% accountability. But such results may not be accurate or reliable. For processes involving both VLE and liquid-liquid extraction (LLE), often separate phase equilibrium expressions will be needed to adequately model each. Although theory suggests that both can be modeled by a single phase equilibrium expression, in practice this does not often work out to be true.

For adsorption, it is important to measure the working capacity of the adsorbent material; that is, the capacity after repeated regeneration cycles. Often, working capacity is considerably lower than the capacity of fresh material, because micropores become filled with adsorbed molecules that are not completely removed during regeneration (either by raising the temperature or reducing the pressure). That is, the effective adsorption capacity may be limited by kinetic limitations within micropores. So, care should be taken to measure the reversible adsorption capacity after repeated cycling between adsorption and desorption steps using the chosen regeneration method – not the adsorption capacity of new, unused material.

For crystallization, the validity of solid-liquid equilibrium data should be checked by measuring solubility and checking the material balance. Solubility should be measured for solvent compositions that will be encountered in the proposed design, not for the pure-component solvents. Phase equilibrium expressions vary in their ability to reliably correlate or predict solubility in mixtures of solvents, so the results must be checked by comparison with mixture data. Furthermore, it is important to establish whether the solid-liquid system you are dealing with is a true eutectic system or whether intermolecular compounds, solid solutions, or regions of partial miscibility in the solid phase are present. This information may be obtained by using differential scanning calorimetry [18]. If the system is eutectic in nature, then highly pure product crystals may be produced in a single stage of crystallization. But in cases involving formation of compounds or solid solutions or partial miscibility, solid-phase purity will be limited. So understanding the type of solid phase behavior is critical to understanding crystallization process options.

6.2 Fluid Flow and Mixing: Imagine the Best Flow Path Through the Equipment

In designing separation equipment, the first thing we do is imagine the path we would like the various phases to take on their way through the equipment – the best path we can imagine that will enhance mass transfer and avoid regions of flow that would degrade the separation. Then we can ask, what options might enable such a path? Which ones might be simplest to implement, yet still be effective? This is a mental exercise and commonsense approach that can help guide conceptualization of appropriate equipment geometry and the necessary internals. These conceptual images of what will be needed for good performance must first be envisioned in the mind before any design details can be calculated and optimized.

A classic example of this is the need for uniform distribution of liquid reflux at the top of a randomly packed distillation column. The importance of uniform distribution was not generally recognized by the industry until the 1980s. In fact, up until that time some designers thought it was necessary to avoid distributing liquid near the column wall because they thought all of it would flow down the wall and bypass the packing, so in many designs liquid was deliberately distributed in a concentrated flow to the center of the column, avoiding the outer area near the wall altogether. But once it was realized that uniform distribution was an important design goal, mass transfer performance (total number of stages achieved by the packed column) could be significantly improved, often by 20% or more, by distributing flow uniformly across the entire cross-section including near the wall. All it took really was the idea that for best performance the flow of liquid to the top of the packing should be uniformly distributed without favoring the center. This then opened up the thinking about how to make this happen, and today many suppliers offer excellent designs for uniform distribution. For more discussion, see Sec. 7.4.1, Column Design. Another example has to do with the flow of gas through fixed-bed adsorbers filled with adsorbent granules. Again, the best path we can imagine involves uniform distribution of the gas at the front of the bed. Some designs involve use of nozzles in a lateral pipe distributor. Ensuring that uniform distribution is achieved involves understanding pressure drop through pipes and nozzles and how to specify each nozzle to ensure all nozzles flow at the same rate over the entire range of total feed rates the unit must handle. But the need for uniform flow must first be recognized as a design goal. Some adsorber designs we have seen do not include a gas distributor at all, using only an inlet pipe with diffuser, so the opportunity for best performance is lost. For more discussion, see Sec. 10.4.2, Adsorber Dimensions, Internals, and Piping Diagram.

Many commercially successful separation processes are designed to emulate and approach the ideals of either countercurrent or chromatographic contacting of two distinct phases. Countercurrent contacting involves movement of the phases in opposite directions to take advantage of favorable concentration gradients as solute transfers from one phase to another. This may involve so-called differential contacting in which individual elements of the dispersed phase move through a bulk continuous phase, both phases flowing in opposite directions. Or it may involve multiple discrete stages of contacting in which the phases are mixed together at each stage, then separated and routed in opposite directions to the next stage of contacting. Either approach magnifies separation power and can maximize the ability to achieve both high solute recovery and purity. Chromatographic processing, on the other hand, normally involves the movement of the feed phase through a second stationary phase. This approach also magnifies separation power and can provide for both high recovery and purity, but now, cyclical operation is required to compensate for the fact that the stationary phase is not moving. Cyclical operation with multiple units may be used to simulate and approach the performance of true countercurrent processing.

There are many ways to approach these ideal flow patterns and many process options. Practical considerations for a given application include the following:
- Batch versus continuous operation or fed-batch operation
- Dynamic versus steady state
- Cyclical steady-state
- Differential processing versus the use of discrete contacting stages
- Binary separation (in two product streams) versus generation of additional products in additional product streams
- Internal recycle in the form of reflux
- The need to purge impurities to avoid accumulation within the equipment
- Large inventory versus small inventory held within equipment
- Short contact time versus long contact time
- Operating period between start-up and shut-down (batch cycle time and continuous operation between maintenance periods)
- Start-up time or step times
- Implications for downstream processing requirements
- The presence of quiescent or inactive zones within equipment
- The degree to which backmixing of phases occurs, which disrupts uniform flow and degrades the separation to some extent

When first introducing the subject of industrial chemical separations, it seems that most examples that people point to involve some sort of large-scale, continuously fed, steady-state process, such as distillation carried out in petroleum refineries or in large-scale petrochemical plants. Continuous, steady-state processing is a large-scale concept not found in the laboratory where batchwise non-steady-state (dynamic) processing is the more convenient way of working at small scale. For large-scale operations, continuously fed, steady-state processing can offer higher productivity in terms of minimizing the required size of equipment and reducing unproductive downtime. When continuous processing is not feasible, the large-scale process often is designed to emulate a continuous one; that is, such a process is designed to handle a continuous flow of feed even though the internal operation of the process involves some form of cyclical (dynamic) operation or cyclical steady state.

With all of this in mind, we have often wondered how countercurrent processing first came about as a concept to be emulated. Drawings of ancient alembic batch distillation apparatus show a boiling pot of liquid connected to another vessel via an un-insulated metal funnel and tube assembly. The tubing is uninsulated and air cooled to allow condensation of the vapor. Most of the resulting condensate is collected in a second vessel (the receiver), but it seems that some of the condensate formed in this way must fall back into the boiling pot countercurrent to the rising vapor. The result is a natural (unintentional?) form of countercurrent vapor-liquid processing taking place within the top cover or the tubing at the connection to the pot (an early form of countercurrent liquid reflux). This must have given somewhat better performance, which people began to appreciate and deliberately mimic at some point. Later versions of batch stills show more elaborate tubing coils with long vertical connections that may have provided some additional internal reflux. Various water-cooled condensers that allowed for deliberate return of liquid to the pot were introduced in the eighteenth century by early chemists such as Justus Baron von Liebig (1803–1873), and some of these involved countercurrent flow of cold water for better heat transfer. For more discussion, see Section 3.2: Pioneers, and the introduction to Chapter 7, Distillation: Rectification, Stripping, and Absorption – the Workhorse Methods. So, perhaps the concept of countercurrent processing began as a fortunate accident that people began to recognize and further develop.

6.2.1 Ideal fluid flow

Here are some examples of design goals related to ideal fluid flow patterns, the details of which are discussed in later chapters dealing with the specific separation method:
- Favorable fluid flow patterns
  - Uniform distribution of liquid reflux and incoming feed within distillation columns (as mentioned above)
  - Uniform flow of the continuous phase in one direction and the swarm of dispersed droplets in the other (in countercurrent fashion) within static liquid-liquid extraction columns (differential contactors)
  - Well-mixed flow in agitated extractors within mixing zones, and uniform flow from one zone to the next – as in a Scheibel rotary-impeller extraction column [19]
  - For crystallizers, the goal often is to achieve well-mixed, high volume circulation, with only low-to-moderate shear, to suspend all the crystals but avoid attrition, and to maintain a uniform distribution of crystals per unit volume (uniform particle suspension)

- Quiescent flow within liquid settlers (decanters), without inducing circulation at drain points that can upset the clarification of the settled layers
- In some membrane modules, fluid flow is directed across the membrane in a direction parallel to the membrane surface (called tangential flow or cross flow) to continuously flush the surface to help reduce fouling
- Favorable hydraulics. High volumetric flow through the equipment with low resistance to flow (low pressure drop).

Hydraulics involves the application of fluid mechanics to understand how fluids flow through equipment, the pressure drop that results, and the maximum flow capacity that can be sustained before something unwanted happens with the flow, such as excessive turbulence, entrainment of one phase into another (flooding), foaming, or emulsion formation. Hydraulic phenomena determine the pressure drop characteristics of fluid traffic within distillation columns and adsorbers, the dimensions needed for uniform flow through feed distributors, the appropriate sizes of packing in packed distillation columns and packed extractors, the dimensions of distillation and extraction trays, and the flow conditions at which the equipment begins to exceed its maximum flow capacity (begins to flood). And because of all this, hydraulics also influences mass transfer performance.

In designing a separator, its dimensions are specified, in part, by using well-known calculations of pressure drop within conduits as a function of fluid physical properties and fluid velocity, and by using criteria quantified in terms of dimensionless groups that correlate with the desired type of flow, such as Reynolds number and Weber number. Classic design methods often focus on fluid velocities or volumetric flow rates per unit area, and these parameters are specified to stay at or below a critical pressure drop, which experience has shown is needed for good mass transfer performance, and if exceeded, leads to poor performance.

The visualization of fluid flow within the equipment is facilitated by the application of computational fluid dynamics (CFD) software. Such software is designed to facilitate the setting up and solving of systems of differential equations that describe a specified geometry or flow field, and then plot the results in a way that helps to visualize the flow. This can be a fairly straightforward tool in cases involving flow of single phases and simple geometries, but it can require experienced specialists in other cases involving complex two-phase flow or complicated geometry. The software is continually being updated, so more and more applications are likely to be examined with this approach. For example, CFD is used to simulate flows within distillation column distributors and packings [20], crystallizers [21], oil/water separators [22], and fixed-bed adsorbers [23]. CFD is especially useful for adjusting the placement of baffles, mixing impellers, and other internals – to adjust flow so it better approaches the desired ideal flow pattern. It is often used to identify several promising geometries and flow patterns before testing prototypes in the laboratory or miniplant. The specific goal of a CFD study may be to identify ways of promoting uniform flow with little backmixing, or well-mixed flow with low shear, or high volumetric flow with low pressure drop – in the course of evaluating various potential equipment geometries to determine which is likely to be most effective.

In using CFD as a tool, there are a number of things to keep in mind that can help ensure good results. Make the geometry as simple as possible; that is, eliminate small details that will not significantly impact the results. This will allow use of a larger calculation mesh size. Also, first model single-phase flow to establish the bulk continuous phase flow pattern before attempting to add a second phase. Modeling bulk fluid flow patterns may be sufficient to establish a suitable design. Modeling two-phase flow is considerably more complex. Also, be sure to check that the converged solution to the simulation closes the material balance reasonably well [24, 25].

6.2.2 Mixing design

In the field of chemical separations, the term mixing design generally refers to specifying the internals of separation equipment in order to obtain near-uniform distribution of dispersed-phase particles or elements within the continuous phase and to obtain near-uniform solute concentrations within the continuous (bulk) phase. Mixing is an important consideration for any process involving discrete stages of contacting, where two phases are mixed together to allow the process to approach equilibrium at that stage, after which the phases are separated and routed on to other contacting stages or discharged from the process. Mixing design is especially important in liquid-liquid extraction for the design of agitated column extractors and mixer-settlers, and in crystallization to ensure adequate circulation and turnover of the contents of the crystallizer and adequate suspension of solids. A number of handbooks are devoted to mixing technology [26–28]. We recommend consulting these resources as well as discussing options with various suppliers of mixing equipment.

In specifying a mixing design, the designer needs to be aware of the scale of mixing that is required (whether blending large or small elements of fluid) and the appropriate type of mixing impeller or other mixing device. Macro-mixing (or bulk blending) refers to bulk flow and circulation of fluid throughout a vessel. Meso-mixing refers to the turbulent dispersion of fluid elements into a flowing fluid. Micro-mixing (also called mixing at the Kolmogorov scale) occurs at the scale of the smallest eddies that form within turbulent flow. For example, in the design of a stirred-tank batch crystallizer or a mixer-settler used for extraction, macro-mixing design focuses on circulation of fluid and is quantified by the rate at which a batch of fluid is completely overturned within a vessel (called turnover). The meso-mixing scale is the focus when ensuring suspension of solid particles within a fluid or when dispersing additives and reactants into a fluid at a submerged injection point. Here, the injection point may be placed just above the impeller for rapid dispersion into the flowing fluid. This can be important in minimizing concentration gradients and avoiding localized high concentrations of the additive. Micro-mixing may be required when dispersing reagents into rapid precipitation reactions or reactive crystalli-zations with rapid kinetics. But micro-mixing is intense and should only be pursued if needed, as it may lead to poor performance due to unwanted emulsion formation (in an extractor) or to excessive secondary nucleation or crystal breakage (in a crystallizer).

The choice of impeller type and its operation will vary depending on the required scale of mixing. Axial-flow (hydrofoil) impellers and pitched-blade impellers may be used when high-volume, low-shear mixing is called for at the macro scale. They may also serve for applications in the meso-scale, depending on the impeller size and speed of operation. But at the micro-scale, a high-shear impeller such as the Rushton turbine is normally called for. The characteristics of the many types of impellers (their geometry, power numbers and power consumption, shear rate, relative volu-metric pumping capacity, and so on) are given in the handbooks. In any case, as scale-up of mixing cannot be exact and to allow flexibility in handling a variety of feeds, we recommend specifying a variable-speed drive to allow for some range of adjustment during operation.

6.3 Interfacial Area

So far, we have focused on various flow patterns that may be considered ideal or most favorable for different types of separators. But there is another related aspect that must be emphasized as well, and that is the need to provide sufficient interfacial area between phases. No matter how the flow of the phases is established and di-rected, separation equipment simply will not function as required unless the interfa-cial area per unit volume is sufficiently high. Simply stated, a mass transfer device is an interfacial-area generator. Product solute must transfer from the feed phase across an interface and into a second phase, and to do so at commercially viable production rates, there must be plenty of area between the phases. To accomplish this, separation equipment is designed to disperse one phase into the other in such a way as to pro-mote the generation of interfacial area. Membrane modules may be considered an ex-ception, but they still must provide sufficient area for mass transfer. So except for membrane-based separations, one phase is chosen to be the continuous phase that fills the equipment (or specific zones within the equipment), and the other phase is chosen as the dispersed phase present as small gas bubbles, liquid droplets, liquid films or rivulets, small solid particles, or some other form – packed as closely together as possible (or practical). And the choice of which phase to disperse can greatly affect how much interfacial area is generated. In this section, we introduce various concepts and techniques used to promote the generation of interfacial area. Additional discus-sions are given in Part III, Engineering Practice, for specific types of separators.

In specifying a separator, the designer must set the equipment geometry, internal flow pattern, and internal flow rates, which may or may not come from mechanical agita-tion, to achieve sufficient interfacial area with only moderate resistance to flow. The choice should also ensure that the internals have little tendency to become fouled or plugged with deposits over time. This requires specifying internals such as packings, trays, or mixing zones with flow channels or orifices that are not too small (perhaps no smaller than 5 mm or so). It is not unusual for equipment to begin performing poorly at some point due to fouling or plugging of the internals, requiring periodic shutdown for cleaning. So the design should tolerate some level of fouling to allow for extended operation, but it should also allow for easy access to internals for rapid and thorough cleaning when needed. And in our experience with well-used equipment, the required internals sometimes become damaged over time and in some cases no longer exist due to corrosion, so the materials of construction should be chosen with mechanical strength and corrosion resistance in mind. It is also important to pay attention to the amount of fluid shear that is generated in the effort to maximize interfacial area, especially for mechanically agitated equipment, making sure it is not too severe. This may be analyzed in terms of Kolmogorov turbulent eddies (produced at the micro-mixing scale), noting that if present, any eddies smaller than suspended liquid drops or crystals may promote formation of liquid-liquid emulsions or cause crystal breakage. See Section 6.2.2: Mixing Design.

Although interfacial area per unit volume is essential for good performance, it can be difficult to measure and calculate, so often the actual interfacial area operating within a separator is not known. Instead, its effect is lumped into other performance measures such as stage efficiency or the height equivalent to a theoretical plate (HETP). Designers know they can increase these performance measures by making design choices that tend to increase interfacial area. Some general concepts can be helpful in thinking about how to go about this. For example, in specifying a distillation packed column, packing is chosen to provide area that promotes formation of liquid films. Designers have worked for decades to specify packings that deliver high packing area per unit volume with large open area for low resistance to flow – over four generations of packing designs (see Section 7.4: Equipment Design and Rating Methods (for distillation)). Efforts to pack more area into the packing per unit volume have resulted in novel shapes of high-surface-area packing pieces that fall together to form open structures that allow flow with low pressure drop. This thinking has also led to the development of various structured packings that deliver high interfacial area in large segments of preformed packing that must be carefully inserted into the column or tower. Interestingly, it was found that the resistance to flow of liquid from one packing segment to another could be high enough to cause liquid backup, and this led to the development of special channel geometries or crimps at the edges of segments that allowed for greater flow from one segment to another. The cost of packing often is a major contributor to the total cost of equipment, so in practice, a packing type and size is chosen that provides a good balance in delivering sufficient area and low pressure drop per unit height along with affordable purchase and installation costs – to achieve satisfactory performance at reasonable cost per length of column height.

Different packings have different packing areas (m^2 per m^3), but the actual interfacial area that is generated will not necessarily be close to the stated packing area. This depends on operating factors such as liquid loading. At low liquid loading (low liquid flow rates per unit cross-sectional area through the packing) with good initial liquid distribution, the liquid tends to disperse in thin layers on the packing surfaces, so interfacial area can be very high and approach that of the packing itself. But as liquid loading increases, the spaces between packing elements can become filled with liquid and interfacial area can then decrease significantly. In conventional trayed contactors at normal operating rates, the liquid typically forms a continuous layer of liquid on top of a tray, and vapor is dispersed as bubbles rising up through the liquid layer. As vapor traffic increases, the liquid can become highly agitated and frothy with increased interfacial area. Beyond a certain point, the interfacial area may increase dramatically as vapor velocity increases to form sprays of liquid droplets rising from the tray deck. At very high vapor rates this may result in the liquid spray height impacting the bottom of the tray above, causing liquid to be entrained into that tray. At that point, significant liquid entrainment can severely reduce separation efficiency.

In liquid-liquid extraction (LLE), normally one liquid is dispersed into another not as films or rivulets, but as dispersed droplets suspended in a continuous liquid. In this case, interfacial area is determined by physical properties and interfacial tension. The packing in a packed LLE column serves to influence the flow and population of droplets and minimize backmixing. When performance is not sufficient, LLE designs often include some form of mechanical agitation to reduce droplet size and increase the number of droplets per unit volume (the population density). For best performance, the extractor may be intentionally operated just shy of a flooding condition where one phase becomes entrained into the outlet of the other or dispersed droplets begin to coalesce and phase inversion begins to occur. Operating near the flooding condition can maximize performance by generating maximum droplet population density and thus maximum interfacial area. In order to find this most effective operating condition, an operator often has to risk an upset, but once the flooding point is found, the agitation intensity and flow rates can be adjusted to back off somewhat to avoid an upset during normal operation and maintain excellent performance. The experimental testing/design method used to identify the flooding curve for a reciprocating-plate extraction column (Karr column) is a good example of how an effective operating regime may be determined, as discussed in ref. [19]. The same kind of approach to *carefully* pushing an operation toward its limits can be important in maximizing the performance of many kinds of separation equipment, not just mechanically agitated equipment. In distillation, this may involve the operation of packed columns or trayed columns near the maximum vapor and/or liquid velocities, that is, running the column just below the flood point. Naturally, this must be done carefully to avoid too great an upset and with some way of dealing with the potential consequences of such a test. The idea is to just touch on the flooding condition and then back off, and then operate the unit at 80% of flooding or somewhat higher to obtain excellent performance. In practice, the flood point may be known for a given packing from data provided by the packing supplier. Aspects of this approach are discussed in more detail in Section 7.4: Equipment Design and Rating Methods (for distillation columns).

In adsorption processes, the interfacial area that resides within each adsorbent particle must be very high to achieve satisfactory adsorption capacity, typically between 100 and 1,000 m^2 per gram. And this surface area must be accessible to the solute, so the internal pore structure must allow for rapid diffusion of the solute into and out of the particles. Beware of adsorbents that claim extremely high internal area, because much of this area may not contribute to mass transfer performance, especially at high throughput, because the pores are too small. See Section 10.1: Adsorbent Materials for more discussion.

In crystallization processes, ensuring satisfactory interfacial area of the growing crystal mass can be very important to achieving desired results. This refers to the surface area of existing crystals within the crystallizer. Sufficient area is needed to provide surface for solute in solution to adhere to and grow the existing crystals. This helps reduce the tendency for the solute to nucleate new (unwanted) crystals or to foul the crystallizer by forming deposits on the crystallizer internal walls. As a result, often, a continuous crystallizer must be operated with a minimum solids slurry density (weight of product crystals per unit volume) for good results. See Sections 11.5, Batch Crystallization from Solution, and 11.6, Continuous Crystallization from Solution.

In membrane-based separations, interfacial area is built into the equipment, so the amount is known exactly. It is constant and independent of feed and product flow rates. This can be a key advantage over other methods when fouling is not an issue or when fouling tendencies can be managed by periodic cleaning. And all of the area is available for mass transfer, unlike the inaccessible microporous area of some adsorbents. The total area is determined by the area built into a single membrane module times the number of modules used. Thus, specifying designs for large-scale production rates is straightforward and reliable, although a large number of modules may be required and other methods may be better suited to a given application or more cost effective. For more discussion, see Section 12.1: General Features (of membrane-based processes).

6.4 Residence Time Effects

In this section, we look at the various ways that residence time is characterized for various types of separators. The average time an element of feed spends within a separator can dramatically affect separation performance. And deciding whether a particular phase should be continuous or dispersed can dramatically affect its residence time in the equipment. The residence time chosen for a particular design often is a compromise. Longer times may increase mass transfer and improve separation. But shorter times are needed to increase productivity and may help reduce or avoid degradation of the feed components at the required operating conditions, which can involve adverse conditions of high temperature, high pressure, or other conditions that may promote undesired reactions (such as acidic conditions). Interestingly, how residence time typically is characterized by designers varies greatly depending on the type of separator. For example, it is rare for residence time to be specifically called out in the design of a distillation column or an adsorption bed, yet residence time is often a specific design variable for a continuous crystallizer. In any case, it is important to be aware of residence time and its potential impact on performance.

Distillation. Liquid residence time in a distillation column is determined by liquid hold-up divided by the average liquid rate within the column. Liquid residence time normally is on the order of minutes or tens of minutes – less time for packed columns with lower liquid holdup, and more time for trays with greater liquid holdup. At normal operating conditions, the liquid is the dispersed phase in the former case (with packing), and the continuous phase in the latter (with trays). Often performance is improved by adding extra packing height or extra trays, and inherent in this is the fact that doing so adds extra residence time for mass transfer. This effect is included in the concept of the transfer unit, as we discuss later in Section 6.7.1: Mass Transfer Units and Mass Transfer Coefficients. Vapor-phase residence time within a distillation column is much shorter, of the order of seconds or a few minutes. It is liquid residence time that we normally focus on, because long times may be needed to promote mass transfer or, for a reactive distillation, to promote a desired liquid-phase reaction. Low liquid-phase residence time in the column and in the reboiler may be needed to minimize product degradation. In general, when thinking in terms of theoretical stages, residence time is not taken into account except that it is assumed (or confirmed by experiment) to be sufficient. It is important to carefully consider whether the assumed number of theoretical stages can actually be achieved; or, in other words, that the appropriate stage efficiency is understood. On the other hand, residence time is inherent in the use of transfer units – residence time increases in direct proportion to the number of transfer units (assuming all else remains the same).

Extraction. Providing long residence time per stage of contacting is an important attribute of mixer-settler type extractors. This type of equipment may provide 10 or 15 min of contacting in the mixing section of a mixer-settler stage, which may be needed due to slow kinetics of a solute-extractant reaction or due to high liquid viscosity. Residence times are lower in other types of extraction equipment. Minimizing residence time may be an important consideration when extracting a sensitive product in harsh conditions, and this may be accomplished by using a centrifugal extractor.

Liquid-liquid phase separator (settler or decanter). The design of a liquid settler is often discussed in terms of providing sufficient liquid residence time, but more precisely, performance depends on the rate of feed flow to the main interface between coalescing layers and the rate of droplet coalescence at the interface [19]. The feed rate must not be so fast as to entrain droplets past the interface.

Adsorption. The feed velocity in an adsorber bed must not exceed a critical velocity. There must be sufficient time for the solute to diffuse into the adsorbent pore structure. If velocity is too fast, then the solute does not have time to diffuse, so little can be adsorbed. The appropriate face velocity normally is determined by measuring breakthrough curves in small-scale tests.

Crystallization. We often think of residence time when designing a crystallizer. For batch crystallization, this is how long we operate the crystallizer from start to finish. For continuous (steady-state) crystallization, this is the average time the liquid spends in the vessel as determined by liquid or slurry volume divided by liquid feed rate, as in a continuous stirred tank reactor (CSTR). Longer residence time generally allows production of larger crystals (up to a point), but this depends on how the crystallizer is operated so that growth dominates nucleation. And spending too much time in a stirred vessel can result in excessive crystal attrition or breakage. This can be a particularly pressing issue when holding a finished slurry within a stirred vessel for extended periods of time; for example, while waiting for a downstream filter to be operational.

Membrane-based separations. The feed rate must be sufficiently low as to allow sufficient time for appreciable solute to permeate the membrane. That is, a lower feed rate (and thus higher residence time spent in the membrane module) may allow for higher recovery of the desired permeating solute. But the selectivity that is achieved may be a function of residence time as well – because more time also affects the permeation of co-solute, and the impact of residence time on co-solute permeation may be different. These factors must be quantified in testing. They depend on the type of operation (chemical system, liquid or gas) and the type of membrane (its chemical composition and structure).

6.5 The Material Balance: Keeping Track of All Components

What goes in must come out, or accumulate, or react. With this simple rule, we already know quite a lot about the main features of a separation process. For example, consider a steady-state, binary separation process such as a standard continuously fed distillation column used to isolate and purify a desired overhead product. Given the flow rates and compositions of feed and product streams, the material balance sets the flow rate and composition of the bottoms stream. This is the fundamental basis for tracking the majority of components in such a process.

By material balance we mean accounting for all components present in all process streams including feed streams, recycle streams, purge streams, and waste streams, as well as the product stream. This includes raw materials and solvents, reaction products, inert components, and all impurities, in both large concentrations and in trace amounts. In professional practice, a detailed material balance spreadsheet almost always accompanies a process flow diagram. Not all of the trace components are likely to be identified in terms of their exact chemical structure, but at a minimum, their presence and their potential impact on the process should be understood.

But for some components, the material balance cannot be calculated because of uncertainties due to the possibility that these components may accumulate or react within the process. This is especially the case for trace impurities. A typical commercial separation process handles feed containing hundreds of components. There may be only two or three main components of commercial interest, but the feed likely will contain many more related components and impurities present in small or trace amounts. Analysis of feed samples using gas chromatography or liquid chromatography often shows hundreds of peaks. And process simulation normally cannot be relied upon to yield the true (exact) material balance in the case of minor components. Data from sampling will be needed to check and adjust the simulation. Given this situation, more than anything else (except safety), the job of the process engineer and process chemist is to determine and monitor the true process material balance. Having a good understanding of where all components flow within the process and their ultimate fate is critical, because success with process simulation, design, and commercial operation often depends on it.

Understanding the true material balance requires a concerted effort to sample and analyze all process streams and then determine how well each component can be accounted for; that is, how well the material balance closes for each component. This statement is certainly true for an existing commercial operation. But for a newly envisioned separation process, much of the required information will be missing. An important focus of process development is to identify or characterize all of the components and to understand their fate. All major components should be accounted for within 99% or better. For components present in smaller amounts, especially trace impurities, the ability to completely close the material balance normally is quite limited, but it is still important to understand their fate. This will require careful attention to sampling and appropriate use of analytical methods, in close collaboration with analytical chemists.

We strongly emphasize the importance of carefully tracking the material balance for another reason as well. That is because we know this is an excellent practice for monitoring process performance and highlighting opportunities for improvement (such as working to reduce the amount of product that ends up in a waste stream) or for gaining early awareness of a problematic trend (such as gradual buildup of an impurity in a recycle stream or within a particular unit operation), allowing time to make needed adjustments *before* they cause failure or lead to a crisis situation. This is also good business practice, as it is important to know what percentage of raw materials can be accounted for in conversion to final products and work to increase this percentage as much as possible. And this is equally important in the development of a new process. Understanding what can happen with each component or type of component is key to successfully developing an appropriate process scheme and final design. All processes need to have outlets for impurities, and if these are not adequately included in the process design, then the commercial design will fail ultimately. With trace impurities, it may take months of operation before they accumulate to levels that cause failure. In some cases, the culprit impurities are present is such low concentrations (parts per million or even lower levels) that they cannot be detected in the feed. It is only after accumulating in the process over an extended time period that they show their presence.

The material balance is calculated on a mass basis normally and not in terms of moles. This is because products are sold on a mass basis, but also because for some components, especially the impurities, their chemical structures and appropriate molecular weights are not known. It is not always possible to identify individual components because there are so many of them in a crude process stream. In that case, specific classes of compounds may be detected and measured, as in lumping all phenolic impurities together or all inorganic ash components. The idea is to track all components critical to process performance (including impurities) as much as possible or practical.

6.5.1 Material balance split and separation power

The performance of a separation process is constrained not only by the overall material balance around the process but also by its ability or power to separate the feed components. The former is a fundamental law of physics and chemistry. The latter depends not only on the laws of nature, but also on economic choices. Economic considerations limit just how powerful the separator can be – only so tall or so big or with so much mass transfer area. A separation process magnifies the fundamental separation factor that characterizes the given chemical system. This magnified separation factor is called separation power. It is determined by measuring the individual material balances for key components to quantify the overall enrichment that is achieved by the overall process. Determining separation power is like determining the power output of a car's engine. Measuring how much horsepower it has using a dynamometer says a lot about the car's capabilities. Similarly, measuring the separation power a separation process has by measuring the material balance for key components says a lot about its capabilities. The concept of separation power also helps in understanding the tradeoffs that must be made in operating a separation process. If changes are made to alter the flow rate or purity of one stream, this must affect other streams, and these effects can be assessed to a large extent simply by knowing the material balance and the limit on separation power.

So, understanding a separation process begins with quantifying the material balance around the process. To illustrate the basic concepts, let us keep things simple by considering a binary feed containing components i and j. The internal details of the process, such as the type of separator and the number of theoretical stages or transfer units or even the chemical system, are not specified. It is a black box as far as we are concerned. Only the overall results from the process are relevant here. The process splits the feed into two fractions (I) and (II). The two components *i* and *j* may be considered the key components of interest for a multicomponent system. The process can either be a continuous one, in which case we are concerned with constant mass flow rates of the feed and each of the fractions, or it can be a batch or a cyclical process, in which case we are concerned with the total mass of feed and the two fractions after the process has been completed or after it has completed a cycle of operation.

Such a process separates a feed mixture into two fractions that differ in their chemical composition. Let us designate component *i* to be the one that is enriched in fraction (I), either due to preferential phase equilibrium or due to a faster rate of mass transfer relative to component *j*. If component *i* is the desired product, then fraction (I) is the product fraction and fraction (II) is a purge fraction that is enriched in component *j*. On the other hand, if component *i* is a key impurity, then fraction (I) is the purge fraction, component *j* is the desired product, and fraction (II) is the product fraction. Also, for now, we need not to be concerned with whether the phases involved are gas, liquid, or solid. The material balance concepts are the same. In this discussion, we focus on the mass balance and mass fractions, but the same approach may be used in terms of moles if desired (for a non-reacting system).

The overall material balance for the process is simply:

$$M_F = M_1 + M_2 \tag{6.13}$$

where M_F = feed mass or mass flow rate
$\quad M_1$ = mass or mass flow rate of fraction (I)
$\quad M_2$ = mass or mass flow rate of fraction (II)

The individual material balances for each component are given by:

$$M_F X_i^F = M_1 X_i^I + M_2 X_i^{II} \tag{6.14}$$

$$M_F X_j^F = M_1 X_j^I + M_2 X_j^{II} \tag{6.15}$$

where X_i^F = mass fraction of a specific component in the feed
$\quad X_i^I$ = mass fraction of a specific component in fraction (I)
$\quad X_i^{II}$ = mass fraction of a specific component in fraction (II)

For a two-component system $(i + j)$, the material balance requires that $X_i + X_j = 1$ in the feed and in each fraction. For a multicomponent system, the overall material balance and the component balances would include the key components plus other components such as a solvent. These are not included here for simplicity.

We now have a system of interrelated algebraic equations that we can solve, given a minimum set of input values. The masses of each fraction are set by the material balance split (MBS), which is defined as the mass ratio of one fraction to the other:

$$MBS \equiv \frac{M_1}{M_2} \tag{6.16}$$

MBS is a fundamental operating parameter for the separation process, because specifying MBS and separation power will set how a given component must partition between the two fractions. This sets the purity and recovery of the purified component. For a binary feed, the purity of component i in fraction (I) is simply its mass fraction, x_i^1. For a multicomponent system (with key components i and j), its purity with respect to j is given by:

$$\text{Purity of key component } i \ (\%) = 100 \left(\frac{X_i^I}{X_i^I + X_j^I} \right) \tag{6.17}$$

Separation power (SP) is the effective (magnified) separation factor achieved by the overall process. The process divides the feed into fractions I and II, and its separation power is defined by:

$$SP \equiv \frac{(X_i/X_j)_{fraction \ I}}{(X_i/X_j)_{fraction \ II}} = \frac{X_i^I/X_j^I}{X_i^{II}/X_j^{II}} = \text{a constant value} \tag{6.18}$$

Manipulating and rearranging of eqs. (6.13) to (6.18) gives an expression for SP in terms of MBS:

$$SP = \frac{\left[\left(\frac{X_i^F}{X_i^{II}} \right)(MBS+1) \right] - 1}{\left[\left(\frac{1-X_i^F}{1-X_i^{II}} \right)(MBS+1) \right] - 1} = \text{constant} \tag{6.19}$$

where

$$X_i^I = X_i^F \left(\frac{(MBS+1)}{MBS} \right) - \frac{X_i^{II}}{MBS} \tag{6.20}$$

The recovery of i into fraction (I) is given by:

$$\text{Recovery, } i \ (\%) = 100 \left[\frac{M_F X_i^F - M_2 X_i^{II}}{M_F X_i^F} \right] = 100 \left[\frac{M_1 X_i^I}{M_F X_i^F} \right] \tag{6.21}$$

In terms of MBS, this is:

$$\text{Recovery, } i \ (\%) = 100 \left(\frac{MBS}{MBS+1} \right) \left(\frac{X_i^I}{X_i^F} \right) \tag{6.22}$$

Similar equations can be written for a multicomponent system. The basic concepts are the same.

So, one can specify X_i^F, MBS, and SP and then calculate the corresponding values of X_i^{II} and X_i^I (and X_j^I and X_j^{II}) from Eqs. (6.19) and (6.20). The solution of these equations is easily automated in an iterative spreadsheet calculation; for example, by using Solver in Microsoft Excel. (In doing so, be sure to calculate the material balance as well, to see that it closes to confirm that the calculations are correct). The desired purity of component i in fraction (I) is obtained by adjusting the MBS. For example, for a binary feed with $X_i^F = 0.3$, MBS = 0.4, and SP = 100, solving eq. (6.19) by trial and error gives $X_i^{II} = 0.068$. Inserting this result into eq. (6.20) gives $X_i^I = 0.88$. The other values are then $X_j^I = 0.12$ and $X_j^{II} = 0.93$. The recovery of component i in fraction (I) is calculated by using eq. (6.21). In this case, recovery = 83.8%. Now, say, we wanted a higher purity of component i in fraction (I). If we assume constant separation power, we can adjust MBS to send less material into fraction (I) with the following results: for $X_i^F = 0.3$, *MBS* = 0.35, and *SP* = 100, then $X_i^{II} = 0.088$, $X_i^I = 0.91$, $X_j^I = 0.094$, $X_j^{II} = 0.912$, and recovery of component i in fraction (I) = 78.3%. So we increased the purity of component i in fraction (I) by sending less total material into fraction (I), but at the expense of somewhat lower recovery of component i. We also reduced the purity of component j in fraction (II).

These calculations illustrate how the concepts of SP and MBS are related. This can be useful in understanding the overall separation capability of a binary separation process – because SP and MBS quantify the limit as to what a given process can achieve in terms of both purity and recovery of the desired product. The same general trends apply to multicomponent feeds. Consider key components i and j in a binary separation, where component i is the one enriched in fraction (I). If we assume that separation power is nearly constant as is often the case for a specific separator operating within a narrow operating range, then the general process trends are as follows:

- An increase in the purity of component i in fraction (I) is obtained at the expense of reduced recovery of component i in fraction (I).
- Similarly, an increase in the purity of component j in fraction (II) is obtained at the expense of reduced recovery of component j in fraction (II).
- Reducing the MBS increases the purity of component i in fraction (I) and it reduces the purity of component j in fraction (II).
- Increasing the MBS increases the purity of component j in fraction (II) and it reduces the purity of component i in fraction (I).

Equations (6.13) to (6.22) and the trends they highlight are particularly useful in characterizing and troubleshooting the performance of an existing separation unit. If a unit is not performing as needed, determining its current separation power to see how far off it is from its original design specification helps characterize the issue. Monitoring separation power over time can identify trends in declining power. A slow decline for whatever reason (fouling over time, for example) can then be detected and dealt with, before it becomes a crisis. This approach can also help identify small changes that may improve performance, as summarized in the above general trends. However, the validity of the assumption that separation power is constant (or nearly constant) will vary. In processes with reflux streams, as in many distillation separations, separation power increases as the amount of reflux increases, which in turn reduces the MBS (for reflux obtained from fraction (I)). SP reaches a maximum at total reflux (when productivity goes to zero). So in that case, the assumption of constant separation power is not valid, especially for large changes in the MBS. It is interesting to note that the concept of SP seems to come from the famous Fenske equation, which relates the separation power of a fractional distillation process to the thermodynamic separation factor (relative volatility) and the minimum number of (countercurrent) theoretical stages at total reflux:

$$\text{Separation Power} \equiv \frac{\left(X_i/X_j\right)_D}{\left(X_i/X_j\right)_B} = (\text{Separation Factor})^{N_{min}} \quad \text{at total reflux} \qquad (6.23)$$

Any consistent units may be used. For more discussion, see Section 7.3.5: Short-Cut Calculations.

Like a dynamometer test of an internal combustion engine, separation power can be obtained from direct measurement of an existing process, by measuring the compositions of all outlet streams. Doing so does not require calculations using a model or simulation, although having a validated model of the process is certainly helpful. Just as engine power is a function of engine rpm measured on a dynamometer, SP can be measured directly in the field by running the process at different reflux ratios, or it can be calculated using a validated model. For a given product purity specification, a certain minimum SP will be needed in order to achieve the required purity and recovery. Typical values of SP required for commercial processes range between 1,000 and 50,000 or so, depending on the chemical system to be separated (since this affects the internal driving force for the separation; that is, the separation factor) and the specifications for purity and recovery, whether stringent or loose. Of course, the SP that can be achieved by a given process depends on the details of the specific equipment, as in the number and type of contacting stages, as well as the chemical system.

This type of analysis is particularly useful for understanding the operation of distillation and extraction processes. For a standard continuously fed distillation column, fraction (I) comprises the overheads stream and fraction (II) the bottoms stream. For a standard batch distillation, fraction (I) is the overheads condensate and fraction (II) is the material left in the pot after distilling the batch. For our purposes here, we do not need to know the path taken by the process, the number of stages achieved, or the internal dynamics – only the material balance results. The analysis is applicable to any type of process that separates the feed into two fractions, even including cyclical binary separation processes such as simulated moving-bed chromatography and pressure swing adsorption (when analyzed over a complete cycle of operation at cyclical steady-state). This is because a cyclical process must satisfy the overall material balance over its cycle of operation for steady operation. And in general, any process may be analyzed in an analogous manner. The number of fractions may be larger and the appropriate equations describing the process may be more extensive, yielding a much more complicated system of equations. But in any case, by taking a look at the overall and individual material balances and by quantifying separation power, the "big picture" becomes clearer. In the case of a distillation column with multiple side draws, there will be two or more material balance splits to consider.

6.5.2 The impact of trace impurities: beware or be sorry

Naturally, we focus on the main components to be separated when discussing a separation process: the desired product and the major impurities to be removed from the product. But as mentioned above, normally there are many more components present in very small concentrations that also need to be taken into account. Trace impurities may enter with the feed in undetectable amounts (at ppm or lower concentrations) and subsequently build up within the process, or they may be produced by reaction or decomposition of some of the main components within the process. Common reactions include oxidation and hydrolysis from reaction with dissolved oxygen and water from infiltration of moist air into a process. For example, exposure of oxygenated organics such as isopropyl ether and isopropyl alcohol to dissolved oxygen forms peroxides (peroxy compounds) over time. This may present a dangerous safety hazard, as peroxides can be highly unstable and even explosive if allowed to build up in concentration, especially under evaporation conditions that leave solid residues [29]. Another example involves hydrolysis of chlorinated organics to form HCl in solution, which can lead to severe corrosion and equipment failure. Yet another example involves processing of acrylic monomer, which forms unwanted dimer over time, so care must be taken to avoid excessively high temperatures and limit processing and storage times. These kinds of reactions may be ongoing during processing, as with dimer formation, or they may occur during periods outside of normal operating conditions; that is, during upsets which may involve leakage of air into a process through failed gaskets, holding liquids in equipment for extended periods of time due to an unscheduled shutdown, or unintended evaporation of liquids in open vessels and formation of solid residues. Or, upsets may occur due to accumulation of water from condensation in relatively cool sections of vent piping, or by exposure of vessel contents to elevated temperatures due to failure of a cooling system, or due to delivery of a key raw material that contains a new unexpected and unidentified odor-causing impurity such as an aldehyde. Some impurities can lead to objectionable odors in a product even at extremely low concentrations below ppm levels, sometimes preventing the sale of the product. These are just a few of many possibilities. Upsets can and will happen, so it is important to understand the potential consequences of upsets and how to deal with them.

Because of all this, the qualification of actual raw materials in terms of their initial trace impurities and any tendencies to undergo trace reaction is an important aspect of design and quality control. The danger of not doing so is that in normal long-term operation trace impurities may have the opportunity to accumulate to concentrations that cause severe problems, and if this is not known and dealt with during design, this potential may lead to failure to perform as needed or even the development of a dangerous safety hazard. It is critical then to recognize possible ways that trace impurities may enter or form within a process, and design the process to allow efficient ways of purging them as needed without emptying everything and starting over. Unfortunately, we have heard of several projects where failure to understand trace impurities led to long delays in the start-up of a new process design (up to a year or more), requiring very expensive modifications in the field. Dealing with this issue requires extensive chemical analysis and testing of raw materials to understand the potential impact of impurities and any potential reactive chemical hazards, as well as judgment as to what measures are warranted to ensure that potential impurities can be purged as needed. Keep in mind that correcting mistakes during design is easier and often less costly than doing so once the commercial equipment has been designed and installed. This is not only because of the added capital cost required for large-scale modifications, but perhaps even more importantly, because of the delay that will result in meeting production and sales commitments.

An example of how important it can be to keep track of trace components and their effects on a separation process is a distillation column that the authors were involved with [30]. This column that was located in a cellulosics production unit used steam stripping at near-atmospheric pressure to remove methanol and other light organics from process wastewater and reduce the organic loading on a downstream treatment facility. As the wastewater was very dilute in methanol, averaging about 2,000 ppm(wt) with about 200 ppm(wt) of other organics lighter than methanol, the column was equipped with a stripping section containing 26 dual flow trays for fouling resistance topped by a packed rectifying section containing 28 feet of 40 mm ring-type packing to concentrate the removed organics for disposal or potential recycling by reducing water content.

With a very dilute feed concentration, if the methanol distillate was to be concentrated with a very low water content (anticipated to be < 0.1 wt% from computer simu-
tions), the column material balance dictated that the distillate-to-feed ratio would
quite low. Taken together with the moderate steam-to-feed ratio needed
mn to give the required methanol removal, this would result
ignificant change in liquid concentration and solvent
d point to mostly methanol at the top of the rec-

, the required methanol removal from the column
higher distillate water content, roughly 3–5 wt%,
to-feed ratio and reflux ratio than was predicted. In

addition, recurring flooding episodes in the packed rectifying section were observed at irregular intervals, with packed section pressure drops as high as 8 psi where < 0.5 psi pressure drops were expected. These flooding episodes resulted in loss of column control with bottoms and reflux drum levels fluctuating so wildly that operators were often forced to shut the column down and initiate a time-consuming restart, after which stable column operation was restored until the next flooding episode occurred anytime from 12 h to several days later.

After a period of time with recurring flooding episodes, an analytical chemist noticed that a minor unknown peak was occasionally present and partially obscured by other peaks in distillate gas chromatograph analyses. Subsequent investigation with mass spectrometry revealed that the unknown was 1,2-dimethoxypropane (DMOP), which had not been detected in the wastewater feed analyses but was deemed likely to be present at trace levels based on process knowledge. This discovery prompted the question of whether the presence of this minor component could, in some way, be responsible for the ongoing column flooding problems.

Laboratory experiments were initiated with a 2-inch diameter glass distillation column to investigate whether DMOP could be accumulating in the column to concentrations that could cause the flooding events. The 2-inch lab column was continuously bottom-fed with drum quantities of the actual wastewater fed to the production column and operated at total reflux to approximate the operation of the top packed section of the column. Operation at total reflux with only the occasional small reflux stream sample for a distillate stream presented an opportunity to observe how DMOP might behave in the top of the production column. Although the 2-inch column did not exhibit signs of flooding during operation, experimental results showed that DMOP indeed accumulated in the top section and was difficult to purge in the distillate, as shown in the following table of sample results:

Elapsed time, hr	Reflux DMOP conc, wt%	Reflux H_2O conc, wt%
2.0	7.6	3.3
7.5	4.1	0.09
10.5	1.6	0.03

It is clear from these results that with the high degree of SP in the lab column, DMOP at first was rectified to significant concentrations but then was held further down in the column as the reflux became drier and higher in methanol content. This was an indication that DMOP is less volatile than methanol when present in the absence of water. This was confirmed by later vapor-liquid equilibrium measurements showed that the relative volatility of methanol to DMOP was 1.4 at atmosphe sure and high methanol concentration. However, since the column bottom showed indetectable levels of DMOP throughout the entire experiment appeared to be easily stripped from the wastewater even as it was being held down out of the top of the column.

As a result, with highly efficient stripping and rectifying sections in the production wastewater distillation column, it appeared that DMOP was acting like a trapped component, not able to be purged at significant rates in the wastewater stream exiting the bottom of the column, nor from the distillate stream exiting the top. In this manner, DMOP was allowed to build up to high concentrations in the rectifying section to the point where phenomena such as foaming or high vapor and liquid recirculation rates in the section would result in the flooding episodes. Based on this information and interpretation, column operation was modified by changing the rectifying section control temperature set point to higher values such that the water content in the distillate was allowed to rise from the 3–5 wt% previously maintained to a range of 5–10 wt% on an ongoing basis. This resulted in significantly higher distillate DMOP concentrations, and the recurring flooding episodes ceased.

This real-world example demonstrates the importance of fully characterizing feed streams to be separated and understanding the thermodynamics of how all of the components, even those at supposedly trace concentrations, interact with each other under the conditions expected in the separation device, and any potential reactions.

6.6 Theoretical Stage Models

Now we come to how the inner workings of a separation process may be characterized and quantified. One of the earliest concepts developed for this purpose is the theoretical stage concept introduced by Ernest Sorel (1889, 1893). Sometimes called an equilibrium stage or an ideal stage, the theoretical stage is an elegant and highly useful construct that allows design of separation equipment with no direct knowledge of the mass transfer rate. Besides the base-case information (compositions of each inlet phase and their relative flow rates), knowledge of phase equilibrium and the number of theoretical stages is all that is needed to calculate the resulting compositions of effluent phases and separation power – because in using the theoretical stage model, we assume that mass transfer is sufficiently fast to allow the system to become equilibrated at certain points or stages within the process. With this approach, we envision mixing two phases together in an idealized contactor. The contacting stage instantly comes to equilibrium with no mass transfer resistance and the newly equilibrated phases exit the stage in separate outlets, each having acquired equilibrated mass and composition, with the total mass being conserved. The phases then move on to other stages in a multistage process scheme (normally a countercurrent scheme) until the desired separation is obtained. No commercial separation process operates exactly in this way. However, many applications can be treated as if they do, because the mass transfer rate is sufficiently fast so that each contacting stage operates sufficiently close to equilibrium.

6.6.1 Theoretical stage relationships

In separation processes involving solute diffusion, the concept of a single theoretical stage involves four main steps:
1. Bring two different phases of matter into contact, or alter the feed conditions to produce two phases out of a single feed phase.
2. Generate sufficient interfacial area (using fluid flow and geometric structures or mechanical agitation) to allow components to diffuse from one phase across the interface into the other.
3. Give sufficient time for the two phases to attain thermodynamic equilibrium (or nearly so).
4. Physically separate the phases to isolate a desired product in one of the phases.

Multiple stages can then be used to achieve a required overall separation. Although commercial separation processes do not work exactly in this way, many approach this ideal, and the theoretical stage concept provides a useful means of quantifying performance. Examples include trayed distillation columns and mixer-settler extraction cascades. The design of such a separator requires some knowledge of phase equilibrium, and this can be described in terms of the slope of an equilibrium line, a partition ratio, or an equilibrium isotherm.

So, the separation achieved by a single theoretical stage is determined by phase equilibrium and the material balance, giving the following relationship:

$$F_R \equiv \frac{x_{in}}{x_{out}} \approx \frac{x_{in} - \left(\frac{y_{in}}{m}\right)}{x_{out} - \left(\frac{y_{in}}{m}\right)} = \frac{\mathcal{T} - \left(\frac{1}{\mathcal{T}}\right)}{1 - \left(\frac{1}{\mathcal{T}}\right)}, \text{ for } \mathcal{T} > 1 \tag{6.24}$$

$$\mathcal{T} = m\frac{S}{F} \tag{6.25}$$

$$y = mx + y_{in}, \text{ at equilibrium} \tag{6.26}$$

where F_R = solute reduction factor
  x = concentration of solute in the feed phase, mole fraction or other consistent units
  y = concentration of solute in the second phase, mole fraction (or other units)
  m = slope of the equilibrium line, dimensionless
  S = second phase flow rate, mass or moles per unit time
  F = feed phase flow rate, mass or moles per unit time
  $\mathcal{T}$ = solute transfer factor

See Section 6.1.3: Partition Ratio, Separation Factor, and Transfer Factor for discussion of the transfer factor. Equations (6.24) to (6.26) describe the performance of an idealized stage for which phase equilibrium has a simple linear relationship. Furthermore, any change in the mass of each phase as solute transfers from one phase to another is ignored, and heat effects are considered negligible. These simplifying assumptions normally are valid for dilute feed mixtures, and the resulting equations serve to illustrate the main relationships. The fraction of solute recovered from the feed phase into the second phase is given by:

$$\text{Solute recovery (\%)} = 100 \left(1 - \frac{1}{F_R}\right) \tag{6.27}$$

The concentration of solute in the second phase outlet is determined by the ratio of the phases $\mathcal{F}/S$ and the reduction factor achieved by the process, $F_R = x_{in}/x_{out}$. For constant flow rates and zero solute in the second phase inlet, this gives:

$$y_{out} = x_{in} \left(\frac{F}{S}\right) \left(1 - \frac{1}{F_R}\right) = x_{in} \left(\frac{F}{S}\right) \left(\frac{\text{Solvent Rec'y(\%)}}{100}\right) \tag{6.28}$$

Now consider a continuously fed, countercurrent, multistage process operating at steady state with the same simplifying assumptions as above. This type of process is aimed at transferring a high portion of solute present in the feed into the second phase (for high solute recovery). In this case, performance is described by the Kremser-Souders-Brown (KSB) equation, so-named after Alois Kremser, who introduced the concept in 1930 [94], and after Mott Souders, Jr., and George Granger Brown, who further developed it in 1932 [31]:

$$N = \frac{\ln\left[\left(\frac{x_{in} - \left(\frac{y_{in}}{m}\right)}{x_{out} - \left(\frac{y_{in}}{m}\right)}\right)\left(1 - \frac{1}{\mathcal{T}}\right) + \left(\frac{1}{\mathcal{T}}\right)\right]}{\ln \mathcal{T}} \tag{6.29}$$

$$\mathcal{T} = m\frac{S}{F}, \mathcal{T} \neq 1 \tag{6.30}$$

$$F_R \equiv \frac{x_{in}}{x_{out}} \approx \frac{x_{in} - \left(\frac{y_{in}}{m}\right)}{x_{out} - \left(\frac{y_{in}}{m}\right)} = \frac{\mathcal{T}^N - \left(\frac{1}{\mathcal{T}}\right)}{1 - \left(\frac{1}{\mathcal{T}}\right)} \tag{6.31}$$

Solute recovery and the second phase outlet concentration are calculated just as in Eqs. (6.27) and (6.28) for single stage operation.

As discussed in Section 6.1.3, the quantity $\mathcal{T}$ is called the solute transfer factor as a general term, or simply the transfer factor. It is called the stripping factor in stripping-mode distillation and the extraction factor in liquid-liquid extraction. The transfer factor $\mathcal{T}$ is a key process variable determining overall performance, because it reflects the solute-carrying capacity of the second phase (or solvent phase) relative to that of the feed phase. On a McCabe-Thiel diagram (discussed in Section 7.3.4: Continuous Fractional

Distillation), it is the slope of the equilibrium line m divided by the slope of the operating line F / S. The transfer factor must be greater than unity if the second phase is to have sufficient capacity to carry a large portion of the key solute present in the feed. Note that adding stages in countercurrent fashion greatly magnifies the impact of the transfer factor. Compare $\mathcal{T}$ in eq. (6.24) with $\mathcal{T}^N$ in eq. (6.31).

When the transfer factor or carrying capacity of the second phase is less than unity, then the maximum possible solute recovery is equal to the transfer factor.

$$\text{Maximum solute recovery } (\%) = 100 \times \mathcal{T}, \text{ for T} < 1 \tag{6.32}$$

and

$$F_R|_{max} = \frac{x_{in}}{x_{out}} = \frac{1}{1 - \mathcal{T}}, \text{ for } \mathcal{T} < 1 \tag{6.33}$$

The question of whether the maximum values in eqs. (6.32) and (6.33) are actually realized by the process depends on the allotted residence time and the number of theoretical stages.

These equations are strictly valid only for dilute solutions, but they may also be applied in other cases to estimate performance by using an average value for the slope of the equilibrium line (the geometric mean). This is given by:

$$m_{ave} = \sqrt{m_{low} m_{high}} \tag{6.34}$$

Additional ways of adjusting the equations to account for curved equilibrium and operating lines are discussed by Kohl [32].

In modeling real processes, stage efficiency can be applied to take into account deviations from equilibrium for real systems, so N becomes $E \times N$, where E is an efficiency factor. For packed columns, efficiency can be expressed in terms of the *height equivalent to a theoretical stage* (*HETS*), also called *HETP*. The shorter the height, the greater is the efficiency. For trayed columns, real behavior is expressed in terms of tray efficiency or the number of theoretical stages achieved per actual tray. (Other definitions of efficiency also may be used. We discuss these in the next section and in later chapters for specific types of equipment). Stage efficiency, as defined here, usually is less than unity; but believe it or not, the calculated efficiency that is obtained assuming a given contacting device acts as a single well-mixed stage can sometimes be somewhat greater than unity because of advantageous crossflow within the device that deviates from perfectly well-mixed flow, as can happen for large single-pass crossflow distillation trays with long liquid flow paths or short-path rotary evaporators (more on this later). The resulting stage model is easy to use, but one must keep in mind that stage efficiency is a strong function of flow rates and residence time in a given device, and if flow through the device becomes too fast, the time allotted for mass transfer becomes insufficient. Then the stage efficiency that matches actual performance will plummet, or the experimentally determined value for *HETS* will increase dramatically. At this point, performance becomes controlled or *limited* by the mass transfer rate, and the design requires a rate-based approach with some way of taking into account the mass transfer rate or allotted contact time. So, it is critically important in using the theoretical stage model to understand the assumptions being made and to take care not to extend or extrapolate the model to conditions where these assumptions are no longer valid; that is, *be sure to design the device with sufficient residence time.* If this aspect of the design is not recognized, then a trap is set that can lead the designer to very poor designs that cannot work as intended. In practice, separation equipment is often evaluated using a stage model whether or not it exhibits fast mass transfer rates or low mass transfer resistance, especially for a binary separation. In this case, a stage model can serve adequately as long as the designer is fully aware of residence time requirements and the corresponding stage efficiencies, which may be low. An informed awareness of the residence time issue and an appreciation for real-world stage efficiencies can prevent a designer from falling into the trap.

Perhaps a more important shortcoming of the theoretical stage model and potential trap involved in its use arises when dealing with multicomponent mixtures, where the thermodynamic drivers and mass transfer rates and resistances can be significantly different for different components. In this case, a theoretical stage model will predict multicomponent separation as if all components exhibit the same fast mass transfer rate. As a result, the use of theoretical stages may correctly model the separation of the key components of interest, but overpredict or underpredict the separation of non-key components. In that case, the material balance numbers obtained from theoretical stage-based process simulations will not be exactly correct for impurity components. Accurately modeling the separation for all components in a multicomponent feed may require manual adjustments to the material balance from the analysis of actual samples or the application of rate-based process simulation.

Calculating the number of theoretical stages is also a common and convenient way to characterize overall separation difficulty for any application, even if the process does not actually operate anywhere near equilibrium and the theoretical stage model will not be used for design purposes – the greater the number of theoretical stages, the greater the separation difficulty for a given application. Since this assessment of difficulty does not take into account how close or far the feed composition is from equilibrium for a given application, comparing the number of required theoretical stages is not particularly meaningful for comparing difficulty across different kinds of separations and chemical mixtures. For example, in the distillation of low viscosity liquids, a given number of theoretical stages is generally easier to attain for solutes with low relative volatility ($\alpha < 2$) compared to those with much higher relative volatilities, because the solutes with higher relative volatility have to go further to reach equilibrium – at least that is one way to look at it.

6.6.2 Stage efficiency

In designing separation equipment with discrete contacting stages such as a trayed distillation column, determining the required number of theoretical stages, even if done rigorously with great accuracy, is not sufficient. The designer must connect the result to a corresponding number of actual stages or trays. The concept of stage efficiency is often used for this purpose and has been the object of much study and discussion. As a result, there are a number of ways that stage efficiencies are defined and used. There are three main types: point (or local) efficiency, Murphree tray efficiency, and overall tray or column efficiency, each of which has its own area of application.

Point or Local Efficiency. The point efficiency is applied to a specific location on a tray or in a froth or film rather than across a full crossflow plate or column. As such, it is usually not used to characterize the difference between an actual and theoretical tray. Instead, point efficiencies are typically used to evaluate differences in efficiency between different locations on a tray with different conditions and determine which factors and areas on the tray have the greatest impact on overall tray performance. It is usually characterized by examining a differential slice of froth across the flow path on a representative tray, as shown in Figure 6.2. In this differential slice, the liquid phase is assumed to be well-mixed vertically.

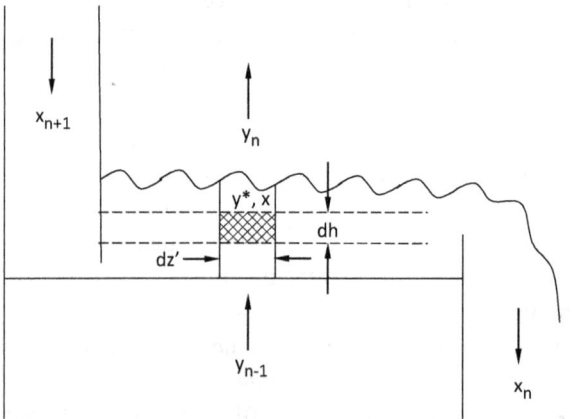

Figure 6.2: Point or Local Stage Efficiency.

Point efficiency, based on local vapor phase concentrations for instance, can then be defined as [33]:

$$E_{pV} = \frac{y_n - y_{n-1}}{y^* - y_{n-1}} \tag{6.35}$$

where E_{pV} = vapor phase point efficiency

y_n = vapor mole fraction exiting the local point

y_{n-1} = vapor mole fraction at the entrance to the local point

y^* = mole fraction of vapor in equilibrium with the liquid at the local point

Point efficiency may be significantly less than unity because of a mass transfer rate limitation. So, point efficiency often is explained by relating it to a rate-based concept called the overall gas-phase mass transfer unit:

$$N_{OG} = \int \frac{dy}{y^* - y} \qquad (6.36)$$

(Mass transfer units are defined and discussed in detail in Section 6.7.1: Mass Transfer Units and Mass Transfer Coefficients.) The integration can be performed from below the plate at point n-1 to above the plate at point n for the differential slice illustrated in Figure 6.2 with plug flow in the vapor phase and a constant value of y^* in the slice:

$$N_{OG} = \int_{n-1}^{n} \frac{dy}{y^* - y} = -\ln(y^* - y_n) + \ln(y^* - y_{n-1}) = -\ln\left(\frac{y^* - y_n}{y^* - y_{n-1}}\right) \qquad (6.37)$$

Combining this result with eq. (6.35) above, we get fairly simple relationships between the vapor-phase point efficiency and the number of overall gas-phase mass transfer units:

$$E_{pV} = 1 - e^{-N_{OG}} \qquad (6.38)$$

and

$$N_{OG} = \ln\left(\frac{1}{1 - E_{pV}}\right) \qquad (6.39)$$

Think of N_{OG} as an effective number of mass transfer stages, each with less than 100% efficiency. It is important to note that although point efficiencies can vary significantly between locations on commercial-scale crossflow trays, they can be very close to the average tray efficiency or Murphree efficiency when the flow path length is very short such as in laboratory scale Oldershaw columns. In addition, in countercurrent flow trays such as dual flow trays and some modern high-capacity trays, point efficiencies may also represent the average or Murphree efficiency.

Murphree Tray Efficiency. Murphree efficiency is named for Eger Vaughan Murphree, who developed the concept in the early 1920s while studying vapor-liquid contacting as a Research Associate in chemical engineering at Massachusetts Institute of Technology [34]. Murphree efficiency is defined in a manner similar to point efficiency, but applied to the entire tray rather than at a specific location on the tray (as in Figure 6.3). The Murphree efficiency is based on concentrations in a single phase (vapor or liquid), with the assumptions that the vapor between trays is well-mixed but passes in plug flow

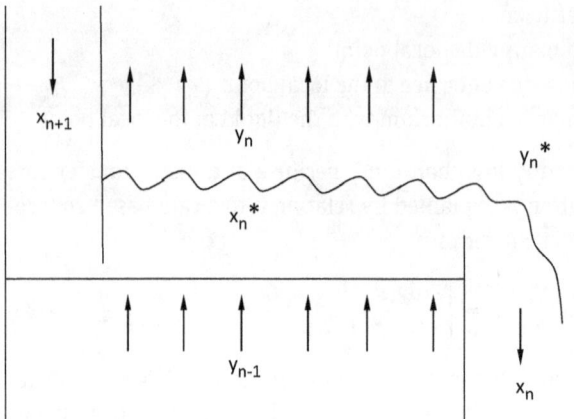

Figure 6.3: Murphree Stage Efficiency.

through the froth on each tray and that the liquid on the tray is well-mixed vertically. For crossflow trays such as sieve plates and valve trays, although the liquid is assumed to be well-mixed vertically, the liquid is also assumed to be in plug flow as it passes across the surface of the tray. A McCabe-Thiele diagram illustrating the concept of the vapor-phase Murphree tray efficiency is given elsewhere [35, 36].

On this basis, the Murphree vapor phase tray efficiency is defined as [36]:

$$E_{MV} = \frac{y_n - y_{n-1}}{y_n^* - y_{n-1}} \tag{6.40}$$

where E_{MV} = Murphree vapor phase efficiency

y_n = vapor mole fraction exiting tray n

y_{n-1} = vapor mole fraction entering tray n

y_n^* = vapor mole fraction in equilibrium with the liquid exiting tray n

If the liquid on the tray is well-mixed horizontally in addition to vertically, the liquid concentration on the tray is constant, and the vapor exiting the liquid would also have a uniform concentration. As a result, the point efficiencies at all locations on the tray would be constant and would be equal to the Murphree efficiency, which, in turn, would be limited to a maximum value of 100%. However, in real world applications of crossflow trays where the liquid flow path across the surface of the tray is long enough, a significant liquid concentration gradient can exist along the flow path so that the liquid concentration at the tray exit, x_n, is appreciably lower than the average liquid concentration on the tray. This can result in a reduced vapor concentration in equilibrium with the exiting liquid, y_n^*, relative to the average vapor concentration exiting the tray, y_n, which subsequently results in an enhanced Murphree vapor efficiency. In cases where the vapor-liquid contact on the tray is very effective and the liquid flow path length is long enough, y_n^* may actually become less than the average vapor concentration exiting the tray y_n, and the Murphree efficiency may in turn become greater than 100% (up to 120% or even more). This does not break any laws of thermodynamics. It simply reflects the fact that such a tray does not function as a true well-mixed single-stage device. Instead, the cross-flowing liquid undergoes successive contacts with up-flowing vapor, which provides greater mass transfer than would be achieved in a well-mixed stage. This is sometimes observed for single-pass trays with liquid crossflow path lengths greater than a meter or so. It may also be observed with multi-pass trays, but normally only in large columns with diameters larger than 2 meters, depending on the number of passes.

When appropriate, the Murphree tray efficiency can also be defined based on liquid concentrations:

$$E_{ML} = \frac{x_n - x_{n+1}}{x_n^* - x_{n+1}} \tag{6.41}$$

where E_{ML} = Murphree liquid phase efficiency

$\quad x_n$ = liquid mole fraction exiting tray n

$\quad x_{n+1}$ = liquid mole fraction entering tray n

$\quad x_n^*$ = liquid mole fraction in equilibrium with the vapor exiting tray n

Liquid phase Murphree efficiencies are mainly useful in stripping sections where changes in liquid compositions are usually more important than changes in the vapor. They are especially relevant in strippers with high relative volatilities operating at elevated transfer factors (stripping factors in this case), since the mass transfer limitations that arise in those applications are manifested almost entirely in the liquid phase, and so the liquid phase Murphree efficiencies will be most representative. However, for most typical distillation applications, the accepted practice is to refer to vapor-phase Murphree tray efficiencies when characterizing tray separation performance.

Overall Column Efficiency. Perhaps the most practical tray efficiency used in distillation design is the overall column (or section) efficiency. This is simply the ratio of the number of theoretical stages or trays in a column divided by the number of actual physical stages or trays in the column:

$$E_O = \frac{N_T}{N_A} \quad \text{and} \quad N_A = \frac{N_T}{E_O} \tag{6.42}$$

where E_O = fractional overall column or stage efficiency

$\quad N_T$ = number of theoretical stages required to achieve the desired separation

$\quad N_A$ = number of actual physical trays required in the column

Clearly, the overall column efficiency is a very useful concept for column design, and estimates of overall efficiencies are frequently used for that purpose. Values normally are obtained using operating data for real world devices under conditions experienced in commercial service. Depending on the application, overall efficiencies may be applied to an entire column or to different sections or groupings of trays within a column, each with its own efficiency according to the process conditions and tray characteristics within that section or grouping. A frequent approach is to identify a representative tray within a column section and use the conditions on that tray to estimate an overall efficiency and apply it to the rest of the trays in that section, for design purposes.

Although many of the traditional rigorous models and more recent mechanistic models for overall efficiency are difficult to use and involve input values that are difficult to quantify, some of the empirical correlations that are available have proven to be relatively straightforward with acceptable accuracies for real world devices, under conditions experienced in commercial applications. As a result, the empirical correlations tend to be more commonly used in practice for estimating actual tray counts.

One of the simplest empirical correlations for overall efficiency is the O'Connell correlation, which was developed in graphical form by H. E. O'Connell in 1946 [37]. O'Connell plotted overall tray efficiency versus key component relative volatility times liquid viscosity, based on operating data from 31 distillation columns using bubble cap trays. This approach is well-described elsewhere [33, 36, 38]. Although it was developed for bubble cap trays, experience has shown that it also yields reasonable estimates of overall efficiency for other types of conventional trays and so has seen wide use over the years, not in the least because of its simplicity and ease of use.

Since a graphical correlation does not lend itself well to computer calculations, analytical expressions for the O'Connell correlation have been developed by a number of investigators. Lockett [95] developed the following expression:

$$E_O = 0.492(\mu_L \alpha)^{-0.245} \tag{6.43}$$

where E_O = Fractional overall column or tray efficiency
$\quad \mu_L$ = Liquid viscosity, cP
$\quad \alpha$ = Relative volatility of the light key component

More recently, Duss and Taylor [39], using the data of Chan [40], developed an improved, more accurate expression for the O'Connell correlation, which includes the solute transfer factor as a parameter rather than the relative volatility and is still quite easy to use:

$$E_O = 0.503\mu_L^{-0.226}\mathcal{T}^{-0.08} \tag{6.44}$$

where $\mathcal{T}$ = solute transfer factor for the column section being considered.

It is important to note that when correlating the parameters in eq. (6.44) above, Duss and Taylor always considered the value of the solute transfer factor to be greater than 1.0. That is, in the stripping section of a distillation column where the stripping factor $mV/L > 1.0$, the stripping factor is used for the solute transfer factor. Conversely, in the rectification section where the absorption factor $L/(mV) > 1.0$, the absorption factor is used for the solute transfer factor. The solute transfer factor is more fully defined and discussed in Section 6.1.3: Partition Ratio, Separation Factor, and Transfer Factor.

Interestingly, Duss and Taylor subsequently [41] developed a further-improved correlation methodology for overall efficiency primarily intended for sieve trays. It relates stage efficiency to mass transfer rate limitations (the subject of our next section). Duss and Taylor used the data of Chan [40] as before and added the data of Garcia [42] to develop the following relationships:

$$N_G = 1.3 \left(\frac{\mu_G}{0.01}\right)^{-0.3} \left(\frac{\mu_L}{0.2}\right)^{-0.1} \left(\frac{FOA}{0.08}\right)^{-0.2} \left(\frac{D_{hole}}{12}\right)^{-0.1} \left(\frac{h_{weir}+5}{55}\right)^{0.05} \tag{6.45}$$

$$N_L = 5 \left(\frac{\mu_L}{0.2}\right)^{-0.6} \left(\frac{FOA}{0.08}\right)^{-0.2} \left(\frac{D_{hole}}{12}\right)^{-0.1} \left(\frac{h_{weir}+5}{55}\right)^{0.05} \tag{6.46}$$

where N_G = number of gas transfer units based on local mass transfer coefficients
   N_L = number of liquid transfer units based on local mass transfer coefficients
   μ_G = gas viscosity, cP
   FOA = fractional open area in tray bubbling area
   D_{hole} = sieve tray hole diameter, mm
   h_{weir} = tray outlet weir height, mm

These parameters are then applied in the form of the sum of gas-phase and liquid-phase resistances:

$$\frac{1}{N_{OG}} = \frac{1}{N_G} + \frac{S}{N_L} \tag{6.47}$$

where N_{OG} = number of overall gas phase transfer units
   S = stripping factor = mV/L

The resulting value of N_{OG} is then used in eq. (6.38) to determine the fractional vapor-phase point efficiency E_{pV}, restated here:

$$E_{pV} = 1 - e^{-N_{OG}} \tag{6.48}$$

The point efficiency is then used in an expression for the Murphree vapor-phase tray efficiency:

$$E_{MV} = \frac{e^{E_{pV}S} - 1}{S}\left(1 - e^{-5.0FPL}\right) \tag{6.49}$$

where E_{MV} = fractional Murphree vapor phase tray efficiency
   FPL = tray flow path length, m

Lastly, since the point and Murphree efficiencies calculated up to this point are equilibrium stage-based, the overall column or tray efficiency must be calculated by taking overall mass transfer rate limitations into account. This is done by using the relationship derived by Lewis [43], which assumes straight operating and equilibrium lines within the tray or column section to which it is applied (not necessarily overall):

$$E_O = \frac{\ln[1 + E_{MV}(S-1)]}{\ln S} \tag{6.50}$$

Although the implementation of this series of equations seems complex, the equations are basically sequential and so are easily implemented in a spreadsheet or other automated calculation. We should also note that all of these expressions for overall column efficiency by Duss and Taylor assume that the column is operated within its design operating range where the efficiency is relatively constant versus throughput. Duss and Taylor describe this as being between 25% and 98% of capacity, with "capacity" likely being the percentage of throughput where the onset of jet flooding occurs. More detailed discussions are given by Duss and Taylor [39, 41, 44].

6.7 Rate-Based Models

As discussed above, the theoretical stage concept is best used to characterize systems for which separation performance is primarily controlled by phase equilibrium such that stage efficiencies are close to unity. Rate-based models, on the other hand, should be used when separation performance is controlled by (or limited by) the mass transfer rate and mass transfer resistance. In this case, a given separation process operates away from equilibrium. The rate of mass transfer for a given solute of interest is normally expressed in terms of a concentration difference (or a partial-pressure difference) between actual and equilibrium values of solute concentration, and the rate of mass transfer is modeled as a linear function of this deviation. Often, the rate is controlled by diffusion of solute across thin films or boundary layers on each side of the interface between two phases, as opposed to transfer of solute within the bulk phases. The actual concentration of solute at the interfacial boundary is not known, so a difference between bulk and equilibrium concentrations is used instead. Characterizing the net or overall rate of mass transfer this way is normally sufficient for design purposes. Knowledge of interfacial concentrations and individual rate-limiting steps is desirable, of course, and may be critical to understanding opportunities for improvement. But the point we make here is that rate-based design methods may be advantageously applied even without detailed knowledge of the actual mass transfer mechanism. Furthermore, mass transfer rates and efficiencies remain difficult to predict with the accuracy required for design, so many design methods require access to some performance data for the same or similar systems, normally obtained in pilot-plant studies or from commercial-scale operations. For distillation columns, this often involves access to data expressed in terms of the height equivalent to a theoretical plate (*HETP*) or the height of a transfer unit (*HTU*) or in terms of tray efficiencies. Or, data may be expressed in terms of overall mass transfer coefficients – for different chemical systems and column internals. In generating such data, various standard chemical systems are used to model typical systems, such as mixtures of hydrocarbons (nearly ideal), mixtures of hydrocarbons + oxygenated hydrocarbons (moderately non-ideal), and aqueous organic mixtures (highly non-ideal). Standard systems are described elsewhere for testing distillation [45] and extraction [19] equipment.

6.7.1 Mass transfer units and mass transfer coefficients

The classic approach to modeling a rate-based mass transfer process employs either mass transfer units or mass transfer coefficients, concepts that are well covered by many excellent books on mass transfer (see Bibliography for Chapter 6 which lists some of these). Our intent here is to clarify several points regarding their definition and use. The first point to make is that mass transfer coefficients and transfer units are directly related; so in principle, the choice of whether to use one or the other is simply a matter of preference. In our experience, mass transfer coefficients (normally combined with interfacial area) tend to be used for process modeling calculations, while mass transfer units tend to be used as a framework for characterizing and correlating process performance and for rating equipment. But again, the two concepts are related and either one may be used for process analysis and equipment design. Mass transfer coefficients are first-order rate constants derived from the two-film theory of mass transfer. They were introduced between 1904 and 1923 by Walther H. Nernst, Warren K. Lewis, and Walter G. Whitman [46–49]. The related mass transfer units were introduced and popularized by Allan P. Colburn and others in the 1930s with reference to the analogy between mass transfer and heat transfer [50–52].

Let us begin with the mass transfer unit because it begs comparison with the theoretical stage. A single theoretical stage is familiar to many: bring two phases together, generate interfacial area, allow sufficient time for the phases to come to equilibrium, and then physically isolate the equilibrated phases. A single mass transfer unit, on the other hand, is not quite as intuitive – but it is related and may be applied in similar ways. It can be regarded as a kind of contacting stage that does not reach equilibrium. Hence, for a given separation, the number of transfer unit "stages" is always greater than the number of theoretical stages. We have already touched on this in Section 6.1.3: Partition Ratio, Separation Factor, and Transfer Factor. We recommend reviewing that discussion and Figure 6.1 as an introduction to the following.

Mass transfer units are used to characterize a process that is driven by a deviation from equilibrium and always operates away from equilibrium. The number of transfer units (NTU) is a dimensionless number that represents the change in solute concentration achieved by such a process relative to the magnitude of the deviation-from-equilibrium driving force. One transfer unit is achieved when ΔC, the concentration change, is equal to the average driving force $C - C^*$ such that

$$NTU \equiv \int \frac{dC}{C - C*} \qquad (6.51)$$

where $*$ denotes equilibrium.

The mass transfer coefficient is derived from two-film theory. It is simply the rate constant for a first-order mass transfer rate expression:

$$\text{mass transfer rate} = \frac{dC}{dt} = ka(C - C*) \qquad (6.52)$$

where C = solute concentration (moles per unit volume)
$C - C*$ = the average concentration driving force (deviation from equilibrium)
k = rate constant (length per unit time)
a = interfacial area per unit volume (m^2/m^3)
t = time

Rearranging gives

$$\frac{dC}{C - C*} = ka\, dt \qquad (6.53)$$

The value of NTU is then given by:

$$NTU = \int \frac{dC}{C - C*} = ka \int dt = ka\tau \qquad (6.54)$$

where τ is the average residence time.

Thus, for a column-type separator the value of NTU is given by:

$$NTU = ka\tau = ka \times \frac{z}{v} = \frac{kaz}{v} = \frac{kaz}{\frac{Q}{A_x}} = ka\left(\frac{zA_x}{Q}\right) \qquad (6.55)$$

where z = total height or length of the column
v = average velocity of the feed phase through the column
τ = average residence time
Q = volumetric flow rate, m^3 per unit time
A_x = column cross-sectional area, m^2

Equations of this general form are used to characterize performance for many types of mass transfer equipment. The units used here are typical, but they can vary depending on how one chooses to represent residence time for a given equipment geometry, as well as the choice of units for solute concentration. But the general concepts are the same [38, 53].

Given the relationship between NTU and ka, the value of NTU may be interpreted as a dimensionless processing time equal to residence time in the separator divided by a characteristic transfer time $t**$, where:

$$t^{**} = \frac{1}{ka} \tag{6.56}$$

and

$$NTU = ka\tau = \frac{\tau}{t^{**}} \tag{6.57}$$

With this interpretation, NTU characterizes how much time a separator allows for transfer of solute relative to a time scale that is characteristic of the chemical system of interest.

So, unlike the theoretical stage concept, the transfer unit concept deals directly with the mass transfer rate in terms of residence time and characteristic transfer time. For example, doubling the length or height of a packed column separator doubles the residence time and doubles the number of transfer units. This assumes that the underlying mass transfer mechanism remains unchanged and interfacial area and all else remain the same. This may or may not be a good assumption for a given separator. That is, the mass transfer rate may be uniform across the entire separator as in a typical stripping-mode distillation column, or it may differ significantly from one section to another as in a typical fractionation column with both stripping and rectification sections. In the latter case, each section of the separator should be analyzed separately.

For separations carried out in countercurrent columns with no distinct stages or trays (differential contactors such as packed columns), the height of the column z is a key design variable, so NTU often is expressed in terms of the height of a transfer unit given by $HTU = z/NTU$. For separations carried out in single-stage mixing devices or in distillation trays, the height of the device or tray is not a key variable used to model performance. Instead, performance is expressed as the number of transfer units achieved per device or per tray.

For a dynamic, batchwise separation process designed to transfer solute out of a feed batch, the concentration of solute remaining in the feed as a function of time can be related to the number of transfer units achieved by the process, by integrating the rate equation, eq. (6.52) as a function of time. This yields a general equation of the form:

$$\exp(-kat) \approx \frac{C_{end}}{C_{start}} = \exp(-NTU) \tag{6.58}$$

where t is processing time and the quantity ka is the time constant for the process. This result is obtained for a dilute feed solution assuming that the mass of the feed batch remains essentially constant (used here for illustration). The fraction of solute recovered from the feed is then given by

$$\text{Recovery (\%)} = 100 \left(1 - \frac{C_{end}}{C_{start}} \right) \tag{6.59}$$

Equation (6.58) represents the characteristic behavior of a batchwise mass transfer process (for a dilute feed). Here, NTU is a lumped parameter that characterizes performance. Its value will vary with batch operating time, but also as the rate of mass transfer changes due to an operational change such as a change in evaporation rate, temperature, or agitation intensity. The NTU value that fits the batch data may be regarded as a characteristic or effective value for the given operation, but only for the specific set of operating conditions and the specific chemical system.

So, for either steady-state or batchwise operation, the number of transfer units is a dimensionless lumped parameter that characterizes mass transfer performance. An increase in the value of NTU reflects greater separation capability, either due to an increase in the operating time or the residence time of the feed phase (the numerator in eq. (6.57) or due to a decrease in the characteristic transfer time needed for mass transfer (the denominator). The latter may be accomplished by increasing agitation intensity in agitated vessels to reduce diffusion path length and increase interfacial area, or by adding baffles in column-type differential contactors to better approach plug flow and reduce backmixing. This interpretation of NTU also helps explain the differences in performance observed between laboratory-scale separators and full-scale separators. On scaling up to a full-scale unit, mass transfer performance often is observed to degrade somewhat due to difficulties in maintaining desired fluid flow patterns. So, the design must be modified to mitigate or counter such effects. This may be done by specifying reasonably efficient internals while compensating for any negative scale-up effects by designing for somewhat longer residence time. As a classic example, often a full-scale packed distillation column must be considerably taller than the packed column in the laboratory to achieve the same separation performance (the same NTU), and this amounts to adding extra residence time on scale-up. Similarly, in batchwise processing, the full-scale unit in the manufacturing plant often must be operated for a somewhat longer time to achieve the same performance as the lab-scale unit.

Reliable values of NTU normally are determined from actual separation performance data, either from operation of a full-scale separator in the manufacturing plant or from pilot-plant data. The value of NTU is calculated from the material balance around the process using a suitable design equation or simulation program. For example, a well-known design equation for a countercurrent steady-state process involving dilute feed is obtained by integration of eq. (6.51), assuming linear equilibrium and operating lines, analogous to the derivation of the KSB equation, eq. (6.29) [38]. This yields the following steady-state design equation, which is sometimes called the Colburn equation after Allan Colburn, who developed the mass transfer unit concept in the 1930s:

$$NTU = \frac{\ln\left[\left(\frac{x_{in} - \left(\frac{y_{in}}{m}\right)}{x_{out} - \left(\frac{y_{in}}{m}\right)}\right)\left(1 - \frac{1}{\mathcal{T}}\right) + \left(\frac{1}{\mathcal{T}}\right)\right]}{1 - \frac{1}{\mathcal{T}}} \tag{6.60}$$

$$\mathcal{T} = m\frac{S}{F}, \; \mathcal{T} \neq 1 \tag{6.61}$$

$$F_R \equiv \frac{x_{in}}{x_{out}} \approx \frac{x_{in} - \left(\frac{y_{in}}{m}\right)}{x_{out} - \left(\frac{y_{in}}{m}\right)} = \frac{\exp\left[NTU\left(1 - \frac{1}{\mathcal{T}}\right)\right] - \left(\frac{1}{\mathcal{T}}\right)}{1 - \left(\frac{1}{\mathcal{T}}\right)} \tag{6.62}$$

As before, solute recovery and the concentration of solute in the second phase outlet are given by eqs. (6.27), (6.28), (6.32), and (6.33), and $\mathcal{T}$ is the transfer factor.

The Colburn equation has the same limiting assumptions as the KSB equation, eq. (6.29), except that now a measure of the rate of mass transfer is included in the value of NTU. As with the KSB equation, estimates can be made for non-dilute feeds by using an averaged value of m eq. (6.34). Both the KSB equation and the Colburn equation have the same basic form, except for the denominator, but this makes for a significant difference. The KSB equation predicts continued improvement in performance with increasing transfer factor (carrying capacity) without limit – because it assumes the mass transfer rate is always more than sufficient. The Colburn Equation, on the other hand, predicts that performance will reach a plateau above a transfer factor of 10 or so, because the transfer factor appears only as the reciprocal value, so its influence becomes vanishingly small at large values. This reflects the limitation imposed by the finite mass transfer rate inherent in the transfer unit model. Normally, designers specify a transfer factor between about 1.3 and 5 or so (but higher values of 10 or more are sometimes used for special reasons to be discussed in later chapters). Below a value of 1.3 the required number of transfer units becomes excessive. The choice of whether to use theoretical stages or transfer units to model a given application hinges on the difference in how the different models extrapolate performance as the transfer factor is varied. As mentioned earlier, if the designer is not aware of the inherent assumptions in the theoretical stage model (where it is assumed that $t^{**} \ll \tau$) and does not consider how stage efficiency changes as a function of residence time in the device, then extrapolation can lead to highly misleading results. The transfer unit model generally is more reliable in this sense, although it must also be used with caution because of its limiting assumptions.

Note that with these simplified design equations, separation performance is independent of the absolute magnitude of the feed concentration, and the calculation of NTU does not explicitly include processing time or scale of operation (absolute production rate). The calculation deals only with change in the relative concentration of solute as a function of position in the equipment (in versus out) (as in eq. (6.62). This is because these equations are derived for steady-state operation and dilute concentrations. Concentrations and flow patterns are considered constant with respect to time, but they vary with position in the equipment. The design procedure involves specifying the ratio of flow rates, one phase with respect to the other (as expressed by the transfer factor). It may seem that the steady-state Colburn equation does not reflect processing time, but it does. Processing time (on a relative basis) is taken into account in the value of NTU (as in eq. (6.57)). The resulting expressions provide a framework for characterizing and quantifying mass transfer performance for design purposes.

Comparing eqs. (6.29) and (6.60) shows that the value of NTU may be related to an equivalent number of theoretical stages by considering the value of the transfer factor:

$$NTU = N\left(\frac{\ln \mathcal{T}}{1 - \frac{1}{\mathcal{T}}}\right), \text{ for } \mathcal{T} > 1 \qquad (6.63)$$

$$NTU = N = \frac{X_{in} - \left(\frac{y_{in}}{m}\right)}{X_{out} - \left(\frac{y_{in}}{m}\right)} - 1, \text{ for } \mathcal{T} = 1 \qquad (6.64)$$

$$F_R \approx \frac{X_{in} - \left(\frac{y_{in}}{m}\right)}{X_{out} - \left(\frac{y_{in}}{m}\right)}\bigg|_{\text{max}} = \frac{1}{1 - \mathcal{T}}, \text{ for } \mathcal{T} < 1 \text{ and large } NTU \qquad (6.65)$$

As mentioned earlier, the number of transfer units required for a given separation is always larger than the number of theoretical stages, except when the transfer factor is equal to or less than unity. Fundamentally, this is the basis for equating theoretical stage efficiency to mass transfer units as discussed in the previous section.

In summary, compared to theoretical stages, transfer units provide a better way of characterizing performance for separation processes that operate away from equilibrium. But the two concepts are related. Values for NTU may be determined directly by integration of the concentration profile within a separator, as indicated by the definition of NTU given in equation (6.51) and illustrated by the Colburn Equation derived for the specific case of straight equilibrium and operating lines (eqs. (6.60) and (6.61)). Or, NTU may be determined by using models involving mass transfer coefficients as indicated in eq. (6.55). Sometimes practitioners use another approach that involves first analyzing the proposed process to rigorously determine N, the number of required theoretical stages, followed by converting N to NTU from knowledge of the transfer factor. This is an approach promoted by Dr. Lanny Robbins at The Dow Chemical Company beginning in the 1980s [54]. This analysis is normally done for different sections of the process for which the transfer factor is different – as in the rectification section of a fractional distillation process (using the absorption factor) and separately in the stripping section (using the stripping factor). This approach allows use of standard process simulators that focus on rigorously determining N. Then, converting N to an equivalent NTU provides a way to determine the number of actual contacting stages or the actual column height that is required for the given separation. In other words, converting N to NTU allows a way to include rate limitations and efficiency into the calculation – in order to determine the actual size of the equipment. This approach can work especially well for designing rate-limited distillation and extraction columns. This is because the measures of transfer unit efficiency (*HTU* and the number of transfer units per tray) are normally fairly constant as a function of transfer factor, much more so than mass transfer efficiency expressed either as the efficiency of each contacting stage (expressed as theoretical stages) or as the *HETP*. Good estimates for *HTU* may be obtained from available *HETP* data (which are normally measured at conditions corresponding to transfer factors near unity), by setting *HTU* = *HETP* and then using the same *HTU* value for all values of the transfer factor (assuming *HTU* is constant). Then, the total required height is given by *NTU* × *HETP*, which is greater than *N* × *HETP*. We discuss this approach in more detail in Section 7.3.4: Continuous Fractional Distillation.

Mass Transfer Coefficients and Solute Diffusivity. As discussed above, mass transfer coefficients are defined in terms of a first-order rate expression for solute diffusion through thin films on each side of the interface between a feed phase and a second phase. The mass transfer coefficient, normally, is a function of solute concentration, solute diffusion coefficients (diffusivities), and fluid flow (velocity and turbulence), and often it includes interfacial area (as *ka*). Mass transfer coefficient correlations have been developed for many types of equipment and may be used to estimate mass transfer rates for initial design studies [38, 53, 55].

Two-film theory invokes a film on each side of the interface through which solute must diffuse. The overall mass transfer coefficient is simply the overall rate constant, and the overall coefficient is a function of the individual film coefficients – just as in heat transfer theory. This relationship is expressed by the rule of additive reciprocal resistances used to sum up the individual in-series resistances encountered as solute diffuses across the two-film interface:

$$\frac{1}{K_{overall}} = \frac{1}{k_{feed\ side}} + \frac{1}{mk_{second\ phase\ side}} \tag{6.66}$$

where *m* is the equilibrium relationship (slope of the equilibrium line). Normally the value of *m* is significantly greater than unity, as most commercially viable applications require a favorable thermodynamic driving force. Then, the term $1/mk_{second-phase-side}$ is relatively small, and *K*overall is dominated by *k*feed-side. Exceptions to this rule may involve applications where the viscosity of the feed phase is much greater than that of a second fluid phase. Various theories concerning the behavior of the individual film coefficients and their fundamental relationship to molecular diffusivity have been developed. Most famous are two-film theory (1924), penetration theory (1935), surface renewal theory (1951), and boundary layer theory, where:

$$k \propto D, \text{ from two – film theory (Lewis and Whitman, 1924)} \tag{6.67}$$

$$k \propto D^{\frac{1}{2}}, \text{ from penetration theory (Higbie, 1935)} \tag{6.68}$$

$$k \propto D^{\frac{1}{2}}, \text{ from surface renewal theory (Dankwerts, 1951)} \qquad (6.69)$$

$$k \propto D^{\frac{2}{3}}, \text{ from boundary layer theory (Spalding, 1964)} \qquad (6.70)$$

Theories relating mass transfer coefficients to diffusivity, film thickness, and fluid velocity have been studied using apparatus with straightforward geometry and well-defined interfacial area and film thickness. The results of this fundamental work provide a framework and guide for correlating mass transfer coefficient data for the more complicated geometries and operating conditions of real equipment. Semitheoretical expressions are developed for real equipment using dimensionless numbers including the Reynolds number, Schmidt number, Sherwood number, Weber number, and Peclet number, among others. These are discussed elsewhere [38, 53, 55, 56]. For some of these applications, interfacial area could be measured and correlations of interfacial area data are available. For other applications, the data are correlated in terms of ka without separating out the interfacial area term, because in these studies it was not feasible to measure interfacial area directly. Theory provides a highly useful framework for correlating data. However, correlations must be used with caution because they apply only to the specific operating conditions and equipment geometries used to develop them (the specific chemical compositions, solute concentrations, and flow conditions).

Mass Transfer with Reaction. As discussed above, for non-reacting systems the main resistance to mass transfer normally resides in the feed phase, or more precisely, in the feed-side film. Exceptions to this general rule include reaction-enhanced separations such as reactive absorption. With the inclusion of reaction, the main resistance to the overall mass transfer may be due to slow reaction relative to physical mass transfer, and this resistance may reside in either phase depending on the chemistry. A well-known example of reaction occurring in the feed phase is the absorption of highly soluble gases such as HCl or NH_3 into water. In this type of process, the main resistance to mass transfer is in the gas phase (the gas-side film resistance), because the kinetics of dissolution and dissociation in the liquid are faster than the rate of mass transfer from the gas phase into the liquid (the gas film resistance) – as is normally expected for non-reactive mass transfer. Another example involves absorption of CO_2 from a feed gas into aqueous NaOH. In this case, the main resistance is in the liquid phase, as the rate of reaction involving CO_2 dissolution and reaction with base in the liquid phase is slow compared to the gas film-side mass transfer rate. See Section 7.3.7: Reactive Absorption for Gas Scrubbing.

6.7.2 Phenomenological and other models

This approach to modeling mass transport and separation processes was championed by R. Bryan Bird, Warren E. Stewart, and Edwin N. Lightfoot with the publication of their classic textbook, Transport Phenomena in 1960. A second edition was published in 2006 [57]. Phenomenological models are mechanistic (or quasi-mechanistic) models developed starting with the fundamental continuity equations of fluid flow (mass and energy conservation equations) plus additional terms for mass transfer driving force and mass transfer resistance. This gives systems of equations representing bulk flow (convective or advective terms), backmixing, accumulation, reaction, and the rate of mass transfer. The mass transfer rate expression includes the concentration driving force (the deviation from equilibrium) and may be a linear expression written in terms of a first-order rate constant (a mass transfer coefficient), or it may be a non-linear rate expression. These models may also include simultaneous solution of mass and energy rate equations and eliminate any need to apply stage efficiencies in the calculations. The full-blown system of equations often must be solved numerically, which can be difficult to do, so in many cases approximate models are developed by removing terms that have small or negligible effects on performance – focusing only on the major terms that dominate performance. This is called scaling, where terms that are likely to have little effect on the results are systematically eliminated. This approach is discussed by William B. Krantz [25].

The development and application of phenomenological models is particularly important for better understanding the mechanisms of mass transfer. Such models may be used to identify and better understand rate-limiting factors and opportunities for improvement. The equations may be written to model multiple discrete contacting stages or the flow of a dispersed phase within a stationary or counter-flowing continuous phase. A version of this general approach is often used in modeling adsorption and chromatography processes as we discuss next, in Section 6.7.3: Adsorption Dynamics: Breakthrough Curves.

For discussion of rigorous distillation models, see refs. [58] and [59] and subsequent articles that cite these key references.

For systems containing solid particles, another approach is called discrete element modeling [60]. This involves modeling of individual particles in two-phase or three-phase flows of solid particles in liquids and gases. This approach may be used to understand the flow and classification of solid particles within conveying systems and storage bins and within crystallizers. Sometimes, systems of multiple particles are approximated as bulk flows and modelled using continuity equations to simplify the calculations and still represent the overall (macroscopic) behavior.

6.7.3 Adsorption dynamics: breakthrough curves

Adsorption processes commonly are carried out using a continuously fed bed of adsorbent particles. The particles are packed into a column (or fixed bed), and the feed (either gas or liquid) is continuously fed to the bed until such time that the bed approaches saturation and solute begins to breakthrough into the effluent stream. Figure 6.4 illustrates a

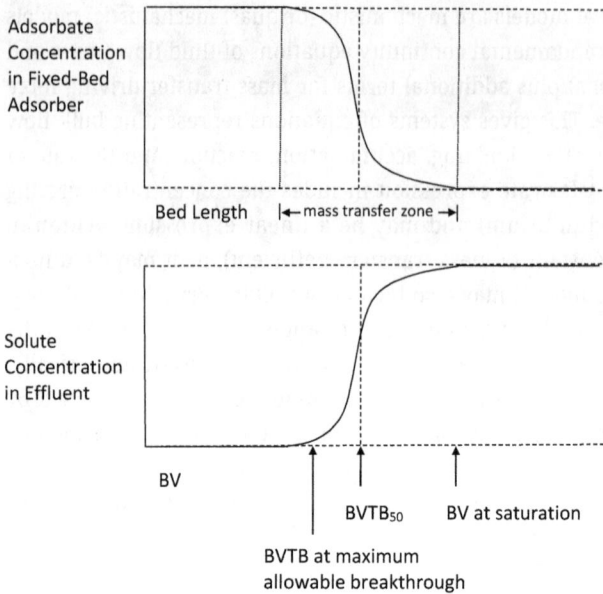

Figure 6.4: Fixed-Bed Adsorption Profiles. Top) Concentration of Adsorbed Solute within the Adsorbent Bed, Showing the Initial Saturated Zone and the Mass Transfer Zone (MTZ); and bottom) Concentration of Solute in the Fluid Exiting the Bed, Showing the Breakthrough Curve.

hypothetical concentration profile of adsorbing components within such a bed (top of figure) and the associated breakthrough curve (bottom).

Typically, breakthrough is characterized by an *s*-shaped line that curves about a center point, with equal areas above and below the curve at this point, as approximated in Figure 6.4. This is the mean breakthrough point, characterized by a theoretical step change in the effluent concentration assuming no mass transfer resistance. That is, the center point is determined solely by the equilibrium adsorption capacity [96]. This is commonly referred to as t_{50}, the mean breakthrough time. It can also be denoted by $BVTB_{50}$ as indicated in Figure 6.4 (and as we discuss below). The spreading of the curve about this point is determined by finite adsorption/desorption kinetics and axial dispersion.

A useful way to think about the capacity of these processes is the concept of bed volumes to breakthrough (BVTB), a popular concept in the ion exchange industry. This concept refers to the number of empty bed volumes that can pass through the bed before breakthrough begins. For a constant feed rate, it is a normalized process time that is independent of the actual bed size, and so data obtained using a small bed may be expressed in a form that applies directly to larger beds as well. The BVTB is given by the material balance equating the amount of solute in the feed with the amount that can be adsorbed, as follows:

$$BVTB\left(\frac{\pi D^2}{4}L\right)\rho_{liquid}\Delta x_i = \Delta q_i\left(\frac{\pi D^2}{4}L\right)\rho_{bed} \tag{6.71}$$

where D = bed diameter, ft

$\qquad L$ = bed length, ft

$\qquad \rho_{bed}$ = adsorbent bulk density, kg per m^3 of bed

$\qquad \rho_{liquid}$ = density of the feed solution, kg/ m^3

$\qquad \Delta q_i$ = effective adsorption working capacity, wt. adsorbed per wt. of adsorbent

Note that eq. (6.71) does not include the volume of liquid needed to fill a dry bed at start-up, but this amount normally is negligible. So, the breakthrough for an idealized, zero-resistance, plug flow adsorber is determined solely by the equilibrium adsorption capacity, given by:

$$BVTB_{50} = \frac{\Delta q_i}{\Delta x_i}\frac{\rho_{bed}}{\rho_{liquid}} \tag{6.72}$$

Popular methods used to analyze and model breakthrough curves include:
- Length of unused bed (LUB) models
- Statistical models
- Kinetic models
- Phenomenological models

The LUB method assumes that the concentration profile within an adsorber bed soon establishes a steady shape and length. That is, the length of the mass transfer zone (MTZ) is constant as it moves down the adsorber bed. This seems to be observed with many adsorption processes. The challenge is to calculate the appropriate MTZ length. Normally this is measured in miniplant tests and confirmed with operating data from full-scale adsorbers. The LUB concept has been attributed to J. J. Collins (from 1967) [61]. Although useful, the LUB method has generally been displaced by phenomenological models. Details of the LUB method are given elsewhere [38, 96].

Statistical models relate the number of theoretical stages achieved by the adsorber to a statistical measure of breakthrough [63–66]. The theoretical stage model with Gaussian normal distribution is essentially the same as the tanks-in-series model used to characterize residence time distribution for reactor design [62]. With this approach, the number of theoretical stages achieved by the adsorber is given by:

$$N_{stages} = \left(\frac{BVTB}{\sigma}\right)^2 \tag{6.73}$$

where σ is the standard deviation for spreading of the curve about the mean value $BVTB_{50}$. Increasingly narrow (sharp) breakthrough (and smaller σ) is characterized by a larger number of theoretical stages. The breakthrough curve can be calculated from the relationship:

$$\frac{BV}{BVTB} = 1 - \left[\frac{1}{\sqrt{N_{stages}}} \times F_{t,\,inv} \left(at\ x_{i,\ effluent} / x_{i,\,feed} \right) \right] \tag{6.74}$$

In this expression, $F_{t,inv}$ is the inverse t distribution function. It is available in many statistical packages and calculation/spreadsheet programs such as Microsoft Excel. At the Dow Chemical Company, this approach, among others, was introduced and led by Dr. Lanny Robbins, primarily for analyzing the performance of commercial-scale adsorbers. This model differs from LUB and phenomenological models in that it does not predict a constant MTZ length. Instead, the MTZ zone is predicted to sharpen as the adsorber length increases (adding more stages for mass transfer), approaching uniform plug flow. Nevertheless, the statistical framework provides a useful way of characterizing adsorber performance in terms of the number of theoretical stages that are achieved, especially when rating the performance of an existing adsorber. The performance of adsorbers in commercial operation can then easily be rated and compared. Once a full-scale unit is in operation, the statistical model provides a way of characterizing the operation and specifying similar adsorbers.

Kinetic models represent an alternative approach to adsorber design that is similar to an analysis of reaction kinetics. This approach is certainly a valid one, but it has generally been displaced in favor of phenomenological models (described below). For example, the Thomas kinetic model assumes adsorber performance is determined by Langmuir isotherm behavior and simplified expressions for the solute uptake rate and desorption rate, without explicit terms for axial dispersion effects. Breakthrough data typically are fit by adjusting a linear adsorption rate constant. Neglecting desorption, an approximate solution to the Thomas model is given by:

$$\frac{C_A}{C_{Ain}} = \frac{\exp[k_a C_{Ain}\theta]}{\exp\left[\frac{k_a \rho_b q_{ref}}{u_s} z\right] + \exp[k_a C_{Ain}\theta]} \tag{6.75}$$

where C_A = concentration of solute A
$\quad$ (CA) = in concentration of solute A at the inlet to the adsorber bed
$\quad$ k_a = linear adsorption rate constant
$\quad$ θ = corrected time (time minus time for liquid to pass through empty bed volume)
$\quad$ ρ_b = bulk density
$\quad$ q_{ref} = adsorption capacity at $(C_A)_{in}$
$\quad$ z = bed length
$\quad$ u_s = superficial velocity (face velocity)

For more detailed discussion, see H. Patel [67] and A. Ghribi and M. Chlendi [68].

Today, it seems that the most popular approach to modeling breakthrough involves the application of phenomenological models. This involves solving systems of differential equations representing the fundamental fluid dynamics and mass transfer phenomena. These are mechanistic models developed starting with the continuity equations of fluid flow (mass and energy conservation equations) plus additional empirical terms for mass transfer driving force and mass transfer resistance. In principle, the complete system of equations will include terms for bulk flow (convection and advection), solute accumulation, and non-linear diffusion of solute within the adsorbent pore structure including separate terms for micropore diffusion, film diffusion, and surface monolayer diffusion. Additional terms are added to represent backmixing (eddy diffusion and axial dispersion). The full-blown system of equations is difficult to parameterize and solve, so in many cases approximate models are developed by removing terms that are judged to have little effect, or by combining terms using lumped parameters – focusing only on the major phenomena that control performance.

One of the most popular of the simplified models is the Linear Driving Force (LDF) approximation. This approach involves modeling adsorption and mass transfer driving forces in terms of linear expressions, by specifying a constant equilibrium K value (or Henry's law constant) and a constant mass transfer coefficient. The linear driving force model has proven to be a good general-purpose model for adsorber design, one which often provides a good fit to performance data without a need to determine a large number of model parameter values [69]. A linear isotherm is assumed and all mass transfer resistance effects including pore diffusion and axial dispersion are lumped into a single mass transfer coefficient k_i. For example, the mass transfer equations to be solved may be written as:

$$\varepsilon_b \frac{\partial C_i}{\partial t} + (1 - \varepsilon_b) \frac{\partial q_i}{\partial t} + v(t) \frac{\partial C_i}{\partial z} = 0 \tag{6.76}$$

$$(1 - \varepsilon_b) \frac{\partial q_i}{\partial t} = k_i \left(C_i - C_i^{eq} \right) \tag{6.77}$$

$$q_i = K_i C_i^{eq} \tag{6.78}$$

where z = position along the length of the adsorber
$\quad$ t = elapsed time
$\quad$ ε_b = void fraction (fraction of adsorber volume occupied by empty space between particles)
$\quad$ q_i = adsorption capacity for solute i
$\quad$ $v(t)$ = bulk fluid velocity
$\quad$ k_i = overall mass transfer coefficient for transfer of solute i form liquid to adsorbent
$\quad$ K_i = partition ratio for solute i distributed between fluid and adsorbent
$\quad$ C_i = concentration of component i in the fluid phase
$\quad$ C_i^{eq} = equilibrium concentration of component i in the fluid phase

Somewhat more complex equations also may be used. For example, see ref. [70]. The LDF model is available for use in free web-based software called FAST [71]. Similar software tools can also be found by searching the internet.

6.7.4 Crystallization rates: nucleation, growth, and the population balance

Solute concentration in excess of the equilibrium or saturation amount (called supersaturation) is the deviation-from-equilibrium driving force for crystallization. The development and design of a crystallization process, whether batch or continuous, involves manipulation of process variables to generate supersaturation in ways that allow control over the relative rates of crystal nucleation and growth. Perhaps the most common way to generate supersaturation is to cool a saturated solution to develop several degrees of sub-cooling. For practical purposes, this ΔT often serves as a convenient way to represent the supersaturation level. Supersaturation may also be defined in terms of solution activities [72], but this is not common practice for crystallizer design, especially for commercial-scale operation.

In general, supersaturation may be generated in several ways:
- Cooling the mother liquor solution or melt below the saturation temperature or freezing point
- Evaporation of solvent from a saturated solution
- Addition of anti-solvent to decrease the solubility of solute in solution
- Addition of a reagent that reacts with the feed solute to form a product with lower solubility
- Combinations of the above, as in processes involving evaporation with cooling (evaporative cooling) and cooling followed by anti-solvent addition

For discussion of various crystallization processes and the types of crystallizers designed to generate and control supersaturation in various ways, see Chapter 11: Crystallization from Solution and from the Melt: Taking Advantage of Eutectic Behavior.

Crystal Nucleation and Growth Rate. Primary nucleation refers to the spontaneous birth of crystal nuclei in the absence of other nuclei or crystals of the same kind. It can be either homogeneous (in a clear solution without influence from other particles or solid surfaces of any kind) or heterogeneous (influenced by solid surfaces such as dust or dirt particles or container walls). Secondary nucleation refers to nucleation in the presence of other crystals of the same kind and normally also includes influences of other particles and solid surfaces plus the influence of fluid shear caused by a mixing impeller. So, the term *primary* refers to the very first nuclei to appear out of a clear solution, and the term *secondary* refers to subsequent nuclei appearing in a crystal suspension or layer of existing crystals. Primary nucleation is also called spontaneous nucleation. Similar concepts are used to describe nucleation of gas bubbles within boiling liquids [73]. Crystallizer design and operation normally deals with secondary nucleation, but heterogeneous primary nucleation is sometimes employed to start-up certain kinds of crystallizers such as an unseeded batch crystallizer or a static melt crystallizer.

Primary nucleation is a stochastic or statistical phenomenon, meaning that the timing of a nucleation event is not precisely determined by the process operating conditions. Rather, in repeated experiments with the same system under identical conditions, nucleation is observed to occur over a range of times with a distribution of probabilities. As a result, primary nucleation is difficult to predict. Nucleation is also highly sensitive to the supersaturation level, which makes it difficult to control. For example, in trying to induce primary nucleation in a clear liquid, it is often observed that as the supersaturation level is increased (by sub-cooling the solution more and more, for example), at first nothing happens. Then, at some level of sub-cooling (which is variable and not precisely predictable), the solution suddenly becomes filled with tiny crystals. It is as if a switch has been flipped or a supersaturation *rubber band* has snapped. This unpredictable, switch-like behavior is often observed for primary nucleation. Secondary nucleation normally is less sensitive, because existing nuclei or larger crystals can catalyze formation of secondary nuclei, reducing the supersaturation level at which significant nucleation takes place, making the process somewhat easier to control. In either case, however, nucleation remains sensitive to the level of supersaturation.

Due to this sensitivity to supersaturation, the rate of nucleation (primary or secondary) is often described in terms of an empirically derived power-law expression, such that

$$B \propto \sigma^n, \ 1 \leq n \leq 3 \tag{6.79}$$

where B = nucleation or birth rate, crystal nuclei per unit time per unit volume
$\quad \sigma$ = a measure of the supersaturation level
$\quad n$ = a numerical constant

As discussed above, primary nucleation tends to be more sensitive than secondary, so the exponent for primary nucleation normally will be significantly larger in value. The crystal growth rate normally is less sensitive than nucleation of either kind, and a typical empirical expression is

$$G \propto \sigma^m, \ 1 \leq m \leq n \tag{6.80}$$

where G = growth rate, crystal length per unit time per unit volume
$\quad \sigma$ = a measure of the supersaturation level
$\quad m$ = a numerical constant

The details of such expressions are discussed elsewhere [28, 72, 74, 75, 97]. Normally, it is expected that the exponent m in eq. (6.80) will be significantly smaller in magnitude than the exponent n in eq. (6.79). But in any case, it is very difficult to generate reliable kinetic data of this kind at crystallizer operating conditions. Instead, the sensitivity of the system to supersaturation level is characterized for design purposes by measuring the width of the metastable zone in laboratory experiments; that is, by determining the level or region of supersaturation where growth of existing crystals dominates nucleation of new ones, as discussed in the next section. Both nucleation and growth are affected by mixing intensity and other crystallizer operating variables, so the metastable zone must be measured under conditions that mimic crystallizer conditions. Secondary nucleation will always be occurring to some extent, but the idea is to find conditions that minimize nucleation and promote growth (when the goal is to grow reasonably large crystals).

Metastable Zone. The metastable zone is the region of process operating conditions at which a solution is supersaturated but the level of supersaturation is too low for significant nucleation to occur, so a solution remains a clear liquid phase for an extended period of time [76]. In this zone, if seed crystals of sufficient size are introduced, the crystals grow at a rate that increases with increasing supersaturation. Also within this region, smaller, metastable crystals may dissolve and contribute solute for the growth of larger crystals – a phenomenon called Ostwald ripening. (A similar phenomenon is observed with colloidal liquid-liquid dispersions with relevance to the formation and persistence of emulsions in liquid-liquid extraction [77].) The concepts of Ostwald ripening and the metastable zone were introduced at the beginning of the twentieth century by Friedrich Wilhelm Ostwald [78] and by Henry Alexander Miers and Florence Isaacs [79–81]. A plot of the solubility curve showing zones of metastable growth and the boundary where nucleation begins to dominate growth is known as an Ostwald-Miers diagram (illustrated in Figure 6.5). Ostwald is also known for residue curve maps used in analyzing ternary distillations (see Section 8.1: Heterogeneous Azeotropic Distillation and Steam Distillation).

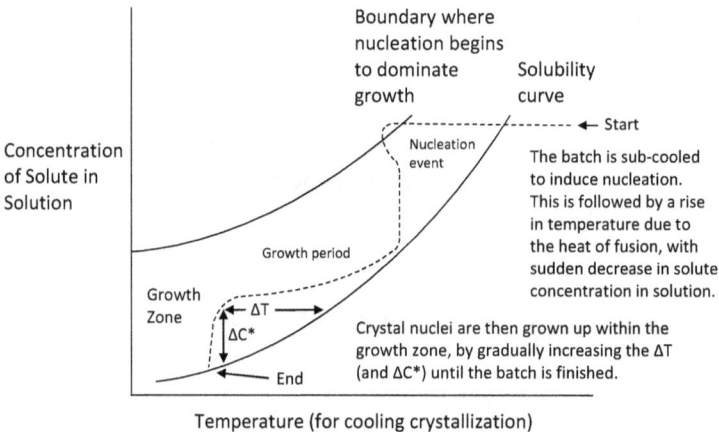

Figure 6.5: Conceptual Ostwald-Miers Diagram with Regions of Saturation, Supersaturation, and Undersaturation. The figure illustrates the path of a hypothetical batch cooling crystallization, showing an initial nucleation event with rise in temperature due to the heat of fusion and sudden decrease in solute concentration as supersaturation is relieved. This is followed by a crystal growth period involving gradual increase in ΔT (a parabolic cooling profile). This gradually increases the supersaturation, ΔC^*.

An Ostwald-Miers diagram shows regions of saturation (solubility curve), supersaturation (labile region including zones for crystal growth, secondary nucleation, and primary nucleation), and undersaturation (no-growth region or crystal dissolution region). The crystal growth zone is the portion of the labile region where growth dominates nucleation, although in most cases nucleation is always occurring to some extent. The classic Ostwalt-Miers diagram plots these regions in terms of liquid composition and temperature, depicting both the ΔT away from the solubility curve and its relationship to the supersaturation level ΔC^* that results (Figure 6.5). This clearly shows why ΔT can be used as a convenient way to represent or infer the supersaturation level. For other types of crystallizations involving different ways of generating supersaturation, other process variables may be used. Instead of ΔT, alternative diagrams might indicate the amount of anti-solvent addition (for a drowning-out crystallization) or the degrees of boiling point elevation (for an evaporative crystallization).

In designing a crystallization process to grow large, well-formed crystals, the goal is to operate at a supersaturation level within the growth zone that avoids excessive nucleation while also allowing for reasonably fast growth for good productivity; that is, by avoiding operation unnecessarily close to the solubility curve. For an unseeded batch process with in situ generation of nuclei, the process will need to start at conditions that favor nucleation. These nuclei may then be grown by operating within the growth zone to finish the batch. Figure 6.5 shows the potential path of a hypothetical unseeded batch cooling crystallization process as it navigates the various regions. The addition of seed crystals is a way to avoid the need for the initial nucleation event, allowing the process to always operate within the crystal growth zone (or at least with less of an excursion into the nucleation zone at the beginning of the process). For a continuous process, crystals are always present and secondary nucleation is always occurring to some extent, but a typical continuous process is designed to operate within the growth zone where growth dominates. We discuss this in more detail in Chapter 11: Crystallization from Solution and from the Melt: Taking Advantage of Eutectic Behavior.

Population Balance. A crystallization process will yield a range of particle sizes. The distribution of sizes can be narrow or wide or it can exhibit multiple peaks and significant shoulder effects. The population balance is the term given to calculation of the particle size distribution by accounting for the birth and death of solid particles during crystallization, over a wide range of particle sizes. The term *particle* is used here because groups of crystals (agglomerates) are treated the same as single crystals (all are particles). Note that a population balance is not a true *balance* in the sense of a mass or energy balance; that is, the total population numbers are not held constant or conserved. In principle, a population balance accounts for all simultaneously occurring rate phenomena that affect the numbers of particles of different sizes. These include the nucleation or birth of new small crystals and the growth of small crystals to larger ones. Other potential rate phenomena, which may or may not be significant for a given application include crystal agglomeration (birth of large particles plus the associated death of smaller particles that combined to form the larger ones), crystal breakage (death of large particles plus the associated birth of new smaller particle pieces), and crystal annealing or Ostwald ripening (dissolution or disappearance of very small crystals and the associated growth of larger ones). The latter may be considered a special kind of crystal growth mechanism. Another possibility involves formation of fouling deposits within the crystallizer, which does not affect crystal particle size directly, but it may do so indirectly by consuming solute that would otherwise contribute to crystal growth. All of these simultaneous rate phenomena determine the overall number density function for the process; that is, the number of particles per unit volume for successive size ranges – for successive bins in which the different sizes fit and are accounted for. And all of this determines a key outcome of the crystallization process – the final particle size distribution.

A population balance calculation does not model the purity of the resulting crystals. However, purity and the particle size distribution are related because both are affected by the same rate phenomena. Crystal nucleation and growth occur simultaneously and are interdependent because they compete for solute from the same limited supply (the feed solution or *mother liquor*). Once a crystal has formed, its properties up to that point are not easily reversed, so the results of the various competing rate phenomena are locked into the crystals as they form and grow. Rapid nucleation may produce defects involving inclusion of trapped mother liquor (solvent and impurities) within the crystal nuclei, which degrade crystal purity. Rapid nucleation also tends to produce large numbers of nuclei that cannot grow to be very large because there are so many crystals competing for the same limited supply of solute. After nucleation, if the growth rate is too fast, many additional defects may be formed and locked into the crystal lattice, also resulting in poor crystal purity (but not necessarily small crystal size). This dependence of both size and purity on rates of change and the path of the process will be discussed in more detail in Chapter 11.

Figure 6.6 is a classic diagram that illustrates how supersaturation generation and mixing affect the various rate phenomena that consume solute from the same limited supply – the phenomena that drive the population balance. These concepts were pioneered by Randolph and Larson [82, 83] and further developed by others [84–86]. A population balance applies not only to crystallization processes of interest here, but also to other particulate processes involving birth and death phenomena, such as liquid droplet dispersions.

For crystallization, the overall population balance, written for a single dimension of crystal growth (L), is given by:

$$V\frac{\partial n}{\partial t} + V\frac{\partial (Gn)}{\partial L} = V(B - D) + \varphi_{in} + \varphi_{out} \qquad (6.81)$$

where n = number density function, volume basis (for a range of crystal sizes)

L = length of a particle or a key crystal dimension

t = time

V = volume of crystallizer

G = growth rate per unit volume of crystallizer (which may be a function of crystal size)

B = birth rate for a specific crystal size range, per unit volume, due to nucleation (at small sizes) and agglomeration (at large sizes)

D = crystal death rate for a specific crystal size range, per unit volume, due to agglomeration (at small sizes) or breakage (at large sizes)

φ_{in} = inlet stream number density per unit time, for continuous operation. Equal to zero for batch operation

φ_{out} = outlet stream number density per unit time, for continuous operation. Equal to zero for batch operation

The measure chosen to represent crystal length will vary depending on the specific application. It is chosen to conveniently represent particle size, and one may choose the largest dimension of a characteristic crystal shape (such as the length of a needle), the effective diameter of a roughly spherical or cubical particle, or the apparent size obtained by using a particle sizing instrument such as a Coulter counter, which tends to size particles according to particle volume.

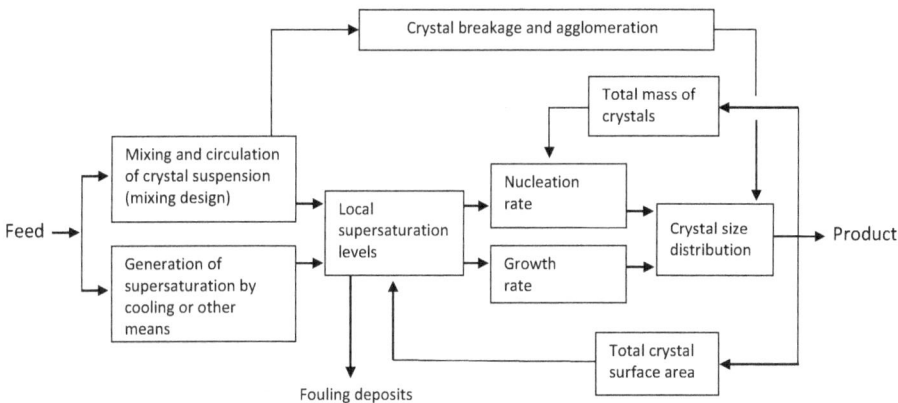

Figure 6.6: Crystallization Population Balance Concept Showing Interrelated, Simultaneous Phenomena. Background refs. [74, 84, 87, 97].

Equation (6.81) represents a set of first-order, non-linear partial differential equations with respect to crystal length and time. The equations are also functions of supersaturation, because the rate phenomena depend on the level of supersaturation, which may be kept constant or varied by manipulating the process operating conditions. Equation (6.81) also assumes uniform distribution of solute throughout the crystallizer vessel (ideal, well-mixed). This is the Mixed Suspension Mixed Product Removal (MSMPR) ideal crystallizer, analogous to the CSTR in reaction engineering (for continuous operation), and the Mixed Suspension ideal crystallizer (for batch operation). The boundary condition is the crystal size distribution at time zero (for a batch process). Given suitable expressions for the various rate phenomena (normally power-law expressions determined experimentally as discussed above) and specifying the supersaturation level, the resulting set of equations may be solved numerically. The population balance can also be evaluated for non-ideal mixing using computational fluid dynamics software [88].

The application of population balance modeling to directly design a crystallization process to achieve a desired crystal size distribution is still a research-intensive activity and not a common design practice. This is due in large part to the time and expertise required to experimentally determine accurate nucleation and growth kinetics (and agglomeration and/or breakage rate expressions if needed) for input to a population balance model, and because the resulting kinetic expressions sometimes do not scale well as the operating conditions necessarily change on scale-up. In particular, the mixing environment and solute concentration (supersaturation) gradients can change significantly on scale-up. The effort put into developing a population balance model also varies depending on the importance of the given application to the business and the difficulty in obtaining desired results; that is, whether the process proves fairly easy to develop and run, or whether it proves to be difficult to run and achieve the required results. And as mentioned above, population balance calculations do not provide information about crystal purity, which has to be determined experimentally. Nevertheless, having a basic understanding of the interconnection between nucleation and growth rates and the impact this can have on the crystal size distribution and the purity of the crystal mass can help guide practical design efforts (as discussed in Chapter 11). The well-known practice of using a parabolic temperature profile in batch cooling crystallization is fundamentally a result of the phenomena described by the population balance.

In practice, the design of a crystallization process relies heavily on experimental methods, and many tools have been developed to characterize a given crystallization application for design purposes. These include real-time (in situ) sensors or instruments generally called Process Analytical Technology (PAT) devices, as well as apparatus for rapid high-throughput experiments. Some of the more common PAT devices are sensors for measuring temperature and turbidity, the Focused Beam Reflectance Measurement (FBRM) device, which measures crystal chord length to characterize crystal size. Also included are various spectroscopy instruments (mainly ATR-FTIR and Raman), and Particle Vision and Measurement (PVM) involving real-time video/imaging of crystals with software for image analysis. Many of these tools may be used to measure solubility curves and characterize metastable zone width. They may also be used to characterize properties of the growing crystal mass as a function of operating conditions during crystallization. Operating trends and useful operating conditions that give good results are identified using directed or statistical process experiments in laboratory or mini-plant apparatus fitted with a number of PAT instruments. This experimental approach is standard practice in the pharmaceutical industry for batch crystallization (see Section 11.2: Types of Processes and Equipment, Section 11.3: Process Analytical Technology, and Section 11.4: Batch Crystallization from Solution).

6.7.5 Membrane permeance and separation factor

Most commercial membrane-based separations are rate-based separations. See Section 6.1.2: Drivers for Membrane-Based Separations. These processes do not operate at phase equilibrium. The membrane serves to selectively enrich one or more of the feed components above what would be achieved by phase equilibrium in the absence of the membrane. The solute permeation rate or flux is driven primarily by pressure drop but enhanced by thermodynamic phenomena occurring within the membrane structure. Both porous and non-porous (ultra-thin film) membranes are used depending on the application. The mechanisms for solute transport through a semipermeable membrane are described in the literature in terms of a solution-diffusion model, a hydrodynamic model, or a molecular-sieving model, depending on the type of membrane and its application. In many cases, the actual mechanism likely is some combination of these phenomena. Detailed discussions of membrane types and mechanisms are given elsewhere [1–3].

<u>Liquid Separations</u>. Liquid separations involve selective transfer of feed components from a liquid mixture on the feed side of the membrane to another liquid on the permeate side. A general expression for the flux across the membrane may be written as:

$$j_i = \frac{\Delta P - \Delta \Pi_i}{(R_m + R_d)\mu z} \approx \frac{\Delta P - (\Delta c_i I_i)RT}{(R_m + R_d)\mu z} \qquad (6.82)$$

where j_i = volumetric flux, cm^3 liquid/cm^2 membrane · s
$\Delta \Pi_i$ = osmotic pressure difference (permeate side – feed side), atm
Δc_i = solute concentration difference (permeate side – feed side), moles per unit volume
ΔP = transmembrane pressure drop (feed side – permeate side), atm
I_i = van't Hoff constant for component i accounting for non-ideal osmotic pressure, dimensionless
R_m = lumped parameter representing mass transfer resistance to flow due to membrane material properties and pore structure, cm^3/(cm · s · atm · cP)
R_d = mass transfer resistance due to fouling of the membrane (increasing over time), cm^3/(cm · s · atm · cP)
μ = liquid viscosity, cP
z = membrane thickness, cm
R = Universal gas constant
T = absolute temperature, K

The osmotic pressure difference derives from a difference in the chemical potential of the permeating component. For reverse osmosis, this is the difference between the chemical potential of water in brine (on the feed side of the membrane) and pure water (on the permeate side). Water dissolved in brine is the lower energy state, so extra pressure is needed to transfer water out of the brine (overcoming the natural osmotic pressure). Analogous to a modified ideal gas law, osmotic pressure is approximated for dilute liquid solutions by the expression:

$$\Pi = \frac{n_i}{V} I_i RT = c_i I_i RT \tag{6.83}$$

$$\Delta\Pi_i = \Delta(c_i I_i) RT \tag{6.84}$$

where Π = osmotic pressure, atm

n_i = moles of component i in liquid solution

V = volume, m^3

c_i = moles per unit volume

The van't Hoff constant I_i accounts for the difference in chemical potentials due to non-ideal interactions of solute with other solutes in solution (such as the attractive interactions between water and salt ions in brine). A similar equation expressed in terms of molality (moles per kg) generally applies at higher concentrations and is known as the Morse equation [89, 90].

Equation (6.82) highlights the main factors contributing to mass transfer resistance: 1) osmotic pressure difference; 2) inherent properties of the membrane itself (material properties and porosity); 3) liquid viscosity; and 4) the thickness of the membrane-discriminating layer (not including the membrane support). Not included in this general equation is a potential phenomenon called concentration polarization that can also increase mass transfer resistance. This is due to the establishment of a co-solute concentration gradient (of salt ions in the case of reverse osmosis for desalination) within a boundary layer next to the membrane, due to selective transfer of solute across the membrane (water in the case of desalination). This added resistance increases the effective osmotic pressure difference that needs to be overcome. Concentration polarization tends to be established within minutes and should not be confused with fouling of the membrane, which develops over a longer time period [91].

The separation factor of a membrane-based liquid separation depends on the ability of the membrane to minimize permeation of unwanted co-solutes. There will always be some co-permeation or leakage, although it may be small. The actual selectivity of the process depends on a non-equilibrium separation factor determined by experiment.

$$\alpha_{i,j}\big|_{non-equilibrium} \equiv \frac{j_i}{j_j} \tag{6.85}$$

where j_i represents the flux of an individual component in the presence of all other components.

Note that the process commonly called "membrane-supported liquid-liquid extraction" differs from membrane-based liquid separation described above. In the case of extraction, the membrane serves only to physically stabilize a liquid-liquid meniscus existing within the pores of the membrane [19]. The membrane serves only to control the amount of interfacial area, and mass transfer occurs within the pores at the liquid-liquid interface. In that case, the membrane does not influence mass transfer, only the available interfacial area. In a true membrane-based liquid separation, the membrane serves to enhance the transfer of desired components between the two liquids; that is, the membrane alters or enhances mass transfer occurring within the membrane structure itself.

Gas Separations. A general equation for flux of gas across a membrane is given by:

$$j_i = \frac{\mathcal{P}_i\left(p_{i_0} - p_{i_l}\right)}{l} = \frac{\mathcal{P}_i^{gas}}{l}\left(P_{feed}\, y_{i_0} - P_{permeate}\, y_{i_l}\right) \tag{6.86}$$

where j_i = volumetric flux, cm^3 (STP) of component i/cm^2 polymer $\cdot$ s

$\mathcal{P}_i^{gas}$ = effective gas permeance. In practice, $\mathcal{P}_i^{gas}$ is a lumped transport parameter that is a function of the pressure drop across the membrane. It includes the effects of solute diffusivity, absorption, and adsorption within the membrane, and the overall resistance to flow across the membrane barrier.

p_{i_0} = partial pressure of component i on the feed side of the membrane

p_{i_l} = partial pressure of component i on the permeate side of the membrane

y_{i_0} = mole fraction of component i on the feed side of the membrane

y_{i_l} = pressure of component i on the permeate side of the membrane

P_{feed} = operating pressure, feed side, atm

$P_{permeate}$ = operating pressure, permeate side, atm

Equation (6.86) illustrates why membranes generally are not well suited to processing feeds that are dilute in the permeating species, because the flux decreases as feed-side concentration decreases.

The non-equilibrium or effective separation factor of a membrane-based gas separation process is given by relative flux:

$$\alpha_{ij}\big|_{non-equilibrium} = \frac{\mathcal{P}_i^{gas}\left(p_{i_0} - p_{i_l}\right)}{\mathcal{P}_j^{gas}\left(p_{j_0} - p_{j_l}\right)} = \frac{\mathcal{P}_i^{gas}\left(P_{feed}\, y_{i_0} - P_{permeate}\, y_{i_l}\right)}{\mathcal{P}_j^{gas}\left(P_{feed}\, y_{j_0} - P_{permeate}\, y_{j_l}\right)} \tag{6.87}$$

As discussed above, eq. (6.87) is derived by considering the effect of relative rates of mass transfer on selectivity, not by assuming equilibrium is achieved (because it is not). It is interesting to note that this non-equilibrium (rate-based) separation factor differs from the separation factor of the membrane material itself, which may be defined by:

$$a_{ij}|_{membrane\,material} = \frac{\mathcal{P}_i^{gas}}{\mathcal{P}_j^{gas}} \tag{6.88}$$

The material's inherent selectivity does not account for the relative partial pressure difference which contributes to the selectivity actually achieved by the process. Also, values of $\mathcal{P}_i^{gas}$ often are measured using pure gases, so they do not include any multi-component effects. This can give highly misleading results because it neglects any co-solute effects such as swelling of a membrane polymer material by absorption of co-solute. For this reason, it is always good practice to use representative feeds to ensure that any co-solute effects are taken into account.

Pervaporation. A membrane-based pervaporation process has characteristics of both liquid and gas-phase membrane separations. Pervaporation employs liquid on the feed side of the membrane and evaporation to vapor on the permeate side. Evaporation may occur to some extent within the pores of the membrane. A general equation for pervaporation flux can be written as:

$$J_i = \frac{\mathcal{P}_i^{gas}}{l}\left[\left(\gamma_i x_i^{liquid} p_i^{sat}\right) - \left(y_i^{permeate} P_{permeate}\right)\right] \tag{6.89}$$

Note the similarity between eqs. (6.86) and (6.89). Normally, the permeate side of the membrane is operated under vacuum. Pervaporation flux is a strong function of temperature because p_i^{sat} increases with temperature. So a pervaporation process may be operated near the maximum temperature the membrane can reliably tolerate to maximize flux. For polymeric membranes, this may be between 80 and 100 °C. Also, the heat of vaporization needed for evaporation commonly is supplied by heating the liquid that is fed to each membrane module. However, keep in mind that operating at the maximum membrane temperature may also reduce the membrane's selectivity due to swelling or other changes in membrane properties. Detailed discussions are available elsewhere [92, 93].

As discussed above, the effective separation factor for a pervaporation process is given by the relative flux across the membrane (for a single membrane stage):

$$a_{ij}|_{non-equilibrium} = \frac{\mathcal{P}_i^{pervap}\left[\left(\gamma_i x_i^{liquid} p_i^{sat}\right) - \left(y_i^{permeate} P_{permeate}\right)\right]}{\mathcal{P}_j^{pervap}\left[\left(\gamma_j x_j^{liquid} p_j^{sat}\right) - \left(y_j^{permeate} P_{permeate}\right)\right]} \tag{6.90}$$

Equation (6.90) shows how the insertion of a selective membrane between phases can boost enrichment of the permeating component from liquid to vapor compared to what can be achieved by a single theoretical stage of liquid-vapor distillation (where $\alpha_{ij} = K_i/K_j = \gamma_i p_i^{sat}/\gamma_j p_j^{sat}$). This far exceeds the thermodynamic separation factor obtainable by distillation at concentrations near the azeotropic composition.

References

[1] A. F. Ismail and T. Matsuura, *Membrane Separation Processes: Theories, Processes, and Solutions*, Elsevier, Amsterdam, 2021.

[2] A. K. Pabby, S. S. Rizvi and A. M. S. Requena Eds., *Handbook of Membrane Separations: Chemical, Pharmaceutical, Food, and Biotechnological Applications*, 2nd ed., CRC Press, Boca Raton, Florida, 2015.

[3] W. S. Winston Ho and K. K. Sirkar Eds., *Membrane Handbook*, Springer, Heidelberg, Germany, 1992.

[4] G. Blandin, F. Ferrari, G. Lesage, P. Le-Clech, M. Héran and X. Martinez-Lladó, "Forward Osmosis as Concentration Process: Review of Opportunities and Challenges," *Membranes*, vol. 10, no. 10, article no. 284, 2020.

[5] W. Suwaileh, N. Pathak, H. Shon and N. Hilal, "Forward Osmosis Membranes and Processes: A Comprehensive Review of Research Trends and Future Outlook," *Desalination*, vol. 485, no. 114455, ISSN 0011-9164, 2020.

[6] R. D. Noble and C. A. Koval, "Chapter 17, Review of Facilitated Transport Membranes," in *Materials Science of Membranes for Gas and Vapor Separation*, Y. Yampolskii, I. Pinnau, and B. Freeman, Eds., Wiley, Hoboken, New Jersey, 2006, pp. 411–435.

[7] S. J. Doong, "Section 7.3.2, Facilitated Transport Membranes, in Chapter 7, Membranes, Adsorbent Materials, and Solvent-Based Materials for Syngas and Hydrogen Separation," in *Functional Materials for Sustainable Energy Applications*, J. A. Kilner, S. J. Skinner, S. J. C. Irvine and P. P. Edwards, Eds., Woodhead Publishing, Sawston, Cambridgeshire, England, 2012, pp. 179–216.

[8] A. Hussain and M.-B. Hägg, "A Feasibility Study of CO_2 Capture from Flue Gas by A Facilitated Transport Membrane," *Journal of Membrane Science*, vol. 359, no. 1-2, pp. 140–148, 2010.

[9] A. A. Kiss and O. M. K. Readi, "An Industrial Perspective on Membrane Distillation Processes", *Journal of Chemical Technology and Biotechnology*, vol. 93, no. 8, pp. 2047–2055, 2018.

[10] B. A. Cinelli, D. M. G. Freire and F. A. Kronemberger, "Membrane Distillation and Pervaporation for Ethanol Removal: Are We Comparing in the Right Way?," *Separation Science and Technology*, vol. 54, no. 1, pp. 110–127, 2019.

[11] "Point-of-use Reverse Osmosis Systems", United States Environmental Protection Agency, [Online]. Available: https://www.epa.gov/watersense/point-use-reverse-osmosis-systems. [Accessed 8 January 2023].

[12] "Membrane Modules for Nitrogen and Oxygen Generator Systems (Catalogue: K3.1.333a)", p. 4, Parker Hannifin Corporation, 2009. [Online]. Available: https://www.parker.com/Literature/Hiross%20Zander%20Division/PDF%20Files/PIS/K3.1.333_Technology_overview_membrane_modules_for_nitrogen_and_oxygen_systems_EN.pdf. [Accessed 8 January 2023].

[13] G. Lebon, D. Jou and J. Casas-Vázquez, *Understanding Non-equilibrium Thermodynamics: Foundations, Applications, Frontiers*, Springer, Heidelberg, Germany, 2008.

[14] L. Wang and J. Min, "Modeling and Analyses of Membrane Osmotic Distillation Using Non-equilibrium Thermodynamics," *Journal of Membrane Science*, vol. 378, no. 1-2, pp. 462–470, 2011.

[15] Y. Demirel and S. I. Sandler, "Non-equilibrium Thermodynamics in Engineering and Science," *The Journal of Physical Chemistry B*, vol. 108, no. 1, pp. 31–43, 2004.

[16] P. L. Jackson and R. A. Wilsak, "Thermodynamic Consistency Tests Based on the Gibbs-Duhem Equation Applied to Isothermal, Binary Vapor-Liquid Equilibrium Data: Data Evaluation and Model Testing," *Fluid Phase Equilibria*, vol. 103, no. 2, pp. 155–197, 1995.

[17] J. D. Raal and A. L. Mühlbauer, *Phase Equilibria: Measurement and Computation*, Taylor & Francis, Abingdon-on-Thames, Oxfordshire, England, 1998.

[18] H. Takiyama, H. Suzuki, H. Uchida and M. Matsuoka, "Determination of Solid–Liquid Phase Equilibria by Using Measured DSC Curves," *Fluid Phase Equilibria*, Vol. 194–197, pp. 1107–1117, 2002.

[19] T. C. Frank, B. S. Holden and A. F. Seibert, "Section 15: Liquid-Liquid Extraction and Other Liquid-Liquid Operations and Equipment," in *Perry's Chemical Engineers' Handbook*, 9th ed., D. W. Green and M. Z. Southard, Eds., McGraw-Hill, New York, 2019.

[20] F. H. Yin, C. G. Sun, A. Afacan, K. Nandakumar and K. T. Chuang, "CFD Modeling of Mass Transfer Processes in Randomly Packed Distillation Columns," *Industrial & Engineering Chemistry Research*, vol. 39, no. 5, pp. 1369–1380, 2000.

[21] M. Chen, J. Wang, S. Li, T. Wang, X. Huang and H. Hao, "CFD–PBE Model and Simulation of Continuous Anti-solvent Crystallization in an Impinging Jet Crystallizer with a Multiorifice at Different Positions," *Industrial & Engineering Chemistry Research*, vol. 60, no. 31, pp. 11802–11811, 2021.

[22] N. Kharoua, L. Khezzar and H. Saadawi, "CFD Modelling of A Horizontal Three-Phase Separator: A Population Balance Approach," *American Journal of Fluid Dynamics*, vol. 3, no. 4, pp. 101–118, 2013.

[23] F. Augier, C. Laroche and E. Brehon, "Application of Computational Fluid Dynamics to Fixed Bed Adsorption Calculations: Effect of Hydrodynamics at Laboratory and Industrial Scale," *Separation and Purification Technology*, vol. 63, pp. 466–474, 2008.

[24] R. Löhner, *Applied CFD Techniques: An Introduction Based on Finite Element Methods*, Wiley, Hoboken, New Jersey, 2001.

[25] W. B. Krantz, *Scaling Analysis in Modeling Transport and Reaction Processes: A Systematic Approach to Model Building and the Art of Approximation*, Wiley, Hoboken, New Jersey, 2006.

[26] S. M. Kresta, A. W. Etchells III, D. S. Dickey and V. A. Atiemo-Obeng, *Advances in Industrial Mixing: A Companion to the Handbook of Industrial Mixing*, Wiley, Hoboken, New Jersey, 2015.

[27] E. L. Paul, V. A. Atiemo-Obeng and S. M. Kresta, *Handbook of Industrial Mixing: Science and Practice*, Wiley, Hoboken, New Jersey, 2004.

[28] D. A. Green, "Chapter 10 – Crystallizer Mixing," in *Handbook of Industrial Crystallization*, 3rd ed., A. S. Myerson, D. Erdemir and A. Y. Le, Eds., Cambridge University Press, Cambridge, England, 2019.

[29] P. Urben Editor, *Bretherick's Handbook of Reactive Chemical Hazards*, 7th ed., Two Volume Set, Academic Press, Cambridge, Massachusetts, 2006.

[30] B. S. Holden, P. H. Au-Yeung and T. Kajdan, "Watch Out for Trapped Components in Towers," *Chemical Processing*, May 2012.

[31] M. Souders and G. G. Brown, "IV. – Fundamental Design of Absorbing and Stripping Columns for Complex Vapors," *Industrial and Engineering Chemistry*, vol. 24, no. 5, pp. 519–522, 1932.

[32] A. L. Kohl, "Chapter 6, Absorption and Stripping," in *Handbook of Separation Process Technology*, *Ronald W. Rousseau, Ed.*, Wiley-Interscience, New York, 1987, pp. 340–404.

[33] H. Z. Kister, *Distillation Design*, McGraw-Hill, New York, 1992, p. 365.

[34] E. R. Gilliland, "Eger Vaughan Murphree 1898-1962, A Biographical Memoir," in *Biographical Memoirs*, National Academy of Sciences (USA), Washington, D.C, 1962, pp. 225–238.

[35] W. L. McCabe, J. C. Smith and P. Harriott, *Unit Operations of Chemical Engineering*, 7th ed., McGraw-Hill, New York, 2005.

[36] J. D. Seader and E. J. Henley, *Separation Process Principles*, 2nd ed., Wiley, Hoboken, New Jersey, 2006.

[37] H. E. O'Connell, "Plate Efficiency of Fractionating Columns and Absorbers," *Transaction AIChE*, vol. 42, pp. 741–755, 1946.

[38] P. C. Wankat, *Separation Process Engineering: Includes Mass Transfer Analysis*, 3rd ed., Prentice Hall, Upper Saddle River, New Jersey, 2012.

[39] M. Duss and R. Taylor, "Predict Distillation Tray Efficiency," *Chemical Engineering Progress*, pp. 24–30, July 2018.

[40] H. Chan, "Tray Efficiencies for Multicomponent Distillation Columns, PhD Dissertation," Univ. of Texas at Austin, Austin, Texas, 1983.

[41] M. Duss and R. Taylor, "A New Simplified Tray Efficiency Model with Improved Accuracy for Sieve Trays," *Chemical Engineering Research and Design*, vol. 146, pp. 71–77, 2019.

[42] J. A. Garcia, "Fundamental Model for the Prediction of Distillation Sieve Tray Efficiency, PhD Dissertation," Univ. of Texas, Austin, Texas, 1999.

[43] W. K. Lewis, "Rectification of Binary Mixtures," *Industrial and Engineering Chemistry*, vol. 28, no. 4, pp. 399–402, 1936.

[44] R. Taylor and M. Duss, "110th Anniversary: Column Efficiency: From Conception, through Complexity, to Simplicity," *Industrial & Engineering Chemistry Research*, vol. 58, pp. 16877–16893, 2019.

[45] Z. Xu, D. R. Summers, T. Cai, B. Holden, R. Olsson, L. Pless, M. Schultes, F. Seibert and S. X. Xu, *AIChE Equipment Testing Procedure – Trayed & Packed Columns: A Guide to Performance Evaluation*, 3rd ed, AIChE/Wiley Press, Hoboken, New Jersey, 2014.

[46] W. Nernst, "Theorie der Reaktionsgeschwindigkeit in Heterogenen Systemen," *Zeitschrift Für Physikalische Chemie*, vol. 47, no. 1, pp. 52–55, 1904.

[47] W. K. Lewis, "Laboratory and Plant: The Principles of Counter-Current Extraction," *Industrial and Engineering Chemistry*, vol. 8, no. 9, pp. 825–833, 1916.

[48] W. G. Whitman, "The Two-Film Theory of Gas Absorption: It Seems to Explain Satisfactorily the Well-Recognized Differences of Absorption Rate for Varying Concentrations," *Chemical and Metallurgical Engineering*, vol. 29, no. 4, pp. 146–148, 1923.

[49] W. K. Lewis and W. G. Whitman, "Principles of Gas Absorption," *Industrial and Engineering Chemistry*, vol. 16, no. 12, pp. 1215–1220, 1924.

[50] T. H. Chilton and A. P. Colburn, "Distillation and Absorption in Packed Columns," *Industrial and Engineering Chemistry*, vol. 27, no. 3, pp. 255–260, 1935.

[51] T. Baker, "Correspondence – Distillation and Absorption in Packed Towers," *Industrial and Engineering Chemistry*, vol. 27, no. 8, pp. 977, 1935.

[52] A. P. Colburn, "The Simplified Calculation of Diffusional Processes. General Consideration of Two-Film Resistances," *Transactions of the American Institute of Chemical Engineers*, vol. 35, pp. 211–236, 1939.

[53] E. L. Cussler, *Diffusion: Mass Transfer in Fluid Systems*, 3rd ed, Cambridge University Press,, 2009.

[54] L. A. Robbins, Interviewee, *Personal Communication with T. C. Frank at The Dow Chemical Company*. [Interview]. ca. 1990.

[55] P. C. Wankat and K. S. Knaebel, "Section 5B, Mass Transfer," in *Perry's Chemical Engineers' Handbook*, 9th ed., D. W. Green and A. Marylee Z., Eds., McGraw-Hill, New York, 2019, pp. 5–42 to 5–74.

[56] C. J. Geankoplis, A. Hersel and D. Lepek, *Transport Processes and Separation Process Principles*, 5th ed, Pearson, London, 2018.

[57] R. B. Bird, W. E. Stewart and E. N. Lightfoot, *Transport Phenomena*, 2nd ed, Wiley, Hoboken, New Jersey, 2006.

[58] R. Krishnamurthy and R. Taylor, "A Non-equilibrium Stage Model of Multicomponent Separation Processes," *AIChE Journal*, vol. 31, no. 3, pp. 449–456, 1985.

[59] R. Taylor and R. Krishna, *Multicomponent Mass Transfer*, Wiley, Hoboken, New Jersey, 1993.

[60] P. B. Umbanhowar, R. M. Lueptow and J. M. Ottino, "Modeling Segregation in Granular Flows," *Annual Review of Chemical and Biomolecular Engineering*, vol. 10, pp. 129–153, 2019.

[61] J. J. Collins, "The LUB/Equilibrium Section Concept for Fixed Bed Adsorption," in *Chemical Engineering Progress Symposium Series*, Vol. 63, AIChE, 1967, pp. 31–35.

[62] H. S. Fogler, *Elements of Chemical Reaction Engineering*, 5th ed, Pearson, London, 2016.

[63] F. C. Denizot and M. A. Delaage, "Statistical Theory of Chromatography: New Outlooks for Affinity Chromatography (Plate Theory/distribution Profile/moments/variance Peak Width/fast Reactions)," *The Proceedings of the National Academy of Sciences USA*, vol. 72, no. 12, pp. 4840–4843, Dec 1975.

[64] R. MacMullin and M. Weber, "The Theory of Short-circuiting in Continuous-flow Mixing Vessels in Series and Kinetics of Chemical Reactions in Such Systems," *Transactions of American Institute of Chemical Engineers*, vol. 31, no. 2, pp. 409–458, 1935.

[65] E. B. Nauman, "Residence Time Distributions," in *Handbook of Industrial Mixing: Science and Practice*, Wiley Interscience, Hoboken, New Jersey, 2004, pp. 1–17.

[66] J. C. Giddings and H. Eyring, "A Molecular Dynamic Theory of Chromatography," *Journal of Physical Chemistry*, vol. 59, no. 5, pp. 416–421, 1955.

[67] H. Patel, "Fixed-bed Column Adsorption Study: A Comprehensive Review," *Applied Water Science*, vol. 9, article no. 45, 2019.

[68] A. Ghribi and M. Chlendi, "Modeling of Fixed Bed Adsorption: Application to the Adsorption of an Organic Dye," *Asian Journal of Textile*, vol. 1, pp. 161–171, 2011.

[69] S. Sircar and J. Hufton, "Why Does the Linear Driving Force Model for Adsorption Kinetics Work?," *Adsorption*, vol. 6, pp. 137–147, 2000.

[70] S. D. Feist, Y. Hasabnis, B. W. Pynnonen and T. C. Frank, "SMB Chromatography Design Using Profile Advancement Factors, Miniplant Data, and Rate-Based Process Simulation," *AIChE Journal*, vol. 55, no. 11, pp. 2848–2860, 2009.

[71] S. Schimmelpfennig and A. Sperlich, "FAST Fixed-bed Adsorption Simulation Tool," [Online]. Available: http://www.fast-software.de/. [Accessed 8 Oct 2022].

[72] D. Erdemir, A. Y. Lee and A. S. Myerson, "Chapter 3, Crystal Nucleation," in *Handbook of Industrial Crystallization*, 3rd ed., A. S. Myerson, D. Erdemir and L. Alfred Y., Eds., Cambridge University Press, 2019, pp. 76–114.

[73] M. Blander and J. L. Katz, "Bubble Nucleation in Liquids," *AIChE Journal*, vol. 21, no. 5, pp. 833–848, 1975.

[74] J. C. G. Moyers and R. W. Rousseau, "Chapter 11, Crystallization Operations," in *Handbook of Separation Process Technology*, Wiley-Interscience, 1987.

[75] A. Y. Lee, D. Erdemir and A. S. Myerson, "Chapter 2, Crystals and Crystal Growth," in *Handbook of Industrial Crystallization*, 3rd ed., A. S. Myerson, D. Erdemir and L. Alfred Y., Eds., Cambridge University Press, Cambridge, England, 2019.

[76] T. L. Threlfall and S. J. Coles, "A Perspective on the Growth-Only Zone, the Secondary Nucleation Threshold and Crystal Size Distribution in Crystallization," *CrystEngComm*, vol. 18, no. 3, pp. 369–378, 2016.

[77] P. Taylor, "Ostwald Ripening in Emulsions," *Advances in Colloid and Interface Science*, vol. 75, no. 2, pp. 107–163, 1998.

[78] W. Ostwald, "Über Die Vermeintliche Isomerie Des Roten Und Gelben Quecksilberoxyds Und Die Oberflächenspannung Fester Körper (On the Supposed Isomerism of Red and Yellow Mercury Oxide and the Surface Tension of Solid Bodies)," *Zeitschrift Für Physikalische Chemie*, vol. 34, pp. 495, 1900.

[79] H. A. Miers and F. Isaac, "The Refractive Indices of Crystallizing Solutions with Especial Reference to the Passage from the Metastable to the Labile Condition," *Journal of the Chemical Society*, vol. 89, pp. 413–454, 1906.

[80] H. A. Miers and F. Isaac, "The Spontaneous Crystallization," *Proceedings of the Royal Society of London*, vol. A79, pp. 322–325, 1907.

[81] H. A. Miers, *Journal of the Institute of Metals*, vol. 37, p. 331, 1927.

[82] A. D. Randolph and M. A. Larson, "Transient and Steady State Size Distributions in Continuous Mixed Suspension Crystallizers," *AIChE Journal*, vol. 8, no. 5, pp. 639–645, 1962.

[83] A. D. Randolph and M. A. Larson, *Theory of Particulate Processes: Analysis and Techniques of Continuous Crystallization*, 2nd ed, Academic Press, Cambridge, Massachusetts, 1988.

[84] Å. C. Rasmuson, "Chapter 6, Crystallization Process Analysis by Population Balance Modeling," in *Handbook of Industrial Crystallization*, 3rd ed., A. S. Myerson, D. Erdemir and A. Y. Lee, Eds., Cambridge University Press, Cambridge, England, 2019, pp. 172–196.

[85] H. M. Hulbert and S. Katz, "Some Problems in Particle Technology. A Statistical Mechanical Formulation," *Chemical Engineering Science*, vol. 19, no. 8, pp. 555–574, 1964.

[86] D. Ramkrishna, *Population Balances: Theory and Applications to Particulate Systems in Engineering*, Academic Press, Cambridge, Massachusetts, 2000.

[87] S. J. Jančić and P. A. M. Grootscholten, *Industrial Crystallization*, Delft University Press, Delft, Netherlands, 1984.

[88] A. Mersmann, B. Braun and M. Löffelmann, "Prediction of Crystallization Coefficients of the Population Balance," *Chemical Engineering Science*, vol. 57, no. 20, pp. 4267–4275, 2002.

[89] G. N. Lewis, "The Osmotic Pressure of Concentrated Solutions and the Laws of the Perfect Solution," *Journal of the American Chemical Society*, vol. 30, no. 5, pp. 668–683, 1908.

[90] K. Janáček and K. Sigler, "Osmotic Pressure: Thermodynamic Basis and Units of Measurement," *Folia Microbiologica*, vol. 41, pp. 2–9, 1996.

[91] R. W. Field and J. J. Wu, "Permeate Flux in Ultrafiltration Processes – Understandings and Misunderstandings," *Membranes*, vol. 12, no. 2, 5 Feb 2022.

[92] H. K. Dave and K. Nath, "Effect of Temperature on Pervaporation Dehydration of Water-Acetic Acid Binary Mixture," *Journal of Scientific and Industrial Research*, vol. 76, pp. 217–222, 2017.

[93] H. Thiess, A. Schmidt and J. Strube, "Development of a Scale-up Tool for Pervaporation Processes," *Membranes*, vol. 8, no. 1, 15 Jan 2018.

[94] A. Kremser, Theoretical Analysis of Absorption Process, *National Petroleum News*, vol. 22, no. 21, pp. 43–49, 21 May 1930.

[95] M. J. Lockett, *Distillation Tray Fundamentals*, Cambridge University Press, Cambridge, England, 1986, p. 119.

[96] C. Tien, *Introduction to Adsorption: Basics, Analysis, and Applications*, Elsevier, Amsterdam, 2019.

[97] W. J. Genck, "Section 2.3, Crystallization from Solutions," in *Handbook of Separation Techniques for Chemical Engineers*, 3rd ed., P. A. Schweitzer, Eds., McGraw-Hill, 1997, pp. 2–141 to 2–172.

Bibliography

Chapter 5

G. N. Lewis and M. Randall, *Thermodynamics and the Free Energies of Chemical Substances*, McGraw-Hill, New York, 1923.

L. Pauling, *The Nature of the Chemical Bond and the Structure of Molecules and Crystals*, 3rd ed., Cornell University Press, Ithaca, New York, 1960.

L. B. Schreiber and C. A. Eckert, "Use of Infinite Dilution Activity Coefficients with Wilson's Equation," *Industrial & Engineering Chemistry Process*, vol. 10, no. 4, pp. 572–576, 1971.

L. H. Horsley, *Azeotropic Data – III, Advances in Chemistry Series*, vol. 116, American Chemical Society, Washington, D.C., 1973.

S. M. Walas, *Phase Equilibria in Chemical Engineering*, Butterworth-Heinemann, Oxford, England, 1985.

J. M. Prausnitz, R. N. Lichtenthaler and E. G. De Azevedo, *Molecular Thermodynamics of Fluid-Phase Equilibria*, 3rd ed., Prentice Hall PTR, Upper Saddle River, New Jersey, 1999.

J. M. Smith, H. C. Van Ness and M. M. Abbott, *Introduction to Chemical Engineering Thermodynamics*, 7th ed., McGraw-Hill, New York, 2005.

J. Richard Elliot and C. T. Lira, *Introductory Chemical Engineering Thermodynamics*, 2nd ed., Prentice Hall, Upper Saddle River, New Jersey, 2012.

S. I. Sandler, *Chemical, Biochemical, and Engineering Thermodynamics*, 5th ed., Wiley, Hoboken, New Jersey, 2017.

J. R. Elliott, C. T. Lira, T. C. Frank and P. M. Mathius, "Section 4, "Thermodynamics"," in *Perry's Chemical Engineers' Handbook*, 9th ed., D. W. Green and M. Z. Southard, Eds., McGraw-Hill, New York, 2018.

P. H. Van Konynenburg and R. L. Scott, "Critical Lines and Phase Equilibria in Binary van der Waals Mixtures," *Philosophical Transactions of the Royal Society of London A*, vol. 298, pp. 495–540, 1980.

R. L. Scott, "Van der Waals-like Global Phase Diagrams," *Physical Chemistry Chemical Physics*, vol. 1, no. 18, pp. 4225–4231, 1999.

H. S. Leff, "Entropy, Its Language, and Interpretation," *Foundations of Physics*, vol. 37, pp. 1744–1766, 2007.

F. L. Lambert, "The Conceptual Meaning of Thermodynamic Entropy in the Twenty-first Century," *International Research Journal of Pure and Applied Chemistry*, vol. 1, no. 3, pp. 65–68, 2011.

B. G. Kyle, "The Essence of Entropy," *Chemical Engineering Education*, vol. 23, no. 4, pp. 250–255, 1989.

P. Atkins, J. De Paula and J. Keeler, *Physical Chemistry*, 11th ed., Oxford University Press, Oxford, England, 2018.

N. Aspiron, B. Rumpf and A. Gritsch, "Work Flow in Process Development for Energy Efficient Processes," *Applied Thermal Engineering*, vol. 31, no. 13, pp. 2067–2072, 2011.

B. Predel, M. Hoch and M. Pool, *Phase Diagrams and Heterogeneous Equilibria*, Springer, Heidelberg, Germany, 2004.

A. Reisman and E. M. Loebl, *Phase Equilibria: Basic Principles, Applications, and Experimental Techniques*, Academic Press, Cambridge, Massachusetts, 1970.

B. Predel, M. Hoch and M. J. Pool, *Phase Diagrams and Heterogeneous Equilibria: A Practical Introduction*, Springer, Heidelberg, Germany, 2004.

S. M. Lai, M. Y. Yuen, L. K. S. Siu, K. M. Ng and C. Wibowo, "Experimental Determination of Solid-Liquid-Liquid Equilibrium Phase Diagrams," *AIChE Journal*, vol. 53, no. 6, pp. 1608–1619, 2007.

https://doi.org/10.1515/9783110695052-010

Chapter 6

P. C. Wankat, *Separation Process Engineering: Includes Mass Transfer Analysis*, 4th ed., Prentice Hall, Upper
  Saddle River, New Jersey, 2016.
E. L. Cussler, *Diffusion*, 3rd ed., Cambridge University Press, Cambridge, England, 2009.
R. E. Treybal, *Mass Transfer Operations*, 3rd ed., McGraw-Hill, New York, 1980.
C. Judson King, *Separation Processes*, McGraw-Hill, New York, 1971, Dover Publications, Mineola,
  New York, 2013.
J. D. Seader and E. J. Henley, *Separation Process Principles*, 2nd ed., Wiley, Hoboken, New Jersey, 2006.
R. W. Rousseau, Ed., *Handbook of Separation Process Technology*, Wiley, New York, 1987.
P. A. Schweitzer, Ed., *Handbook of Separation Techniques for Chemical Engineers*, 3rd ed., McGraw-Hill,
  New York, 1997.
R. Bryan Bird, W. E. Stewart and E. N. Lightfoot, *Transport Phenomena*, 2nd ed., Wiley, Hoboken,
  New Jersey, 2006.
W. B. Krantz, *Scaling Analysis in Modeling Transport and Reaction Processes: A Systematic Approach to Model
  Building and the Art of Approximation*, Wiley-AIChE, Hoboken, New Jersey, 2006.
C. J. Geankoplis, A. Hersel and D. Lepek, *Transport Processes and Separation Process Principles*, 5th ed.,
  Pearson, London, 2018.
T. L. Bergman, A. S. Lavine, F. P. Incropera and D. P. DeWitt, *Fundamentals of Heat and Mass Transfer*,
  8th ed., Wiley, Hoboken, New Jersey, 2017.
W. L. Luyben, *Process Modeling, Simulation, and Control for Chemical Engineers*, 2nd ed., McGraw-Hill,
  New York, 1989.
S. M. Kresta, A. W. Etchells III, D. S. Dickey and V. A. Atiemo-Obeng, Eds., *Advances in Industrial Mixing:
  A Companion to the Handbook of Industrial Mixing*, Wiley, Hoboken, New Jersey, 2015.
E. L. Paul, V. A. Atiemo-Obeng and S. M. Kresta, Eds., *Handbook of Industrial Mixing*, Science and Practice,
  Wiley, Hoboken, New Jersey, 2004.
B. Andersson, R. Andersson, L. Håkansson, M. Mortensen, R. Sudiyo and B. Van Wachem, *Computational
  Fluid Dynamics for Engineers*, Cambridge University Press, Cambridge, England, 2012.
A. S. Myerson, D. Erdemir and A. Y. Lee, eds., *Handbook of Industrial Crystallization*, 3rd ed., Cambridge
  University Press, Cambridge, England, 2019.
C. Wibowo and K. M. Ng, *Conceptual Design of Crystallization Processes*, De Gruyter, Berlin/Boston, 2021.
C. Tien, *Introduction to Adsorption: Basics, Analysis, and Applications*, Elsevier, Amsterdam, 2019.
E. L. Cussler, "Membranes Which Pump," *AIChE Journal*, vol. 17, no. 6, pp. 1300–1303, 1971.
M. C. Yang and E. L. Cussler, "Designing Hollow-Fiber Contactors," *AIChE Journal*, vol. 32, no. 11,
  pp. 1910–1916, 1986.
B. W. Reed, M. J. Semmens and E. L. Cussler, "Membrane Contactors," *Membrane Science and Technology*,
  vol. 2, pp. 467–498, 1995.
A. F. Ismail, M. A. Rahman, M. H. D. Othman and T. Matsuura, eds., *Membrane Separation Principles and
  Applications*, Elsevier, Amsterdam, 2018.

Part III: **Engineering Practice**

In Part II we covered various fundamentals governing chemical separation. Here in Part III, we focus on how that understanding may be applied in designing a separation process. At times, we and others have thought about trying to program all such work into a computer for a virtual design method – allowing rapid development and visualization of a newly proposed process without any laboratory or miniplant studies. After all, the automotive industry is moving in this direction, specifying more and more of a new car design in a computer. And electronic circuits are designed by using a computer, although the fabrication of the actual microchips involves considerable experimentation and validation.

A virtual design method certainly is an appealing vision for the future of separation process design, and progress is being made in that direction. Certain applications already are designed in a computer without need of experiments, such as the design of standard heat exchangers and heat-exchange networks, standard designs for drying gases, and standard distillation applications involving well-characterized chemical systems and standard column configurations and internals. Many new distillation columns are almost exact duplicates of existing columns, just in new locations. But these well-understood applications benefit from many years of design practice and have extensive databases to draw upon. And they represent only a fraction of the great variety of chemical systems and separation methods. When dealing with a novel separation application, certain computer simulations such as process simulators, computational fluid dynamics (CFD) programs, and molecular modeling software can greatly reduce the need for experimentation, especially for initial design activities such as outlining equipment geometries and identifying useful process conditions and candidate solvents. And these tools are becoming more and more capable. But in many cases certain physical, thermodynamic, and transport properties still need to be measured, and novel designs still need to be tested in miniplant or pilot-plant studies. This is required for many adsorption and crystallization applications and for many other applications as well. And, by its very nature, exploratory research will always involve some experimentation. So, although virtual design is already practiced in many cases (especially for distillation design) and no doubt will become more and more feasible over time (a good thing), a combination of design calculations, computer simulations, and key experiments will continue to be required for many new applications for the foreseeable future.

But there is more to it than that. Design work also involves a good deal of creative thinking – in identifying most-effective separation methods and in evaluating opportunities for process simplification, for process intensification, or for integrating two or more separation methods into a hybrid process scheme. This is more than doing calculations or running experiments. It involves devising a mental picture of what is needed and a plan to put it all together into a working process scheme. It takes the kind of ingenuity and drive discussed in Section 3.2: Pioneers and in Section 3.3: The Engineering Method. We are not aware of a computer program that can do all that (at least not yet). So, we begin Part III by describing design methods for various individual separation operations, many of which involve a combination of design calculations and key

https://doi.org/10.1515/9783110695052-011

experiments (in Chapters 7–13). We end the discussion (in Chapter 14) by commenting on ways to think "outside the box" in the approach to process design, by using creative thinking to strike a balance between process simplicity, intensity, and complexity, one that is most appropriate for the application at hand.

7 Distillation: Rectification, Stripping, and Absorption – The Workhorse Methods

In its simplest form, distillation is the method used to separate a crude liquid mixture by boiling it in a closed kettle or pot. The pot is heated and a portion of the feed liquid is vaporized. The resulting vapor (called the overheads) is directed through an overhead tube or pipe into a condenser. The condenser may simply be an air-cooled section of coiled piping. The piping is bent away from the pot so the overheads condensate drains into another container called the receiver. A very simple process. In many cases, the condensate collected in the receiver is the desired product. But sometimes the product is the liquid left in the pot. This simple process, which dates from before the Middle Ages, is known as alembic distillation or pot distillation. Today, a single-pot distillation process like this is also known as Rayleigh distillation (more on this later). Such a process normally gives a noticeable separation; that is, the composition of the liquid in the overheads receiver differs from that of the initial feed and that of the liquid left in the pot, because on boiling, the overhead vapor is enriched in the more volatile components (called the lights). An alembic process is fairly easy to set up and practice. It was used in ancient times to collect overhead products such as drinking water from seawater, enriched alcohol from beer or wine, and perfumes and essential oils from plant extracts. It is still in use today, often in the form of copper or wood-lined kettles used to produce specialty alcoholic drinks such as cognac and rum as overhead products, or in the form of single-stage batch evaporators used to remove light solvents and other volatiles from heavy liquid products, a process called devolatilization. Pictures of typical alembic equipment can easily be found on the internet. Small units can be fine works of art, with finely hammered copper kettles, onion-shaped kettle tops, and swan-neck-shaped overhead tubes.

Simple alembic distillation generally is not used for industrial-scale chemical separation because it is inefficient and tedious to practice whenever greater separation power is needed beyond what a single-stage process can deliver. For example, when further enrichment of an overhead product is needed, alembic distillation has to be repeated in multiple batches of vaporizing/condensing. A sequential process involves taking the overheads condensate from a first distillation, vaporizing a portion of it in a second distillation, collecting the overheads condensate from the second distillation, vaporizing a portion of it in a third distillation, and so on – a fairly tedious proposition. Furthermore, although this scheme allows for greater enrichment and concentration of lights in the overheads, now another problem arises: The amount of overhead product recovered after many distillations becomes diminishingly small. This can be alleviated to some extent by recycling the pot residue from the second distillation to the feed entering a subsequent first distillation and likewise recycling the pot residue from the third distillation to the feed for a subsequent second distillation, and so on in a three-stage, countercurrent batch processing scheme. Such a scheme may be a useful laboratory

https://doi.org/10.1515/9783110695052-012

technique or a way to make a specialty alcoholic liquor (a craft spirit), but the handling of the various pot residues and overhead products is tedious and time-consuming, so generally it is not recommended for industrial processing.

We can imagine that at some point in the history of alembic distillation someone decided to deliberately cool the overhead piping at its vertical connection to the top of the pot, in addition to cooling the section ahead of the condensate receiver. The removal of heat from this initial section of piping caused partial condensation of the rising vapor, and some of the resulting condensate drained back into the pot (not the receiver). Now, in effect, the process involved use of an overhead pipe or "column" to contact rising vapor with falling liquid (called reflux), and this was found to further enrich the lights in the overheads. Whether deliberate or not, some internal condensation of vapor and reflux of condensate likely occurred to some extent whenever the overhead piping was not insulated. This gave better separation because the falling liquid condensate absorbed some of the less volatile components (called heavies) out of the rising vapor. The technique we now call batch rectification was born (or so we imagine).

This contacting of rising vapor with falling liquid is somewhat akin to conducting a series of vaporizing/condensing steps as described above for multiple countercurrent alembic distillations. But now, these steps are carried out in countercurrent fashion within the column of a single distillation operation (as in Figure 7.1). With the addition of countercurrent reflux, a temperature gradient is established within the column due to differences in the boiling points of lights and heavies (cooler at the top of the column and hotter toward the bottom). As the hot rising vapors intermingle with the cooler falling liquid, they exchange heat, causing some portion of liquid to vaporize and about the same amount of vapor to condense – a phenomenon called constant molar (or molal) overflow. This occurs repeatedly along the entire length of the column. In effect, the column forms a countercurrent cascade of partially condensing vapor and partially vaporizing liquid. As a result, the composition of the overhead vapor leaving the top of the column is further enriched in the light components *and* the recovery of lights into the overhead product is increased – all in a single distillation operation. The increase in recovery is due to sharper separation of lights from heavies and the fact that the reflux liquid returns to the pot carrying unrecovered lights with it, allowing additional recovery of lights as the distillation proceeds.

It appears that it was not until the early nineteenth century that people began in earnest to exploit this phenomenon called rectification (see Section 3.2: Pioneers, and ref. [1]). Improvements made at that time include the deliberate addition of external reflux (instead of relying on internal condensation) by metering a portion of the overheads condensate back to the top of the column. Also, the use of specially designed column internals (distributors, trays, and packings) further improved vapor-liquid contacting by directing vapor and liquid traffic in desired (uniform) paths within the column and by promoting generation of interfacial area. With these advances, the term "overheads" refers to the total material, usually all vapor, leaving the top of the

Overheads Vapor

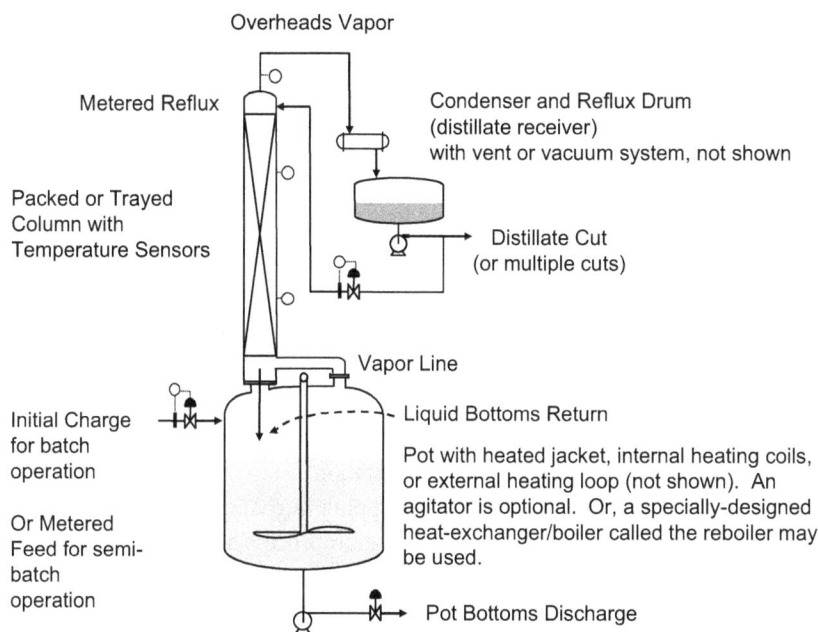

Metered Reflux

Packed or Trayed
Column with
Temperature Sensors

Condenser and Reflux Drum
(distillate receiver)
with vent or vacuum system, not shown

Distillate Cut
(or multiple cuts)

Vapor Line

Initial Charge
for batch
operation

Or Metered
Feed for semi-
batch
operation

Liquid Bottoms Return

Pot with heated jacket, internal heating coils,
or external heating loop (not shown). An
agitator is optional. Or, a specially-designed
heat-exchanger/boiler called the reboiler may
be used.

Pot Bottoms Discharge

Figure 7.1: Traditional Batch Distillation with Reflux (Dynamic Batch Rectification).

column. Reflux is the portion of the overheads condensate that is returned to the top of the column, and distillate is the portion taken out of the process as the overhead product. The reflux ratio is defined in two ways: as the ratio of reflux flow rate to overheads flow rate (called the internal reflux ratio) and as the ratio of reflux flow rate to distillate flow rate (called the external reflux ratio). Common practice is almost exclusively to refer to the external reflux ratio which can range from zero to infinity. The maximum concentration of lights in the overheads occurs when all of the overheads condensate is returned to the column as reflux, resulting in an infinite external reflux ratio, a condition called total reflux. Of course, this cannot be a productive operating condition, but it is commonly used in starting up a distillation column.

The concept of the middle-fed distillation column was introduced at about the same time as the batch rectification column, leading to a continuously fed column with top and bottom sections above and below the feed point, as illustrated in Figure 7.2. The top section is called the rectification section (or the enriching section or absorption section) and the bottom section is called the stripping section. The entire column is called a fractionation column. The rectification section employs liquid reflux as described above to purify the overheads and produce distillate. In the stripping section, vapor rising up from a boiler at the bottom (called the reboiler) serves as a kind of vapor reflux (called boilup) to purify the liquid exiting the bottom of that section (and entering the reboiler). The bottom product (or simply the bottoms) is the liquid discharged from the reboiler after some fraction has been evaporated to serve as the boilup vapor (analogous to the liquid distillate discharged from the process after removing the reflux liquid from the overheads condensate). As vapor from the reboiler rises up through the stripping section, it removes or strips light components out of the descending liquid. The rising vapor becomes more and more enriched in lights, which must come from the liquid, so the descending liquid contains fewer and fewer lights on its way down the column. The boilup ratio is defined in two ways analogous to how the reflux ratio is defined: as the ratio of the boilup vapor rate to the rate of liquid that exits the stripping section and enters the reboiler (called the internal boilup ratio) and as the ratio of the boilup vapor rate to the bottoms flow rate (called the external boilup ratio). In practice, we often refer to the internal boilup ratio which can range from zero to one when expressed as a fraction, or from 0% to 100% when expressed as a percentage. However, the external boilup ratio is often more easily calculated in real-world operations because the boilup rate usually can be calculated from the reboiler duty, and the bottoms discharge rate usually is measured directly with process instrumentation.

With this combination of rectification and stripping introduced in the early nineteenth century, the basic features of modern distillation processing were now in place. Furthermore, the basic concepts of countercurrent processing, use of reflux, and middle-fed fractionation have subsequently been used in the design of many other separation technologies as well, as we discuss in subsequent chapters.

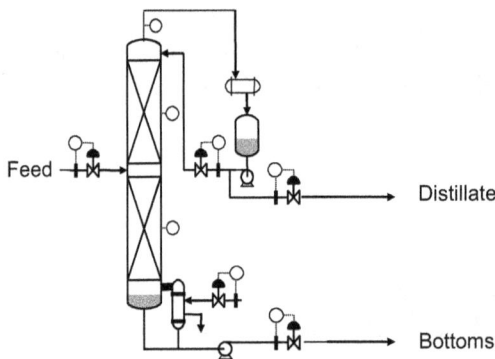

Figure 7.2: Continuously Fed, Steady-State Fractional Distillation (Middle-Fed Distillation Column).

7.1 General Features and Utility

Batch distillation is a dynamic process involving change in the composition of the pot contents with time. In the case of batch rectification (Figure 7.1), this does not necessarily involve change in the composition of the overheads with time, because the overheads composition can be held constant by manipulating the flow of reflux (more on this later). On the other hand, the operation of a continuously fed fractionation column (Figure 7.2) is a steady-state process with constant flow rates and compositions throughout, varying only with a deliberate change in operating conditions or control parameters. Along with limited use of simple alembic distillation, these distillation configurations and operating schemes are the classic or traditional ways of practicing distillation for industrial chemical separation.

But other configurations and operating schemes are also possible and may be used to advantage depending on what products need to be isolated – light, heavy, or intermediate. Traditional batch rectification is well suited to producing multiple light products by removing some of the overheads condensate in successive distillate portions called cuts. On the other hand, if the desired product is a much heavier component (not an intermediate) and the goal is to remove light contaminants, then batch rectification is not well suited to the job. It can still be effective to some degree, but not optimally: Distillate is removed overhead until the residual pot contents are sufficiently concentrated in the desired heavy components, yielding a heavy product in the pot. But this approach can take a long time which can cause significant thermal degradation of the pot contents. In addition, operating at a low reflux ratio to speed removal of lights and reduce thermal degradation can also carry considerable amounts of the desired heavy product into the distillate due to reduced separation power, resulting in poor product recovery. This is because the traditional batch rectification configuration has only one stage of stripping – in the pot. The column provides rectification stages for lights purification but no stripping stages for heavies purification.

To produce purified heavies, the inverted batch configuration shown in Figure 7.3 may be used for genuine multi-stage batch stripping. In this configuration, the feed is loaded into a vessel from which liquid is fed to the top of the column with the condensed overhead vapor from the column being returned to the feed vessel. The liquid exiting the bottom of the column is sent to a reboiler in which a portion is vaporized and returned to the column as boilup. The unvaporized remainder of the bottoms liquid is then discharged as purified heavy product (the bottoms product). In this manner, the extent of volatiles stripping from the desired heavy product is controlled by the boilup ratio rather than a reflux ratio. Here, the maximum separation power and minimum lights concentration in the bottoms is achieved at total boilup, with all of the liquid exiting the bottom of the column being vaporized and returned to the column as boilup. As in the case of total reflux in batch rectification, this would result in zero productivity as there would no longer be any bottoms product available for discharge. But it can be a useful strategy in starting up such a process or when servicing downstream equipment with a need to temporarily suspend the column bottoms discharge.

A batch stripper as described above can be much more productive and deliver greater separation power whenever the desired products are heavies. Such a process is fairly easy to practice. In fact, the traditional batch column configuration may be modified to operate in either rectification or stripping mode, or in both modes at the same time by using two columns as indicated in Figure 7.4. The resulting middle-vessel dual-column batch scheme allows for production of both light distillate cuts and heavy bottoms cuts – a kind of general purpose operation. It is the batchwise

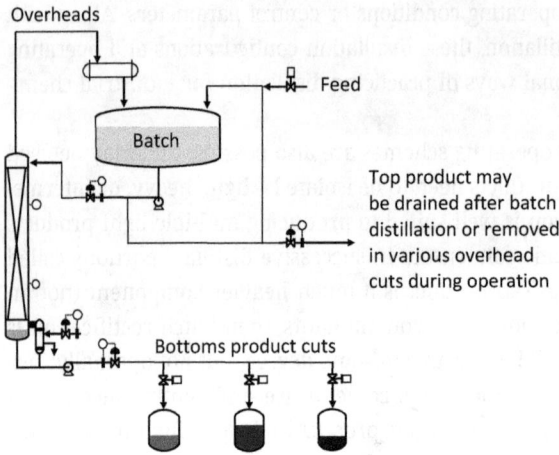

Figure 7.3: Batch Stripping (Inverted Mode).

version of a continuously fed fractional distillation column (Figure 7.2). In this scheme, the pot serves as a middle vessel to collect liquid from the top rectification column and distribute it to the bottom stripping column. The pot (middle vessel) is filled with a batch of feed liquid on start-up. During operation, the liquid from the top column mixes with the batch in the pot before being fed as a mixture to the bottom column. Flowrates and boilup are adjusted as needed depending on feed composition and desired products. The vapor from the bottom column is sent directly to the top column, and an auxiliary heater may be added to the middle vessel (not shown) to offset the effects of a cold feed and heat losses or to serve as a reboiler for operation in standard batch rectification mode. Further discussion of middle-vessel batch distillation, its background, and various modes of operation is given in ref. [2].

To our knowledge, the middle-vessel dual-column batch scheme shown in Figure 7.4 is not widely practiced in industry. Batch rectification remains the most popular batch distillation configuration, even when a middle-vessel or inverted batch stripping configuration would be more efficient. But a change to one of these other batch schemes may allow a significant reduction in energy consumption, depending on the application. See Section 7.6: Energy Utilization.

Batchwise operation is used when flexibility is needed, as when processing small quantities in the laboratory or pilot plant or a specialty production plant or whenever producing multiple products from a single feed or when processing a variety of feeds in multiple campaigns. Batchwise processing may also be used to maintain batch lot integrity as may be required in pharmaceutical production. Continuously fed operation may be preferred for the production of large quantities or when steady-state operation is needed to improve process productivity, by reducing variability in operation and by eliminating non-productive periods such as fill and drain times. It may also eliminate or reduce the need for large distillation vessels when processing large quantities.

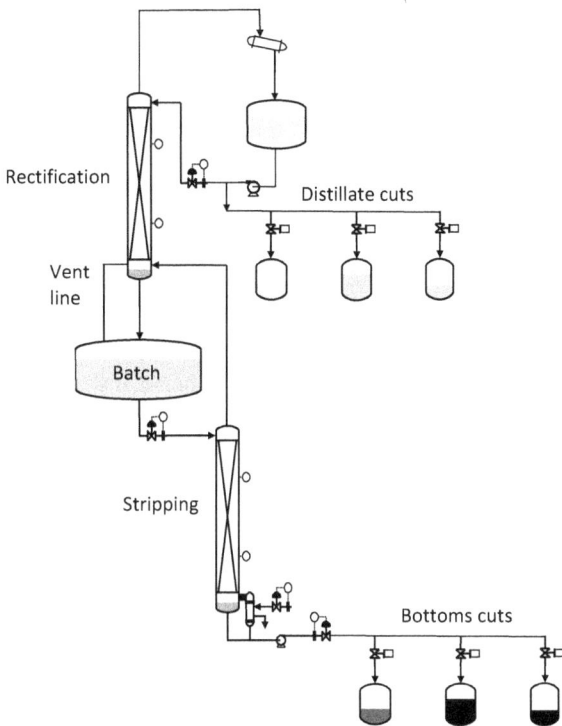

Figure 7.4: Middle-Vessel Dual-Column Batch Fractionation (Dynamic Rectification + Stripping).

Today, the various process configurations are practiced using many different styles of equipment with many distinguishing characteristics and optional features, some of which are listed in Box 7.1. With so many features and options, distillation has proved commercially successful over many decades, even centuries, and new applications continue to be introduced. As a result, distillation is by far the most common industrial chemical separation method. No other method comes close in terms of the number of applications and the total volume of material processed. It has been estimated that roughly 40,000 distillation columns are in operation in the U.S. alone [3].

Box 7.1: Distillation Equipment Features and Options

– Few moving parts (except for pumps used to pump liquids)
– Many different types of internal packings and trays
– Liquid and vapor distributors and collectors
– Reboilers and total or partial condensers
– Multiple feed points and side draws
– Top pasteurization section for removing lights above the main overheads removal point
– Cross exchange of feeds and discharges from the column for energy recovery
– Coupled reboiler/condenser with heat pump cycles for energy savings

- Internal dividing walls for multiple high-purity side-draw products with lower capital and energy costs
- Mechanically complex columns such as rotating packed beds and trays with integrated heat exchangers
- Evaporators: thermosyphon, forced circulation, wiped film, thin film, falling film, short path
- Multi-column distillation trains for multiple products
- Reactive distillations with heterogeneous catalysts located within the column or in external circulation loops, or homogeneous catalysts dissolved in the liquid feed or injected into the column

Distillation is effective in separating many different kinds of liquid mixtures, so it can be used for a wide variety of applications. This is because most liquids are comprised of compounds whose volatilities in solution vary significantly with chemical structure. Exceptions to this rule include liquids comprised of components with very similar structures (such as isomers), liquids that contain ionized species which form strong ionic interactions in liquid solution (and so are not very volatile), and ultra-low-volatility liquids such as polymer melts. The presence of an azeotrope (where liquid and vapor compositions become equal) due to non-ideal interactions between components also limits the effectiveness of standard distillation. However, if an azeotrope is present, enhanced distillation methods are available to overcome (or break) the azeotrope or to deliberately exploit the azeotrope to enhance separation of other components (see Chapter 8: Enhanced Distillation). In other words, an azeotrope does not necessarily present an insurmountable obstacle, and it may even be used to enhance performance.

And distillation often is highly effective. A single distillation column may be designed to achieve many theoretical stages or transfer units for high separation power. Typically, this is about 10 or 15 stages but in special cases it can be as many as 100 or even more. As a result, it is not uncommon for distillation separation power to be between 10,000 and 50,000. In special cases, separation power can reach 100,000 or even higher. And distillation often is effective across an exceptionally wide concentration range, successfully producing a highly concentrated product out of a dilute feed with high product recovery. In many cases, the driving force for transfer of solute between vapor and liquid is adequate whether the desired solute is present in low or in high concentration in the feed – simply because a single column can be designed with so many separation stages and high separation power. As a result, distillation may be used in making crude cuts or in isolating high-purity products, as needed. In addition to pure products, distillation can be used to process crude feeds containing hundreds or even thousands of components, separating them into multiple fractions or boiling point ranges.

As mentioned earlier, a single distillation unit may be used for multiple applications in the same facility, processing a variety of feeds in separate campaigns. This allows for high utilization and overall productivity of the installed equipment with little idle time. Distillation can also allow isolation and recovery of multiple products from a single feed. Batch operated columns can isolate multiple products by controlling the boilup rate, reflux rate, and the timing of when products are removed from the over-heads (or the bottoms). Single continuously operated columns can have multiple side draws with different compositions, with the option of employing internal dividing walls to sharpen the composition of the side draws. And continuously operated distillation trains of multiple columns can be designed to isolate any number of products (but commonly up to five).

Distillation is also well suited for commercial operation because it may be practiced over a wide range of production scales, from small laboratory units, to pilot plants, to very large commodity-scale units with very high production rates. As a result, a distillation column may be scaled up as a single unit with good economy of scale; that is, with lower and lower cost per unit of product as column size increases (including capital and operating costs). The largest distillation columns are as big as 15 meters (50 feet) in diameter and 110 meters (360 feet) tall or even bigger. Packing may be used, or the column may be built with as many as 160 large-diameter, specially designed trays (or more). At this large scale, some designers prefer to call them *towers* instead of columns.

Distillation can also be practiced across a wide range of pressures, from deep vacuum (normally down to about 2 kPa (15 mmHg) or somewhat less, with lower pressures obtainable with specialized equipment such as short-path evaporators) to pressures significantly above atmospheric, typically up to 3000 kPa (about 30 atmospheres) and sometimes higher. This allows adjustment of operating pressure for operation at temperatures within a practical range for heating and cooling and to avoid or minimize product degradation or freezing from exposure to extreme temperatures. Adjustment of operating pressure sometimes also allows optimization of the thermodynamic driving force by allowing operation within the temperature range that maximizes relative volatility. In special cases, distillation may be economically operated at extreme temperatures, as in cryogenic distillation of air to separate oxygen and nitrogen at around −200 °C, and crude-oil distillations running at bottoms temperatures of 400 °C or even higher [4].

Distillation equipment generally is robust in operation, being resistant to wear and plugging in normal use. Fouling of distillation columns is not uncommon, but normally it can be managed with proper design and periodic cleaning. In severe fouling service, as in processing heavy crude oils or reactive monomers, specialized designs with large orifices and large open areas for vapor and liquid traffic allow for extended operating time before internal deposits interfere with performance. Also, filtration of the feed normally is not needed, as many types of distillation equipment can handle some limited amounts of solids entering with the feed. However, in some cases the liquid feed must be pre-filtered, as when stripping out volatile contaminants from a stream containing solids that have absorbed a portion of the contaminants.

And with distillation's long history, the many types and styles of distillation equipment are well developed and readily available from commercial suppliers. Furthermore, design methods and control schemes are well known and generally reliable. Control schemes include various ways of controlling distillate and bottoms compositions in batch and continuous processing by manipulating such process variables as heating and cooling (reboiler and condenser duties), reflux ratio, boilup ratio, and flow rates (feed, distillate, and bottoms). Common control schemes manipulate a process variable to control the overheads temperature or a temperature within the column (often the temperature that is most responsive to changes in composition). This provides an inferential indicator of chemical composition that often is fast-responding, robust, and cost-effective. In special cases, online chemical analyzers may be needed [5]. In addition, various schemes have been proposed to enhance mass transfer by pulsing the flow of vapor and liquid within a column [6–8]. And many hybrid process schemes involving distillation plus another separation method have been developed to further improve separation performance. Hybrid schemes increase separation capability and efficiency across a wider composition range (as we discuss in Chapter 14: Inventive Engineering: Process Simplification, Intensification, and Hybrid Processing). Each separation method is used within that portion of the composition range for which it is best suited.

The energy required for distillation can sometimes be obtained from sources that would otherwise go unused. Examples include distillation/evaporation of sea water to make fresh water aboard seagoing ships, taking heat from the ship's diesel-engine cooling system. Modern ships may utilize reverse osmosis membranes in port, but distillation is still a popular option when underway at sea. The evaporator is operated at reduced pressure (about 350 mmHg or 47 kPa) to allow boiling of sea water at the sub 100 °C temperatures of the engine's cooling water discharge (normally about 80 °C). Analogous examples involve distillation carried out within large, integrated manufacturing sites, using heat recovered from other units such as reactors and furnace exhaust gases to make steam needed for distillation. An example is in the Andrussow process for hydrogen cyanide production where steam generated from the heat recovery unit on the catalytic HCN reactor is used to run the ammonia recovery column [9]. Special designs allow reduced energy consumption, as in cross-exchange of heat between streams (called pinch technology), mechanical vapor recompression, hybrid processing, and inverted batch stripping as mentioned above. Optimization of multiple-column configurations also may reduce energy consumption, as in dividing-wall column designs. General design considerations of this kind are covered elsewhere [10–15].

Of course, there are also reasons why distillation is not well suited for some applications and why we need many other separation methods:

– When dealing with non-volatile or very-close-boiling materials, as well as many non-condensable gases. Some non-condensable gases are processed by using gas-liquid absorption or stripping, a subcategory of distillation, as discussed below.
– When a solid is the desired product in the form of specific kinds of crystals.
– When other methods offer advantages (alone or in combination with distillation) – such as greater driving force, milder operating conditions, higher selectivity, lower energy consumption, or lower total cost, as we will see when discussing other methods and various hybrid process schemes.

A major issue with distillation is its dependence on thermal energy. A distillation process may be viewed as a type of heat engine or energy-transformation process that transforms thermal energy into separation work, and as such its thermodynamic efficiency is limited (see Section 7.6: Energy Utilization). It is estimated that distillation consumes about 40% of all the energy consumed by chemical manufacturing [16, 17], much more than any other separation method. This fact derives not only from distillation's inherent energy consumption (which can be high, but this depends on the application), but also from the popularity and much wider use of distillation compared to other methods. An important opportunity for energy savings is in the replacement of older inefficient distillation operations with newer more-efficient ones and the implementation of energy recovery technologies such as cross exchange (pinch technology) [15, 16].

Some have suggested that the use of distillation will decline with further development of alternative methods such as adsorption and the use of membranes [18]. And indeed, since the late 1970s, alternative methods have made significant progress toward replacing certain distillation and gas processing applications (see Chapter 10: Adsorption-Based Processes and Chapter 12: Membrane-Based Separations). But even so, we feel distillation is likely to remain the dominant separation method for the foreseeable future. For one thing, the cost of energy consumed by distillation per unit of product often is low compared to the value of the product delivered by the process. But perhaps the more pressing reason is that finding alternatives that can reliably handle crude, multicomponent, industrial feeds is difficult, especially for large volumes. Also, although it is often assumed that other methods can save energy, in practice this depends on the specifics of the application [19]. For membrane-based processes, this depends on the required product purity, the concentration of the permeating species in the feed to the membrane (the higher the better), membrane flux and resistance to flow (pressure drop that needs to be overcome by compressors and pumps), and the membrane's selectivity, which along with the feed composition and product purity specification determines how many membrane contacting stages are needed. In addition, operational issues like fouling over time and difficulty cleaning membrane units can come into play and must be considered. Given all this, one study has concluded that commercial polymeric membranes available today are unlikely to replace distillation for most applications [20]. These authors conclude that better membranes with higher permeability and selectivity will be needed to enable cost-effective separation with few membrane stages, and this is an ongoing area of research.

Nevertheless, some membrane applications have found considerable commercial success replacing distillation. These are often ones that require only three or fewer stages of membrane separation. A good example is reverse osmosis for desalination of seawater. A modern reverse osmosis process with pressure-recovery capability consumes between 8% and 40% of the energy required by distillation with mechanical vapor recompression (the state of the art) [21]. Another successful application involves pervaporation of certain mixtures of polar + relatively non-polar components such as water + alcohols. In this application, relatively polar (or non-polar) components selectively permeate a membrane and evaporate on the permeate side depending on the relatively hydrophilic (or hydrophobic) nature of the membrane. Up to three membrane stages are typical. Another application area now under development involves use of membranes to separate olefin/paraffin mixtures, notably ethylene/ethane and propylene/propane [22, 23]. Researches into other potential alternatives to distillation are ongoing, including greater use of adsorption as well as membranes. Membrane-based methods surely will continue to improve as new membranes are introduced with more favorable combinations of selectivity, flux, pressure drop, and resistance to fouling – for lower cost in use. And the development of pressure recovery technology (pressure-exchangers) has significantly reduced the energy consumption of reverse osmosis for desalination of seawater. For more discussion of these and other commercially successful membrane processes, see Chapter 12: Membrane-Based Separations.

But instead of replacing distillation altogether, in many cases membranes may best be utilized within a distillation + membrane hybrid process. An excellent example involves the incorporation of pervaporation membranes into a distillation process designed to remove water from alcohols that form an alcohol + water azeotrope. The distillation column can easily handle the main load, and the added membrane deals with the concentration range where distillation becomes ineffective. The same may be said for distillation + adsorption (see Section 12.5: Pervaporation; Section 10.6: Pressure Swing Adsorption; and Chapter 14: Inventive Engineering: Process Simplification, Intensification, and Hybrid Processing.)

7.2 Process Types

The term "distillation" includes many types of processes, all of which are driven in some way by deviation from vapor-liquid or gas-liquid equilibrium. In special cases, distillation can also involve chemical reaction, primarily in the liquid phase. In addition to batchwise, fed-batch, and continuously fed modes of operation, distillation processes are classified according to the following general functions. A number of these have already been introduced, and here we give a little more explanation.

- Rectification (Figure 7.5). Removal of heavy (less-volatile) impurities to obtain a purified overhead light product. This involves transfer of heavies from the vapor phase into the liquid phase, increasing the concentration of lights in the vapor. Rectification is also called enrichment or absorption. The liquid phase in rectification is called reflux liquid or absorption fluid. The term "reflux" refers to material that exits the process with the overheads vapor and is returned to the process (or is refluxed back) as liquid condensate.
- Stripping (Figure 7.6). Removal of light (more volatile) impurities to obtain a purified heavy liquid bottoms product. This involves transfer of lights from the liquid phase into the vapor phase, increasing the concentration of heavies in the bottoms liquid.

- Fractional distillation. Production of both light and heavy product streams involving both rectification and stripping. An example is the continuous middle-fed distillation column discussed earlier (Figure 7.2), having an upper rectification section (with liquid reflux used to absorb heavies from the rising vapor) and a lower stripping section (with the rising vapor stripping light impurities from the falling liquid). The batchwise version of fractional distillation is illustrated in Figure 7.4. Multi-product batch rectification is another kind of fractionation process.
- Absorption. Absorption is another name for rectification, and the rectification section of any distillation process may be called the absorption section. However, the name "absorption process" most often refers to a process in which impurities are absorbed out of a non-condensable gas. In such a process the absorption liquid (or scrubbing liquid) is supplied to the top of the column from an external source, while in a distillation column with rectification, the "absorption" liquid is supplied as condensed overheads liquid (reflux) returned to the top of the column. We classify absorption from gas to liquid as a special type of distillation process because it is driven by deviation from gas-liquid equilibrium which is analogous to vapor-liquid equilibrium. The design equations are essentially the same or very similar.
- Stripping with a non-condensable gas (non-condensable at process conditions). Analogous to stripping-mode distillation, but carried out using an added stripping gas instead of vapor. The gas acts as a carrier gas for evaporated liquid that saturates the gas phase, allowing stripping to be conducted a temperatures below the liquid boiling point, but sometimes at the cost of reduced effectiveness. The relative volatility driving force is reduced by the ratio of liquid vapor pressure to total pressure. In effect, the gas is simply a carrier for the liquid's vapor, and the liquid's partial pressure is the fraction of the total gas phase that does the actual thermodynamic stripping (as discussed in Section 7.3.3: Continuous Stripping and Absorption).
- Enhanced distillation. A distillation process involving addition of a material separating agent (an entrainer or solvent) to enhance the separation capability by altering the relative volatility of key components. This is the subject of Chapter 8: Enhanced Distillation: Taking Advantage of Molecular Interactions.
- Reaction-enhanced distillation. A distillation process involving use of chemical reaction to enhance a desired separation. An example is reactive absorption of acid gases. Here, acidic impurities in a feed gas react with basic components in a scrubbing liquid to form non-volatile salts, thereby enhancing their removal from the feed. Another example is the transfer of CO_2 out of a hydrocarbon feed into an aqueous alcohol-amine where it reacts with the amine functional group to form a non-volatile molecular complex or adduct. In this case, the complex is thermally reversible, so regeneration of the aqueous alcohol-amine scrubbing fluid is done by heating it in a stripping column to strip out CO_2 gas which is then vented to the atmosphere or compressed for further treatment or disposal. Typical alcohol-amines are diethanolamine and monoethanolamine. For more details, see Section 7.3.7: Reactive Absorption for Gas Scrubbing. Similar technology is

also used to remove CO_2 from liquids in a reactive liquid-liquid extraction process (see Chapter 9). The same general approach may be used to remove CO_2 from industrial flue gases to reduce CO_2 emissions to the atmosphere (see Chapter 4: Addressing Universal and Changing Needs).

– Reactive distillation. A type of hybrid process scheme in which both a desired reaction and a required separation are accomplished within the same process. For example, this type of process is used to improve equilibrium-limited reactions, as the *in-situ* separation (removal) of a reaction product from the reaction zone to drive further conversion.

– Hybrid distillation. As mentioned above, distillation is very often used in close association with another separation process. This is termed a hybrid process scheme when the resulting operation allows for enhanced overall separation and recovery of desired components or improved overall efficiency.

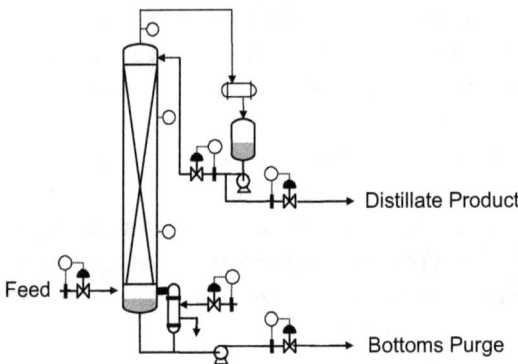

Figure 7.5: Continuously Fed Steady-State Rectification-Mode Distillation Column (Rectifier).

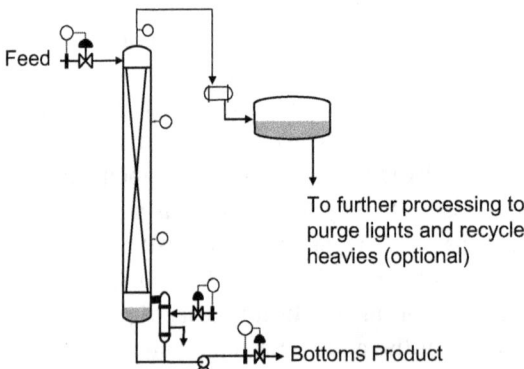

Figure 7.6: Continuously Fed Steady-State Stripping-Mode Distillation Column (Stripper).

7.3 Distillation Operations and Design Considerations

In this section we review the design and operation of various types of distillation processes, drawing on the concepts discussed earlier in Part II, Design Fundamentals. Non-reactive, stand-alone rectification and stand-alone stripping processes are special cases that often involve dilute feeds. A dilute feed allows for relatively straight-forward design calculations involving straight (or nearly straight) equilibrium and operating lines. Like stripping operations, non-reactive absorption is often conducted at dilute conditions. Absorption is analogous to rectification, so analogous design equations may be used. Fractional distillation design is more complex because a fractionation column includes both rectification and stripping sections usually operating at conditions well beyond the dilute (linear) range. Design calculations require simultaneous solution of model equations for both sections, matching the material balance at the feed point where the two sections come together. Processes combining reaction with separation require additional knowledge of chemical reaction equilibrium (for processes operating close to equilibrium) or knowledge of the actual reaction kinetics (for reaction-rate-limited processes).

In practice, the design and specification of a distillation process involves a number of work processes, each interconnected and interdependent. These include:

- Quantifying the thermodynamics. Developing an appropriate phase equilibrium model.
- Conceptualizing the process scheme. This involves deciding on the type of equipment and the arrangement of equipment in an overall process scheme or configuration that best fits into the larger, integrated manufacturing operation. For example, this might involve deciding whether to use regular batch rectification or inverted batch stripping, or whether to use a dividing-wall distillation column or a series of columns.
- Determining operating conditions and the required number of theoretical stages or transfer units (the required separation capability or separation power). This requires knowledge of the thermodynamic driving force (from the model development) and depends on choices in equipment operation such as operating pressure (which determines the operating temperature range) and reflux ratio. This is the step normally emphasized in undergraduate textbooks.
- Deciding how to control the process and integrate its operation into the overall process scheme.
- Determining energy requirements. This is a function of feed composition, the required separation performance (required product recovery and purity), the potential use of energy-saving design options, and how the distillation process fits into an overall process scheme.
- Specifying process equipment type and size needed to achieve the required separation capability. This involves specifying equipment internals and determining such things as column diameter and the required number of actual trays or

packed column height. Doing so requires knowledge of hydraulics and mass transfer resistance.

– Adjusting the process for cost-effective design and operation. This involves balancing such things as capital and operating costs, operating conditions and available utilities, and ease of operation and robustness of the equipment.

In the following sections, we touch on certain aspects of these design steps to illustrate and explain various design considerations and practices.

7.3.1 Single-stage distillation: Rayleigh distillation and continuous evaporation

Single-stage batch distillation without column reflux is known as Rayleigh distillation. It is essentially the same as simple alembic distillation. Its separation power is limited to only one theoretical stage (at best). The quantifying theoretical-stage equation, known as the Rayleigh equation, was introduced in 1902 in Britain by John William Strutt, known as Lord Rayleigh [24]. Lord Rayleigh is famous for describing fundamental physical principles – of buoyancy in fluid flow, of acoustics, and of many other physical phenomena including distillation. The Rayleigh distillation equation is commonly applied in scientific studies of chemical, biological, geological, and ecological processes involving distillation-like transfer of key components or tracers within the system of interest. For example, see ref. [25]. In addition, Rayleigh distillation is sometimes used to measure infinite-dilution relative volatilities of binary pairs of components in order to determine the corresponding infinite-dilution activity coefficients or Henry's law constants [26].

In the chemical laboratory, Rayleigh distillation is commonly used to evaporate light solvents or other volatiles from less-volatile liquids. For small quantities, this may be done using a simple round bottom flask placed in a heating mantle, using a magnetic stir bar and optionally operating under vacuum with a light flow of sweep gas. Or it may be done using a glass rotary evaporator such as the well-known Rotovap®. This type of batch evaporator often works well in the lab, so at times it is scaled up for use in the pilot plant for devolatilizing liquid products. This is sometimes done using large 20-liter (or even larger) rotating glass flasks. But this direct scale-up of laboratory equipment is not recommended. For one thing, glass construction should be avoided as it can break unexpectedly, and this can present significant safety hazards from sudden release of the hot contents with potential exposure to personnel and potential for fire, not to mention the poor heat transfer characteristics of glass. For the pilot plant and commercial production, a continuously fed shell-and-tube type evaporator (falling-film, rising-film, or calandria style) or an engineered mechanically agitated rotary evaporator (wiped-film or thin-film evaporator) should be used instead. Other types of evaporators also are available and may have advantages [27, 28]. Any of these units may be used alone or as the evaporator for a Rayleigh batch distillation. In that case, liquid is repeatedly circulated from the pot containing the batch through the evaporator and back to the pot until the lights content has been reduced to the desired low level. In other cases, these units may be used as the reboiler for a refluxed distillation column, yielding a process with much greater separation power (as we discuss in the next several sections).

For a single-stage Rayleigh distillation, mass transfer performance is given by:

$$\frac{C_{final}}{C_{initial}} \approx (1 - \text{ø})^{a-1} \left[\frac{1 - C_{final}}{1 - C_{initial}}\right]^{a} \tag{7.1}$$

where $C_{initial}$ = initial concentration of light component in the feed batch
$\quad C_{final}$ = final concentration of light component in the batch after evaporation
$\quad$ ø $\quad$ = fraction of feed liquid evaporated
$\quad a \quad$ = constant (or the mean) relative volatility of light key to heavy key

For a constant or average evaporation rate,

$$t_{batch} = \frac{\text{ø}M_{feed}}{\dot{M}} \tag{7.2}$$

where M_{feed} = mass of the initial feed batch
$\quad \dot{M} \quad$ = evaporation rate, mass per unit time
$\quad t_{batch}$ = batch elapsed time

This version of the Rayleigh equation is derived for a binary mixture and may easily be solved using an iterative calculation in a spreadsheet, as in using Solver in Microsoft Excel. The bases for this and other versions of the Rayleigh equation are discussed elsewhere [29–31]. They all assume that evaporation is sufficiently slow such that the boiling liquid is in equilibrium with the vapor (or nearly so). Furthermore, eq. (7.1) is best used to analyze applications with only low to moderate relative volatility. In that case, it may be used to estimate evaporator performance or to determine an effective average relative volatility by fitting batch evaporation data. An alternative model for systems with higher relative volatility is the rate-based *NTU* model, as expressed in eq. (6.58) and restated here:

$$\frac{C_{final}}{C_{initial}} = \exp(- kat_{batch}) = \exp(- NTU) \tag{7.3}$$

Equation (7.3) is applicable when the fraction evaporated is small, as is normally the case when relative volatility is high.

Continuously fed rotary evaporators and shell-and-tube evaporators are designed to evaporate a specific fraction of the feed liquid and have limited separation power to isolate lights from heavies. They are primarily heat transfer devices. Because of this, they are not generally used as stand-alone devices for mass transfer unless the relative volatility of the light component to be removed is very high. And they are not single-stage devices. Instead, the feed liquid is distributed as a film that continues to evaporate as liquid moves through the device, for a kind of differential contactor. In that case, mass transfer performance may be correlated in terms of the number of mass transfer units (*NTU*) achieved by the process. Similar to eq. (7.3), this has the basic form:

$$\frac{C_{out}}{C_{in}} \approx \exp(-NTU) \qquad (7.4)$$

when relative volatility between the solute and solvent is high. Here, C_{in} and C_{out} are the concentrations of solute in the liquid before and after passing through the evaporator. Equation (7.4) may be used to correlate experimental data, with *NTU* being a function of the amount of feed evaporated and other operating variables (for dilute feeds). For more discussion, see Section 6.7.1: Mass Transfer Units and Mass Transfer Coefficients. Most continuously fed evaporator designs involve co-current flow or crossflow of vapor and liquid, but some designs operate with countercurrent flow. In any case, such devices are not expected to deliver more than about 2 mass transfer units depending on feed viscosity, operating conditions, internal design details, and feed rate. Two or more passes through the evaporator may be needed to reduce the lights content to the desired low level.

Production-scale rotary evaporators generally are used to devolatilize liquid feeds that are particularly difficult to process otherwise. This may be because of high liquid viscosity as with specialty polymer melts or because the feed contains temperature-sensitive, high-boiling compounds such as pharmaceutical actives. These devices normally operate under vacuum with short contact time to minimize thermal exposure. Mechanical agitation serves to control liquid film thickness and improve heat transfer. There are many types of rotary evaporators, and normally performance must be measured in miniplant tests. Laboratory-scale glass units may be used in initial tests to allow direct observation of any sputtering, fouling, foaming, or entrainment of liquid into the vapor. Tests carried out in metal units are needed to accurately determine simultaneous heat-transfer and mass transfer characteristics, suitable operating parameters, and capacity. Units are designed to mechanically spread a uniform liquid film onto a cylindrical heated wall for rapid heat transfer and evaporation. Both vertical and horizontal units are available with various types of internals (liquid distributors, multiple wipers or thin-film blades, one or more heated surfaces, and so on) and some with internal screw-type positive-displacement pumps to allow processing of highly viscous liquids and even solid residues (as with the Rototherm-P® unit by Artisan Industries). A wiped-film unit has rotating wipers in direct contact with the heat-transfer surface. A thin-film unit has rotating blades that maintain a specified distance or clearance from the heat transfer surface, forming a turbulent bow wave ahead of the rotating blade. Rotary evaporators may have the condensing surface located within the body of the unit for operation under deep vacuum, as in short-path evaporation (also called molecular distillation). Or, the condenser may be located outside the unit. Various designs and analysis methods are described elsewhere [32–34].

Falling-film evaporators utilize a liquid distributor to evenly distribute feed liquid onto a heated surface, most often the inside surface of tubes in a shell-and-tube configuration. Gravity serves to pull the liquid film down the heated surface and out the bottom of the exchanger body. It is important that the liquid film wet the entire surface all along the length of a tube with no dry regions. Vapor typically flows upward, countercurrent to the falling liquid. Rising-film evaporators are fed at the bottom of the tube bundle and both vapor and liquid rise up through the tubes as evaporation proceeds. This avoids dry regions (unless a phenomenon called vapor blanketing occurs) and may offer improved heat transfer characteristics. Design methods and processes are described elsewhere [27].

7.3.2 Batch distillation with reflux

As discussed earlier, adding column reflux to a batch distillation process increases the separation power considerably compared to simple Rayleigh distillation. Batch distillation with reflux (batch rectification) is a very common laboratory technique. Typical laboratory apparatus include glass Vigreux columns, glass columns packed with glass beads or wire mesh, glass condensers, and intricate distillation heads designed to divide the flow of overheads condensate between reflux return and distillate take-off [35]. Such apparatus are often used to remove unwanted contaminants and isolate desired components from crude liquids and reaction mixtures, as when developing a new chemical process in the lab. And they often do a satisfactory job, so the temptation is to scale up what works in the lab into the pilot plant and into production (as is also the case with the Rotovap mentioned earlier). This approach is sometimes teasingly (or annoyingly?) called the jumbo-lab approach to production. It is easy to conceptualize and demonstrate. And it has to be said that it sometimes gives good results. But it can also lead to less-than-optimal processing in the production plant. Other options, such as the use of an inverted batch stripper (Figure 7.3), a continuously fed stripping column (Figure 7.6), or another separation method altogether, may provide considerably better results. So, instead of automatically assuming that it is a good idea to scale up the particular method used in the lab, other engineering options should also be considered for the production scale.

But certainly batch rectification has its place. It can be an excellent process for isolating light overhead products (Figure 7.1). These are collected in sequential quantities of distillate called cuts. Each successive cut contains a heavier product or larger amounts of the heavier components. This is a dynamic process, so the concentration of lights in the pot steadily decreases as the distillation proceeds. If the process is operated with a constant distillate take-out rate, the concentration of lights in the overheads will steadily decrease as well. But the overheads composition can be maintained at a desired level – by steadily increasing the amount of overheads condensate returned to the column as reflux, steadily reducing the amount taken out of the process as distillate. In this way, the column's separation power is steadily increased to maintain constant (or nearly constant) overheads composition – until all of a given cut has been removed and the column is at or near total reflux (with near-zero distillate flow). The process is then ready and waiting to begin taking the next (heavier) cut. This control scheme is easily automated using overheads temperature as an indirect (inferential) indicator of the desired composition for each cut. The first cut corresponds to a specific overheads temperature set-point, the second cut to a higher temperature set-point (indicating fewer lights and a heavier product), and so on. This approach often is used to first remove unwanted lights, then isolate a desired middle cut, and later purge a heavier cut. Artisan craft distillers refer to this general approach as removing the heads, collecting the heart, and removing the tail [36]. Referring to Figure 7.1, the control scheme we recommend works like this:

- The column is conveniently operated at constant pressure, normally atmospheric pressure, but vacuum and elevated pressures also may be used.
- The pot heater is turned on for a constant initial boilup rate, and the condenser temperature is adjusted to completely condense all of the overheads vapor without too much sub-cooling of the condensate. Reflux is returned to the column near its bubble point for best performance. The distillate can be cooled later on if needed.
- During start-up, often all the overheads condensate is initially sent back to the column as reflux, the condition described earlier as total reflux. This allows the column to develop the needed concentration profile along the column length. Once the overheads temperature reaches the desired set-point temperature indicating a desired amount of lights in the overheads, the operator will begin to collect the first cut, at first using a relatively high distillate rate with modest reflux.
- As the distillation proceeds and lights are removed with the distillate, the overheads temperature will begin to climb above the set-point as more heavies are taken overhead. At this point, the controller will begin to increase the amount of condensate sent back to the column as reflux. This will increase the column's separation power, causing the amount of lights in the overheads to increase again, and the overheads temperature to fall back, returning to the set-point. For a given rate of heat input to the pot, the total amount of overheads condensate will be fairly constant (but may decline over time). So increasing the amount of reflux requires reducing the distillate flow.
- If the overheads temperature happens to go below the set-point, the controller will compensate by slightly reducing the amount of reflux (taking a little more distillate). This will reduce the amount of lights in the overheads, causing the temperature to climb back up toward the set-point.
- If the operating pressure is somewhat variable, then a pressure-compensated temperature set-point may be used to better indicate composition as a function of temperature and pressure [37]. This is especially helpful for operation under vacuum.

In the lab, this scheme is commonly accomplished by using a solenoid-actuated diverter valve to send overheads condensate either back to the column (its normal POWER OFF position) or to the distillate receiver (the diverter valve POWER ON position). This involves time-proportioned ON/OFF control: the diverter valve is switched ON for some fraction of a given time period, say for 5 s out of a 30 s time period, to divert flow to distillate. During the remaining 25 s it is switched OFF to divert flow back to reflux. In the next 30 s time period, the controller changes the time fraction (the proportion of time spent ON versus OFF) as needed to control the overheads temperature. And so it continues, with the controller adjusting the time fraction every 30 s to minimize the deviation between the measured overheads temperature and the set-point temperature.

This control scheme is analogous to using a standard proportional-integral-derivative (PID) temperature controller to control a heater. In the lab set-up, as lights are removed with the distillate and the overheads temperature begins to climb above the set-point, the controller responds by increasing the fraction of time the diverter valve is switched OFF, reducing distillate take-out (analogous to turning OFF a heater to lower the temperature). If the temperature falls below the set-point, the controller responds by increasing the fraction of time the diverter valve is switched ON, increasing distillate take out, analogous to turning ON a heater to increase the temperature. Of course, this analogy does not refer to the distillation-pot heater which remains ON at all times. On scale-up to production, this thermostatic type of control logic is implemented by proportional metering of reflux and distillate flow rates instead of by using time-proportioned ON/OFF control. Either way, the process is set up to automatically manipulate reflux to remove a desired distillate cut – steadily increasing reflux and reducing distillate take off until all of the desired cut is removed (at total reflux). This provides an automatic and convenient stopping point. The operator then begins collecting the next cut simply by setting a new, higher set-point temperature.

The batch cycle time and productivity of batch rectification for a given column depends on the boilup rate and the required number of distillate cuts, as well as the separation power of the particular column. Because reflux and distillate flow rates change with time and the boilup rate can change with time, calculating overall cycle time is not a simple task. Furthermore, as discussed earlier, batch distillation configurations other than batch rectification may allow for shorter cycle times and should also be considered. In general, batch cycle time is determined by:
–   The types of products to be isolated (light, intermediate, or heavy) and the number of required product cuts.
–   Choice of batch configuration: regular rectification, inverted stripping, or middle-vessel fractionation, among other possibilities. Which one is most appropriate depends on the types of products to be isolated.
–   The ability of the column to make a sharp separation (its separation power). This depends on specific details of the column design (diameter, height, and type of internals) as well as the difficulty of separation and the required reflux flow rate.

- Choice of control scheme. The control scheme and any adjustments in boilup rate over the course of the distillation will impact reflux and distillate flows over time. Holding reflux flow constant or distillate flow constant are potential options, but other schemes (as in the previously mentioned manipulation of reflux to control overheads temperature) may give significantly better results.
- Non-productive times. Fill and drain times. Pre-heating, start-up, cooling, and cleaning times.
- Heat duty and boilup rate. This depends on heat transfer capability and may change over time, especially for jacketed distillation pots whose heat transfer capability diminishes as liquid level goes down and heat-transfer area decreases. This may be avoided or mitigated by using fed-batch operation.
- Fed-batch operation. Continuous or periodic refilling of the pot. This semi-continuous approach to batch processing often allows for shorter cycle time. It can also allow for efficient solvent exchange in which a new higher-boiling solvent is added to the pot to chase the existing solvent overhead.

A batch distillation simulator may be used to evaluate these options and identify a suitable operating scheme. Detailed discussions are available elsewhere [38–45].

Sometimes a batch distillation process involves addition of an azeotropic entrainer to aid in stripping sparingly soluble components overhead. Normally, the entrainer is chosen so the overheads condensate splits into two liquid layers, with the layer containing the majority of the entrainer automatically overflowing or underflowing (depending on the relative density of the entrainer) back to the column. The other layer is removed as distillate. This kind of process is often called a Dean-Stark distillation. It is another common laboratory technique. We discuss this kind of process in Section 8.1: Heterogeneous Azeotropic Distillation and Steam Distillation.

7.3.3 Continuous stripping and absorption

A common use of a continuously fed stripping column is to strip organic impurities from water, a distillation application called steam stripping [46–50]. The organic-laden aqueous feed is fed to the top of the stripping column, and steam is generated in a bottom reboiler or injected directly into the bottom from an outside source such as a utility boiler (called live steam or open steam injection). The latter eliminates the need for the reboiler. The overhead vapor is condensed, and in many cases the condensate splits into an organic layer and an aqueous layer (Figure 7.7). The organic layer may be decanted and processed for recycle or it may be disposed of in a suitable incinerator. The aqueous layer normally is sent back to the stripping process by mixing it with the incoming feed. The decanter actually serves the function of a rectification section by separating the two liquid layers. An analogous process involves stripping water out of a wet organic feed by reboiling the organic phase.

Typical stripping applications include:
- Steam stripping volatile organics from wastewater or brine prior to biotreatment or discharge to the environment (for example, reducing the organic concentration from 1000 ppm → 1 ppm).
- Air stripping volatile organic impurities from groundwater or wastewater (for example, 100 ppm → 1 ppm).
- Drying a reaction solvent to remove traces of water prior to conducting a water-sensitive organic reaction (for example, 5000 ppm or 500 ppm → 50 ppm or less).
- Stripping dissolved oxygen out of liquids (deoxygenation). Example liquids are reaction media such as toluene, glycerol, glycol, or some other organic liquid of moderate viscosity.

Table 7.1 lists some specific contaminants that can be stripped this way. These applications often involve use of a packed column, which is our focus here. Trayed columns may also be used. The same general considerations apply.

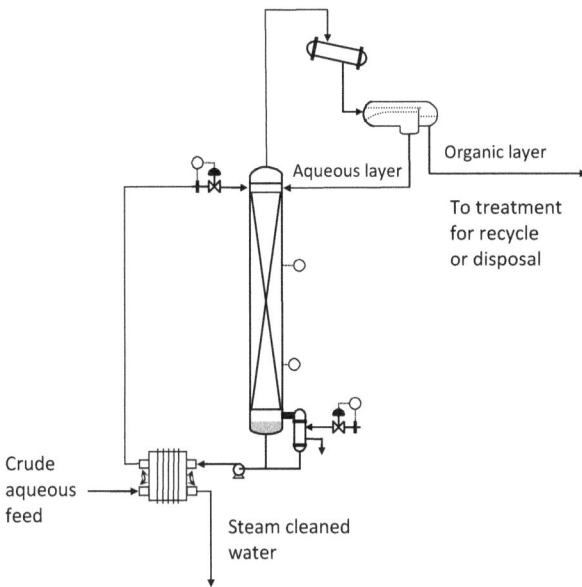

Figure 7.7: Steam Stripper for Removing Sparingly Miscible or Partially Miscible Organics from Water or Brine. The condenser and decanter serve the function of rectification – the overheads condensate spontaneously splits into two liquid phases. The bottoms stream heats the incoming feed to reduce energy consumption.

Stripping-mode distillation is an alternative to other separation methods that also may be used to remove contaminants from liquids, such as liquid-solid adsorption, liquid-liquid extraction, bio-treatment, and the use of membranes. It is a relatively simple

Table 7.1: Example Stripping Applications*.

Steam stripping sparingly soluble organics out of water (estimated values)		
Organic solute	Solubility of organic in water at 25 °C, ppm(wt)	$\alpha_{organic, water}$ near infinite dilution at 100 °C
toluene	500	8000
benzene	1130	7000
cyclohexane	63	130000
ethyl acetate	75000	120
o-dichlorobenzene	150	420
nitrobenzene	2500	100
naphthalene	20	1500
methyl isobutyl ketone	19000	180
1-hexanol	5700	75
n-decane	1	2×10^7

Stripping water out of hydrophobic solvents (estimated values)			
Solute	Solvent	Solubility of water in solvent at 25 °C, wt%	$\alpha_{water, organic}$ near infinite dilution at 100 °C
water	toluene	0.055%	475
water	benzene	0.072%	588
water	cyclohexane	0.007%	5500
water	ethyl acetate	3.2%	15

*Values are estimated from various sources of solubility and vapor pressure data, for illustration. For design purposes, values should be obtained from reliable databases or other suitable sources.

process that does not require regeneration, and it can enable recovery and recycle of the stripped components, especially when these components are sparingly or partially miscible in the feed solvent (through phase splitting of the overheads condensate).

At first glance, it may seem that designing such a simple-looking process should be much simpler than designing a dynamic batch distillation process or a continuously fed fractional distillation process. And in some respects it is. But stripping is far from straightforward in its fundamentals, and it serves as a good introduction to important design considerations for distillation in general, considerations that if not recognized and adequately addressed can easily lead to poor designs:

– The significant impact of non-ideal phase equilibria on distillation performance.
– The need to take into account both thermodynamic and kinetic factors (relative volatility and mass transfer resistance).
– Assessing which of these factors controls performance under what conditions; and related to this, appreciating differences in the use of theoretical stage versus transfer unit models.
– The significance of the transfer factor at low and at high values (the stripping factor in this case).

- The importance of phase diagrams and the benefit of knowing where on the diagram the process operates.
- The importance of uniform liquid and vapor distribution and the need to deal with a certain amount of backmixing of vapor and liquid within the column.
- The potential effect of salts (electrolytes).
- The impact of using a non-condensable stripping gas (air or N_2 stripping).
- How these factors can affect energy consumption.
- Points of comparison when considering other separation methods.

In the case of a stripping column, this boils down to knowing how to deal with two main design issues:
- How to set the stripping rate. How much boilup vapor is needed and when is using more vapor no longer effective? In other words, what is the most appropriate vapor to liquid ratio?
- How to provide sufficient vapor-liquid contacting. In particular, how tall should the column be?

To begin this discussion, consider the way stripping performance typically is modeled at dilute conditions. A stripping column operating at steady state normally is designed with constant vapor/liquid flow within the column to reduce the concentration of impurity in the feed by a constant ratio or reduction factor. This ratio, $F_R = x_{in}/x_{out}$, may be modelled using either theoretical stages or transfer units or the equivalent mass transfer coefficients:

$$F_R \equiv \frac{x_{in}}{x_{out}} \approx \frac{S^N - \left(\dfrac{1}{S}\right)}{1 - \left(\dfrac{1}{S}\right)} \approx S^N \text{ at large } S, \text{ in terms of theoretical stages} \qquad (7.5)$$

Equation (7.5) is not recommended whenever $S > 2$. In that case, the preferred model is

$$F_R \equiv \frac{x_{in}}{x_{out}} \approx \frac{\exp\left[NTU\left(1 - \dfrac{1}{S}\right)\right] - \left(\dfrac{1}{S}\right)}{1 - \left(\dfrac{1}{S}\right)} \approx \exp(NTU), \text{ at large } S \qquad (7.6)$$

The stripping factor is given by $S = m\frac{V}{L}$, and $m \approx a_{i, H_2O}$ at dilute conditions. So, in discussing stripping-mode distillation we focus on the stripping factor and the ratio V/L. This is the internal boilup ratio. At stripping factors greater than 2, a stripping column operates away from equilibrium and performance is dominated by the mass transfer rate. For this reason, relying on theoretical stages to characterize performance can give highly misleading results, because doing so assumes that stage efficiencies remain high as the value of S increases, which normally is not a realistic assumption.

Additional explanation is given in Section 7.3.5: Short-Cut Calculations. See Examples 7.7 to 7.9.

Equation (7.6) is derived from the Colburn equation discussed in Section 6.7.1: Mass Transfer Units and Mass Transfer Coefficients. An example of stripping column design is discussed in Section 7.3.5: Short-Cut Calculations (see Examples 7.6 to 7.9). Performance depends on the stripping factor (but only up to a value of 10 or so) and on the design of column internals and the column height. As a rule, beyond $S = 10$ or so, performance is solely determined by the number of transfer units the column can achieve, with essentially no improvement from further increase in stripping factor. That is, stripping performance no longer improves as more and more vapor is used – unless a change in the physical flow of vapor somehow impacts vapor-liquid contacting and the number of transfer units achieved by the column. Furthermore, when relative volatilities are very high as in Table 7.1, the practical minimum vapor/liquid ratio is determined not by considering the stripping factor, but by considering the ability of the column internals to uniformly distribute vapor within the column and for vapor flow to overcome pressure drop in the column and balance any heat losses. This may allow operation in reboiled or direct steam-injected columns at a vapor/liquid ratio as low as 0.05 or perhaps as low as 0.02 in special cases. At very low vapor/liquid ratios in columns with condensable stripping vapor, fluctuations in feed flow rate or temperature can easily result in the entire stripping vapor stream being condensed so that no overhead vapor leaves the column and no reduction in solute concentration occurs. Minimizing backmixing is an important consideration, as well. It does not take much backmixing to destroy separation performance when the vapor rate relative to the liquid rate is very small – like chimney smoke rising up against Niagara Falls. This may be an exaggeration, but it brings to mind a picture of just how challenging it can be to maintain performance at very low vapor to liquid rates, but a challenge worth pursuing to minimize energy consumption (as we discuss below).

Even considering all that can affect performance, a stripping column often operates with a constant reduction factor for a given vapor/liquid ratio and column internals (generally as indicated in eq. (7.6)). A common control method involves setting a constant steam to feed rate; that is, adjusting the boilup rate as the feed rate changes to maintain a constant ratio. Feed composition and feed rate normally are fairly constant, but may vary somewhat in operation, and this control scheme helps ensure sufficient boilup at all times. As the concentration of solute in the feed varies, so will the residual solute concentration in the bottoms, according to $F_R \approx$ constant. A typical column will be designed to give a reduction factor between 100 and 1000. Note that this is much higher than the single-stage distillation methods discussed earlier. So, a stripping column may be able to reduce an impurity concentration from 100 ppm in the feed down to 1 ppm or perhaps even as low as 0.1 ppm in the bottoms. Higher reduction factors may be achieved by using extra tall columns, greater than 10 meters or so (greater than 30 ft), and by paying very careful attention to the design of internals to minimize backmixing of vapor with liquid.

Sometimes, other factors can affect performance, especially when stripping to very low contaminant levels:

- Saturated feeds containing droplets of sparingly or partially soluble contaminants. Such feeds must first be processed through a coalescer/decanter to scrupulously remove all second phase liquid. A stripping column is useful only for removing *dissolved* contaminants. Second phase droplets can go in the column and come right out with little change because of low residence time and inadequate heat transfer for evaporation.
- The presence of dirt or other solid particles that slowly release contaminants during and after stripping. The residence time within a column usually is not sufficient to release all contaminants from suspended solids. Often liquid feeds must be pre-filtered in order to achieve ultra-low contaminant levels.
- Trace chemistry effects that tie up solute in a less strippable chemical form, or trace reactions that continue to produce solute due to chemical decomposition or other reactions.
- Exposure to contaminants after stripping, such as moisture in the air when trying to dry solvents or trace impurities left in equipment used for transport and storage. These factors can put a practical limit on how low contaminant levels can be taken and maintained without extraordinary measures.

All these possibilities should be evaluated when attempting to strip a feed down to very low contaminant levels.

A stripping column can be an attractive option compared to alternative technologies because it is simple and often robust and reliable in operation. It is sometimes used with other methods to obtain better performance. This may involve further treating the bottoms in an adsorption column or with biotreatment to further reduce contaminant levels. In this case, the stripping column removes the majority of the contaminant, and the adsorption column or biotreatment plant removes remaining trace amounts which distillation has difficulty removing. With adsorption, this approach can provide reliable removal down to very low contaminant levels without requiring frequent regeneration of the adsorbent.

Strippability. What makes a contaminant strippable? This has to do with both the inherent volatility of the contaminant (its pure component vapor pressure) and the magnitude of non-ideal molecular interactions in solution. Both factors contribute to strippability. In extreme cases, non-ideal interactions result in limited solubility of the contaminant in the feed solvent. In fact, many organic compounds and even some solids that are high boilers and much less volatile than water (as pure components) may be stripped out of water because they have very limited solubility in water (as in Table 7.1). For the same reason, water may be stripped out of non-water-miscible organic solvents.

To understand why, consider a sparingly miscible organic compound dissolved in water. The following discussion corresponds to a steam stripper as depicted in Figure 7.7.

It is not directly applicable to air or inert gas stripping where the bulk liquid phase is not at its boiling point at the column operating pressure, although the underlying principles do apply in general (more on this later).

A sparingly miscible system is a two-liquid-phase system that forms a heterogeneous azeotrope. This corresponds to a vapor + liquid + liquid phase diagram as shown in Figure 5.6. For background, see Section 5.6.3: Heterogeneous Azeotropes. The ability to strip an organic compound out of water is represented by its relative volatility with respect to water which depends on both the ratio of its pure component vapor pressure to that of water and the activity coefficient of the dissolved compound at dilute conditions:

$$a_{organic,\,water} = \gamma_{organic,\,water} \frac{p_{organic}^{sat}}{p_{water}^{sat}} \approx \gamma_{organic,\,water} \frac{p_{organic}^{sat}}{P_{total}} \tag{7.7}$$

where $\gamma_{organic,\,water}$ = activity coefficient of organic dissolved in water
$\quad\quad p_i^{sat}$ = pure component vapor pressure for component i
$\quad\quad P_{total}$ = total pressure

The activity coefficient for the organic dissolved in the aqueous phase can be related to solubility through liquid-liquid equilibrium. For a heterogeneous aqueous + organic system at equilibrium, the activity of an organic component distributed between the two liquid phases is the same in each phase:

$$a_i^I = a_i^{II} \tag{7.8}$$

$$(x_i\gamma_i)^I = (x_i\gamma_i)^{II} \tag{7.9}$$

where the superscripts I and II denote the organic and aqueous phases. For a sparingly miscible system, the organic phase is comprised mostly of organic. The small amount of dissolved water can be neglected for our purposes here. That means that the mole fraction and the activity coefficient for the organic component in the organic phase are nearly unity:

$$(x_i)^I \approx 1.0 \text{ and } (\gamma_i)^I \approx 1.0, \text{ so } (\gamma_i x_i)^I \approx 1.0 \tag{7.10}$$

Then, according to eq. (7.9), the activity coefficient for the organic component in the aqueous phase is obtained from:

$$(\gamma_i x_i)^{II} = 1.0, \text{ so } (\gamma_i)^{II} = (1/x_i)^{II} \tag{7.11}$$

We can then write the expression for relative volatility in a form explicitly showing the impact of limited solubility:

$$a_{organic,\,water} \approx \frac{1}{x_{organic,\,water}^{sat}} \frac{p_{organic}^{sat}}{P_{total}} \tag{7.12}$$

We can further relate this quantity to the vapor-liquid-liquid phase diagram by recognizing that the vapor phase composition of the heterogeneous azeotrope is determined by the vapor pressures exerted by both liquids, as described by Robbins [51]. For sparingly miscible systems, both liquids are almost pure components, so the vapor phase composition in terms of the fraction of organic in the vapor is given by

$$\alpha_{organic,\,water} \approx \frac{1}{x_{organic,\,water}^{sat}} \frac{P_{organic}^{sat}}{P_{total}} \tag{7.13}$$

Then,

$$\alpha_{organic,\,water}^{sat} \approx \frac{y_i^{AZ}}{x_{organic,\,water}^{sat}} \tag{7.14}$$

This is the relative volatility of organic dissolved in the aqueous phase at saturation; that is, at the solubility limit where the aqueous liquid is in equilibrium with the azeotropic vapor. For sparingly miscible systems, this limit is very low, so this relationship also approximates relative volatility at lower concentrations of organic in the aqueous phase. With the above expressions in hand, we can obtain a value for the relative volatility directly from the phase diagram (as shown in Figure 5.6), by locating two points on the diagram: (1) the heterogeneous azeotrope composition (point E in Figure 5.6); and (2) the solubility of the organic in water at the solubility limit (point C in Figure 5.6). This sparingly miscible system represents an extreme case, but it serves to illustrate the factors that affect strippability. Note that we have focused on the solubility limit for the aqueous liquid, not the organic liquid, because normally the organic liquid is not actually present in a steam stripper. The solubility limit for organic dissolved in the aqueous phase is all that matters. Of course, the same fundamental concepts apply when focusing on the organic phase; that is, when stripping dissolved water out of an organic liquid. In that case:

$$\alpha_{water,\,organic} \approx \frac{1}{x_{water,\,organic}^{sat}} \frac{P_{water}^{sat}}{P_{total}} \approx \frac{y_{water}^{AZ}}{x_{water,\,organic}^{sat}} \tag{7.15}$$

The properties of a heterogeneous azeotrope are discussed in more detail in Section 5.6: Phase Diagrams: Equilibrium Curves, Azeotropes, and Eutectics.

The same concepts are in play for partially miscible systems, although the activity coefficient must be extrapolated from the solubility limit to the composition of interest using a suitable activity coefficient correlation equation (as discussed in Section 5.5: Key Expressions of Phase Equilibria). Near infinite dilution, the appropriate expression is:

$$\alpha_{organic,\,water}^{\infty} = \gamma_{organic,\,water}^{\infty} \frac{P_{organic}^{sat}}{P_{water}^{sat}} \approx \exp\left(\frac{A}{RT}\right) \frac{P_{organic}^{sat}}{P_{total}} \tag{7.16}$$

where R is the ideal gas law constant, T is absolute temperature, and A is a constant that can be determined directly from azeotropic composition [52]. For completely miscible systems, solubility is not a factor, of course, but a non-ideal activity coefficient still is. The same expression as eq. (7.16) applies. If a homogeneous azeotrope is present, the constant A may be determined directly from the azeotropic composition. If not, the constant A may be determined by regression of available VLE data. It can also be estimated from the corresponding vapor-liquid phase diagram, as the relative volatility at dilute conditions is represented by the initial slope of the equilibrium line.

Note that the actual solution chemistry involved in stripper operation can be more complex due to the presence of electrolytes, sometimes causing severe fouling of the equipment during operation. Fouling is a common issue for air and steam strippers used to strip organics from ground water containing calcium carbonate and other minerals. In such cases, it may be necessary to adjust the pH of the feed water, as in conditioning boiler water to minimize scale formation [27, 53–55].

Henry's Law and Infinite Dilution Activity Coefficients. Henry's law constants often are used to model and analyze stripping applications. In this application, the Henry's law constant can be thought of as an effective vapor pressure:

$$H_i = \gamma_i^{\infty} p_i^{sat} = a_{i,j}^{\infty} P_{total} \tag{7.17}$$

Then, the stripping factor is given by

$$S = \frac{H_i}{P_{total}} \times \frac{V}{L}, \; at \; T_{strip} \tag{7.18}$$

Values of Henry's law constants have been reported for a large number of compounds dissolved in water including dissolved gases [56, 57]. Values of infinite dilution activity coefficients (also called limiting activity coefficients) also are available [49, 58, 59].

Salting-Out Effect. The addition of salt to an aqueous feed liquid generally increases the value of γ_i^{∞} and of H_i. The result is a significant increase in the strippability of organic contaminants. This effect is often expressed with the use of a Setschenow constant, as follows:

$$\ln \frac{\gamma_{i, brine}}{\gamma_{i, water}} = k_S \times C_{salt} \tag{7.19}$$

where $\gamma_{i,j}$ = activity coefficient for component i dissolved in j (brine or water)
$\quad\quad\; k_S$ = Setschenow constant (liter/g-mol)
$\quad\quad\; C_{salt}$ = salt concentration (g-mol/liter)

This relationship was first reported by Setschenow in 1889 [60]. The Setschenow constant is a function of the effective molar volume of the salt. Data on Setschenow constants for various common salts are available elsewhere. The rule of thumb for NaCl is that γ_i^{∞} doubles with every 7 wt% addition of salt. This explains why it is often considerably easier to strip contaminants from brine compared to stripping them from water [61].

Column Height. For steam stripping designs employing popular types and sizes of commercial column packing, the height of a transfer unit (H_{OL}) typically is in the range of 24 to 36 inches or about 0.6 to 0.9 meters. So, a typical column with 20 ft of packing (about 6 meters) can achieve between 6 and 10 transfer units. Since tray spacings in trayed columns are often in the range of 24 inches (0.6 m) and crossflow tray mass transfer efficiencies are often in the range of 0.6 to 1.0 *NTU* per tray, stripping columns with a 20 ft (6 m) trayed section can also typically achieve between 6 and 10 transfer units. This number is affected by liquid and vapor flow rates (liquid residence time and vapor velocity) and whether steps are taken to insure good (uniform) liquid distribution and minimal backmixing in packed columns and regular flow patterns and minimal entrainment in trayed columns. For more discussion, see Section 7.4: Equipment Design and Rating Methods.

Hydraulics. A steam stripper is a highly productive device in terms of liquid throughput per unit volume (liquid loading) because of the relatively low vapor to liquid ratios frequently employed. A typical liquid rate is 20 gallons per minute per ft^2 of packing (or about 50 m^3/h per m^2). Even higher rates are sometimes used. This is higher than the liquid loading in most liquid-liquid extraction columns (typically 5 to 15 gpm/ft^2 or 12 to 38 m^3/h per m^2). And it is much higher than a typical liquid loading in solid-liquid adsorption beds (about 1 or 2 gpm/ft^2 or up to about 5 m^3/h per m^2). The liquid flow rate that can be used in liquid-solid adsorption is limited by significant pressure drop for flow through the bed. But perhaps more importantly, good adsorption performance requires slow liquid flow to allow sufficient residence time for solute to diffuse into the adsorbent's internal pore structure, and this limits its volumetric throughput considerably compared to steam stripping. For this reason, when both technologies are effective, the use of steam strippers may be favored over the use of adsorbers, especially for commodity-scale operations, simply because many fewer strippers may be needed for the same job.

For trayed strippers, liquid loading can be similar to packed columns, but this depends on the type and design of tray as well as operating characteristics such as foaming tendencies. Designers think about the liquid carrying capacity of a tray, which is reflected in weir loading and downcomer capacity which can vary significantly, although weir loadings in the range of 10 gpm/inch of weir length (0.025 m^3/s/m) are quite typical in strippers. Weir length on a tray is the chordal width of the downcomer open area. Additional explanation is given elsewhere [62]. Unlike full fractional distillation, in stripping-mode distillation the vapor-traffic pressure drop typically is not a limiting factor affecting liquid flow rate, although this might become a constraint when stripping under vacuum to minimize operating temperature, as when stripping reactive monomers. We touch on the hydraulics of distillation columns in general in Section 7.3.4: Continuous Fractional Distillation. The liquid loading in the stripping section of a fractional distillation column is no higher and normally is less than that in a stripper because vapor loading usually is more significant.

Energy Consumption. Figure 7.8 shows two design curves for stripping a hypothetical contaminant at dilute conditions (calculated using eq. (7.6)). For illustration, here we focus on a sparingly miscible contaminant that has a relative volatility with respect to the feed solvent equal to 50. The resulting design curve is a plot of the number of transfer units (N_{OL}) versus V/L, for two different design goals (100× and 1000× reduction factors). This illustrates the trade-off between column height and the amount of stripping vapor needed by the process. As discussed earlier, due to hydraulic factors the practical lower limit on V/L for reboiled or direct injection strippers occurs at a V/L ratio of about 0.02. This might be feasible with very stable operating conditions (feed flow rate and temperature), but a somewhat higher value of V/L normally is used. At the other extreme, the height of the required stripping column can be minimized by operating at a V/L ratio ten times higher, at around 0.2, which is much too high to be economical. The practical optimum exists somewhere in between these limits. In this example, it appears to be about 0.05.

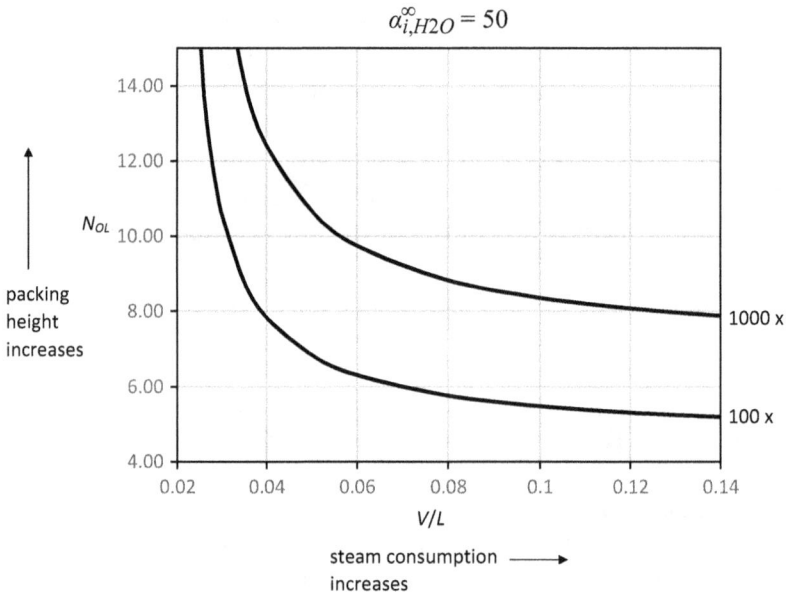

Figure 7.8: Typical Design Curves for Steam Stripping Sparingly Miscible or Partially Miscible Contaminants. The curves shown here illustrate approximate relationships.

For a steam stripper designed to remove sparingly miscible contaminants, the amount of steam required by the process is approximated by:

$$\frac{steam}{feed} = \frac{V}{L}\left(\frac{V}{L}+1\right)$$

(7.20)

where V/L is determined from analysis of relative volatility, stripping factor, and number of transfer units achieved by the column, as described above. The steam to feed ratio is somewhat greater than V/L because, in this case, the aqueous condensate from the overheads decanter is returned to the process (as shown in Figure 7.7). Here, we neglect any additional sensible heat needed to heat the incoming feed, which is normally a reasonable assumption when the incoming feed is cross-exchanged with the bottoms and heated by returning the overheads. The concentration of contaminant in the feed to the stripper is then given by:

$$x_{in} = \frac{x_{i,feed}F + x_{i,water}^{sat}V}{L}, \quad L = F + V \tag{7.21}$$

As stated earlier for the example design curves in Figure 7.8, we could choose to operate with $V/L = 0.2$ to minimize column height. But this corresponds to 0.24 kg steam/kg feed, which is too high. Increasing the height of the column can dramatically reduce this amount. As mentioned above, the economical optimum may be about $V/L = 0.05$, which corresponds to 0.0525 kg steam/kg feed, a considerable reduction in energy consumption. This corresponds to 33 kWh/metric ton of treated feed (51 Btu/lb) not including pumping costs or the cost of condensing the overheads. For minimum energy consumption, the overheads should not be subcooled too much below the condensing temperature. For discussion of energy consumption and efficiency for distillation in general, see Section 7.6: Energy Utilization.

As a rough general rule, a stripping application may be economical whenever relative volatility is greater than about 10. For systems with relative volatilities greater than 50, the practical minimum V/L ratio is determined by hydraulic considerations and may be about 0.05 or so, as discussed earlier. Of course, affordability is relative to what other methods can deliver. When the goal is to remove small amounts of contaminants from liquids, liquid-solid adsorption is a worthy alternative, one that normally also requires thermal energy for regeneration. The required amount of energy will depend on the concentration of contaminants and the adsorption capacity of the adsorbent. The use of membranes requires compression energy to overcome the mass transfer resistance of the membrane. Furthermore, because the permeating species is present in such low concentrations, achieving a reasonably high flux of contaminant species with high selectivity is a real challenge. So steam stripping can be a good option for purifying liquids, especially when dealing with sparingly miscible contaminants.

Using a Stripping Gas. Stripping with a non-condensable stripping gas such as air (or nitrogen) allows operation below the boiling point of the feed solvent. It has the potential to achieve the same performance as steam stripping while reducing energy consumption, but only under specific conditions involving stripping factors above $S = 10$ or so. At other conditions, air stripping will be significantly less effective than steam stripping. This all stems from the fact that a non-condensable gas essentially acts as a carrier gas for water vapor. Performance approaches that of steam stripping as the operating temperature increases and approaches the boiling point of water. The partial pressure of water is key to the effectiveness of the stripping carrier gas. The same concepts apply when using air to strip water from organics (substituting organic vapor for steam in that case).

The appropriate equations for air stripping follow from those described above for steam stripping. At dilute conditions:

$$S = \frac{p_{water}^{sat}}{P_{total}} \, \alpha_{organic, \, water}^{\infty} \, \frac{V}{L}, \quad at \; T_{strip} < T_{bp} \tag{7.22}$$

$$S = \frac{p_{water}^{sat}}{P_{total}} \left[\gamma_{organic, \, water}^{\infty} \, \frac{p_{organic}^{sat}}{p_{water}^{sat}} \right] \frac{V}{L}, \quad at \; T_{strip} < T_{bp} \tag{7.23}$$

$$S = \gamma_{organic, \, water}^{\infty} \, \frac{p_{organic}^{sat}}{P_{total}} \, \frac{V}{L}, \quad at \; T_{strip} < T_{bp} \tag{7.24}$$

So the equations are the same, but the specific vapor pressures used in the equations differ according to temperature. This is why air stripping can be less effective than steam stripping, and also why its effectiveness increases with increasing temperature, up to the boiling point.

For example, let's compare the relevant stripping factors at 100 °C and at 25 °C. This can be expressed as follows (all else the same):

$$\frac{S_{steam, \, 100 \, C}}{S_{air, \, 25 \, C}} = \frac{H_{organic, \, water} \, at \; 100 \; C}{H_{organic, \, water} \; at \; 25 \; C} = \left(\frac{\gamma_{organic, \, water} \, at \; 100 \; C}{\gamma_{organic, \, water} \; at \; 25 \; C} \right) \left(\frac{p_{organic}^{sat} \; at \; 100 \; C}{p_{organic}^{sat} \; at \; 25 \; C} \right) \tag{7.25}$$

$$\sim 0.5 \qquad\qquad\qquad \sim 10 \; to \; 100$$

So for many organics, steam stripping can be 5 to 50 times more effective (in terms of the stripping factor), depending on temperature. Of course, as discussed above, beyond a stripping factor of 10 or so, any further increase in stripping factor is no longer effective, and the two processes will then perform about the same (assuming no significant difference in mass transfer efficiency). It is at these very high stripping factors that air stripping can provide virtually the same separation performance as steam stripping and achieve a reduction in energy consumption.

But the process is not complete yet. The organic-laden stripping gas will need to be treated before discharge to the atmosphere. If a nearby furnace or other combustion device is able to accommodate it, the discharged stripping gas may be able to be used as supplemental combustion air with subsequent destruction of the removed organics. But if a suitable use for the discharged stripping gas is not available, it will be necessary to recover the contaminants as a concentrated liquid for recycle or for incineration (as with condensation of the overheads from a steam stripper). This might be accomplished by using pressure swing adsorption or by using steam-regenerated activated carbon.

Although these additional units also require energy, they will be processing a relatively small air stream (small compared to the liquid feed to the air stripper), so the resulting hybrid process scheme has the potential to reduce the total amount of energy consumed. Most likely, the hybrid process will also be significantly more expensive in terms of the initial capital cost. For more discussion of PSA and steam-regenerated activated carbon adsorption, see Section 10.5: Temperature Swing Adsorption and Section 10.6: Pressure Swing Adsorption.

In summary, designing a stripping column involves:

- Specify the required reduction factor (typically between 100 and 1000).
- Determine relative volatility for the contaminant (from database or measurement).
- Calculate the required number of transfer units, for a range of stripping factors.
- Determine the height of a transfer unit for the selected packing (with proper liquid distributor and bottom vapor distributor) or the number of transfer units per tray if trayed internals are selected.
- Choose a vapor-liquid ratio that gives good mass transfer performance with good economy (the lowest feasible V/L that gives the required separation). The minimum practical V/L may be on the order of 0.05 or perhaps somewhat lower for a very steady process.
- Specify the total height and diameter of the column with the goal of reducing the contaminant concentrations while also achieving low energy consumption.
- Compare air stripping with steam stripping. If the stripping factor is above 10, then air stripping may allow significant reduction in energy consumption (in combination with another method to treat the contaminant-laden air).
- Evaluate hybrid process schemes for advantages compared to using steam stripping alone.

The height of a transfer unit normally is constant for a given stripper design (for a dilute feed). A suitable value for H_{OL} may be estimated from various mass transfer correlations; however, it is best to measure it directly. Typical commercial H_{OL} values for packings at commercial scale are between 18 and 36 inches (0.45 and 0.9 m) when using a proper liquid distributor. This varies depending on the type and size of packing and the contaminant species of interest.

We end this section by noting another category of distillation design that is analogous to stripping column design in many ways. This is the design of an absorption column to scrub contaminants from a gaseous feed into a liquid. The solute transfers from the gas phase into the liquid phase instead of the other way around. The approach to design for dilute gas concentrations is analogous in that the relevant equilibrium and operating lines are often linear and the design equations have essentially the same form – expressed in terms of the Colburn Equation (see Section 6.7.1: Mass Transfer Units and Mass Transfer Coefficients). Although analogous in that sense, the equations are opposite in another sense because mass transfer is in the opposite direction. The relevant transfer factor is then the absorption factor, which is the reciprocal of the stripping factor we discussed earlier for air stripping (or gas stripping). The absorption factor A is given by:

$$A = \frac{1}{S} = \frac{P_{total}}{P_{water}^{sat}} \left(\frac{1}{\alpha_{organic,\,water}^{\infty}} \right) \frac{L}{V} \qquad (7.26)$$

The term "absorption" in the current discussion refers to transfer of solute from a non-condensable gas into a liquid. However, absorption is essentially the same as rectification in the design of a fractional distillation column which includes both stripping and absorption (rectification) within the same column. In the rectification section, solute transfers from the vapor phase into the liquid phase, analogous to transfer from gas to liquid. The relevant solute transfer factor used in analyzing the rectification section is in fact the absorption factor (Eq. 7.26). We will discuss absorption in the context of fractional distillation in the next section. We return to stand-alone absorption column design in Section 7.3.7: Reactive Absorption for Gas Scrubbing.

7.3.4 Continuous fractional distillation

Fractional distillation may be used to simultaneously recover and purify one or more components present in the feed. The term "continuous fractional distillation" refers to a continuously fed distillation column in which a feed stream is separated into a light overhead fraction and a heavy bottoms fraction. In many cases, one or more additional intermediate fractions are also removed from the column in one or more side draw streams. Such a column, often called a fractionator, may also have multiple feed points introduced at various locations along the length of the column. On the other hand, the term "batch fractional distillation" normally refers to batch rectification where different overhead cuts or fractions are taken off separately to generate products of different light/heavy concentrations, especially when the feed contains many components. That process is covered in Section 7.3.2: Batch Distillation with Reflux.

Although a wide variety of process simulation programs such as Aspen Plus, PRO/II, CHEMCAD, and others are commonly used today for the design of continuous fractional distillation, stripping, and absorption columns, long-established simplified methods such as the McCabe-Thiele graphical method for binary distillation can still offer visualization opportunities, providing valuable insights into distillation design and operation and highlighting key factors controlling performance. In the discussion below, many of the considerations and decisions that need to be made in conceptualization and design using modern simulation programs are illustrated and explained (visualized) by using the McCabe-Thiele method. An example McCabe-Thiele diagram is shown in Figure 7.9 (to be discussed later). Examples of McCabe-Thiele diagrams and instructions on how to prepare and use them are given in many textbooks. The same general concepts may also be applied in the analysis and design of liquid-liquid extraction and other processes

involving diffusion of solute between phases, driven by a deviation from equilibrium. In the discussion of the McCabe-Thiele method that follows, we offer recommendations concerning its use, but we assume the reader already is familiar with
the basic method.

The McCabe-Thiele method generally is considered a teaching method for illustrating factors involved in designing a distillation column, rather than a rating method. Its
most frequent use is to determine how many theoretical stages are required to achieve
a specified separation rather than how much separation a given number of theoretical
stages can achieve. However, using the McCabe-Thiele diagram to determine what distillate and bottoms concentrations will result from a given number of theoretical stages
is certainly possible, so it may also be used for rating purposes. Its usefulness in analyzing a distillation design is underscored by the practice of some engineers to construct a
McCabe-Thiele diagram from the output of a computer process simulation to visually
check the design.

The McCabe-Thiele diagram was introduced in a 1925 paper by Warren L. McCabe
and Ernest W. Thiele [63]. It followed the introduction of another graphical method by
Worth H. Rodebush in 1922 [64]. Prior to this, the calculation of distillation theoretical
stages was generally done analytically using vapor-liquid equilibrium relations and
plate-to-plate heat and material balances, as in methods introduced by Warren. K. Lewis
and Ernest Sorel (generally known as the Lewis-Sorel method, ca. 1915). However, in the
pre-computer era these calculations were tedious and time-consuming, and simplified
calculation methods were developed with the assumption that the degree of separation
in each stage was relatively small so that differential equations and integration could be
used to arrive at a number of required theoretical stages for a given separation. A shortcoming of these simplified methods was that as long as the number of theoretical stages
was large, the error in the required stage count was of little consequence, but for separations requiring few stages with larger stage-to-stage concentration changes the error
became appreciable. In the early 1920s detailed graphical methods were proposed by
Ponchon [65] and by Savarit [66], but these were quite complex and difficult to carry
out. The advantage of the McCabe-Thiele graphical method was that being an inherently
stepwise method, it did not rely on the assumption of a continuous relationship between
the equilibria and heat and material balances and thus was not subject to the constraint
of small stage-to-stage concentration changes – and it was fairly easy to use.

An assumption the McCabe-Thiele diagram *is* based on is that of constant molar
overflow (also called constant molal overflow), which maintains that the molar heat
of vaporization of the liquid in the column does not change appreciably with composition. As a result, the number of moles of vapor rising up the column (and those of
liquid going down) may be considered constant except for changes due to the introduction of feeds or the removal of intermediate streams (side draws) from the column. It then follows that the molar liquid-to-vapor ratio (L/V) of bypassing streams in
each section of the column is also constant, meaning that the slope of the operating
line representing the concentrations of those bypassing vapor and liquid streams is

constant, and the operating line is straight on a plot of vapor and liquid mole fractions with linear axis scales. Keep in mind that the mechanism of vapor-liquid contacting still involves differential vaporization of cooler falling liquid with concurrent condensation of hotter rising vapor, but the overall ratio of vapor to liquid is constant (or nearly so).

Another underlying requirement for the assumption of constant molar overflow is for the column environmental heat losses and the heats of mixing to be minimal with respect to reboiler and condenser duties. Environmental heat losses often are small in commercial-scale distillation columns where the column surface-area to volume ratio is relatively small and insulation systems are of high quality. And negligible heats of mixing are typical of mixtures that do not stray too far from ideality such as with components that are chemically similar (for example, homologous series and hydrocarbon separations).

In the McCabe-Thiele method, a diagram is constructed on a graph of vapor concentration on the y-axis versus liquid concentration on the x-axis. Although the method is primarily intended for use with binary mixtures only, it can be applied successfully to multicomponent mixtures through the use of non-standard axes or the consideration of only the light and heavy key components in the feed, distillate and bottoms streams. However, care must be taken to understand the limitations and constraints when using alternative approaches with the diagram for multicomponent applications. Here are ten important steps and considerations to keep in mind when using the McCabe-Thiele method:

1. Generate x-y axes with desired scale
2. Plot equilibrium line
3. Locate distillate, feed, and bottoms points
4. Plot rectifying operating line
5. Determine feed stream quality
6. Plot the feed stream q-line
7. Plot stripping operating line
8. Step off theoretical stages
9. Calculate total number of theoretical stages
10. Determine if mass transfer limitations are significant

Generate x-y Axes with Desired Scale. In the most straightforward case of a simple distillation of two components with percentage residual concentrations of the light in the bottoms and the heavy in the distillate, the diagram x and y axes are linear and constructed from 0 to 1 mole fraction of the light component in the liquid and vapor respectively. In cases of low relative volatility or operation close to minimum reflux ratio where some or most of the stage steps cover small concentration changes, it may be advantageous to construct a large diagram to make the stage steps more accurate and easier to draw.

However, there are also occasions where splitting the diagram into multiple plots with different scales is desirable such as when very pure distillate and/or bottoms streams are specified. When residual concentrations get into the fractional percentage or ppm range, magnifying the scales may be necessary in those sections of the column where concentrations become very low. But in the portions of the column where larger concentration changes occur in each stage, a conventional scale and diagram size is often appropriate. In the case of some strippers or rectifiers where low concentrations in the liquid (strippers) or vapor (rectifiers) exist throughout the entire column, a single diagram with a magnified scale from 0 to the feed concentration may be all that is needed. It should be noted that in some stripper and rectifier cases, the phase into which the removed component is transferring may exhibit significantly different concentrations of the removed component than in the feed phase. As a result, in these cases the x and y axes may need significantly different scales.

Another consideration for construction of the x and y axes arises in the case of strippers or absorbers where more than two components are present (e.g., air strippers, drying absorbers, gas scrubbers, etc.). In these cases, the use of mole ratios for the axes is generally more useful than mole fractions, particularly when the predominant components of the liquid and gas do not transfer appreciably between phases. For example, in a deoxygenating stripper where nitrogen is used to strip oxygen from a hydrocarbon, the x-axis might best be chosen to represent moles oxygen per mole hydrocarbon, and the y-axis might best be chosen as moles oxygen per mole nitrogen. Similarly, in a drying column or tower where sulfuric acid is used to remove water from chlorine gas, the x-axis could be moles water per mole sulfuric acid with the y-axis as moles water per mole chlorine. Using these axes with the liquid and gas flows on a solute-free basis, the L/V ratio will be essentially constant and the operating line will be essentially straight, allowing a conventional McCabe-Thiele diagram construction.

Plot Equilibrium Line. In plotting the vapor-liquid equilibrium (VLE) line on the x-y diagram, either equilibrium data or a VLE correlation can be used. In either case, the quality of the data or the correlation should be scrutinized and validated for the application being evaluated, especially for the operating conditions and concentration range over which the separation is to be performed. For instance, a particular VLE correlation for a pair of components may have been developed from data measured over a limited concentration range. Subsequent application of that correlation to a case covering a wider or possibly entirely different concentration range may, and often will, yield invalid results since solution characteristics and non-idealities (i.e., liquid phase activity coefficients) can change dramatically with concentration resulting in significantly different relative volatilities between components from one end of the column to the other. Conversely, a VLE correlation developed from full-range VLE data may not accurately represent relative volatilities at either end of the column where dilute solutions result in elevated or suppressed liquid phase activity coefficients compared to those in the middle of the column. As a result, in plotting the equilibrium line it may be necessary to draw on different sets of VLE data and/or correlations in different concentration ranges to properly evaluate the separation requirements with a McCabe-Thiele diagram.

In the case of mixtures exhibiting little or no liquid phase non-idealities, however, simplified VLE correlations, assumption of Raoult's law, or even the use of a constant relative volatility may be sufficiently accurate for preliminary design or range finding purposes. These types of mixtures are typically homologous series of hydrocarbons or compounds in the same chemical class close to the same molecular weight (such as aliphatic compounds and isomer separations). And in the case of strippers and absorbers where the component being removed is always dilute in the feed phase, the use of VLE K-values ($K = y/x$ for the component being removed) may be the most effective way to represent the equilibrium line which is usually nearly straight in these situations. Once the equilibrium line is plotted on the diagram, the $y = x$ diagonal line is plotted for reference purposes. This line becomes the operating line at total reflux and is used to locate the concentration points for the streams entering and exiting the column.

Locate Distillate, Feed, and Bottoms Points. The next step in the construction of a McCabe-Thiele diagram involves plotting the feed, distillate, and bottoms concentrations on the $y = x$ diagonal line on the diagram. These are the key plot points that define the range of the intended separation, the material balance envelope for the column, and the ends of the operating line constructions. For a total condenser, the distillate concentration point represents the concentrations of the bypassing vapor and liquid streams at the top of the column where the exiting vapor and the returning reflux have the same composition. As a result, it also represents the start of the operating line for the rectifying (or enriching or absorption) section of the column. Similarly, the bottoms concentration point on the $y = x$ diagonal line represents the start of the operating line for the stripping section of the column. But the feed concentration point on the $y = x$ diagonal, rather than lying on one of the operating lines, instead defines the beginning of the feed or "q" line, which allows the intersection of the rectifying and stripping section operating lines to be determined.

Plot Rectifying Operating Line. The rectifying section operating line is determined from the distillate (also the reflux) liquid composition and the reflux ratio chosen for design. The reflux ratio (external or internal) is used to calculate the molar liquid to vapor ratio (L/V) inside the rectifying section of the column which corresponds to the slope of the rectifying operating line. The internal reflux ratio is defined as the ratio of the molar reflux rate (L) to the molar vapor rate (V) leaving the top of the column which, for constant molal overflow, is also the L/V everywhere inside the rectifying section. However, since the internal reflux ratio is not usually known directly in practice, it is more customary to determine the L/V from the external reflux ratio (R_E), defined as the ratio of the molar reflux rate (L) to the molar distillate rate (D):

$$\frac{L}{V} = \frac{R_E}{R_E + 1}$$

(7.27)

If a partial condenser is used, the molar vapor distillate rate must be included in D when calculating the reflux ratio and the resulting L/V ratio in the rectifying section. With the rectifying section L/V as the slope of the rectifying operating line, the line is constructed by starting from the distillate concentration point on the $y = x$ diagonal and drawing the operating line to lower concentrations at the designated L/V slope. It is apparent from the L/V equation given above that because the denominator $R_E + 1$ is always greater than the numerator R_E, in almost all distillation cases the rectifying section L/V ratio and slope will be less than 1.0. However, at total reflux the slope will be 1.0 but will also result in zero productivity since the distillate and bottoms discharge rates as well as the column feed rate will be zero.

Determine Feed Stream Quality. The vaporization state or quality of the feed stream (or streams) determines where on the diagram the rectifying and stripping section operating lines intersect and subsequently, the plot of the stripping section operating line. Alternatively, the stripping section operating line can be determined directly from the reboiler boilup ratio in a similar manner as with the reflux ratio for the rectifying section operating line so that determining the feed stream quality is not necessary and can be skipped. However, for a given fixed reflux ratio, the required reboiler duty and boilup ratio are usually dependent on the feed stream quality, so it is normally more straightforward to use the feed stream quality to determine the intersection of the operating lines from which the stripping section operating line is plotted.

For a McCabe-Thiele diagram, the feed quality, q, is defined as the molar fraction of feed that consists of saturated liquid at the column operating pressure at the feed point. Therefore, if the feed stream is 100% saturated liquid at column conditions, the quality q is 1, and if it is 100% saturated vapor at column conditions, the quality q is 0. Then, if the feed is say, 75 mole% liquid and 25 mole% vapor, the quality q is 0.75. Now, if the feed is outside of the range between saturated liquid and vapor; that is, if it is subcooled liquid or superheated vapor, q is > 1 or < 0 respectively. In that case it must be calculated according to the feed enthalpy relative to the saturated liquid and vapor enthalpies. For the case of constant molal overflow where the molar heat of vaporization of the liquid is constant with respect to composition, the feed quality q can be calculated from material and enthalpy balances around the feed point:

$$q = \frac{H_V - H_F}{H_V - H_L} = \frac{H_V - H_F}{\lambda} \tag{7.28}$$

where H_V = molar enthalpy of vapor
H_F = molar enthalpy of feed evaluated at feed conditions
H_L = molar enthalpy of liquid
λ = molar latent heat of vaporization

This equation for feed quality indicates that q can be understood as the heat required to vaporize one mole of feed at its entering condition to a saturated vapor divided by the molar latent heat of vaporization. An alternative equation can be derived for "cold" subcooled liquid:

$$q = \frac{C_{pL}(T_{bp} - T_F) + \lambda}{\lambda} \tag{7.29}$$

and for "hot" superheated vapor:

$$q = \frac{C_{pV}(T_F - T_{dp})}{\lambda} \tag{7.30}$$

where C_{pL}, C_{pV} = specific heat of liquid and vapor, respectively
$\quad T_F \qquad$ = feed temperature
$\quad T_{bp} \qquad$ = feed bubble point
$\quad T_{dp} \qquad$ = feed dew point

Plot the Feed Stream q-Line. The next step in constructing a McCabe-Thiele diagram is to plot the feed line, also known as the q-line. The q-line is the locus of points representing the intersections of the rectifying and stripping section operating lines at various reflux ratios for the fixed set of feed conditions that results in the fixed value for the feed quality q calculated previously. With that value of q and the addition of component balances around the rectifying and stripping sections, the equation of the q-line can be determined and plotted on the diagram. However, since the q-line starting point on the y = x diagonal is known from the feed concentration, the full equation is not necessary – only the slope of the line is needed which is calculated as:

$$q_{slope} = \frac{q}{q-1} \tag{7.31}$$

It follows from this equation that for a saturated liquid feed where $q = 1$, the q-line slope is infinite so that the q-line is vertical on the x-y diagram, and for a saturated vapor feed where $q = 0$, the q-line slope is also 0 so that the q-line is horizontal. Then, if the feed is partially liquid and partially vapor so that q is between 0 and 1, it follows that the q-line slope will be negative and between 0 and minus infinity. Now, if the feed is outside of the range between saturated liquid and vapor, that is, subcooled liquid or superheated vapor, q is > 1 or < 0 respectively and the q-line slope will be positive. With the calculated q-line slope, the q-line is then plotted on the diagram starting from the x_F point on the y = x diagonal, and the intersection of the q-line with the rectifying section operating line is located graphically.

Plot Stripping Operating Line. Starting at the intersection of the q-line with the rectifying section operating line, the operating line for the stripping section is easily plotted as the straight line connecting the intersection point with the point on the y = x diagonal representing the column bottoms discharge concentration x_B. The slope of the stripping section operating line is again the molar L/V ratio in the section, although it is different from that in the rectifying section because of the additional liquid resulting from the introduction of the feed stream at the feed location at the top of the stripping section. With the additional liquid from the feed, the L/V in the stripping section will normally be greater than 1.0.

As stated previously, the stripping section operating line can optionally be plotted independently of the q-line and its intersection with the rectifying operating line with a knowledge of the stripping L/V ratio and x_B. In most cases, however, the stripping L/V is usually not easily calculated from the typical column specifications on which a McCabe-Thiele diagram is based. But if the conditions around the reboiler and column bottom happen to be specified rather than the conditions and reflux ratio around the top of the column, it may be more convenient to plot the stripping operating line in a similar manner as the rectifying operating line. This can be done from a determination of the stripping L/V ratio calculated from the external boilup ratio (R_B), defined as the ratio of the molar vapor rate leaving the reboiler (V) to the molar bottoms discharge rate (B):

$$\frac{L}{V} = \frac{R_B + 1}{R_B} = \frac{V + B}{V} \tag{7.32}$$

With the stripping section L/V as the slope of the stripping operating line, the line is constructed by starting from $x = x_B$ on the $y = x$ diagonal and drawing the line to higher concentrations at the calculated L/V slope until it intersects with the rectifying operating line. In a similar manner as with the rectifying operating line, it is apparent that because the numerator $R_B + 1$ in the final equation above is always greater than the denominator R_B, the stripping section L/V and slope will normally be greater than 1.0. At total reflux the slope will be 1.0 as in the rectifying section, but the column productivity will drop to zero since the feed, distillate, and bottoms rates will necessarily also be zero at total reflux. In whatever manner the stripping section operating line is constructed, now that the two operating lines and the equilibrium line are plotted on the diagram, the individual theoretical stages in the column can be drawn in terms of the concentration changes that each one accomplishes.

Equations for McCabe-Thiele Graphical Construction. General equations for the various lines used in constructing the McCabe-Thiele diagram can be derived in terms of parameters normally specified for a column design such as reflux ratio, distillate, feed, and bottoms concentrations, and feed quality. These equations can be helpful in computations and in developing automated McCabe-Thiele diagrams.

Rectifying Operating Line:

$$y_{ROL} = \frac{R_E}{R_E + 1} x_{ROL} + \frac{x_D}{R_E + 1} \tag{7.33}$$

Feed Stream q-Line:

$$y_q = \frac{q}{q - 1} x_q - \frac{x_F}{q - 1} \tag{7.34}$$

or,

$$x_q = \frac{y_q(q-1) + x_F}{q} \qquad (7.35)$$

q-Line/Rectifying Operating Line Intersection:

$$y_{int} = y_{ROL} = \frac{R_E x_F + q x_D}{q + R_E} \qquad (7.36)$$

and

$$x_{int} = x_{ROL} = \left(\frac{R_E x_F + q x_D}{q + R_E}\right)\left(\frac{q-1}{q}\right) + \frac{x_F}{q} \qquad (7.37)$$

Stripping Operating Line:

$$y_{SOL} = \frac{[R_E(x_F - x_B) + q(x_D - x_B)]x_{SOL} + x_B(x_F - x_D)}{x_F(R_E + 1) + x_D(q-1) - x_B(q + R_E)} \qquad (7.38)$$

where q = mole fraction of feed that is saturated liquid
R_E = external reflux ratio, L/D
x_D = mole fraction light component in distillate
x_B = mole fraction light component in bottoms
x_F = mole fraction light component in feed
x_q = liquid mole fraction light component along q-line
y_q = vapor mole fraction light component along q-line
x_{ROL} = liquid mole fraction light component along rectifying operating line
y_{ROL} = vapor mole fraction light component along rectifying operating line
x_{int} = liquid mole fraction light component at q-line/rectifying operating line intersection
y_{int} = vapor mole fraction light component at q-line/rectifying operating line intersection
x_{SOL} = liquid mole fraction light component along stripping operating line
y_{SOL} = vapor mole fraction light component along stripping operating line

Step Off Theoretical Stages. The theoretical stages on the McCabe-Thiele diagram are delineated by marking off the concentration changes across the stages with successive transitions or "steps" between the operating line and the equilibrium line where the points on the operating line represent the concentrations of bypassing vapor and liquid, and where those on the equilibrium line represent the concentrations of the vapor and liquid leaving each theoretical stage. These transitions model the operation of the column as a cascade of stages where the liquid and vapor enter, mix, reach equilibrium, and then separate with the vapor going up to the next stage and the liquid going down.

The first (top) stage step begins at the $x = x_D$ point where the rectifying operating line meets the $y = x$ diagonal. This point represents the vapor and liquid streams bypassing at the top of the top stage where, for a total condenser, the vapor entering the condenser and the liquid returning to the top stage have the same concentration. From the $x = x_D$ point on the diagonal, a horizontal line is drawn to the equilibrium line. This point on the equilibrium line represents the streams exiting the top stage, the vapor rising from the top stage to the condenser and the liquid at a lower light component concentration exiting the top stage and dropping to the stage below. To get the concentration of the vapor coming up from the stage below and bypassing the down-coming liquid from the top stage, a vertical line is drawn from the point on the equilibrium line back down to the rectifying operating line which is now separate from the y = x diagonal.

This construction of lines or "step" from the operating line to the equilibrium line and back to the operating line represents the performance of one theoretical stage at the specified column operating conditions. Subsequent stages in the column are represented with successive steps between the operating and equilibrium lines that progress down to lower and lower light component concentrations until the optimum feed point is reached at the intersection of the stripping and rectifying operating lines. For the case where the actual feed location matches the optimum feed location, when the next stage's horizontal line meets the equilibrium line at a liquid concentration lower than the concentration at the intersection of the operating lines, the vertical line going down to the concentration of the vapor coming up from the stripping section transitions to the stripping operating line rather than the rectifying operating line. The stage steps then progress down between the equilibrium and stripping operating lines until the final stage step results in a down-coming liquid concentration that is less than the concentration of the bottoms liquid discharged from the column, x_B.

Calculate the Total Number of Theoretical Stages. If the final stage step happens to fall right on or very close to the bottoms concentration x_B, the total number of theoretical stages required to perform the given separation is merely the sum of the full stage steps constructed on the diagram. However, if the final stage step results in a concentration appreciably less than x_B, it will be necessary to do a calculation of the final stage step to arrive at the number of theoretical stages required for the separation. If the diagram results in a large number of stages required like when the equilibrium and operating lines are close together, accounting for a fractional stage out of many is perhaps inconsequential. However, for cases where the number of stages required is small, the bottom fractional stage may represent a significant portion of the column being evaluated.

It is tempting to perform a simple linear interpolation of x_B with the liquid concentrations at the start and finish of the final stage such as:

$$N_B = \frac{x_{N-1} - x_B}{x_{N-1} - x_N} \tag{7.39}$$

where N_B = fractional theoretical stage required at column bottom (the last stage or partial stage)

x_B = liquid mole fraction exiting column bottom

x_{N-1} = liquid mole fraction entering final stage at column bottom

x_N = liquid mole fraction exiting final stage as a full theoretical stage

However, the relationship between theoretical stages and concentration is logarithmic rather than linear. This is demonstrated by the Kremser-Souders-Brown (KSB) equation for straight equilibrium and operating lines (Eq. 6–29.), restated here:

$$N = \frac{\ln\left[\left(\dfrac{x_{in} - y_{in}/m}{x_{out} - y_{in}/m}\right)\left(1 - 1/\mathcal{T}\right) + 1/\mathcal{T}\right]}{\ln \mathcal{T}} \tag{7.40}$$

where N = number of theoretical stages required for separation

x_{in} = mole fraction in feed phase entering stage cascade

x_{out} = mole fraction in feed phase exiting stage cascade

y_{in} = mole fraction in stripping phase entering stage cascade at feed phase exiting end

m = slope of equilibrium line

$\mathcal{T}$ = solute transfer factor, mV/L for stripping operations

For further discussion, see Section 6.6.1: Theoretical Stage Relationships.

In many distillation cases, the concentration of the light component in the bottoms stream leaving the stripping section is low (in the dilute linear range) and the stripping operating line is fairly straight, so the KSB equation may be used to quantify the last stage (or partial stage) such that:

$$N_B = \frac{\ln\left[\left(\dfrac{x_{N-1} - x_B/m}{x_B - x_B/m}\right)\left(1 - 1/\mathcal{T}\right) + 1/\mathcal{T}\right]}{\ln \mathcal{T}} \tag{7.41}$$

Here, because we need some measure of the mole fraction of theoretical vapor entering the fractional bottom stage, we have set $y_{in} = x_B$. This is a reasonable approximation in cases where the bottoms product is nearly pure. Although one may point out that a simple linear interpolation of x_B with x_N and x_{N-1} will give an answer that might be good enough for determining the last fractional stage on a McCabe-Thiele diagram, eq. (7.41) will give a more accurate and defensible estimate, especially for cases of high solute transfer factor. Then, the sum total of the full theoretical stages on the diagram plus the fractional bottom stage is the number of theoretical stages required to achieve the given separation of x_D and x_B from x_F.

Determine If Mass Transfer Limitations Are Significant. For systems with low relative volatilities, each theoretical stage only achieves a small change in concentration, and mass transfer rate limitations on separability are usually not significant. As a result, in those cases the separation is limited by the equilibrium driving force rather than the rate of mass transfer. However, if a column handling a system with high relative volatilities has a sufficiently high L/V in the rectifying section or a sufficiently low L/V in the stripping section such that the difference between the slopes of the equilibrium and operating lines is sufficiently large, individual theoretical stages are predicted to achieve quite large concentration changes, and the separation will be limited by the rate of mass transfer rather than the equilibrium driving force.

This difference in the slopes is quantified by the solute transfer factor which for stripping operations is the ratio of the slope of the equilibrium line, m, to the slope of the operating line, L/V, resulting in a solute transfer factor (or stripping factor) of mV/L. However, for rectifying (or absorption) operations, where the dilute component is the heavy component being transferred in the opposite direction from the vapor to the liquid phase, the solute transfer factor is the ratio of the slope of the operating line, L/V, to the slope of the equilibrium line, m. In this case, the solute transfer factor (or absorption factor) is $L/(mV)$, the reciprocal of the stripping factor. In both cases, however, the equations for accounting for mass transfer rate limitations in terms of solute transfer factor are the same.

Since at non-dilute concentrations of the components such as is typically found around the middle of the column there are ample amounts of both components present at the vapor-liquid interface for diffusion through the vapor and liquid films, mass transfer rates are unlikely to be limiting and departures from equilibrium limitations are likely minimal for non-dilute conditions. As a result, evaluation of the column for mass transfer rate limitations is mainly important around the ends of the column where one or the other of the components is dilute and potentially subject to vapor or liquid film diffusion resistance.

If mass transfer rate limitations are significant enough to affect column performance, the number of transfer units required to achieve the separation may differ significantly from the number of theoretical stages required. In cases of high solute transfer factors, the separation represented by a single theoretical stage may actually require several transfer units to achieve. As a result, in sections of a column where mass transfer limitations are significant, the number of transfer units must be considered rather than the number of theoretical stages, and the transfer unit calculation can be approximated by the equation:

$$NTU = N \frac{\ln \mathcal{T}}{1 - 1/\mathcal{T}} \qquad (7.42)$$

where NTU = number of mass transfer units in affected column section

$\quad N$ = number of theoretical stages in affected column section

$\quad \mathcal{T}$ = solute transfer factor for conditions in a given column section, mV/L or $L/(mV)$

For additional discussion, see Section 6.7.1: Mass Transfer Units and Mas Transfer Coefficients.

Evaluating this equation for different values of solute transfer factor, we see that a 10% difference between number of stages and number of transfer units arises with a solute transfer factor of only 1.215. But since mass transfer rate limitations are often only significant in the stages at the ends of the column as stated earlier, a solute transfer factor of 2.0 (which results in an NTU/N of 1.386, a 39% difference) is often used as a criterion for when mass transfer limitations need to be accounted for. However, different applications with different requirements and different design factors may require accounting for these limitations at solute transfer factors greater or less than 2.0. Depending on how relative volatility changes from the top to the bottom of the column, the respective L/V ratios, and the acceptable levels of purity of the distillate and bottoms products, accounting for mass transfer limitations may be required at both ends of the column, neither end, or the entire column.

Accounting for mass transfer limitations usually involves adjusting the height of packing equivalent to a theoretical stage (HETS or HETP) in a packed column section or the number of actual trays per theoretical stage (the reciprocal of the overall tray efficiency E_O) in a trayed column section using a variation on the equation above. Since there is the same height of packing or number of trays in the column section for which N and NTU are calculated, it follows that for packed columns:

$$\frac{NTU}{N} = \frac{HETP}{HTU} = \frac{\ln \mathcal{T}}{1 - 1/\mathcal{T}} \tag{7.43}$$

where $HETP$ = height of packing equivalent to a theoretical plate (stage) in affected
column section

HTU = height (H_{OL} or H_{OG}) of packing equivalent to a transfer unit (N_{OL} or N_{OG})
in affected column section

And, for trayed columns:

$$\frac{NTU}{N} = \frac{N_{TRAY}/N}{N_{TRAY}/NTU} = \frac{1/E_O}{N_{TRAY}/NTU} = \frac{\ln \mathcal{T}}{1 - 1/\mathcal{T}} \tag{7.44}$$

where N_{TRAY}/N = number of actual trays per theoretical stage in affected column
section

N_{TRAY}/NTU = number of actual trays per transfer unit (N_{OL} or N_{OG}) in the affected column section

E_O = fractional stage efficiency of actual tray

A general observation in practice is that the height of packing per transfer unit, HTU (that is, H_{OL} and H_{OG}) and the number of actual trays per transfer unit (N_{TRAY}/NTU) are much more constant and well behaved over a range of solute transfer factors than are $HETP$ values and tray efficiencies. Since at a solute transfer factor of unity ($\mathcal{T} = 1$), $N_{OL} = N_{OG} = N$ so that $H_{OL} = H_{OG} = HETP$ and similarly $N_{TRAY}/N_{OL} = N_{TRAY}/N_{OG} = 1/E_O$, an approximation is sometimes used such that the HTU at a high solute transfer factor is about the same as an $HETP$ measured at a low solute transfer factor in a low relative volatility system which is usually the case for published *HETP* data. Likewise, the approximation that the N_{TRAY}/NTU at a high solute transfer factor is about the same as the N_{TRAY}/N (or $1/E_O$) at a low solute transfer factor is sometimes used. Similar to HETP data, tray efficiencies are usually measured in low relative volatility systems that yield low solute transfer factors. These approximations lead to the following relationships:

$$\frac{HETP_{High}}{HETP_{Low}} = \frac{(N_{TRAY}/N)_{High}}{(N_{TRAY}/N)_{Low}} = \frac{(E_O)_{Low}}{(E_O)_{High}} = \frac{\ln \mathcal{T}_{High}/\left(1-\frac{1}{\mathcal{T}_{High}}\right)}{\ln \mathcal{T}_{Low}/\left(1-\frac{1}{\mathcal{T}_{Low}}\right)} \qquad (7.45)$$

where $HETP_{High}$ = *HETP* at the higher of two solute transfer factors

$HETP_{Low}$ = *HETP* at the lower of two solute transfer factors

$(N_{TRAY}/N)_{High}$ = number of actual trays per theoretical stage at the higher of two solute transfer factors

$(N_{TRAY}/N)_{Low}$ = number of actual trays per theoretical stage at the lower of two solute transfer factors

$(E_O)_{High}$ = fractional stage efficiency of an actual tray at the higher of two solute transfer factors

$(E_O)_{Low}$ = fractional stage efficiency of an actual tray at the lower of two solute transfer factors

$\mathcal{T}_{High}$ = the higher of two solute transfer factors

$\mathcal{T}_{Low}$ = the lower of two solute transfer factors

With these relations, an *HETP* or tray efficiency for a mass transfer limited portion of a column can be estimated from *HETP* or tray efficiency data from an equilibrium limited portion of the column or even, for that matter, measured on a different chemical system with a different solute transfer factor.

It should also be noted that although the solute transfer factor is usually evaluated for the key components to identify when corrections for *NTU/N* need to be made, there are situations where this is not sufficient. In many cases, the relative volatility between key components is low enough that the solute transfer factor for those components never rises above 2.0 so that *NTU/N* is always close to one and the entire column is equilibrium limited. However, there may be non-key minor or trace components which have significantly higher relative volatilities that result in solute transfer factors greater, sometimes much greater than 2.0, so the separation of those minor components is mass transfer limited. If there are important specifications for the separation of those components which need to be met, then adjustments to the actual stage count or height of packing may still need to be done for the segments of the column where the solute transfer factor for the minor components is greater than 2.0. Although this may result in over-separation of the key components, it would still be a necessary adjustment to ensure that all design specifications are met.

Example McCabe-Thiele Diagram.

Example 7.1. Methanol is to be recovered by distillation at a concentration of 95 mol% from a feed stream consisting of 55 mol% methanol/45 mol% isopropanol with a residual 5 mol% methanol in the bottoms discharge. The distillation is to be operated at atmospheric pressure and an external reflux ratio of 2.5 with a flashing feed resulting in 75 mol% liquid and 25 mol% vapor at column conditions. Determine the number of theoretical stages required.

Vapor-liquid equilibrium data for the methanol-isopropanol binary at atmospheric pressure [67, 68] is available online from a subset of the Dortmund Data Bank [69] made freely available for demonstration purposes:

Table 7.2: Methanol-Isopropanol Vapor-Liquid Equilibrium Data, 101.325 kPa.

T [deg K]	xMeOH [mole frac]	yMeOH [mole frac]
352.35	0.06800	0.14600
351.65	0.10600	0.20600
351.65	0.14100	0.26800
349.65	0.18400	0.33600
347.25	0.31000	0.50100
345.35	0.42500	0.61200
343.65	0.52400	0.68400
341.95	0.65200	0.78500
340.25	0.81100	0.88600
339.65	0.82400	0.88900
339.05	0.87800	0.92400
338.45	0.94700	0.96700

These data are plotted on the McCabe-Thiele x-y diagram as the equilibrium line. The methanol-isopropanol VLE exhibits a relative volatility of about 2, ranging from 1.6 at the methanol end to 2.3 at the isopropanol end of the concentration range. This reasonably good relative volatility allows for the required separation to be achieved without an excessive number of stages or a high reflux ratio, so it is well suited to a McCabe-Thiele analysis.

Next, the feed, distillate, and bottoms concentrations are located on the $y = x$ diagonal line. These points give the locations of one end of each of the feed q-line, the rectifying operating line, and the stripping operating line. If the boilup ratio is stated as a design basis then starting the diagram construction with the stripping operating line is a reasonable approach, but this is not often the case. Much more often the reflux ratio is given as the design basis which favors starting by constructing the rectifying operating line.

The rectifying operating line is constructed by starting at the x_D distillate concentration point on the $y = x$ diagonal and drawing the line toward the feed concentration of x_F at the L/V slope of $R_E/(R_E + 1)$, or 2.5/(2.5 + 1) = 0.714. The q-line is then plotted by starting at the x_F feed concentration point on the $y = x$ diagonal and proceeding to the left/top side of the diagonal at the q-line slope of $q/(q-1)$ where q is the feed quality. For the methanol-isopropanol example with the feed consisting of 75 mol% liquid, the slope is then 0.75/(0.75–1) or – 3.0.

With the q-line plotted, its intersection with the rectifying operating line can be located which allows the stripping operating line to be plotted without a determination of its slope by connecting that intersection with the x_B bottoms concentration point on the diagonal. The intersection can be located graphically by inspection of the diagram to this point, or its coordinates can be calculated according to eqs. (7.36) and (7.37). Then,

$$y_{int} = \frac{R_E x_F + q x_D}{q + R_E}$$

$$= \frac{2.5 \times 0.55 + 0.75 \times 0.95}{0.75 + 2.5} = 0.642$$

$$x_{int} = \left(\frac{R_E x_F + q x_D}{q + R_E}\right)\left(\frac{q - 1}{q}\right) + \frac{x_F}{q}$$

$$= \left(\frac{2.5 \times 0.55 + 0.75 \times 0.95}{0.75 + 2.5}\right)\left(\frac{0.75 - 1}{0.75}\right) + \frac{0.55}{0.75} = 0.519$$

With both operating lines constructed, the cascade of stages required for the separation can be plotted by laying out individual stage steps starting at the x_D distillate concentration point on the diagonal, moving horizontally to the equilibrium line, and then vertically back down to the operating line. This sequence is repeated for each subsequent theoretical stage until the transition to the equilibrium line results in an x-value that is less than the x-value of the rectifying operating line/q-line intersection point. For that stage, the transition back to the operating line goes to the stripping operating line rather than the rectifying operating line, and subsequent stage steps use the stripping line. The stage step sequence ends when the ending x-value of the next stage step becomes less than x_B, the bottoms concentration as shown in Figure 7.9.

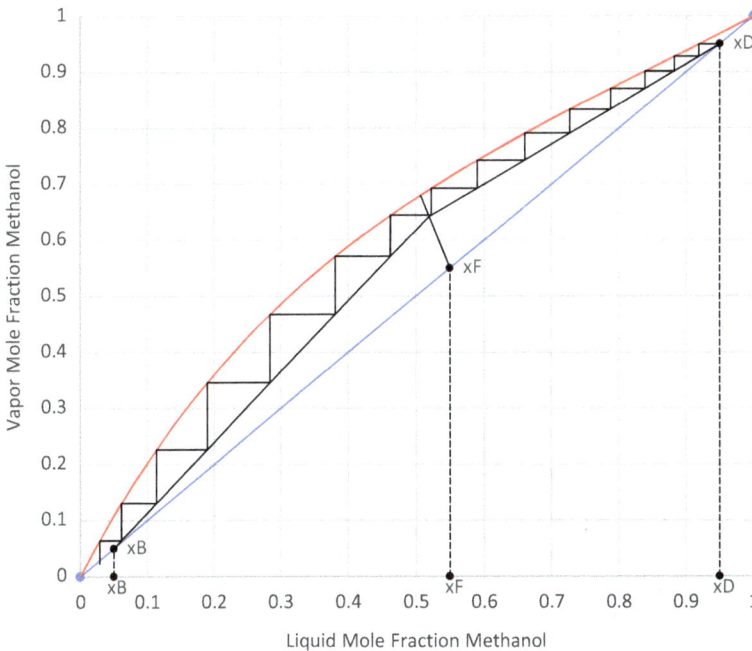

Figure 7.9: McCabe-Thiele Diagram for Example 7.1, Methanol + Isopropanol at 101.3 kPa.

Counting the stage steps, we can see that the recovery of 95% methanol with a residual 5% methanol in the bottoms discharge at a reflux ratio of 2.5 requires 14 theoretical stages plus a fractional stage. The fractional stage is calculated with eq. (7.41), restated here:

$$N_B = \frac{\ln\left[\left(\frac{x_{N-1}-x_B/m}{x_B-x_B/m}\right)(1-1/\mathcal{T})+1/\mathcal{T}\right]}{\ln \mathcal{T}}$$

where N_B = fractional theoretical stage required at column bottom
x_B = liquid mole fraction exiting column bottom
x_{N-1} = liquid mole fraction entering final stage at column bottom
m = slope of equilibrium line
$\mathcal{T}$ = solute transfer factor, mV/L for stripping operations

For this example, x_{N-1}, m, and $\mathcal{T}$ need to be calculated before N_B can be calculated. The liquid concentration x_{N-1} leaving the next-to-last stage at the bottom of the column may be measured off of the stage step diagram just generated as 0.061 mole fraction. The slope of the equilibrium line m can be closely approximated in the dilute concentration region of the bottoms concentration as the equilibrium K-value, or y/x. Using the data point with the most dilute concentration from the equilibrium data in Table 7.2, we get:

$$m = K = \frac{y}{x} = \frac{0.146}{0.068} = 2.15$$

This value for m can also be used to calculate the solute transfer factor $\mathcal{T}$, but we also need a value for L/V, the slope of the stripping operating line. This can be calculated from the x and y values of the intersection point and the x_B point on the y = x diagonal:

$$\frac{L}{V} = \frac{y_{int}-x_B}{x_{int}-x_B} = \frac{0.642-0.05}{0.519-0.05} = 1.26$$

And,

$$\mathcal{T} = \frac{mV}{L} = \frac{K}{L/V} = \frac{2.15}{1.26} = 1.70$$

The fractional bottom stage can then be calculated as:

$$N_B = \frac{\ln\left[\left(\frac{0.061-0.05/2.15}{0.05-0.05/2.15}\right)(1-1/1.7)+1/1.7\right]}{\ln 1.7} = 0.295$$

The total number of theoretical stages required for the methanol-isopropanol separation is then the sum of this fractional bottom stage and the integral number of whole theoretical stages stepped off prior to the fractional bottom stage, or 14.3.

Uses of the McCabe-Thiele Diagram. With the development of process simulation computer programs, McCabe-Thiele diagrams have fallen into disuse for general column design. However, even in the age of simulators and spreadsheets, the McCabe-Thiele diagram can be a valuable tool in visualizing and analyzing column performance and potential operating issues. Potential uses include:

1. Identifying pinch points: A pinch point in a column occurs due to insufficient reflux and reboiler boilup that results in a close approach to or crossing of the operating line and the equilibrium line. With a close approach, a very large number of stages is required for the desired separation, but with a crossing of the lines the separation cannot be achieved even with an infinite number of stages. Where it can be difficult to diagnose a pinch point from the output of a simulation, a pinch point is quite obvious on a McCabe-Thiele diagram even before the stage steps are drawn by looking at the construction of the operating and equilibrium lines and evaluating their orientation. In addition, how much the reflux ratio needs to be increased to eliminate the pinch can be estimated by drawing trial operating lines and adjusting them to allow an appropriate distance between the equilibrium and operating lines.

2. Identifying improper feed location: When the actual feed location differs significantly from the optimum feed location, the number of stages required can again become excessive or even infinite for the desired separation. If the actual feed location is say, 2 stages below the optimum location, the stage steps continue along the rectifying operating line for the actual stage count, extending beyond the intersection between the rectifying operating line and the q-line. Since the operating line will eventually cross the equilibrium line if extended far enough, the improper feed location can result in a pinch point if it varies from the optimum location by a large enough margin.

3. Identifying insufficient or excessive reflux ratios: Examination of the x-y diagram, the distance between the operating and equilibrium lines, and the number of stage steps constructed can often give an idea of the relation between reflux/boilup ratio and number of stages, particularly the relative decrease/increase in number of stages for a corresponding increase/decrease in reflux or boilup ratio. In this manner, an approximate judgment can be made of whether a small increase in stages can result in a significant reduction in reflux ratio and therefore energy consumption, or whether a small increase in reflux ratio would result in a significant reduction in stages and therefore capital cost.

4. Evaluating the need for or potential benefit from intermediate reboilers: Another evaluation that can be made is whether the addition of an intermediate reboiler in the column would result in a modification to the shape of the operating lines and subsequently a significant reduction in the required number of theoretical stages required.

5. Identifying other opportunities for column optimization: McCabe-Thiele diagrams have a number of other uses for column optimization and worker education. Included in these are: use in spotting non-optimum feed thermal conditions such as with a severely subcooled or superheated feed which could result in a pinch; demonstrating the effects of changing feed or product concentrations; evaluating the addition of side draws or multiple feed locations; use as screening tools

to understand the general aspects of and potential issues with proposed applications prior to performing computer simulations and providing starting estimates; and use as training aids for new engineers and operators.

Other Graphical Design Methods. Where the McCabe-Thiele diagram is based on the assumption of constant molar heats of vaporization and constant molal overflow, a Ponchon-Savarit diagram allows for varying heats of vaporization and mixing with the result that it is constructed on a vapor-liquid enthalpy-concentration diagram rather than a simple x-y concentration diagram. However, both the Ponchon-Savarit and McCabe-Thiele methods are based on the assumption of adiabatic column operation.

As a result of including enthalpy balances in its formulation, the Ponchon-Savarit diagram is a more rigorous methodology than the McCabe-Thiele and can give more accurate results, presuming that the enthalpy-concentration relationships are also represented accurately. However, that improved rigor and accuracy is accompanied with increased complexity and difficulty, to the point where its use has become obsolete with the advent of powerful computer simulation software with multiple rigorous calculation algorithms and extensive physical property databanks. On the other hand, its use may be helpful for some applications in visualizing the enthalpic effects of column operation and opportunities for energy conservation and management such as intermediate reboilers or condensers. Ponchon-Savarit diagrams have also been proposed for use in the design of reactive distillation columns [70].

In the Ponchon-Savarit diagram, enthalpies of the vapor and liquid are represented on the vertical axis, and vapor and liquid mole fractions are represented on the horizontal axis. The vapor-liquid equilibrium relationship is plotted as two separate curves: the enthalpy-concentration curve for the saturated liquid and another one for the saturated vapor at equilibrium conditions. These two curves are connected by a series of tie lines between saturated liquid enthalpy-concentration points and the corresponding saturated vapor enthalpy-concentration points in equilibrium with the liquid. The enthalpy-concentration relationships of the bypassing streams in the column are generated through the use of difference points (also known as net flow points or pivot points) for the column reboiler and condenser. The reboiler difference point is located at $(x_B, h_B\text{-}q_B)$ and the condenser difference point is located at $(x_D, h_D\text{-}q_D)$ where h is the enthalpy per unit mole or mass of the stream indicated by the subscript and where q is the amount of heat added to the column at the location indicated by the subscript. As a result, q_B is the reboiler duty, a positive number, and q_D is the condenser duty, a negative number.

In a Ponchon-Savarit diagram, column stages are stepped off by starting with the concentration at one end of the column, for example, the condenser or distillate end. In that case, a straight line would be drawn from that concentration on the liquid enthalpy line up to the condenser difference point and the intersection on the vapor enthalpy line (at the same concentration as the liquid distillate in the case of a total condenser) would be located. The corresponding equilibrium tie line would then be followed from that vapor enthalpy-concentration point to the liquid enthalpy-concentration point in equilibrium with it. A subsequent line would then be drawn from that liquid point back up to the condenser difference point to establish the next stage's bypassing streams and its vapor enthalpy-concentration point at the intersection on the vapor enthalpy line. The tie line from that point back to the liquid enthalpy-concentration point in equilibrium with it is drawn, followed by another line back up to the condenser difference point to establish the next theoretical stage in the column. This sequence continues until the feed concentration is crossed by a tie line after which the reboiler difference point is used to draw the lines corresponding to the concentrations of the bypassing vapor and liquid lines. When the bottoms concentration is crossed by an equilibrium tie line, the diagram construction is complete and if necessary, the fractional remaining stage is calculated from the last tie line.

In addition to the difficulty of determining the equilibrium enthalpy-concentration relationships noted earlier, when high reflux ratios are required because of tight relative volatilities, the condenser and reboiler difference points can be located far from the equilibrium lines and each other, and the diagram construction can become physically ponderous which can compromise accuracy. As a result of this and the ready availability of rigorous computer process simulators, Ponchon-Savarit diagrams are currently rarely used.

7.3.5 Short-cut calculations

Often situations arise in distillation design or troubleshooting where a quick answer is needed to a question or problem that doesn't warrant or allow the time and effort involved in developing a computer process simulation or a graphical construct like a McCabe-Thiele diagram. In these cases, a shortcut calculation using pencil, paper, and calculator or a spreadsheet or simple single-purpose computer program may be the best approach to arriving at an answer to questions regarding required number of theoretical stages, reflux ratio, column diameter, etc. Although shortcut methods are not rigorous and only yield approximate results and estimates, that may be all that is needed to allow decisions to be made regarding design approach, need for deeper analysis or more accurate calculations, or preliminary equipment selection.

Minimum Number of Theoretical Stages (Fenske Equation). One of the first questions to be addressed in distillation design is how many theoretical stages will be required to perform a given separation, and a quick answer can often be provided by applying the Fenske equation for the minimum number of theoretical stages, such as would be achieved for a given separation at total reflux (infinite external reflux ratio). This equation, in principle directly applicable to simple binary mixtures, was developed in 1932 during studies of the compositions and distillation characteristics of Pennsylvania crude oils by Merrell Fenske (1904–1971), a professor of chemical engineering at Pennsylvania State University [71]:

$$N_{min} = \frac{\ln\left[\frac{(x_1/x_2)_D}{(x_1/x_2)_B}\right]}{\ln \alpha_{12}}$$

(7.46)

where N_{min} = minimum number of theoretical stages
$\qquad x_1, x_2$ = mole fraction Component 1 and 2
$\qquad D, B$ = labels for distillate and bottoms streams respectively
$\qquad \alpha_{12}$ = relative volatility of Component 1 to Component 2

This famous equation relates the overall separation power of a fractionation process to the thermodynamic separation factor (relative volatility) and the minimum number of stages:

$$\text{Separation Power} \equiv \frac{(x_1/x_2)_D}{(x_1/x_2)_B} = (\text{Separation Factor})^{N_{min}}$$

(7.47)

It shows how multiple stages of countercurrent processing magnify the separation factor. For more discussion, see Section 6.1.3: Partition Ratio, Separation Factor, and Transfer Factor and Section 6.5.1: Material Balance Split and Separation Power.

Typically, the light or lowest boiling point component of the two is chosen as Component 1 and the heavy component is chosen as Component 2. However, the same numerical result is obtained if the component assignments are reversed as long as the concentration ratios and relative volatility are calculated on a consistent basis. Furthermore, in cases where the relative volatility of the two components to be separated is fairly constant from the top to the bottom of the distillation column, the equation can be applied as-is, but when the relative volatility varies significantly, a geometric mean of the relative volatilities should be used, for example:

$$(\alpha_{12})_{avg} = \sqrt{(\alpha_{12})_{top}(\alpha_{12})_{btm}}$$

(7.48)

It should be noted that the Fenske equation delivers only an estimate of the number of theoretical stages at total reflux and is really only applicable to well-behaved zeotropic mixtures; that is, non-azeotropic mixtures that do not form two liquid phases or that are not otherwise highly non-ideal. However, because of its simplicity and ease of use, it is often a valuable tool in the initial evaluation of the difficulty of a proposed separation.

Example 7.2. In the atmospheric pressure methanol-isopropanol separation described in Example 7.1, estimate the minimum number of theoretical stages required to achieve the desired distillate quality of 95 mol% methanol with 5 mol% residual methanol in the bottoms.

From the methanol-isopropanol vapor-liquid equilibrium data given in Table 7.2, we can see that the relative volatility of methanol to isopropanol varies from 2.34 at the lowest methanol concentration to 1.64 at the highest methanol concentration. Because of this significant difference in relative volatility, a geometric average relative volatility should be used in applying the Fenske equation:

$$(a_{12})_{avg} = \sqrt{(a_{12})_{top}(a_{12})_{btm}} = \sqrt{(2.34)(1.64)} = 1.96$$

Substituting this value along with the distillate and bottoms concentrations into the Fenske equation then yields:

$$N_{min} = \frac{\ln\left[\frac{(x_1/x_2)_D}{(x_1/x_2)_B}\right]}{\ln a_{12}} = \frac{\ln\left[\frac{0.95/0.05}{0.05/0.95}\right]}{\ln 1.96} = 8.75$$

Although the Fenske equation is clearly a tool for two-component systems, it is also applicable in evaluating the separability of multicomponent mixtures. The most common method of application to multicomponent mixtures involves identifying two components which represent the most critical desired separation in the mixture and thus constitute a controlling factor in column design. These two components are then identified as the light key and heavy key components and are used as Components 1 and 2 in the equation. In cases where there are more than two components with critical separation requirements, the different binary component pairs among the critical components can be identified and applied in the Fenske equation. A comparison of those results can determine which pair represents the most demanding separation requirement and the largest minimum number of theoretical stages.

Minimum Reflux Ratio (Underwood Equation). When evaluating a proposed distillation, in addition to estimating the minimum number of theoretical stages at infinite reflux (total reflux), it is also useful to estimate the minimum reflux ratio that would be required if an infinite number of theoretical stages were available. This is the purpose of the equation published by Arthur J. V. Underwood in 1946 and 1948 [72, 73]. The derivation of the Underwood equation is quite involved, but its application is fairly straightforward in the following general form for multicomponent mixtures:

$$R_{E,min} + 1 = \sum_{i=1}^{n} \frac{a_{iH}X_{i,D}}{a_{iH} - \theta} \tag{7.49}$$

where $R_{E,min}$ = minimum external reflux ratio
  n = number of components
  a_{iH} = relative volatility of component i with respect to the heavy key component

$x_{i,D}$ = mole fraction component i in the distillate
θ = Underwood constant, as calculated below

The value of θ to use in the Underwood equation lies between the values of the relative volatility of the light key component and that of the heavy key component (that is, 1.0). It was calculated by Underwood from:

$$\sum_{i=1}^{n} \frac{\alpha_{iH} x_{i,F}}{\alpha_{iH} - \theta} = 1 - q \tag{7.50}$$

where $x_{i,F}$ = mole fraction component i in feed stream
q = feed stream thermal quality ($q = 1$ for saturated liquid feed)

For a simple binary system, these equations reduce to:

$$R_{E,min} + 1 = \frac{\alpha_{12} x_{1,D}}{\alpha_{12} - \theta} + \frac{x_{2,D}}{1 - \theta} \tag{7.51}$$

and,

$$\frac{\alpha_{12} x_{1,F}}{\alpha_{12} - \theta} + \frac{x_{2,F}}{1 - \theta} = 1 - q \tag{7.52}$$

One issue with the Underwood equation is that the non-linear expression for θ in eq. (7.50) does not have an analytical solution for θ, so it must be solved numerically. This does not pose a significant problem to anyone with ready access to a computer and software such as a spreadsheet or math software, but it does mean that a general and direct solution to the equation is not possible. The Underwood equation must be applied with the parameters for a specific case and then solved numerically for that case.

Example 7.3. In the atmospheric pressure methanol-isopropanol separation described in Example 7.1, estimate the minimum external reflux ratio required to achieve the desired distillate quality of 95 mol% methanol from a 75 mol% liquid flashing feed containing 55 mol% methanol with 5 mol% residual methanol in the bottoms.

The first step is to determine the value of the Underwood constant q for the case. Since the relative volatility in the column varies from 1.64 to 2.34 as noted in Example 7.2 earlier, it is again appropriate to use the geometric average of these values for the relative volatility, 1.96. Substituting this and the feed concentration and quality values into the equation for θ we get:

$$\frac{(1.96)(0.55)}{1.96 - \theta} + \frac{0.45}{1 - \theta} = 1 - 0.75$$

$$\frac{1.078}{1.96 - \theta} + \frac{0.45}{1 - \theta} - 0.25 = 0$$

Applying a simple spreadsheet solver function to this equation to find its root yields a value of $\theta = 1.316$. Subsequent substitution of this value and the other parameters into the minimum reflux equation then results in:

$$R_{E,min} + 1 = \frac{(1.96)(0.95)}{1.96 - 1.316} + \frac{0.05}{1 - 1.316}$$

$$R_{E,min} = \frac{1.862}{0.644} + \frac{0.05}{-0.316} - 1 = 1.733$$

Estimating Actual Number of Theoretical Stages and Reflux Ratio (Gilliland Correlation). In addition to estimating the minimum number of stages and minimum reflux ratio required for a separation, it is also valuable to be able to estimate the actual number of theoretical stages required for a separation at a specified non-minimum reflux ratio without going through the time and effort involved in developing rigorous vapor-liquid equilibrium parameters and computer simulations. Perhaps the most popular method for doing so is the correlation developed by Edwin R. Gilliland (1909–1973) and first published in 1940 [74]. Gilliland analyzed more than 50 distillation cases including both binary and multicomponent systems, and the resulting correlation is still used today. Although the Gilliland correlation has been implemented numerically as an estimation method in some process simulation computer programs, it is often applied graphically as displayed on a logarithmic plot as shown in Figure 7.10.

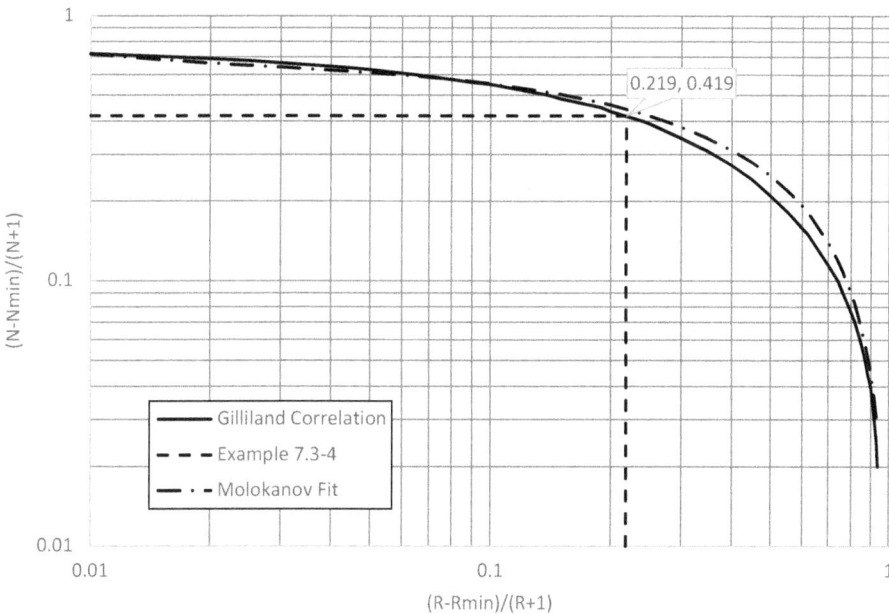

Figure 7.10: Gilliland Correlation for Stages vs. Reflux Ratio. Developed from the plot in Separation Processes, by C. J. King [75].

Examining the abscissa and ordinate axis functions in Gilliland's correlation, it is apparent that the actual reflux ratio and actual number of theoretical stages are correlated with respect to their minimum values such that the axis functions will always conveniently fall between 0 and 1. In this manner, a wide range of distillation applications with differing relative volatilities, separation factors, and operating conditions can be compared to each other and represented on the same plot. It should also be noted that although the Gilliland correlation is usually presented as a single universal relationship between stages and reflux ratio, Gilliland himself noted that the correlation, in addition to being limited by errors in the estimation methods used to determine minimum stages and minimum reflux ratio, is also affected by differences in the feed quality among other factors. In fact, representations of Gilliland's correlation in different sources sometimes differ from each other by small but noticeable amounts depending on whether authors use Gilliland's original curve drawn through the data points he used or whether authors draw their own curves through the same data points. For instance, the resulting ordinate value at an abscissa value of 0.06 from the Gilliland correlation plot in King's Separation Processes [75] from which Figure 7.11 is derived is 0.61. This ordinate value is also reflected in the Gilliland correlation plot in Kister's Distillation Design [62]. However, the ordinate value at the same abscissa on the Gilliland plots in Perry's 4th edition [76] as well as McCabe, Smith, and Harriott's Unit Operations of Chemical Engineering (7th ed.) [77] is only 0.58, a 5% difference. As a result, the accuracy of the Gilliland correlation is limited in a number of aspects and should only be used for preliminary design estimates or for determining starting values for more detailed and rigorous design methods.

In practice, the Gilliland correlation is often applied together with the Fenske and Underwood equations for minimum stages and reflux ratio as a combined methodology for preliminary distillation design estimates and is sometimes referred to as the "FUG" shortcut method. Requiring only values for the relative volatilities, feed, distillate, and bottoms specifications, and feed quality, this can be a very quick method of exploring the feasibility of proposed separations. Although other correlations for stages versus reflux ratio are available and computer process simulation programs can give similar or better results, they usually are more complex and time-consuming to apply effectively. Because the FUG method is so easily applied, it is still used today, even though its origins are decades old.

Example 7.4. In the atmospheric pressure methanol-isopropanol separation described in Example 7.1, estimate the actual number of theoretical stages required if the distillation is operated at an external reflux ratio of 2.5.

Using the Fenske equation results from Example 7.2 and the Underwood equation results from Example 7.3, the minimum number of theoretical stages required is 8.75, and the minimum reflux ratio is 1.733. The design point on the abscissa of the Gilliland correlation plot in Figure 7.10 corresponding to the stated design reflux ratio of 2.5 is then:

$$\frac{R - R_{min}}{R + 1} = \frac{2.5 - 1.733}{2.5 + 1} = 0.219$$

Locating this value on the abscissa of the plot and finding the corresponding point on the correlation curve yields a value on the ordinate of 0.419 as shown in Figure 7.10. The required number of theoretical stages N can then be calculated from the ordinate result Y for the stated reflux ratio of 2.5 as follows:

$$Y = \frac{N - N_{min}}{N + 1}$$

$$YN + Y = N - N_{min}$$

$$N - YN = N_{min} + Y$$

$$N = \frac{N_{min} + Y}{1 - Y} = \frac{8.75 + 0.419}{1 - 0.419} = 15.8 \text{ stages}$$

This result compares favorably with the result from the McCabe-Thiele result in Example 7.1 of 14.3 stages, in that the estimated number of stages from the Gilliland correlation is greater than that from the McCabe-Thiele analysis, a conservative result.

Although the Gilliland correlation was originally developed as a graphical technique, an analytical representation of the correlation would be a convenient way to implement it in computer-based calculations. While early attempts to regress the data points used in the Gilliland correlation to an analytical expression were not particularly accurate, Molokanov et al. [78] developed an equation that fits the data reasonably well:

$$Y = \frac{N - N_{min}}{N + 1} = 1 - \exp\left[\left(\frac{1 + 54.4X}{11 + 117.2X} \right) \left(\frac{X - 1}{\sqrt{X}} \right) \right] \tag{7.53}$$

Here, X represents the abscissa function of the Gilliland correlation, $(R-R_{min})/(R + 1)$, and the plot of the Molokanov equation is shown in Figure 7.11. While the Molokanov equation deviates somewhat from the Gilliland correlation as originally published, it is reported to fit the data points well and is a reasonable method of projecting the number of theoretical stages required for a given reflux ratio or vice versa.

Applying the values for the abscissa X and minimum theoretical stages N_{min} from Example 7.2 (0.219 and 8.75 stages, respectively) to the Molokanov equation yields:

$$Y = \frac{N - N_{min}}{N + 1} = 1 - \exp\left[\left(\frac{1 + 54.4X}{11 + 117.2X} \right) \left(\frac{X - 1}{\sqrt{X}} \right) \right]$$

$$= 1 - \exp\left[\left(\frac{1 + 54.4(0.219)}{11 + 117.2(0.219)} \right) \left(\frac{0.219 - 1}{\sqrt{0.219}} \right) \right] = 0.444$$

$$N = \frac{N_{min} + Y}{1 - Y} = \frac{8.75 + 0.444}{1 - 0.444} = 16.5 \text{ stages}$$

This result compares favorably to the result of the graphical solution of the Gilliland correlation for this case, 15.8 stages.

Theoretical Stages for Straight Equilibrium and Operating Lines (Kremser-Souders-Brown Equation). In this section we provide some examples of the use of the KSB equation (eq. (6.29)). It may be used in cases involving dilute solutions or relatively small changes in gas and liquid compositions, where both the equilibrium and operating lines are nearly straight [79, 80].

Cases involving dilute solutions typically include stripping of low concentration volatiles from liquids and absorption of dilute components from gas streams into liquids such as in scrubbers. Some examples include: steam stripping of organic contaminants from groundwater; nitrogen stripping of oxygen from monomers and solvents; and absorption of isopropyl alcohol from air with water; and absorption of chlorinated organics from air with oil. In some cases where the system equilibria, properties, and operating conditions are constant enough, application of the KSB equation is all that is needed for the mass transfer design of the separation without needing to resort to a rigorous computer process simulation.

Example 7.5. It is desired to recover ethanol from fermentation broth exiting a bioethanol reactor using distillation at atmospheric pressure with direct steam injection. If the fermentation broth enters the distillation column as a saturated liquid containing 4.0 mol% ethanol, how many theoretical stages below the feed are required to reduce the ethanol concentration in the bottoms effluent to 0.01 mol% at a steam-to-feed ratio of 0.15?

In this example, the concentration variables in the KSB equation are:

x_{in} = 0.04 mole fraction ethanol in the feed
x_{out} = 0.0001 mole fraction ethanol in the bottoms effluent
y_{in} = 0.0 mole fraction ethanol in the direct injection steam at the bottom of the column

The other KSB equation variables to be resolved are m, the slope of the equilibrium line and $\mathcal{T}$, the solute transfer factor. The slope of the equilibrium line m, which is approximately $\Delta y/\Delta x$, can be determined from vapor-liquid equilibrium data for the system in question, and data for the ethanol-water system at 101.3 kPa can be found in the literature [81] and are shown in Table 7.3 with concentrations in mole fractions.

Table 7.3: Ethanol-Water Vapor-Liquid
Equilibrium Data, 101.325 kPa.

T, deg C	xEtOH	yEtOH
100.0	0.0001	0.0016
100.0	0.0002	0.0020
100.0	0.0002	0.0025
100.0	0.0002	0.0025
98.6	0.0040	0.0592
98.55	0.0043	0.0598
98.5	0.0048	0.0727
93.0	0.0315	0.2582
92.95	0.0370	0.3625
87.95	0.0740	0.3881

An average m for the separation can be approximated by calculating a $\Delta y/\Delta x$ value between the data points closest to the feed and bottoms liquid concentrations:

$$m_{avg} = \frac{\Delta y}{\Delta x} = \frac{0.3625 - 0.0016}{0.0370 - 0.0001} = 9.78$$

Since the feed to the column is a saturated liquid, little of the directly injected steam will be condensed on the liquid fed to the column, and the solute transfer factor $\mathcal{T}$ can then be calculated by using the stated steam-to-feed ratio of 0.15 as the effective second phase-to-feed ratio, where the second phase in this case is the direct injection steam:

$$\mathcal{T} = \frac{m_{avg}S}{F} = (9.78)(0.15) = 1.47$$

The number of theoretical stages required below the feed can then be estimated with the KSB equation as:

$$N = \frac{\ln\left[\left(\frac{0.04 - (0.0/9.78)}{0.0001 - (0.0/9.78)}\right)\left(1 - \frac{1}{1.47}\right) + \frac{1}{1.47}\right]}{\ln 1.47} = 12.6 \text{ stages}$$

With $\ln \mathcal{T}$ in the denominator of the KSB equation, it is clear that as $\mathcal{T}$ becomes very large, the denominator in the equation also becomes very large so that the number of stages N becomes very small, approaching zero as $\mathcal{T}$ approaches infinity. Large values of $\mathcal{T}$ can occur in stripping of high relative volatility systems such as chlorinated organics from aqueous streams or scrubbing of reactive species from gas streams such as acid gases with alkaline liquids. In these cases, the KSB equation would predict that only a very few theoretical stages would be required to achieve large, even orders of magnitude, reductions in concentration. And a small number of required stages suggest that achieving large reductions under these conditions is trivial, requiring only short columns or small equipment sizes which is not reasonable when taking mass transfer limitations into account.

Example 7.6. A reboiled top-fed wastewater stripping column is proposed to reduce aqueous chloroform concentrations from 1000 ppm(wt) to less than 1 ppm(wt). If the feed stream is preheated to nearly its bubble point at conditions in the column at the feed point, and the stripping vapor is generated in an external forced circulation reboiler operating at 25% vaporization to produce vapor at 10% of the feed rate, how many theoretical stages will be required to achieve the desired reduction?

Assuming that the wastewater feed stream is essentially a binary mixture of water and chloroform, the liquid mole fraction concentration parameters in the KSB can be calculated from the stated mass concentrations (X_{in}, X_{out}) and the component molecular weights, 18.0153 for water (M_W) and 119.377 for chloroform (M_C):

$$x_{in} = \frac{X_{in}/M_C}{X_{in}/M_C + (1E6 - X_{in})/M_W} = \frac{1000/119.377}{1000/119.377 + (1E6 - 1000)/18.0153} = 0.000151$$

$$x_{out} = \frac{X_{out}/M_C}{X_{out}/M_C + (1E6 - X_{out})/M_W} = \frac{1/119.377}{1/119.377 + (1E6 - 1)/18.0153} = 1.51E{-7}$$

In this example, the entering second phase (vapor) concentration y_{in} is not equal to zero as in Example 7.5 since it is generated in the reboiler from the bottoms liquid exiting the column. However, if we assume that virtually all of the chloroform in the liquid entering the reboiler ends up in the vapor exiting the reboiler, then since 25% of the liquid entering the reboiler is vaporized, the size of the vapor stream is 0.25 times the size of the entering liquid stream. As a result, the concentration of chloroform in the exiting vapor is 1/0.25 or 4 times the concentration of chloroform in the bottoms liquid:

$$y_{in} = (4)(1.51E{-7}) = 6.04E{-7}$$

As in Example 7.5, the slope of the equilibrium line m in the KSB equation can be determined from vapor-liquid equilibrium data at 101.3 kPa for the chloroform-water system found in the literature [82] and are shown in Table 7.4 with concentrations again in mole fractions.

Table 7.4: Chloroform-Water Vapor-Liquid Equilibrium Data, 101.325 kPa.

T, deg C	xCHCl3	yCHCl3
99	0.00002	0.012
98	0.00003	0.095
97	0.00005	0.165
94	0.00006	0.338
93	0.00008	0.439
90	0.00011	0.472
84	0.00012	0.577
82	0.00015	0.619
78	0.00018	0.647
76	0.00025	0.688
75	0.00031	0.71

An average m for the separation can be approximated by calculating a $\Delta y/\Delta x$ value between the data points closest to the feed and bottoms liquid concentrations:

$$m_{avg} = \frac{\Delta y}{\Delta x} = \frac{0.619 - 0.012}{0.00015 - 0.00002} = 4670$$

We should note here that the conversion of the inlet and outlet liquid concentrations to mole fractions at the beginning of this example results in the same 1000-fold reduction as indicated in the problem statement in terms of mass fractions. As indicated earlier, although technically the KSB equation is written in terms of mole fractions, mass fractions may also be used as long as the equilibrium line slope m and solute transfer factor $\mathcal{T}$ are also expressed in terms of mass units. However, since in this case the equilibrium data are given in terms of mole fractions, it was more convenient to convert the inlet and outlet liquid concentrations rather than the rest of the example data.

Since the feed to the column is a saturated liquid, little of the vapor generated in the reboiler will be condensed on the feed liquid, and the solute transfer factor $\mathcal{T}$ can be calculated using the stated reboiler vapor-to-feed ratio of 0.10 as the effective second phase-to-feed ratio, where the second phase in this case is the stripping vapor:

$$\mathcal{T} = \frac{m_{avg}S}{F} = (4670)(0.10) = 467$$

The number of theoretical stages required below the feed can then be estimated with the KSB equation as:

$$N = \frac{\ln\left[\left(\frac{0.000151 - (6.04E-7/4670)}{1.51E-7 - (6.04E-7/4670)}\right)\left(1 - \frac{1}{467}\right) + \frac{1}{467}\right]}{\ln 467} = 1.12 \text{ stage}$$

The result of this calculation is that only a single theoretical stage is required to achieve a 1000-fold reduction in chloroform under the stated operating conditions. This seems to be unrealistic since a single theoretical stage may involve only a brief contact between the feed liquid and the stripping vapor. A more appropriate approach for this and other cases involving high solute transfer factor values would be one which takes mass transfer limitations into account, such as the mass transfer unit approach.

Number of Transfer Units for Straight Equilibrium and Operating Lines. As we have stated previously, compared to the KSB equation (eq. (6.29)), the Colburn transfer unit equation (eq. (6.60)) is better able to model separations with mass transfer-rate limitations indicated by values of the transfer factor larger than $\mathcal{T} = 2$ or so. In this section, we provide some examples and further explanation. The basic difference may be summarized as follows: With the Colburn equation, as $\mathcal{T}$ increases, the denominator approaches unity and the number of transfer units approaches a constant finite value. With the KSB equation (eq. (6.29)), as $\mathcal{T}$ increases, the denominator approaches infinity and the number of theoretical stages approaches zero, which is not a realistic result at large $\mathcal{T}$ values. We continue to emphasize this difference simply because it seems that many in the profession continue to invoke theoretical stages when a rate-based approach would be better.

Example 7.7. For the reboiled top-fed wastewater stripper described in Example 7.6, estimate the number of mass transfer units required to achieve the reduction in aqueous chloroform concentration from 1000 ppm(wt) to 1 ppm(wt).

Substituting the parameters of the case into the Colburn equation results in:

$$NTU = \frac{\ln\left[\left(\frac{0.000151-(6.04E-7/4670)}{1.51E-7-(6.04E-7/4670)}\right)\left(1-\frac{1}{467}\right)+\frac{1}{467}\right]}{1-\frac{1}{467}} = 6.93 = 7.0 \text{ transfer units}$$

This result, when compared to the 1.12 theoretical stages predicted for the same case from Example 7.6, appears to be much more reasonable considering that a 1000-fold reduction in concentration is involved, and this takes time (residence time). It is also helpful to observe that when $\mathcal{T}$ becomes very large the Colburn equation reduces to:

$$NTU_{\mathcal{T}\to\infty} = \ln\left[\frac{x_{in}-(y_{in}/m)}{x_{out}-(y_{in}/m)}\right] \tag{7.54}$$

Or, for cases where $y_{in} = 0$ (such as with direct injection steam in a stripper):

$$NTU_{\mathcal{T}\to\infty} = \ln[x_{in}/x_{out}] \tag{7.55}$$

So, we see that in cases with large values of $\mathcal{T}$ and zero or minimal amounts of the removed component in the entering second phase, the required number of transfer units can be estimated as the natural logarithm of the reduction ratio in the feed phase, a very simple calculation.

Example 7.8. For the reboiled top-fed wastewater stripper described in Example 7.6, use the simplified Colburn equation to estimate the number of mass transfer units required to achieve the reduction in aqueous chloroform concentration from 1000 ppm(wt) to 1 ppm(wt).

Substituting the parameters of the case into the simplified Colburn equation results in:

$$NTU = \ln\left(\frac{0.000151}{1.51E{-}7}\right) = 6.91 = 7.0 \text{ transfer units}$$

The difference between this result and that of the full Colburn equation used in Example 7.7 is insignificant, so it is clear that for systems with solute transfer factors in the range of hundreds, the simplified Colburn equation gives results with adequate accuracy.

As was shown in Section 6.7.1: Mass Transfer Units and Mass Transfer Coefficients (eqs. (6.63 to 6.65)), the relation between the number of theoretical stages and the number of mass transfer units when equilibrium and operating lines are straight is given as:

$$\frac{NTU}{N} = \frac{\ln \mathcal{T}}{1-1/\mathcal{T}} \tag{7.56}$$

Evaluating this relation for different values of T results in the *NTU/N* values given in Table 7.5.

Table 7.5: Relation between Number of Transfer Units and Number of Theoretical Stages for Different Values of Solute Transfer Factor with Straight Equilibrium and Operating Lines.

$\mathcal{T}$	NTU/N
1.1	1.048
1.5	1,22
2	1.39
3	1.65
4	1.85
5	2.01
10	2.56
50	3.99
100	4.65
500	6.23
1000	6.91

Example 7.9. For the reboiled top-fed wastewater stripper described in Example 7.6, estimate the number of mass transfer units required from the number of theoretical stages calculated in Example 7.6 using Table 7.5. The number of theoretical stages required for the stated chloroform reduction in Example 7.6 is N = 1.12. Since the solute transfer factor of 467 is closest to the $\mathcal{T}$ value of 500 in Table 7.5, the NTU/N value is about 6.23. As a result, the estimated number of transfer units for the separation is:

$$NTU = 1.12(6.23) = 6.98 = 7.0$$

This compares quite favorably with the calculated NTU value of 6.93 from Example 7.7. This shows that when values of the solute transfer factor are very high, even an approximate value for $\mathcal{T}$ is acceptable for mass transfer calculations.

In considering when it is appropriate to use a mass transfer unit approach to a problem rather than a theoretical stage approach, a common rule of thumb is to use mass transfer units when the solute transfer factor is greater than 2. Since the required number of mass transfer units is 39% greater than the required number of theoretical stages at a solute transfer factor of 2, the use of mass transfer units at that point certainly seems warranted and may also be appropriate at even lower solute transfer unit values.

Choosing a Reflux Ratio. In distillation design, one important task is to determine the design reflux ratio, boilup ratio, or liquid-to-vapor ratio. This usually involves a tradeoff between operating and capital costs since necessarily, as reflux ratio is reduced, the number of theoretical stages increases which results in a taller distillation column and increased capital cost with the benefit of lower energy consumption and correspondingly lower operating costs. This inverse relationship between reflux ratio and number of stages presents an optimization problem which can be addressed through detailed computer simulations and optimization methods or even repeated applications of the Gilliland correlation, followed by preliminary equipment sizing and cost estimates.

However, a quick range-finding method involves the application of a general rule of thumb for the ratio of optimum reflux ratio to minimum reflux ratio, R_{opt}/R_{min}. Recommended R_{opt}/R_{min} values range from 1.05–1.5, depending on whether energy consumption and costs or capital costs are significant. Lower values of R_{opt}/R_{min} of about 1.05–1.2 are usually indicated for high energy consumption cases like columns with refrigerated condensers or cryogenic separations, and higher values of R_{opt}/R_{min} of about 1.2–1.5 are typical in cases where the minimum number of stages is already quite high such as with high product purity requirements or very low relative volatilities. In times and circumstances when the cost of energy is high, lower values of R_{opt}/R_{min} have tended to be preferred, and when the cost of energy has fallen, higher R_{opt}/R_{min} values have tended to prevail. On the balance however, using an appropriate rule-of-thumb value for R_{opt}/R_{min} does provide a very simple, quick, and useful method of getting close to an optimum tradeoff between reflux ratio and number of stages so that preliminary equipment sizing, design, and cost estimates can be done in the early stages of a project. Additional discussion of the utility of shortcut methods is available elsewhere [83].

7.3.6 Computer calculations

Except for the specialized purposes described earlier, graphical and manual distillation design and analysis methods have been almost entirely supplanted by computer-based calculations and simulations which have become the predominant tools for distillation design today. These programs may be in the form of spreadsheet calculations, special purpose-built programs designed for the solution of specific problems, or commercial process simulation programs that allow one to perform a wide range of analysis and design tasks. The advantage of computer-based calculations is the greater speed, rigor, and accuracy which they can often bring to bear on distillation design problems compared to that which is possible with manual graphical design methods. However, when significant effort is required to regress and characterize physical properties and equilibria and/or resolve difficult column convergence issues, the benefits of applying manual graphical methods should not be overlooked. Graphical or shortcut methods can help visualize the problem and provide an approximate result that is useful for checking the results of more-rigorous computer calculations. Furthermore, knowledge of graphical and shortcut methods can help the user of a computer-based tool understand what the computer is doing, so the computer may be viewed as more than a black box calculator.

Although complex chemical process simulation computer programs are often used in industry and academia for the rigorous design and analysis of complex distillation and other separation processes, many distillation calculations can be computerized with spreadsheets or mathematical analysis programs such as MATLAB, Maple, GNU Octave, and many others. Often, these spreadsheets and other applications are developed by individual professionals to perform a calculation to solve a specific problem such as a KSB equation calculation or to estimate the capacity of a type of column internal. However, if used repeatedly, the professional may invest the time and effort to expand the calculation to produce a generalized tool to solve similar problems. Some individuals and organizations have made a number of these types of tools available on the Internet which can be located with search engines. A few examples of publicly available calculational tools which can be found online at the time of this writing are:

- Ternary flash spreadsheet on the LearnChemE.com website [84]
- McCabe-Thiele binary distillation Google Sheets worksheet and a databank of binary VLE Wilson parameters and component physical properties on the CheSheets.com website [85]
- Web-based app for calculating packed column pressure drop, flooding, and column sizing on the CheCalc.com website [86]
- Distillation Stage Calculation spreadsheet using FUG and McCabe-Thiele on the cheresources.com website [87]
- Web-based app for converting between different concentration units on the EngineeringToolBox website (under "Miscellaneous" tools) [88]

In addition, many manufacturers of distillation equipment such as trays, packings, and other internals have developed specialized proprietary programs for sizing and rating their products which they may make available to engineers and designers upon request. Also, some universities with chemical engineering departments provide publicly available online resources and calculational tools to support research and development in distillation and other separation technologies.

When a complex distillation process is planned or a distillation is intended to be integrated into a full or partial flowsheet model, a full-featured chemical process simulation package is usually the best choice for carrying out the design and performance calculations, especially if multiple components or non-ideal thermodynamic and physical property models are required. Although process simulation programs have been available for decades, the computationally intensive nature of the programs at first required the use of mainframe computers which often required significant effort to program and apply to real-world distillation problems. However, the advent and development of microprocessor-based computers and accompanying advances in software, property databases, and computing power over the years have resulted in the ability to quickly set up and analyze multiple simulation cases in process simulators, and this has made their use popular and even indispensable for powerful analysis of some problems.

Although many engineers and practitioners are at least acquainted with one or more of the most popular steady-state process simulation programs such as Aspen Plus® and Aspen HYSYS® by Aspen Technology, PRO/II™ by AVEVA, and UniSim® Design Suite by Honeywell, there are a host of other commercial and open-source process simulation programs available which fill a wide range of process design, analysis, and operational needs. In addition to steady-state process flowsheet modeling, there are programs that provide dynamic process simulation, optimization, batch process modeling, and costing functions among others. A list of many of the available process simulation programs is given in Table 7.6.

Table 7.6: List of Chemical Process Simulation Programs (abstracted from ref. [89]).

Software	Developer	Applications	Operating system	License
Aspen HYSYS	Aspen Technology	Process simulation and optimization	Windows	Closed-source
Aspen Plus	Aspen Technology	Process simulation and optimization	Windows	Closed-source
BatchColumn	ProSim	Simulation and Optimization of batch distillation columns	Windows	Closed-source
BATCHES	Batch Process Technologies, Inc.	Simulation of recipe driven multiproduct and multipurpose batch processes for applications in design, scheduling and supply chain management	Linux	Closed-source

Table 7.6 (continued)

Software	Developer	Applications	Operating system	License
CHEMCAD	Chemstations	Software suite for process simulation	Windows	Closed-source
CHEMPRO	EPCON	Process Flow Simulation, Fluid Flow Simulation, & Process Equipment Sizing	Windows	Closed-source
COCO simulator + ChemSep	AmsterCHEM	Steady state process simulation based on CAPE-OPEN Interface Standard	Windows	Open-source
Design II for Windows	WinSim Inc.	Process simulation	Windows	Closed-source
DWSIM	Daniel Medeiros, Gustavo León and Gregor Reichert	Process simulator	Windows, Linux, macOS, Android, iOS	Open-source, freemium
gPROMS	PSE Ltd	Advanced process simulation and modelling		Closed-source
METSIM	Metsim International	General-purpose dynamic and steady state process simulation system	Windows	Closed-source
OLI Analyzer	OLI Systems, Inc.	Chemical phase equilibrium simulation featuring electrolytes	Windows	Closed-source
PD-PLUS	Deerhaven Technical Software	Steady-State Modeling of Chemical, Petrochemical, and Refining Processes	Windows	Closed-source
PEL Suite	PEL Software	Steady state process simulation	Windows	Closed-source
PRO/II	AVEVA	Steady state process simulation	Windows	Closed-source
ProMax	Bryan Research & Engineering	Process simulator capable of modeling oil & gas plants, refineries, and many chemical plants	Windows	Closed-source
ProSimPlus	ProSim	Steady-state simulation and optimization of processes	Windows	Closed-source
SimCentral	AVEVA	Steady state, Fluid flow and Dynamic process simulator.	Windows	Closed-source
SuperPro Designer	Intelligen	Batch and continuous process simulation.	Windows	Closed-source

Table 7.6 (continued)

Software	Developer	Applications	Operating system	License
SysCAD	KWA Kenwalt Australia	Steady-state and dynamic process simulation	Windows	Closed-source
UniSim Design Suite	Honeywell	Process simulation and optimization	Windows	closed-source

Several of the programs listed in Table 7.6 are more than capable of integrating and modeling the full range of unit operations in an entire chemical process in all its complexities. However, for this work, we will limit the discussion to modeling of individual distillation unit operations. Furthermore, this discussion is not intended to be a full treatment of every aspect of distillation process modeling but rather an overview of some of the knowledge and insights the authors have gained over the course of their careers. To that end, we will begin with the steps and issues involved with developing a distillation process simulator model:

1. Specify the distillation
2. Develop property and thermodynamic models
3. Set up the simulation
4. Determine starting values for column operating conditions
5. Run the simulation
6. Analyze the simulation results
7. Develop a report
8. Deal with convergence problems

Specify the Distillation. Define the separation requirements, including any constraints such as key and other components, including relevant trace components, inlet and required outlet concentrations, flow rates, available utilities, temperature limitations, space limitations, client expectations, and so on. It is also important to understand how strict the requirements are and what limitations there are on tradeoffs such as separation power versus energy consumption.

Develop Property and Thermodynamic Models. This is perhaps the most critical step since if the physical properties and thermodynamic model are not representative of the actual separation system, then the simulation results will not be representative either. This is especially true with complex multicomponent systems when poor-quality characterizations of the thermodynamics of minor components obscure their effect on the behavior of the key components. As a result, it's important to fully characterize the binary relationships between all component pairs, even the relationships between minor components as well as those between minor and key components.

Additionally, inaccurate physical properties such as density and viscosity can result in inaccurate pressure drop, capacity, and flooding calculations for column internals which can lead to underperformance of final column or tower designs.

Set Up the Simulation. Enter the column- and case-specific data in the chosen process simulation package and enter the specifications for the problem. Depending on the program, this may involve specifying a process flowsheet in addition to column data such as number of theoretical stages, feed and side draw (if applicable) locations, condenser type, and other details as needed. Some combination of operating parameters must also be specified such as operating pressure, pressure drop, and whatever means are used to control the separation power and material balance split. Often a reboiler duty or a ratio such as reboiler duty-to-feed or reflux ratio is used to control separation power, and a distillate-to-feed ratio or bottoms-to-feed ratio is used to control the material balance split. The physical property and thermodynamic model parameters must also be entered in a format compatible with the options available in the chosen simulation program, and so it may be necessary to regress the parameters from property data. Some simulators provide regression capability within the simulator itself or in a companion program that will generate property parameters in the format that is compatible with the simulator. In most of the popular simulation programs, physical property and thermodynamic databases are incorporated with parameters for many components and chemical systems that are commonly encountered, and sometimes the ability to link with external databanks like the Dortmund Data Bank (DDB) by DDBST GmbH [69] is provided.

Determine Starting Values for Column Operating Conditions. In some cases, the intent of the distillation simulation may be simply to determine what the performance of a given column will be when operated at a specific set of operating conditions such as condenser pressure, column pressure drop, reflux ratio, and distillate-to-feed ratio. This type of simulation is sometimes called a rating calculation. However, in cases where the intent is to determine what operating conditions are required to achieve certain performance objectives such as distillate and bottoms concentrations, it is usually necessary to vary one or more column operating parameters until the objectives are achieved, often called a design calculation. This can be done by manually adjusting the parameters and manually examining results over multiple simulation runs, but this can become quite time-consuming. It is usually more convenient to use iterative solvers within the simulator itself (referred to as design specifications, controller blocks, or other names) to converge quickly to the desired results by automatically varying operating parameters and applying convergence algorithms such as the Newton-Raphson, secant, or bisection methods. With the use of iterative solvers, it is important to provide reasonable starting values for the operating parameters to be varied (reflux ratio, distillate-to-feed ratio, etc.) to allow a good opportunity for the solver to converge properly. Traditionally, shortcut calculation methods such as the

Fenske-Underwood-Gilliland (FUG) method have been used to arrive at guesses for initial parameter values, but with the ready availability of high-powered computers able to efficiently handle computationally intensive tasks, often it is practical to develop good starting values just by running a very few rating simulations with manually chosen operating parameter values. However, if difficulties arise in running range-finding rating simulations, it may be advisable to perform some FUG calculations or even a McCabe-Thiele graphical analysis to better understand constraints in the problem and choose operating parameters and/or specifications more wisely.

Run the Simulation. Once the simulation has been set up with property and operating parameters specified, the process simulation program will likely give an indication that the problem specifications are complete and the simulation is ready to be run. This may be just a matter of clicking the "Run" button in the simulator or selecting the corresponding menu item to start the simulation program. And for straightforward problems that are well characterized beforehand, this may be all that is required, with a swift program completion and initial calculation results yielding the answers required by the problem specification. However, the program may finish with errors, maximum iterations exceeded, or other irregular results which must be resolved.

A common error is when the program indicates that one or more stages in the column have "dried up" which means that the liquid flow or even the vapor flow has decreased to zero or close to zero on those stages. This is most often a problem caused by specifying a reboiler duty or discharge flow that is too high or too low to provide sufficient liquid or vapor on the stages showing the error. The duty and/or flow specifications can usually be changed to alleviate the problem, but in many cases, changing from absolute flow specifications to ratio specifications will rectify the error. For instance, changing from specifying the reboiler duty to specifying the boilup ratio or reflux ratio, or changing from specifying the distillate discharge flow rate to specifying the distillate-to-feed ratio.

When the maximum iterations are exceeded, it may be a simple matter of increasing the limit on iterations, but many times this is an indication that something is wrong with the column specifications. Although in most cases leaving the convergence algorithm specified as "Standard" is the best choice, sometimes it's necessary to change the algorithm to an alternate method such as Sum-rates, Non-ideal, or Newton or other possible options depending on the simulation program. In other cases, the convergence tolerance may have been specified too tightly and subtle mathematical issues between the physical properties, thermodynamics, and mass and energy balances do not allow the error/tolerance ratio to be reduced to unity or less. But other times, specifications may be set to cause iteration difficulties that are not great enough to cause stages to dry up but nonetheless result in excessive iterations. This may not be a problem with a single simple column simulation, but in a complex simulation with multiple unit operations and many recycle loops and/or design specification loops, a high degree of nesting in the simulation sequencing may require the column calculation to be solved dozens, hundreds, or even thousands of times. In these cases, even a modest increase in the number of iterations needed for the column calculation may result in a significant increase in the amount of computation time for the entire flowsheet simulation.

Other irregular or spurious results may arise with complex columns or highly non-ideal systems. These may take the form of an unclosed material balance, dried up stages not attributable to an improperly specified discharge rate, or unrealistic column profiles and concentrations due to invalid thermodynamics among other issues. In addition, some non-ideal multicomponent systems can demonstrate multiple independent stable steady state solutions that can appear in simulations as well as actual practice. All of these issues require adjustments to column and/or simulation parameters to resolve the problem results, and the required adjustments may not be obvious. As a result, some experimentation with parameters and values may need to be done before a valid solution to the problem can be obtained.

Analyze the Simulation Results. Once any errors or irregularities are sorted out, the simulation results may be reviewed to compare them against the stated separation requirements and constraints. In most cases the initial results will indicate that some adjustments to the column specifications are needed, such as changing the operating pressure to maintain reboiler temperatures within the desired range or adjusting the number of stages or feed location to reduce the required reflux ratio or internal column liquid and vapor flow rates. It may be necessary to carry out some initial column internals sizing calculations based on these liquid and vapor flow rates to estimate the resulting column diameter and height which may help guide the adjustments needed to the number of stages and column specifications and parameters. In addition, it is important to examine the column energy consumption and understand the utility requirements in guiding the adjustments. Condenser duties and cooling media temperatures should not be overlooked as part of the energy analysis since these can be significant drivers in the column design process.

In a wide variety of applications where the component relative volatilities are low (e.g., less than about 2), the separation is probably equilibrium limited so that published tray efficiencies and HETP values can likely be used without accounting for mass transfer limitations. However, when relative volatilities are high, the column internal flow rate and vapor-liquid concentration profiles need to be examined to determine if mass transfer limitations need to be taken into account when considering how many actual trays or how much actual packing height needs to be provided. The methodology for accounting for mass transfer limitations is discussed in detail in Sections 6.7.1 and 7.3.3 through Section 7.3.5. But in general, it involves calculating the solute transfer factor for each stage or a few representative stages and determining if there are any zones with high solute transfer factors. Mathematically, a solute transfer factor of only 1.2 will result in a 10% difference between the number of mass transfer units and the number of theoretical stages which seems significant at first glance.

However, mass transfer limitations usually occur at the ends of the column where component concentrations can become dilute so that only a portion of the overall column length is affected. As a result, often a higher solute transfer factor value is used as a rule-of-thumb for when adjustments need to be made, and for many practitioners a value of 2.0 is used which results in a ratio of mass transfer units to theoretical stages of 1.39.

Even though key component relative volatilities may be low with their separation being equilibrium limited throughout the entire column, the relative volatilities of minor or trace components may be significantly higher. In that case, only those components would exhibit mass transfer limitations so that achieving any concentration specifications involving them would require adjusting the number of actual trays or the actual height of packing. Of course, if those minor or trace components are unimportant or lack specifications, adjustments on their behalf could be unnecessary.

In evaluating column profiles for mass transfer limitations, it is also important to look for regions where compositions and/or temperatures remain nearly constant over a range of stages. This is an indication of a pinch point in the column, represented on a McCabe-Thiele diagram as a location where the equilibrium and operating lines approach each other and nearly touch, thus requiring a large number of stages to accomplish only a small change in composition. This pinched condition is an indication that the column is operating close to the minimum reflux ratio required to accomplish the separation, and a significant reduction in the number of required theoretical stages may be realized by increasing the reflux ratio or boilup ratio (and therefore increasing the energy consumption), sometimes by only a small amount.

Develop a Report. Once all simulations are complete, free of errors, and validated, a report of the simulation results should be developed and communicated to other project team members and clients. This may be as simple as printing out a report file in a default format that the simulation program may generate automatically. Or, it could involve putting together a custom document with tables of the results of multiple simulation runs (e.g., case studies or sensitivity analyses), plots, and examples of column operating conditions with stream results. The exact nature of the report will depend on the nature of the problem statement and the results desired by the project team or client.

In many cases, a primary result desired from the simulation work is profiles of column hydraulic and physical property data to be extracted and entered into design programs for column or tower internals. Some internals manufacturers provide the means to import data files from popular simulation programs into their proprietary design programs and reduce time requirements and the possibility of data entry errors. And conveniently, some internals manufacturers provide information on design and performance correlations for their products on a confidential basis to the simulator company for direct inclusion in their process simulation program distributed to users. These tools, properly utilized, can significantly reduce the time required to produce a column design and improve design accuracy.

Deal With Convergence Problems. Often the cause of convergence problems is a poor set of initial values for the distillation operating parameters. Other causes include tight limits on operating parameter values (which may actually exclude the parameter values necessary for convergence to a solution), selection of an inappropriate calculation algorithm, overly demanding design specifications, or overly tight convergence tolerances.

In cases of complex distillation systems where convergence difficulties arise from poor initial values for operating parameters, it may be difficult to guess at values close enough to the final values to result in successful convergence. In addition to the methods of arriving at good initial values discussed earlier, another approach is to simplify the distillation problem into a problem that can be more easily solved, and use the results of that simplified problem to provide initial parameter values for the complex problem. One way to simplify a complex distillation problem might be to reduce the number of components or even limit the problem to only the key components to converge the problem and using those results to arrive at better choices for the operating parameter initial guesses.

Another way to simplify the problem might be to modify the vapor-liquid equilibria specifications to reduce non-idealities or possibly even use Raoult's Law (fully ideal solutions). In cases where the VLE is characterized using an activity coefficient model with interaction parameters, the degree of non-ideality might be reduced by adjusting those parameters (often by bringing them closer to unity) to achieve convergence and then incrementally adjusting them back to their correct values in small enough steps to maintain convergence in each step. One should realize, however, that intermediate results while the interaction parameters are being adjusted have no real-world meaning since the parameters do not represent the actual VLE behavior.

Lastly, simplification might be achieved by eliminating recycle loops or associated equipment like decanters, side strippers, or sister columns. Once the simplified configuration is made to converge to a solution, the results might give better guesses for the operating parameter initial values in the full process configuration. Other means of simplification may present themselves in specific circumstances, but the intent should be to eliminate factors that could cause large fluctuations in the column conditions from iteration to iteration and thus, the convergence problem.

If recycle streams are included around the distillation column (e.g., from pre- or post-processing units upstream or downstream of the column), it is usually advisable to specify the distillation convergence error tolerance tighter than the recycle tear stream tolerance. It also may be appropriate to tighten the distillation error tolerance when trace components are present, important, and being followed, when two liquid phases are present and one phase is small, or when greater accuracy is desired.

As discussed earlier, increasing the maximum allowable number of iterations is often an immediate response to convergence problems, but if the error/tolerance values initially decrease but end up oscillating, it may be possible to specify a damping function in some simulators to settle the oscillations down and aid convergence. However, an increased damping level almost always slows convergence, so extra iterations may still be necessary. In some simulators, other adjustments can be made to the distillation convergence algorithm that may be needed to achieve convergence or reduce the iterations required.

7.3.7 Reactive absorption for gas scrubbing

As discussed earlier in Section 7.3.3: Continuous Stripping and Absorption, an important vapor-liquid separation technique that is somewhat different from classical distillation is gas absorption, also known as gas scrubbing. It consists of a vapor-liquid contacting device in which a major or minor component in a usually non-condensable gas stream is transferred into an absorbent liquid. The absorbent liquid can act on the absorbed component either by non-reactive (solubility) means or by reactive (chemical reaction) means. Non-reactive absorption driving forces result strictly from the VLE behavior between the components in the gas and the scrubbing liquid at the operating conditions in the scrubber (temperature, pressure, concentration). Examples of non-reactive absorption are removal of alcohols from process vents with water and absorption of chlorinated organics from air into low-volatility, low-viscosity mineral oil.

Reactive absorption driving forces result from beneficial chemical reactions between the component being absorbed and components in the scrubbing liquid in addition to the VLE behavior between the absorbed component(s) and the scrubbing liquid. Because of the usually much greater capacity of the scrubbing liquid for the absorbed component with reactive absorption compared to non-reactive absorption, reactive absorption typically sees greater use in industrial applications. Some examples of reactive absorption are natural gas and refinery gas sweetening with aqueous alkanolamines, CO_2 removal from combustion stack gas with aqueous caustic solutions, and sulfur dioxide scrubbing from flue gas with lime/limestone aqueous slurry (involving reaction in the presence of slurry solids). Note that in many of these applications the ability of the absorption fluid to deal with the heat of absorption without excessive temperature increase is a key design concern, as well as the mass transfer characteristics. This is a major reason why many absorption fluids are aqueous solutions with high heat capacity. Discussion of thermal effects is beyond the scope of our discussion here, but keep in mind the need to control temperature during absorption. For more discussion of this point, see refs. [90] and [91].

The type of contacting device chosen often depends on the quantity and operating conditions of the gas stream to be scrubbed. For small volume streams with modest mass transfer requirements, eductor scrubbers with pumped scrubbing liquid as the motive fluid are often used, which are compact and provide positive suction or draft on the incoming gas stream. This can often negate the need for a blower on that stream and as a result be helpful for process simplicity. For large volume streams with modest mass transfer requirements, venturi scrubbers may be a viable option, where the high-volume gas stream flows at high velocity through the throat of the venturi where it is contacted with the scrubbing liquid being introduced at that point. Because the gas stream is the motive fluid for venturi scrubbers, significant forced draft on the venturi inlet gas and/or induced draft on the venturi outlet is required. This pressure drop requirement usually dictates the use of large fans somewhere in the gas stream, with the scrubbing liquid usually needing to be pumped into the throat of the venturi and often dispersed as a spray through a spray bar or radial nozzles. On the other hand, both eductor and venturi scrubbers are quite forgiving of solids content in the gas stream and scrubbing liquid which can be an important factor in some applications.

For scrubbing applications with significant mass transfer requirements, a packed or trayed scrubbing column or tower is usually required which allows multiple countercurrent stages or mass transfer units to be provided for high removal efficiencies. Scrubbing columns can be designed with large diameters to handle very large gas streams with modest or low pressure drop requirements which reduce or eliminate the need for supplemental blowers or fans and reduces the energy requirements for the system. With the proper choice of column internals, scrubbing columns can also be quite tolerant of solids in the entering gas and scrubbing liquid streams although usually not as tolerant as eductors or venturi scrubbers.

For non-reacting systems, the main resistance to mass transfer normally resides in the feed (gas) phase, or more precisely, in the feed-side (gas) film rather than the liquid film. Because of generally larger mass transfer coefficients in the gas phase compared to those in the liquid phase, gas-phase-controlled systems generally exhibit relatively low heights of a transfer unit for packed applications and high stage efficiencies or high numbers of transfer units per actual tray for trayed applications. However, in reactive absorption, reaction of the absorbed component in the liquid phase is a necessary sequential step in the absorption process. With the inclusion of reaction, the main resistance to mass transfer may be due to slow reaction in the liquid relative to mass transfer through the gas film which then results in the mass transfer in the system becoming controlled by the overall liquid phase resistance. In these cases where the equilibrium concentrations of the absorbed component in the liquid are negligible and the liquid phase reaction kinetics are considerably slower than the gas phase diffusion and therefore controlling the rate of mass transfer, we see significantly elevated heights of a transfer unit in packed columns and low stage efficiencies or low numbers of transfer units per actual tray in trayed columns. An example of slow reaction kinetics resulting in liquid phase resistance-controlled mass transfer is the absorption of CO_2 from a feed gas into aqueous NaOH [92]. In this case, the rate of reaction involving CO_2 dissolution and reaction with NaOH in the liquid phase is slow compared to the gas film-side mass transfer rate, so the main mass transfer resistance is in the liquid phase and heights of a mass transfer unit are high.

In the case of scrubbing applications for dilute non-reacting systems, the number of transfer units required for the desired gas phase component reduction can often be calculated by a version of the Colburn equation given as eq. (6.60) in Section 6.7.1: Mass Transfer Units and Mass Transfer Coefficients:

$$N_{OG} = \frac{\ln\left[\left(\frac{y_{in} - mx_{in}}{y_{out} - mx_{in}}\right)\left(1 - \frac{1}{A}\right) + \frac{1}{A}\right]}{1 - \frac{1}{A}} \tag{7.57}$$

where y_{in} = inlet vapor phase composition

y_{out} = outlet vapor phase composition

x_{in} = inlet liquid phase composition

m = slope of the equilibrium line, $y_i = mx_i + b$

A = $L/(mV)$, solute transfer factor for absorption called the absorption factor

This equation may also be applicable for reactive absorption with reaction kinetics that are so slow that no significant reaction of the absorbed component with the scrubbing liquid occurs within the average retention time of the liquid in the absorbing device. In these cases, the reaction in the scrubbing liquid does not significantly affect the vapor-liquid equilibrium between the absorbed component in the gas and liquid phases, so that equilibrium is the predominant factor in determining the effective value of the absorption factor. That VLE relation may be somewhat difficult to accurately measure experimentally because of the slow reaction, but if it can be determined, the above equation can be used to calculate the number of transfer units.

However, not all liquid phase reaction kinetics are slow. In cases where the relevant reaction is an ionic dissociation reaction, the rates can be quite fast, to the point where the reaction rate is considerably faster than the liquid phase diffusion. As a result, the overall liquid phase resistance is not limiting and the system is once again controlled by gas phase diffusion. A well-known example of reactive absorption with fast liquid phase kinetics is the absorption of highly soluble gases such as HCl or NH_3 into water. In this type of process, even though absorption of the removed component is dependent on the liquid phase reaction, the kinetics of dissolution and dissociation are fast enough so that the main resistance to mass transfer is still in the gas phase (the gas-side film resistance), and mass transfer efficiencies are still high – as would normally be expected for non-reactive mass transfer.

In cases where the highly soluble gas is the major or maybe even the only component in the entering gas stream (such as with hydrogen chloride absorption into water for aqueous HCl production), a significant element of heat transfer is usually involved because of the large amounts of heat released with the transfer of much of the gas phase into the liquid phase. In these cases, because there is very little gas phase mass transfer resistance, heat transfer resistances become the controlling factor in absorber design and as a result is beyond the scope of our discussion here.

In addition, since the VLE for systems such as these greatly favors the absorbed component being in the liquid phase, the relative volatility and therefore the slope of the equilibrium line, m, is quite low. As a result, with m in the denominator, the absorption factor in these systems with fast liquid phase kinetics is quite high which reduces the number of transfer units required to achieve a given separation. In fact, when the scrubbing application includes a fast irreversible reaction in the liquid that converts the absorbed component to a stable non-volatile species such as a salt, the equilibrium partial pressure of the absorbed component over the liquid becomes essentially zero. The slope of the equilibrium line, m, then also becomes essentially zero, and the absorption factor A becomes essentially infinity so that the Colburn equation for absorption as given in eq. (7.57) reduces to:

$$N_{OG} = \ln \frac{y_{in}}{y_{out}} \qquad (7.58)$$

Examples of scrubbing applications where eq. (7.58) is applicable include HCl scrubbing with aqueous caustic and NH_3 absorption with aqueous sulfuric acid.

Another reactive absorption process that is widely used is the removal of acid gases from a variety of gas streams with aqueous alkanolamines such as monoethanolamine (MEA), diethanolamine (DEA), methyldiethanolamine (MDEA), diisopropanolamine (DIPA), diglycolamine (DGA), and piperazine (PZ), among others. The process concept was initially introduced by Robert R. Bottoms who was awarded a U.S. patent in 1930 [93]. Since then, the technology has undergone significant advancement over many years of commercial practice. Frequent applications include H_2S and CO_2 absorption from refinery gas and raw natural gas (gas sweetening).

In the case of alkanolamines, the amine group in the molecule provides basicity which reacts with H_2S and/or CO_2 and results in absorption capacity, and the OH group reduces vapor pressure and provides affinity with water molecules to enhance water solubility over that of the unsubstituted amines. Furthermore, reactivity and absorption capacity as well as thermal stability vary with each type of alkanolamine, with absorption capacity and thermal stability generally increasing as the degree of substitution on the nitrogen atom increases, but with H_2S and CO_2 reaction rates decreasing. Thus, the primary alkanolamines (MEA, DGA, etc.) typically have higher reactivities than secondary alkanolamines (such as DEA) but lower absorption capacities and thermal stabilities. Likewise, the secondary alkanolamines generally have higher reactivities and lower absorption capacities and thermal stabilities than the tertiary alkanolamines such as MDEA and TEA. As a result, blends of different alkanolamines are sometimes considered and can be used to tailor performance for individual applications. And since the alkanolamines exhibit less basicity than aqueous caustic, the liquid phase reaction rates of CO_2 with the alkanolamines are no better than those with caustic, again resulting in relatively large heights of a transfer unit and low tray efficiencies compared to absorption applications where molecular diffusion is the only mechanism resulting in mass transfer resistance.

However, unlike the absorption of acid gases with aqueous caustic which is non-reversible, the absorption of acid gases with alkanolamines results in the formation of amine adducts which are reversible with the use of thermal regeneration, normally accomplished in a hot stripping unit associated and usually integrated with the absorption unit. A typical alkanolamine gas sweetening process with an absorber operating at elevated pressure and a near-atmospheric pressure stripper is shown in Figure 7.11.

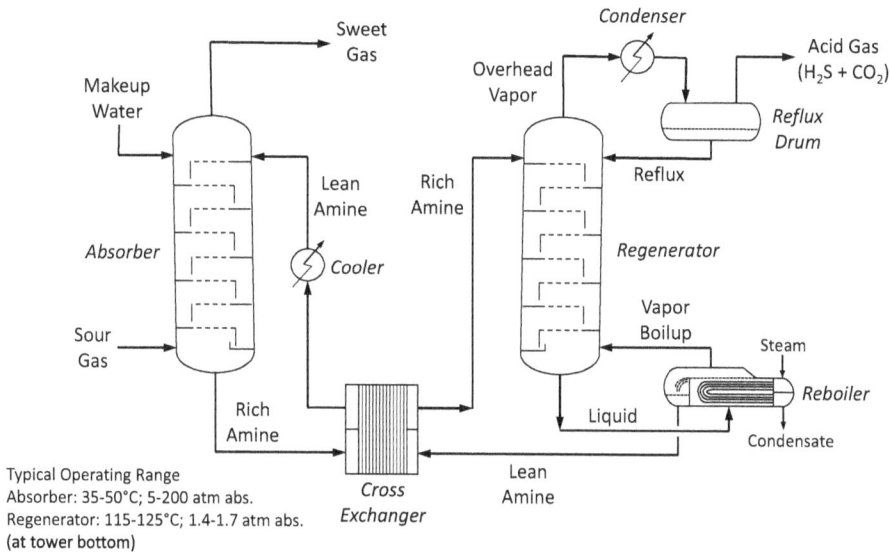

Figure 7.11: Alkanolamine Gas Sweetening Process with Absorber and Stripper. Adapted from drawing by "Raminagrobis" from Wikimedia Commons [94].

This ability to regenerate rich (spent) alkanolamine solutions with little degradation makes the use of difficult-to-produce and high-value alkanolamine compounds feasible and cost-effective in many sweetening and recovery applications, to the point where alkanolamine absorption is even being investigated and considered for carbon capture and sequestration in fossil fuel power generation facilities. However, the eventual need to replace or externally treat and regenerate the alkanolamine absorption media after some period of time due to buildup of thermally stable, high-boiling impurities may impose significant cost [95].

7.4 Equipment Design and Rating Methods

In the study of distillation, most graphical and shortcut methods focus on calculations used to quantify the required number of theoretical stages or transfer units. This may be done very rigorously by using validated phase equilibrium and physical property models as a function of temperature and by including thermal effects. But of course, knowing the exact number of theoretical stages or transfer units is not sufficient. We also need to know something about the efficiency of the mass transfer equipment in terms of the actual height of a stage or transfer unit or the efficiency of a tray as a function of the production rate. In many cases, the major uncertainties in the design of a distillation column have to do with characterizing mass transfer efficiency and the required size of the equipment, not in representing the thermodynamics or the physical properties. And adequately characterizing mass transfer efficiency normally cannot be done with calculations alone. It requires generation of key mass transfer data or access to an appropriate database.

Please note that in all of the following discussions, no preference or endorsement of any specific manufacturer or supplier over any other is intended or given. The analyses and recommendations below are intended for guidance and general instruction for engineers and other practitioners and should be applied with good judgment in the many and varied circumstances that arise in real-world applications. In our experience, most equipment suppliers are conscientious, highly skilled, and deliver high-quality products that perform well in the applications for which they are suited.

7.4.1 Column design

Selection of Column Internals. Once the required separation, mass transfer, throughput, and operating conditions for a distillation application have been determined, the selection of separation equipment such as column internals can proceed. Although unique equipment such as a rotating packed bed or falling film column might be chosen for specific applications, the primary equipment choice decision is usually between packing or trays in a conventional column or tower. Historically, trays have been specified in distillation columns in many more instances than packing, but with significant advances in packed column technology such as improved liquid distributor designs, in recent years packings have been seeing much broader application. And although some applications can make use of either one effectively, often only packing or only trays is the clear choice. However, even when some factors clearly favor the choice of one or the other, other factors in specific cases such as prior experience and corporate preference may result in a different choice. The following are examples of factors to evaluate in an application that would point toward the use of trays versus packing.

Factors favoring trays

High Pressure Operation. Distillation applications operating at high pressures, typically around 1000 kPa (150 psia) or higher, are usually the domain of trays since their higher pressure drops are usually not an issue. Also, their more robust construction generally provides better mechanical strength than packed beds and better resistance to the buffeting and surging that occurs during upsets with high vapor densities at high pressures. In addition, many structured packings under these conditions exhibit elevated HETPs in the middle of their operating range. The cause of this "hump" in the efficiency curve for some structured packings has been attributed to vapor back-mixing due to the high vapor densities and low pressure drops of these packings, but the root cause has not been definitively identified.

Fouling Services and Presence of Solids. Because of the greater turbulence and surface scouring action generally exhibited by trays, they are usually preferred over packings in systems where solids are present. With lower pressure drops than trays, the interaction between vapor and liquid in a packed bed is typically gentler than on trays, giving any solids present more of an opportunity to drop out, deposit, and build up in the interstices between packing pieces to cause bed fouling. In addition, the ability to enter a trayed column or tower to clean out specific areas of fouling is beneficial as opposed to the need to completely remove a packed bed to clear fouling materials. However, when fouling is severe and pressure drops are constrained such as in wash sections, vacuum reformers, flue gas scrubbers, quench sections, and asphalt sections, grid-type packings are usually preferred because of their very open structure, low pressure drop, and lack of horizontal surfaces for solids and fouling materials to deposit and build up.

High Retention Time Requirements. Because of their higher pressure drops than packings, trays generally exhibit larger liquid holdups than packings and therefore, longer liquid retention times. Also, certain tray types such as bubble cap trays can be designed with mechanical configurations to provide specific amounts of holdup volume for specific retention time requirements. So, for distillation applications involving slow mass transfer or depending on chemical reactions, trays provide a better opportunity to achieve the required residence times and reaction conversions and the associated boost in separation factor.

Complex Columns. For columns or towers requiring side draws, multiple feed points for varying feed compositions, and side reboilers or condensers, trays are almost always preferred because they offer a lower cost method of providing those functions than a packed column would allow. In a packed column, each side draw, feed point, reboiler, or condenser requires another packing support, liquid collector, and liquid distributor to be installed which can become prohibitive if several locations are needed. In a trayed column, the whole principle of discrete tray locations offers a ready-made provision for adding those features to the column at little additional cost.

Multiple Liquid Phases. In packed columns, because of the low pressure drops and low turbulence typically present within a packed bed, multiple liquid phases (if present) tend to separate from each other with one phase preferentially flowing down the column and getting to the reboiler ahead of the other. This can sometimes cause severe bumping of the reboiler with a resulting surge of vapor back up through the column, potentially causing mechanical damage to the packing and internals. Conversely, the usually greater liquid holdup on trays with greater turbulence in the liquid promotes mixing and contact between multiple liquid phases, thus minimizing the potential for phase separation and divergent movement of the different phases through the column. As a result, the presence of two or more liquid phases in a column, even just in a portion of the column, usually makes trays the preferred type of internal (in the part of the column where two liquid phases may exist). This is a design issue for heterogeneous azeotropic distillation (see Section 8.1: Heterogeneous Azeotropic Distillation and Steam Distillation.)

Factors favoring packing

Vacuum Operation. The main reason for designing a column or tower to operate under vacuum is usually to keep the operating temperatures, especially reboiler temperatures, low enough to prevent thermal degradation of the column contents or to allow the use of low-temperature heating media. Then, keeping the overall pressure drop across the column to a minimum is often a major criterion in designing for vacuum service. Because packings almost always deliver lower pressure drops per theoretical stage compared to trays, packings are typically specified as internals for vacuum service. Furthermore, structured packings often can deliver even lower pressure drops and higher capacities for a given HETP range compared to random packings, so structured packings may be preferred depending on their cost.

Low Pressure Drop Requirements. Sometimes temperatures need to be kept below critical levels where polymerization or thermal degradation could occur, so operation at reduced pressure may be necessary. These are often situations where the lower pressure drops of packings, especially structured packings, come into play and favor their use. However, in some severe fouling applications where packing can become severely fouled with deposits, the need to provide convenient access to the column internals for periodic removal of deposits may favor the use of trays, at least in those places within the column where most of the fouling occurs. Whether to design for minimum pressure drop to minimize operating temperature (and the rate of fouling), or design to allow for convenient cleaning at the expense of somewhat higher operating pressure and temperature, is a judgment call.

Corrosive Environments, Expensive Materials of Construction. Packings usually require significantly less mass of material per unit column or tower volume than trays, so when corrosive conditions require high alloys or other expensive materials of construction, packings are generally preferred over trays. In addition, when non-metals such as plastics, ceramics, graphite, or engineered materials are required, packings are almost always specified because of the difficulty of fabricating effective tray designs from non-metals. Even when relatively cheap plastics such as polypropylene or CPVC (chlorinated polyvinyl chloride) are used, packings are preferred since the construction techniques required for trays from those plastics are expensive and result in trays that are not as robust as those fabricated from metals.

Low Residence Time Requirements. In cases where reactivity concerns require the minimization of residence times in a column, the lower liquid holdups of packings usually make them a preferable choice over most tray designs. Trays can sometimes be designed with very low liquid holdups, but this often comes at a sacrifice of tray efficiency and/or capacity or operating range, so this rarely favors them over packings for these cases.

Small Diameter Columns. In specialty chemical, pharmaceutical, pilot plant, or other applications where relatively small quantities are produced and required column diameters are small, i.e., less than 1 meter (3 feet) in diameter, packings are generally preferred because of their easier installation and maintenance over trays in small diameters. Since trays typically require personnel entry and access to the inside of columns for installation, the cramped conditions in small diameter columns usually preclude their use. Cartridge tray designs can be specified where an assembly of a number of trays is fabricated outside of a column in a shop and then slid into the column as a unit, but these designs are mechanically complex, usually more expensive, and require special sealing methods at the column walls to prevent liquid bypassing, so they are not a panacea.

Systems Prone to Foaming. With their lower pressure drops, packings generally produce less vapor-liquid turbulence under most operating conditions than trays do, so applications with foaming tendencies tend to favor packings over trays. Tray designs can often accommodate foaming conditions by operating well below their maximum capacity, but this requires larger column diameters and often results in reduced tray efficiencies and higher installed cost than a packed column.

Wide Range of Operating Rates. In years past, packed columns were typically restricted in operating range and turndown capability due to the quality of the liquid distributor designs available then. But in recent years, high-quality liquid distributor technology which has been fully demonstrated has advanced to the point where design HETP values for packings can be maintained over a much wider range than in the past, to the point where turndown ratios of 2:1 are considered easily achievable, 4:1 is reasonable, and 10:1 and higher is possible with modern high-efficiency distributor designs. False loading, which involves maintaining vapor and liquid traffic in a column by recycling portions of the distillate and/or bottoms to the feed, can be used to reduce the need to operate at a high turndown ratio, but this comes at the expense of reduced energy efficiency and higher operating cost. As a result, in cases where a wide range of operating rates is necessary without false loading the column (with its attendant increased cost), packings with good-quality liquid distributors may offer a favorable alternative to trays.

Need for Frequent Internals Replacement. Some column or tower applications require frequent service or replacement of internals due to repeated corrosion or fouling issues. In these cases, use of a random packing which can be quickly removed through an accessway at the bottom of a bed and reloaded through another at the top of the bed, sometimes without a vessel entry, and can result in very minimal downtime and a quick return to normal operation. Conversely, replacement of trays almost always requires vessel entry, dismantling of old trays, and reinstallation of new trays. This usually requires a much longer shutdown and greater expense than a random packing replacement, so packings may be favored in these situations.

Neutral factors
Sometimes there is no clear line of demarcation between conditions when trays are recommended versus packings, so there is a middle ground where either trays or packings can be used effectively. For instance, in cases where mild materials of construction such as carbon steel or low-alloy stainless steels can be used at near-atmospheric pressure with only moderate turndown requirements, often both tray and packing designs can meet process requirements. In these cases, cost comparisons and other factors like prior experience or business preference become the important drivers for decisions about the type of internals to be used.

Just one or two of the factors favoring trays or packings do not necessarily dictate the choice of column internals. For instance, in acrylic acid distillation columns, vacuum operation is required to keep temperatures low enough to allow polymerization inhibitors to function properly and avoid excessive polymer formation, but even with low temperatures, over time sufficient polymer does form, requiring shutdown to remove buildup and restore efficient operation. However, that slow polymer buildup would quickly cause random or structured packings to foul with resulting increased pressure drop and reduced efficiency. As a result, those columns are sometimes provided with dualflow tray internals which have higher pressure drops than clean unfouled packings but are much less prone to polymer fouling and maintain an acceptable pressure drop even when partially fouled.

Expressing Column Capacity. Whether packing or trays are used for column internals, understanding what fraction of its full or maximum capacity a column is operating at is a critical issue for column design and plant operations. If a distillation column is designed to operate (or sometimes operate) close to its maximum capacity, minor fluctuations in operating rates or small upsets can cause the column to exceed its maximum capacity, resulting in significant reductions in separation efficiency and even flooding. Minor exceedances of maximum capacity might be able to be recovered from relatively easily by cutting back on rates or adjusting operating conditions, but a significant upset or a protracted minor exceedance can cause the column to transition into full-fledged hydraulic flooding. Although there are a number of types of abnormal column operation (entrainment, foaming, downcomer backup, etc.) that cause limits on capacity and are referred to as flooding, during hydraulic flooding a column becomes abnormally or completely filled with liquid to the point where pressure drop skyrockets, the column can no longer pass vapor and/or liquid properly, and mass transfer degrades or essentially ceases. In that case, it usually takes quite a while, sometimes even days in the case of very large columns in integrated facilities, to allow excess liquid to drain and reestablish steady state and normal operation. Thus, establishing a means of characterizing column capacity and what fraction it operates at is a very important part of the design process. In discussing limits on capacity, we often refer to liquid or vapor rates in terms of loading. The term "loading" refers to the volumetric flow rate per unit cross-sectional area available for flow and can refer to either liquid or vapor traffic.

No matter what type of column internal is being considered, trays or packing, both gas and liquid rates (gas and liquid loadings) and fluid properties affect capacity. Certainly, as gas rate increases with everything else constant, pressure drop and the fraction of maximum capacity both increase, meaning that column operation gets closer to flooding. Likewise, as liquid rate increases at constant gas rate and properties, pressure drop and fractional capacity again increase. But from a property standpoint the inverse occurs, with fractional capacity increasing as gas and liquid densities decrease.

Based on these interactions, a superficial parameter based on the entrainment correlation of Souders and Brown was developed called the capacity factor (usually denoted as C_S). The capacity factor is intended to reflect the magnitude of the column gas throughput while taking into account different vapor and liquid densities and is defined as:

$$C_S = \frac{G}{\sqrt{\rho_G(\rho_L - \rho_G)}} = v_S \sqrt{\frac{\rho_G}{\rho_L - \rho_G}} \qquad (7.59)$$

where C_S = capacity factor, m/s
$\qquad G$ = superficial gas mass rate, kg/h-m^2
$\qquad \rho_G$ = gas or vapor density, kg/m^3
$\qquad \rho_L$ = liquid density, kg/m^3
$\qquad v_S$ = gas velocity, m/s

Because in many distillation columns the gas and liquid rates are fairly close in magnitude so that the V/L does not stray too far from 1.0, the capacity factor can be considered to represent the absolute magnitude of the column throughput. Also, since the liquid density for many organic chemicals is relatively constant on a mass basis and gas densities are usually orders of magnitude less than liquid densities, the ρ_L-ρ_G

term in the denominator does not vary much and is almost a constant. As a result, a simplified C_S parameter called the "F factor" is often used to reflect the magnitude of column throughput and is defined as:

$$F_S = v_S \sqrt{\rho_G} = \frac{G}{\sqrt{\rho_G}} \tag{7.60}$$

where

$$F_S = \text{F factor}, kg^{0.5}/m^{0.5} - s$$

Also, it follows that:

$$F_S = C_S \sqrt{\rho_L - \rho_G} \tag{7.61}$$

And,

$$C_S = \frac{F_S}{\sqrt{\rho_L - \rho_G}} \tag{7.62}$$

In packed columns, the packing occupies the entire cross-sectional area of the column with no downcomers to take up area and exclude it from participation in active vapor-liquid contact and mass transfer. As a result, the superficial gas mass rate G in the equations for C_S and F_S above should clearly be based on the full column cross-sectional area. However, in the case of trays with downcomers where only a portion of the column area is active bubbling area for mass transfer, the capacity factor is usually based on the superficial gas mass rate through that bubbling area since the approach to flooding will be governed mostly by the gas velocity through the bubbling area rather than the cross-sectional area. The capacity factor for trays with downcomers is then usually designated by C_B to distinguish it and its calculation from the C_S for packings and trays without downcomers.

Inspecting the equations defining the capacity factor and F factor, it is interesting to note that they are essentially gas phase kinetic energy terms, actually the square root of kinetic energy, as kinetic energy is defined as $(1/2)mv^2$. Furthermore, it appears that distillation column throughput is primarily dependent on the kinetic energy of the gas stream, and although the liquid rate plays a part, variations in the gas rate and density will have the greatest effect on column fractional capacity.

Rating Methods for Packing and Trays [96]. With the transition from the conceptualization and preliminary design stage of engineering to an equipment design stage comes the need to determine the diameter and height of a distillation column. And the diameter and height of a column is strongly influenced by the choice of column internals, particularly whether it contains packing or trays and then what type and size of packing or what tray design will be used. In most design or retrofit cases there is not a single type of packing or tray that will fulfill the column preliminary design requirements but a range of types and sizes, with different choices resulting in different combinations of column diameter and height. As a result, the selection of column internals usually presents something of an optimization exercise involving tradeoffs between cost, energy use, and utilization issues such as controllability and ease and cost of maintenance.

In general, the packing size or tray design details (e.g., type of tray, tray spacing, open area, downcomer cut) will determine the column diameter required to handle the specified feed rate and internal vapor-liquid flows based on the hydraulic capacity of the particular packing or tray type chosen. In addition, that same choice will determine the column height required based on the theoretical stages and/or mass transfer units dictated by the separation power specifications and the tray efficiency or packing efficiency (HETP or HTU). As a result, in rating a column's performance, the capacity and efficiency of the chosen packing or tray needs to be determined for the specific operating conditions of the application.

Packing Efficiencies. For packed columns, separation efficiency is characterized in terms of HETP, or height (of packing) equivalent to a theoretical plate. This is the length of packed bed required to deliver one theoretical stage of separation for the given conditions and rates in the packed bed. The HETP in a packed bed is difficult to predict because it is dependent on many factors: the type and size of packing, the liquid and vapor flowrates in the bed, physical properties including liquid-vapor interfacial tension, the thermodynamic driving forces for separation, and the quality and state of liquid distribution in the packed bed. As an inverse parameter, when the value of HETP is high, the separation efficiency is low, and when the value of HETP is low, the separation efficiency is high.

In every case of packed column or tower design, an HETP or HTU value must be chosen which represents the most reliable value for determining the packing height that will deliver the required separation power over the required range of operating rates. If test data are available for the chemical system, the specific packing, and the operating conditions of interest, then those should be used in setting a design value of HETP or HTU. But in many cases, test data on the actual chemical system with the desired packing type and size are not available, so alternative sources must be used. Sometimes data on a similar chemical system or a standard test system such as cyclohexane/n-heptane, isobutane/n-butane, o/p-xylene, or others are available which can be adjusted appropriately. But if no other data are available, or if only a rough estimate of packing efficiency is required for preliminary range-finding purposes, the authors have found in their experience that random packings will often exhibit an HETP on a low relative volatility test system (less than 1.5–1.8 relative volatility or so) that is equivalent to about 12 times the diameter of the packing pieces. This rough HETP value can be adjusted for systems with higher relative volatility using methods described earlier, as in eq. (7.43), but only for estimation or preliminary design purposes.

A typical packing efficiency versus capacity curve is shown in Figure 7.12. This curve is very generic and is intended for illustrative purposes since the shape, magnitude, and range of the HETP capacity curve for a particular application will depend on a number of factors such as type and size of packing, type of system being separated, design of the column internals, how the internals were installed, degree of mass transfer rate limitation, operating pressure, and other factors. But most HETP capacity curves exhibit a number of features that are a result of the hydraulic and mass transfer performance of the packing in the application being studied.

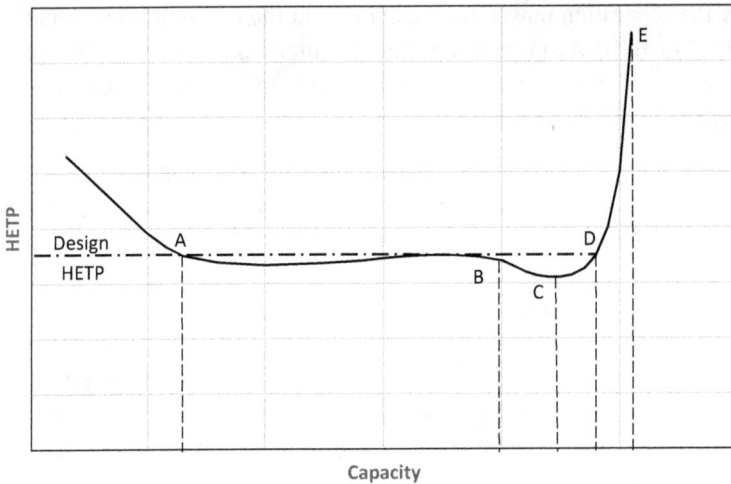

Figure 7.12: Typical Packing HETP versus Capacity Performance Curve.

The relatively flat portion of the curve between points A and B represents the primary range in which the column is intended to operate, with Point A representing the lowest required throughput and Point B representing the design capacity of the column. As the operating rate drops below Point B but prior to reaching Point A, the HETP may drop slightly because as the liquid loading decreases, film thicknesses decrease and mass transfer improves to cause the liquid and vapor streams to operate closer to being in equilibrium with each other. But at some point, as the liquid loading continues to decrease the ability of the liquid distributor to uniformly divide the liquid across the entire cross-sectional area of the packed bed deteriorates, the benefit of thinner films is overcome, and the result is the rise in HETP shown on the curve below Point A. Although this rise in HETP at low liquid loadings is mostly attributed to poor liquid distribution, other factors such as packing wettability may also come into play.

In many cases, as column throughput increases above Point B, the HETP goes through a minimum (Point C) as the vapor and liquid phases undergo an increase in pressure-drop-driven turbulence prior to flooding. This increased turbulence can cause froth generation with extended interfacial area and thinner liquid and vapor films which subsequently result in the reduction in HETP due to improved mass transfer. Beyond Point C, HETP increases progressively as the packing flood point is approached. The flood point is represented by Point E, above which hydraulic stability is lost, column pressure drop skyrockets, and operation is no longer able to be sustained, either due to massive entrainment, foaming, dumping, or filling of the column with liquid.

An exception to the behavior illustrated by Point C are wire gauze packings like Sulzer BX, Intalox Wire Gauze Packing, and Montz A3 which exhibit their best HETP performance at low liquid loadings and progressively higher HETPs as throughput increases right up to flooding. With these types of packings, the wicking nature of wire gauze allows significant spreading of liquid throughout the gauze structure at low liquid loadings with very thin liquid films and significantly enhanced interfacial area. As long as adequate liquid distribution quality is maintained, the result is enhanced mass transfer and significantly reduced HETPs at loadings where random packings and structured sheet metal packings would be delivering at best only slightly reduced HETPs from their design HETPs between points A and B. However, at higher throughputs and loadings characteristic of more typical column design conditions, the advantage that wire gauze packings provide of liquid spreading and thin films is lost, and HETPs are no better and can be even worse than those provided by structured sheet metal packings with equivalent specific surface areas. The liquid loadings at which wire gauze packings lose their HETP advantage tend to be in the $2–4$ gpm/ft^2 ($5–10$ m^3/h/m^2) range.

The design HETP is generally the maximum value between points A and B and can occur at any point in that range, not necessarily the HETP right at the design capacity. In this manner, the required separation efficiency will be achieved or exceeded at all planned operating rates, not just at the design operating rate. Point D is the Maximum Usable Capacity (MUC), the highest throughput at which the design HETP can be achieved while still maintaining stable column operation. Note that there is unused operating capacity between points B and D where acceptable HETPs can still be achieved. However, routinely operating in this range is generally not advisable since small variations in column operating conditions and rates can push the column into regimes where separation performance is compromised and possibly even into flooding where column operation must be curtailed or interrupted to allow stable operation to be restored. A common rule-of-thumb for choosing the Point B design capacity is roughly 90% of the Point D MUC and 80% of the Point E flooding capacity which often puts it immediately prior to the drop in HETP leading to Point C as shown in Figure 7.12. However, it should be noted that there may be valid reasons for choosing another point along the HETP curve for the design capacity of a particular application such as the need for an expanded turnup-turndown ratio.

As mentioned earlier, the shape of the HETP capacity curve depends on a number of factors, and perhaps the most important factor is design and choice of liquid distributor. Even in the early years of packed column development, the need was recognized to evenly distribute liquid feeds and reflux across the surface of a packed bed to achieve desirable levels of packing separation efficiencies. However, the types of liquid distributors available prior to the 1980s were not capable of delivering consistent distribution quality over a wide enough operating range with sufficient distribution points per unit area to allow any but the largest packings (>50 mm) to match the design HETPs achieved in small (<0.5 m diameter) test columns. As a result, most packed column applications were traditionally limited to small diameter columns, those operating at high liquid loadings (above 10 gpm/ft^2 or 25 m^3/h/m^2), corrosive services requiring non-metallic packings and internals, and undemanding separations where high HETPs were acceptable. Examples include high loading stripping and absorbing applications and high relative volatility separations.

Some pre-1980 distributor designs attempted to deliver a large enough number of distribution points in larger diameter commercial towers but suffered from performance and reliability issues due to design challenges. For example, liquid-full circular orifices such as are used in orifice plate and pipe distributors were susceptible to fouling when provided with sufficient point densities to achieve good HETPs because of the small orifice diameters required. And distributors using V-notch or rectangular weirs as metering orifices were quite resistant to fouling problems but were very susceptible to leveling issues because of the high dependence of flow rate on liquid head with the V-notch geometry. As a result, large diameter commercial packed towers were hampered by compromised performance of high-efficiency packings or were restricted to relatively large packings (50 mm or larger) which did not require a large number of distribution points to achieve their design efficiencies.

When novel structured packings were developed in the 1970s and later which offered good packing efficiencies at significantly lower pressure drops per theoretical stage than random packings, it was recognized that high-quality liquid distribution at moderate to low liquid loadings was required to take advantage of the significant benefits of modern structured packings. Much development work by a number of companies and organizations followed which resulted in innovations in distributor design and greatly improved performance. Some of these innovations included the use of side discharge orifices, multiple parting boxes for pre-distribution, and quiescent zones to reduce maldistribution due to liquid velocities. The comparative HETP performance of modern distributors relative to other styles is shown in Figure 7.13.

The base style of distributor for comparison is the same style used for Figure 7.12, which for comparative purposes is termed the standard quality distributor. These types of distributors are typically bottom-orifice type devices such as orifice plates and perforated pipe distributors and are generally effective down to roughly 4 gpm/ft^2

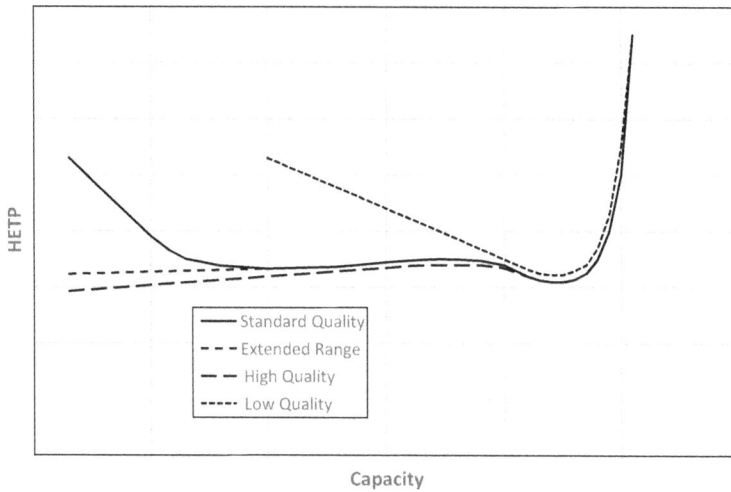

Figure 7.13: Effect of Distribution Quality on Packing HETP vs. Capacity Performance Curve.

($10 \ m^3/h/m^2$). Standard quality distributors may have 4–9 distribution points per square foot (44–100 pts/m²) and can have turndown ranges of 2:1 to 4:1.

Low-quality distributors are usually older styles such as spray nozzle arrays and rectangular or V-notch weir trough distributors that are compromised on turndown, distribution point densities, and leveling sensitivity. As seen in the Figure 7.13, the flat portion of the HETP capacity curve for these devices is very short or non-existent, and the HETP deteriorates rapidly as liquid loading decreases. These types of distributors have only a short range of throughputs at high loadings close to flooding where design HETPs can be achieved. However, they often can handle high liquid loadings and may provide better fouling resistance than higher-quality devices which may qualify them for some extreme service applications today. They might have an effective distribution point density of 1–4 pts/ft² (11–44 pts/m²), a turndown range of perhaps 1.5:1, and may be appropriate for applications with liquid loadings of 10 gpm/ft² ($108 \ m^3/h/m^2$) or more, especially cases where the column is planned to run routinely and consistently at high liquid loadings or conditions close to but below flooding. An example is recirculated caustic scrubbers which typically operate at high fixed liquid recirculation rates.

Extended range distributors also may have 4–9 pts/ft² (44–100 pts/m²) like standard quality distributors but employ design features such as multilevel parting boxes and segregated final distribution zones to achieve turndown ranges of 10:1 or more at liquid loadings as low as 0.1 gpm/ft² ($0.2 \ m^3/h/m^2$). Their HETP capacity curves are usually very close to those of standard quality distributors at high and moderate liquid loadings but do not exhibit a kickup of HETP at low loadings. In fact, because liquid film thicknesses are usually reduced at low loadings, the HETP values at low loadings can be lower than at any other point on the curve.

High-quality distributors employ features such as independent troughs, multiple parting boxes, side orifices, and distribution point spreading devices to achieve turndown ratios of 10:1 or more with effective distribution point densities of more than 10 pts/ft^2 (108 pts/m^2) and the ability to handle liquid loadings down to 0.05 gpm/ft^2 (0.1 m^3/h/m^2). Because of their high distribution point densities, high-quality distributors can often deliver lower HETPs than the other classes of distributors across the full operating range of loadings, but the greatest improvement in HETP is in the lower range of loadings. With the proliferation of fourth-generation random packings and high-efficiency structured packings in recent times, more and more applications have taken advantage of extended range and high-quality liquid distributors. Although standard quality and low-quality distributors are still used in many applications, they are becoming less prevalent in recent years.

Tray Efficiencies. For trayed columns, separation efficiency is characterized by the tray efficiency; that is, the fraction or percentage of the change in composition achieved by the actual tray compared to that which could be achieved by a theoretical equilibrium stage operating at the same conditions as the actual tray. In practice, tray efficiencies are less than 100% for countercurrent trays, but can be greater than 100% for almost all crossflow trays. Tray efficiencies for crossflow trays with long liquid flow path lengths, such as with single-pass trays in large diameter columns, are sometimes significantly greater than 100%. For discussion of the reasons why, see the discussion of Murphree tray efficiency in Section 6.6.2: Stage Efficiency.

A representative tray efficiency versus capacity curve is shown in Figure 7.14. Again, this curve is very generic and is intended for illustrative purposes. The actual shape, magnitude, and range of the tray efficiency capacity curve for a particular application will depend on a number of factors such as tray type, type of system being separated, tray open area, hole size or tray deck device size, degree of mass transfer rate limitation, operating pressure, and other factors. Most tray efficiency capacity curves exhibit a number of similar features that are a result of the hydraulic and mass transfer performance of the specific tray in the application being studied.

In the tray efficiency curve, at low levels of capacity tray efficiencies generally fall off as liquid tends to leak through tray openings due to insufficient vapor velocity through the openings, reduced froth heights, and lower levels of mixing and turbulence on the tray deck. These factors all result in less vapor-liquid contact area and time and as a result, a decrease in mass transfer. However, as tray throughput increases into the broad middle range of the curve, vapor-liquid contacting improves and stabilizes due to proper liquid holdup on the tray deck and good vapor-liquid mixing and contact in the froth on the deck. As throughput increases beyond a certain point however, vapor velocities increase to cause liquid entrainment upward, froth heights increase to encroach on the trays above, and downcomers fill with liquid and back up onto the tray feeding them. These phenomena cause a degradation in tray

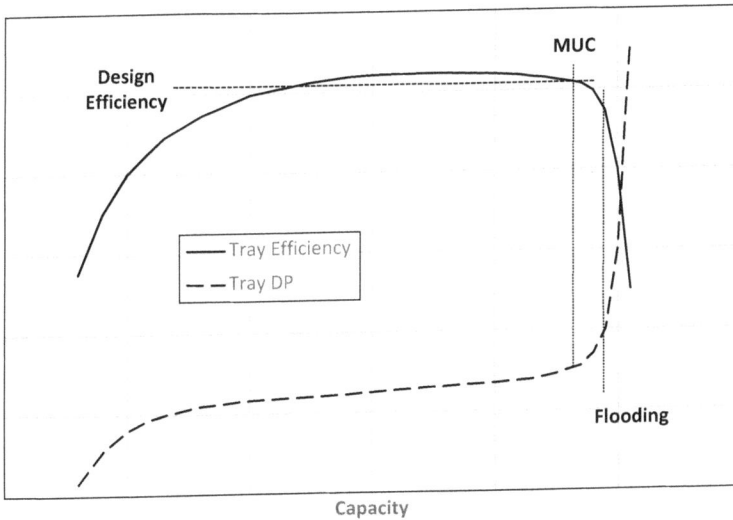

Figure 7.14: Typical Tray Efficiency and Pressure Drop versus Capacity Performance Curves.

efficiency and eventually a progression into flooding with a severe loss of efficiency and a sudden increase in tray pressure drop which can be difficult to reverse.

So in Figure 7.14, the flooding point is identified as the capacity (throughput) at which tray efficiency sharply decreases and pressure drop sharply increases. The tray design efficiency is usually chosen as the best efficiency that can be met or exceeded over the desired range of operation, often characterized as a fairly broad middle range of capacities where the efficiencies do not change very much. Then, the maximum usable capacity is identified (similarly to how it is identified for packings) as the maximum capacity at which the tray design efficiency is still maintained before the onset of flooding and within the range of stable operation.

Now, with the wide variety of tray types and design details (hole or valve size, percent bubbling area, downcomer area, weir height, etc.) available to the column designer, the classic efficiency curve in Figure 7.14 may not be applicable. In many cases, the efficiency curve may deviate significantly from the ideal, lack some of the defining characteristics, and make identifying a reliable design efficiency somewhat subjective as shown in Figure 7.15. In this example, there is no broad range of throughputs where the efficiency is relatively constant, with gradual transitions from low efficiencies at low throughputs to a peak efficiency in the midrange and then into higher pressure drops and reduced efficiency as the column loads up prior to flooding. In the Figure 7.15 example, even the transition to flooding is more gradual, making the determination of the flood point more subjective and problematic. These cases usually require a degree of extra conservativeness in choosing the design range and operating points.

In establishing the design efficiency, operating range, MUC, and flood point for a particular application, a designer may be able to take advantage of published tray

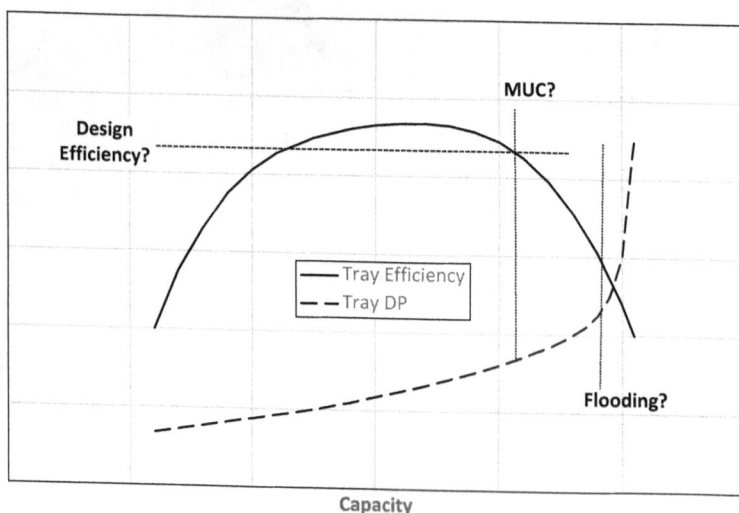

Figure 7.15: Tray Efficiency and Pressure Drop versus Capacity Curve with Less-Clear MUC.

efficiency diagrams and correlations. Or, they may have access to proprietary data and correlations from equipment manufacturers and industry consortia such as Fractionation Research, Inc., the Separations Research Program at the University of Texas at Austin, or others. But it's also possible to take advantage of proprietary computer programs for tray design and rating from some of the major equipment manufacturers which allow a designer to determine tray design and performance information using experimentally verified correlations incorporated into the program code itself. Furthermore, some manufacturers have worked with process simulation program providers to incorporate their correlations into the simulation program itself so that tray rating and design calculations can be done right along with flowsheet simulation execution. This can greatly expedite the distillation design process and improve designer productivity.

7.4.2 Packings

Purpose of Packings. Column packings, ever since they were first introduced and whether random or structured, provide surfaces for liquid to form films in a liquid dispersed/vapor continuous contacting environment. The packing increases interfacial area beyond that which would be available merely from internal baffling or wetted column walls. In general, higher packing specific surface area (surface area per unit packed volume) results in higher separation efficiency (i.e., lower height equivalent to a theoretical plate or HETP) although usually at the cost of higher pressure drops and lower column throughput capacity. Desired attributes for an effective column packing include high surface area per unit volume, high open area (high void volume) to promote high vapor traffic and low pressure drop, and a geometry that minimizes horizontal surfaces to result in a low tendency for liquid to accumulate and cause flooding.

In addition, an effective packing needs to have enough mechanical robustness to resist deformation due to crushing in tall beds or turbulent forces at high throughputs and pressure drops. However, this robustness needs to be accomplished without compromising the open area or introducing structure that would cause excessive liquid holdup or impediments to vapor and liquid flow. Corrosion resistance of packing materials of construction in the process environment must be considered so that the packing geometry and mechanical robustness can be maintained over the design lifetime of the packing. Often metals, particularly less-costly grades of stainless steels, meet these requirements, but when aggressive process environments must be dealt with, ceramics, plastics, or other non-metals may need to be considered usually at the expense of higher pressure drops, lower capacities, and shorter allowable bed heights due to thicker surfaces and reduced open area. High metal alloys with good corrosion resistance and better throughput characteristics may provide an alternative to non-metals in corrosive environments, but high costs of materials may turn out to be prohibitive for many applications. These are all tradeoffs that need to be considered when choosing the best packing type, size, and material of construction for a process application.

Random (or Dumped) Packing. Random packings are the earliest type of packing used in industrial practice. They consist of individual packing pieces that are typically smaller than the column diameter by a factor of 8 or more. Random packings have been in use industrially for around 200 years, with the first packings consisting of gravel, coal or coke, shards of broken glass, or glass or ceramic spheres. By the mid-nineteenth century, other materials came into use such as pieces of pumice and coke with increased porosity which contributed to somewhat better separation efficiencies due to improved liquid adhesion and spreading. However, throughput capacities using these types of packings were severely limited as a result of poor free (open) volume and low surface area per unit of packed volume. These limitations gave rise to work to develop improved geometries and configurations to provide better free volume and surface area and consequently, better capacities and efficiencies, which resulted in what are termed the first generation of manufactured random packings [97, 98].

First-Generation Random Packings. First generation random packings are characterized by attempts to increase the void fraction of the packing and as a result, open up vapor and liquid access to the interior of the individual packing pieces for improved performance. The simplest solution to this problem, and therefore the first, was a hollow cylindrical ring developed at the end of the nineteenth century by Friedrich Raschig with a length equal to its diameter and made of metal, ceramic, or other non-metals such as glass or carbon. Called the Raschig ring (Figure 7.16a), it was patented in 1907 and is still in use in a few applications in modern times [99].

Modifications to the basic shape of the Raschig ring followed with the inclusion of a flat dividing wall across the inside diameter to produce the Lessing ring. Following this development, a second dividing wall was added perpendicular to that in the Lessing ring in an "X" configuration to produce the Cross Partition ring. Both the Lessing ring and Cross Partition ring were intended to increase the internal surface area and thus the wetted surface area available to the vapor and liquid flows inside a column. Another modification in this vein was the spiral ring, in which an internal spiral surface was added, either as a single, double, or triple helix oriented along the central axis of the Raschig ring. All of these modifications were primarily manufactured of ceramic materials because of the difficulty of producing them from metals.

Also usually classed among first-generation packings are Berl saddles (Figure 7.16d), a classic symmetrical saddle shape invented and patented in the 1930s by Ernst Berl which had the advantage of providing increased surface area per unit packed volume and lower pressure drop than Raschig rings. Manufactured in ceramic materials in a variety of sizes from 6 mm to 50 mm, Berl saddles were widely used in research and industry. Although highly effective for their time, Berl saddles had a tendency to nest together in interlocking stacks inside a packed bed which reduced their easily accessible surface area and separation efficiency and provided impetus for further developments in packing shapes [100].

Second-Generation Random Packings. With the success of the Raschig ring and Berl saddle, developments continued on providing greater surface area accessible to gas and liquid flows and lower impediments to flow for lower pressure drops. A significant improvement to the Berl saddle was the Intalox® saddle (Figure 7.16e) developed by Max Leva and licensed to the former U.S. Stoneware Company which utilized a toroidal saddle shape to minimize nesting tendencies when installed in a packed bed and increase the available surface area. The Intalox saddle shape was further refined as the Super Intalox® saddle by the Norton Chemical Process Products Company (now Saint-Gobain NorPro which markets the packing as the Proware™ Super Saddle in ceramic) with the addition of surface perforations and scalloped edges to provide less resistance to gas flows, drip points for greater liquid dispersion, and greater turbulence for improved gas-liquid contact. Although the classic Intalox saddle shape has only been available in ceramic, the Super Intalox saddle was also developed in plastic materials such as polypropylene, CPVC, polyvinylidene fluoride, and others and is available under that name from Koch-Glitsch, LP (Figure 7.16f). Packings with the same general geometry are also available from multiple manufacturers under different names in ceramic as well as plastics.

The Raschig ring was likewise improved in several versions over time from the 1920s to the 1960s by introducing slots in the packing walls to increase the accessibility of gas and liquid to the inside of the ring. BASF AG developed the most well-known version called the Pall ring (Figure 7.16b,c) during World War II and introduced it commercially in the 1950s. The Pall ring geometry includes multiple internal fingers which are made by punching rectangular slots in the sides of the ring with the material from the slots being left attached at one end. This provided internal drip points to better disperse the liquid flow and increased open area in the ring to promote gas-liquid mixing and reduce pressure drop. These improvements resulted in the Pall ring achieving 50% to 80% greater capacity and separation efficiency over that of the Raschig ring. Pall rings are still in wide use today and are available in sizes from 16 to 90 mm from several manufacturers in a variety of metals and plastics as well as ceramic.

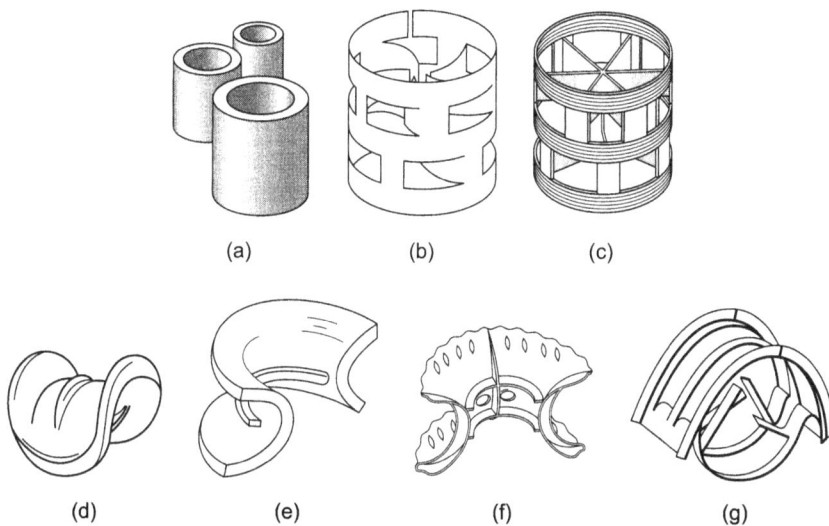

Figure 7.16: Comparison of Random Packing Generations: (a) Raschig rings; (b) metal Pall ring; (c) plastic Pall ring; (d) ceramic Berl saddle; (e) ceramic Intalox saddle; (f) plastic Super Intalox® saddle; and (g) IMTP® (Intalox Metal Tower Packing). Taken from ref. [77], W. L. McCabe, J. C. Smith and P. Harriott, Unit Operations of Chemical Engineering, 7th ed., McGraw-Hill, copyright 2005. Reprinted with permission from McGraw-Hill Higher Education.

A further development of the Pall ring by Norton Chemical Process Products Corporation in the late 1960s was the Hy-Pak® ring in which larger windows were punched in the walls with internal fingers being formed at both ends of the windows for a doubling of the number of fingers. Circumferential ribs were also formed in the walls during manufacturing to provide better rigidity and mechanical strength and allow the use of thinner gauge metal. The increased open area of the packing pieces resulted in increased capacity and reduced pressure drop with only a marginal reduction in efficiency, if any. The Hy-Pak packing (now by Koch-Glitsch LP) and others with similar geometry are available from multiple manufacturers in metals only.

<u>Third-Generation Random Packings</u>. With the success and many applications of the second-generation random packings, a wider variety of packing shapes and modifications were developed to further increase the packing void fraction and porosity. One of the most significant third-generation developments was the Intalox Metal Tower Packing (IMTP®, Figure 7.16g), introduced by Norton Chemical Process Products in 1977 [101] and now produced by Koch-Glitsch, LP. The IMTP packing represented a departure from previous developments in that it was neither a strictly Raschig ring/Pall ring modification nor a saddle-type packing but rather a hybrid type with some characteristics of both. Its open geometry with very high void fraction when packed in a column resulted in significantly lower pressure drops and higher capacities than second-generation packings with the same or better separation efficiencies. In addition, the IMTP geometry lends itself well to relatively straightforward production techniques which keep manufacturing costs low and allow for competitive pricing, so once the IMTP patent expired, many other manufacturers began supplying geometrically identical (or nearly so) IMTP "clones," although marketed under different names. IMTP and its clones are still very widely used today.

Following the success of IMTP, Dale Nutter of Nutter Engineering developed a random packing in 1984 with similar but slightly different geometry called the Nutter Ring™ (Figure 7.17). The Nutter Ring again shares characteristics of both ring-type and saddle-type packings with an open geometry (in a similar fashion to IMTP and its clones) which results in lower pressure drops/higher capacities than second-generation packings with good separation efficiencies. Because their geometries are slightly different, the pressure drops, capacities, and efficiencies of the various sizes of Nutter Rings do not necessarily line up exactly with those of the various sizes of IMTP, but they do tend to fall on the same efficiency vs. capacity trend line. Nutter Rings are currently supplied by Sulzer Ltd. in a wide range of sizes and metal materials.

Figure 7.17: Sulzer Nutter Ring™. Taken from ref. [102]. Reprinted with permission from Sulzer, Ltd.

Another third-generation packing is the Cascade Mini-Ring® (CMR, Figure 7.18), essentially a high aspect ratio Pall ring with a diameter-to-height ratio of approximately 3:1, which was introduced in 1971 by Mass Transfer Limited. The high aspect ratio of the CMR allows better access of gas and liquid to its interior surfaces than the Pall ring and results in the packing pieces preferentially orienting themselves with their diameters horizontal when packed in a bed in a column. This horizontal orientation provides less resistance to gas flow and therefore lower pressure drops and higher capacities than Pall rings or similar packings with roughly the same separation efficiency. Although the packing name is trademarked and currently owned by Koch-Glitsch, the same general packing geometry (with some variations) is available under different names from a variety of suppliers in a number of metal and plastic materials.

Figure 7.18: Koch-Glitsch Metal Cascade Mini-Ring®. Taken from ref. [103]. Reprinted with permission from Koch-Glitsch, LP.

In a departure from the classical ring and saddle type packings, the spherical Jaeger Tri-Packs® packing (Figure 7.19) was developed in the early 1980s as a plastic packing with a filamentous geometry that provided low resistance to gas flow and minimal nesting tendencies. Primarily intended for absorption and scrubbing applications, Tri-Packs have exhibited advantaged performance over second-generation and some third-generation packings. Still in wide use today, Tri-Packs are available from Raschig, and geometrically similar packings are available from multiple other manufacturers under a variety of names, as the original name is a trademark owned by Raschig.

Fourth-Generation Random Packings. Candidates in the latest fourth generation of random packings exhibit mostly evolutionary improvements over third- and earlier-generation packings rather than significantly different geometries. The trend toward more open, filamentous geometries continued with emphasis on reducing nesting tendencies and lowering pressure drops while maintaining separation efficiencies. The results of these developments often are packing designs that deliver higher throughput

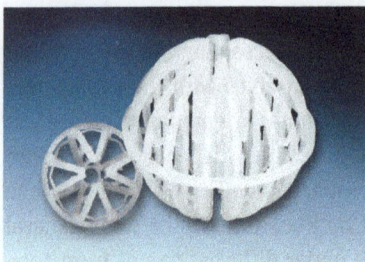

Figure 7.19: Raschig GmbH Hacketten® / Jaeger Tri-Packs®. Taken from ref. [104]. Reprinted with permission from Raschig GmbH.

capacities at the same separation efficiency as packings from earlier generations or conversely, higher separation efficiencies at the same throughput capacity.

An example of this is Intalox® Ultra packing (Figure 7.20) by Koch-Glitsch which is similar in general concept to IMTP but with its arches skewed slightly outward (and with other changes) to allow each packing piece to occupy a greater volume than the equivalent size IMTP. The result is a packing with good mechanical strength exhibiting lower pressure drop and higher capacity when compared with IMTP packing delivering roughly the same HETP.

Another fourth-generation packing example is the NeXRing™ (Figure 7.21) by Sulzer which incorporates fourth-generation principles of an open, filamentous shape with reduced nesting. In a similar fashion to other advanced packings, the NeXRing can deliver higher capacity than third-generation packings like the Sulzer I-Ring™ and Nutter Ring with similar HETPs. Not every size of NeXRing lines up with an I-Ring or Nutter Ring with a similar HETP, but the efficiency vs. capacity trend line of the NeXRing falls noticeably above that of the third-generation packings.

Other packings falling in the general range of performance of fourth-generation packings are the Raschig Super-Ring® (Figure 7.22) and Super-Ring® Plus (Figure 7.23) which achieve improvements with a multi-arch geometric design. The Super-Ring, introduced around 1995, exhibited generally improved performance over Pall rings and other second- and third-generation packings, and the Super-Ring Plus, introduced later, delivers somewhat lower pressure drops and higher capacities than the equivalent-sized Super-Ring with the same or slightly better mass transfer efficiency.

Interestingly, AMACS also provides a packing with generally improved performance over second- and third-generation packings called SuperBlend™ 2-Pac which consists of a blend of two different sizes of IMTP-like packing (a third-generation packing). AMACS maintains that it performs with the capacity of the larger size of the blend but with the HETP efficiency of the smaller size of the blend which is supported by testing by a 3rd party research organization, the Separations Research Program at the University of Texas at Austin. As a result, although made up of third-generation packing pieces, the performance of the blend of packing sizes generally falls in the range of a fourth-generation packing.

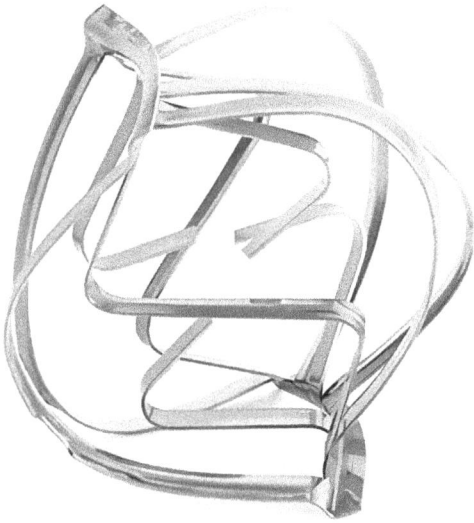

Figure 7.20: Koch-Glitsch Intalox® Ultra Packing. Taken from ref. [105]. Reprinted with permission from Koch-Glitsch, LP.

Figure 7.21: Sulzer NeXRing™ Metal Packing. Taken from ref. [106]. Reprinted with permission of Sulzer, Ltd.

A rather unique packing which might be classified as either third- or fourth-generation and available only in plastic is the Intalox® Snowflake® packing (Figure 7.24), developed by Norton Chemical Process Products Company in the early 1990s and currently owned and produced by Koch-Glitsch. This packing is unique in that it again is filamentous in geometry but in the form of a very high aspect ratio circular ring with

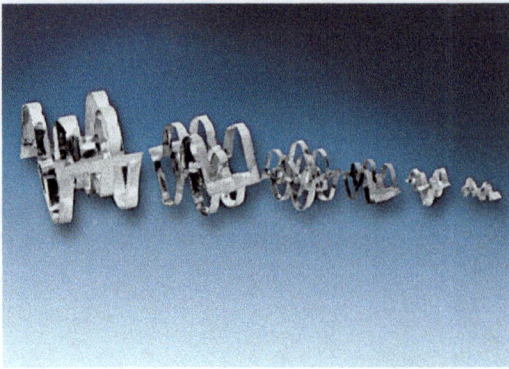

Figure 7.22: Raschig Super-Ring® Metal Packing. Taken from ref. [107]. Reprinted with permission of Raschig GmbH.

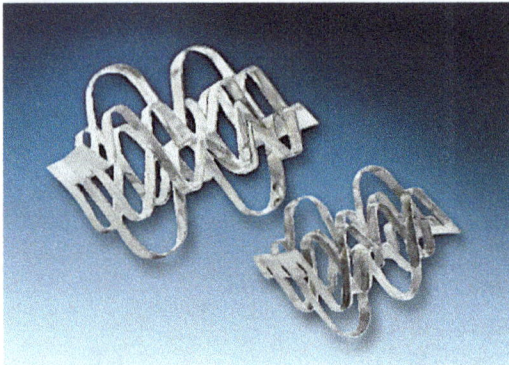

Figure 7.23: Raschig Super-Ring® Plus Metal Packing. Taken from ref. [108]. Reprinted with permission from Raschig GmbH.

intersecting internal filaments in aggregations of 6 – hence the name "Snowflake." The high diameter-to-thickness ratio and complex internal structure results in the packing pieces preferentially orienting themselves horizontally in a packed bed (similar to the way CMR does) so that resistance to gas flow is very low, resulting in low pressure drops and high capacities. Intalox Snowflake has been used very successfully in absorption and scrubbing where it often outperforms other earlier plastic packings in these types of applications such as Pall rings and Super Intalox saddles.

The random packings mentioned above represent only a portion of the many types available from a wide range of manufacturers. The industry has a long and colorful history of invention and development which continues to unfold [110]. Although some new packing designs are sometimes described as fifth-generation packings, our opinion is that most represent incremental improvements and refinements of previous geometries. However, considering the dramatic improvements seen over the first

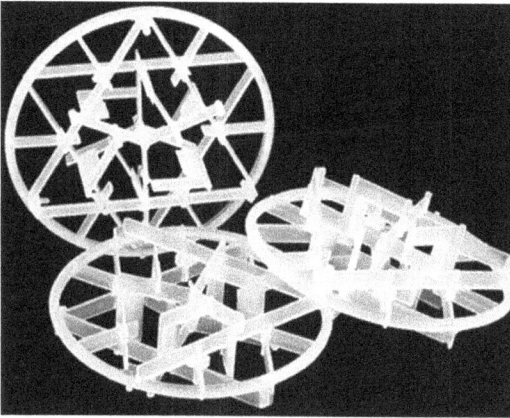

Figure 7.24: Koch-Glitsch Intalox® Snowflake® Plastic Tower Packing. Taken from ref. [109]. Reprinted with permission from Koch-Glitsch, LP.

four generations of random packings, further developments resulting in yet more improvements in performance and representing a true fifth generation of random packings are not out of the question.

Structured Packing. Where random packings consist of discrete individual pieces that are much smaller than the column diameter, structured packings are fabricated in uniform blocks or segments called elements that are roughly 100–300 mm in height and cover the full cross-sectional area of a column. Structured packing elements are most often fabricated by assembling alternating layers of sheet metal or plastic, wire gauze, or mesh with diagonal corrugations and sometimes surface treatments and/or perforations. However, there are also alternative form factors for special applications such as grid-type structured packings in which 25–100 mm thick lattices of fluted, tabbed, or otherwise modified strips forming a grid are stacked in layers to fill a column. Because the alternating layers of corrugated sheet in most structured packings are very similar in configuration to some static mixers, radial mixing of bypassing liquid and vapor are promoted, helping to minimize maldistribution and enhancing mass transfer.

The blocks or elements of structured packings are fabricated in different shapes that, when installed in a column, fit together to completely cover the cross-sectional area of the column with no gaps other than a small gap around the periphery next to the inside wall. This peripheral gap is usually filled with a tabbed strip of sheet metal or wire gauze called a wall wiper to capture liquid wall flow and direct it back into the general volume of structured packing. Because structured packing must be fabricated in particular shapes for the particular column for which they are intended, their manufacturing cost and therefore purchased cost is almost always higher than that of random packing giving the same capacity. And because they must be manually fitted into the column in specific locations in a specific order, their field installation cost is typically higher than that of random packings. Contributing to this high installation cost is the tendency for structured packings to be fabricated from very thin sheets of metal or gauze which can sometimes be quite delicate and require extra care during installation, often requiring the use of walking boards for installers to stand on to prevent damage to the surface of the elements.

In addition to somewhat higher cost, one drawback is in applications where fouling can occur over time such as in monomer services requiring the lower pressure drops of packings versus trays. In these applications, buildup of solids can result in a conglomerated mass of structured packing elements and fouling matter that can be extremely difficult to remove, often much more difficult than random packing in the same service. However, structured packings can often deliver lower pressure drops and higher capacities than random packings for the same or better separation efficiencies. This typically results in lower pressure drops per theoretical stage for structured packings than random packings, a very desirable characteristic.

The first structured packings were crude arrangements of wooden slats or angle iron lengths stacked in a crosshatched fashion and used in absorption columns. However, in the early 1950s a more refined structured packing was developed called Panapak which used corrugated expanded metal sheets to form packing elements and which significantly improved on the performance of bubble cap trays, the predominant column internal at the time, with pressure drops roughly 8 times lower than those of trays. However, separation efficiencies were limited in commercial-scale columns because of the limited quality of liquid distributors available at the time.

Other early structured packings were Goodloe® and Hyperfil™ knitted wire mesh packings made of multiple thin strands of wire knitted together into flat sheet-like pads into which corrugations were formed. The corrugated pads were then fabricated into structured packing elements by stacking the pads in adjacent layers with the corrugations oriented diagonally and alternating from layer to layer. Although the knitted mesh packings were able to exhibit excellent separation efficiency in some applications, the same limits on liquid distribution quality usually restricted their primary use to small diameter columns of roughly 300 mm or less.

In the 1960s, Sulzer Brothers, Ltd. (now Sulzer Ltd.) introduced Sulzer BX packing (Figure 7.25), a structured packing fabricated from corrugated sheets of wire gauze woven in a Dutch Twill weave to promote liquid spreading. The corrugated gauze sheets were fabricated into packing elements in alternating layers similar to the way knitted mesh packings were fabricated but also included perforations in the sheets to allow better passage of liquid and gas from one layer to the next and improve radial mixing. The result was excellent separation efficiency when operated well below the maximum capacity of the packing, but efficiency degraded significantly as loadings were increased close to maximum. However, at low loadings the pressure drops for the packing were also very low which lent itself well to vacuum operations where the much higher pressure drops of trays were infeasible. With the development of high-quality liquid distributors in the 1980s, 1990s, and into the twenty-first century, the high separation efficiencies of wire gauze packings at low loadings were extended to larger diameter commercial-scale towers, and other manufacturers developed their own gauze packings such as Intalox® Wire Gauze Packing by Koch-Glitsch and Montz-Pak Type A3 by Julius Montz GmbH which delivered similar performance. These products have been followed in recent years by wire gauze packings from a range of manufacturers in Asia, Europe, and the Americas, although the performances of some products may vary from what would be expected from the original BX packing. Additional sizes of wire gauze packings with different crimp heights such as Sulzer CY packing are available for applications also requiring low pressure drops per theoretical stage but with higher separation efficiency needs.

Figure 7.25: Sulzer BX Wire Gauze Packing. Taken from ref. [111]. Reprinted with permission from Sulzer, Ltd.

In the 1970s, Sulzer followed up the development of Sulzer BX wire gauze packing with its Mellapak™ structured packings (Figure 7.26), conceptually similar to Sulzer BX but with its corrugated layers made from thin sheet metal rather than wire gauze. With the lower material costs, greater mechanical strength, and low pressure drops per theoretical stage offered by structured sheet metal packings, they saw significant adoption in a wide range of applications in columns up to 10 m diameter and greater (with modern high-quality liquid distributors) and are still in wide use today. Many major column internals manufacturers today offer their own structured sheet metal packings such as Flexipac® by Koch-Glitsch and Montz-Pak Type B1 by Julius Montz GmbH.

The most popular style of these packings has corrugations oriented at a 45-degree angle from horizontal, often designated as "Y" type packings (for example, Sulzer Mellapak 250Y and Koch-Glitsch Flexipac 250Y), but other corrugation angles are available such as a 60-degree angle from horizontal which usually carries an "X" designation.

In general, the "Y" type packings deliver a higher separation efficiency than an "X" type packing with the same surface area per unit volume, but the "X" type packings deliver higher capacities and lower pressure drops.

Figure 7.26: Sulzer Mellapak™ Structured Sheet Metal Packing. Taken from ref. [112]. Reprinted with permission from Sulzer, Ltd.

Research into the flow dynamics occurring within operating distillation columns in the late 1990s showed that in many cases, as flooding was approached on sheet metal packings with straight diagonal corrugations, frothy liquid would start building up at the transitions from one packing element layer to another. As operating rates got closer and closer to flooding, the liquid buildup grew larger and larger until it completely filled the packing volume at the flood point. This was an indication that the capacity of structured packings with straight corrugations was limited by the restrictions caused by the change in flow direction at the tops and sometimes bottoms of the packing elements and not necessarily by the flow patterns within the corrugation channels. As a result of these observations, some structured packing manufacturers modified the shape of the sheet metal corrugations to alleviate the bottleneck. In some cases, this involved a transition from the designated corrugation angle to a nearly vertical orientation at the tops and bottoms of each packing element. An example of typical modifications to the sheet corrugations is shown in Figure 7.27. In some other cases, the corrugations were flattened at the tops and/or bottoms of the elements. In all cases the intent was to reduce the flow disruption at the transitions from one element to the next and reduce the restrictions causing flooding. These modifications to the corrugation geometry resulted in significantly increased capacities with the same or better separation efficiencies than the packings with the same surface area and straight corrugations with little or no difference in manufacturing cost. Sulzer designated these modified packings as their MellapakPlus™ line, and Koch-Glitsch designated their offering as the Flexipac® HC® line. Other structured packing manufacturers have followed with products offering similar sheet corrugation modifications for improved packing throughput and pressured drop performance. These modified structured packings have seen wide adoption in distillation applications with a significant number of retrofits to replace structured packings having straight unmodified corrugations.

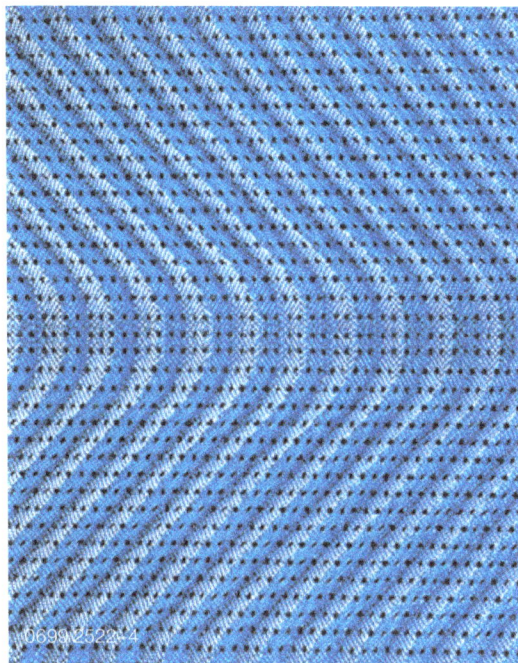

Figure 7.27: Example of Modified Structured Packing Sheet Corrugations (Sulzer MellapakPlus™). Taken from ref. [112]. Reprinted with permission from Sulzer, Ltd.

Grid type packings are sometimes used for high fouling applications requiring very low pressure drops such as vacuum column wash sections, Coker fractionator quench sections, and Fluid Catalytic Cracker flue gas scrubbers. Simplistic versions of grid packings such as those made from wooden slats and lengths of angle-iron mentioned earlier, or some made from repurposed materials such as stacked layers of floor grating, have been used since the early years of packed columns.

7.4.3 Trays

Tray devices have been used in distillation for centuries, and there is some indication that primitive versions of sieve trays were used by the Greeks in the early first millennium. However, it was not until the early 1800s that present-day tray designs were developed with some work encouraged by a competition sponsored by Napoleon to improve the production of alcohol from sugar beet fermentation and distillation. This led to the development by Jean-Baptiste Cellier-Blumenthal in 1813 of the first continuous distillation column with a vertical arrangement (or cascade) of trays similar to modern-era bubble cap trays. And in 1830, Aeneas Coffey developed the Coffey still, a two-column distillation process using trays similar to modern sieve trays to produce alcohol at greater than 90% concentrations and which is still in use today. The production of grain alcohol and the growth of the chemical industry in the nineteenth century and particularly the development of the petrochemical industry in the twentieth century drove improvements in distillation tray design from these early beginnings to the wide range of highly effective offerings available today.

Although there are a wide variety of tray designs available today, they can usually be divided into two general classes: trays with downcomers and trays without downcomers. Among the types of trays with downcomers are bubble cap trays, sieve trays, and valve trays with a number of varieties within these types. For instance, Thormann® trays and tunnel trays are modifications of classical bubble cap designs, and valve trays are divided into designs with either fixed or movable valves. Trays without downcomers include dual-flow-type trays and baffle trays, again with different varieties within those types. For instance, dual-flow-type trays include the classic FRI-developed Dual Flow tray consisting of a flat plate with relatively large perforations, as well as Turbogrid trays with slots and vanes in the plate, and the Ripple Tray™ where the perforated plate is formed with transverse corrugations to give the plates a rippled appearance.

In addition, in recent years a number of modern high-capacity tray designs have been developed that do not neatly fall within one of the classical tray types because of their mechanical complexity and atypical configuration. Some examples of these high-capacity trays include the Koch-Glitsch Ultrafrac® tray, the Shell ConSep™ tray, and the Raschig USA CoFlo tray. Some high-capacity trays are even capable of consistently operating with reasonable tray efficiencies well above the FRI system limit which was once thought to be a strict limit above which stable efficient operation could not be sustained. In general, although they may make use of more-or-less conventional tray deck devices such as fixed or movable valves, these trays accomplish this performance by utilizing inertial separation principles achieved by means such as cyclonic action, vanes, torturous flow paths, and other techniques. These techniques bring about pressure drop-driven vapor-liquid phase separation to reduce or eliminate entrainment at very high vapor velocities. As a result, it has become clear that distillation tray development is not a stagnant art, and continuing research and development may provide further advances in distillation capacity and efficiency.

Trays with downcomers
Bubble Cap Trays. The predominant tray design in industrial use up until the 1960s, bubble cap trays consist of an array of cylindrical risers in the bubbling area of the tray deck between downcomers with the risers topped by inverted round cup-like caps (Figure 7.28). The caps are suspended far enough above the risers to allow ample open area for gas to flow under the caps and bubble out through the liquid on the tray. Because bubbles are promoted by cap features such as serrated edges and rows of slots in the cap skirts, interfacial area can be enhanced, and tray efficiencies can be quite reasonable. In addition, since the risers can be designed tall enough to provide a positive liquid seal, bubble cap trays can perform well over a very wide range of flow rates and are still used today in services requiring a wide operating range or the ability to operate at extremely low liquid rates such as in chlorine drying columns. However, geometric limits on the number of bubble caps that can fit on a tray deck means that open area is likewise limited which causes tray capacity to be limited compared to more modern designs. And the mechanical complexity, parts count, and installation requirements of bubble caps usually cause them to be more expensive than simpler trays like sieve trays or valve trays (Figure 7.29).

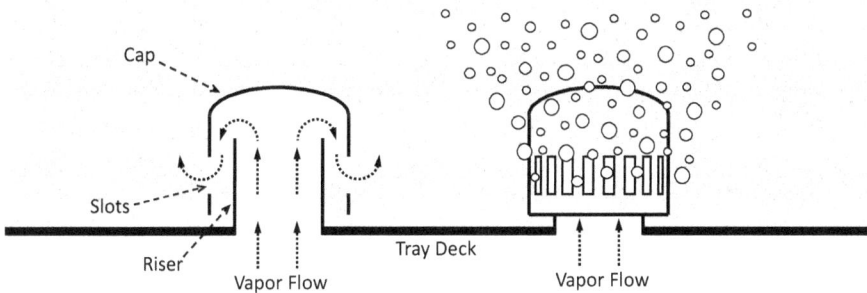

Figure 7.28: Schematic Diagram of Bubble Cap Configuration and Operation.

Related to the generic round bubble cap tray are tunnel trays which use long rectangular risers and caps to provide more open area for vapor flow and better capacity than generic round bubble caps. The rectangular tunnels can be arranged either parallel to the liquid flow path for higher liquid loadings or perpendicular to the flow path for low liquid loadings and longer retention times. Baffles and vanes can also be installed on the tray deck and in the sides of the rectangular caps to direct the liquid flow path and provide less short circuiting and more liquid retention time on each tray and thus achieve better efficiencies in low liquid loading applications. Thormann® trays, a trademark of

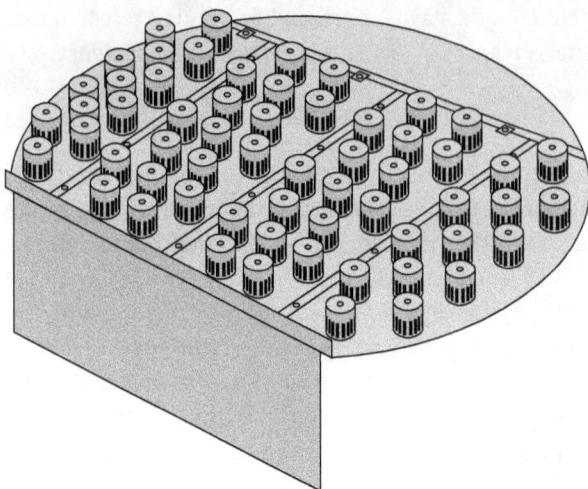

Figure 7.29: Bubble Cap Tray.

Julius Montz GmbH, are a special type of tunnel tray designed primarily for low liquid loading applications such as vacuum distillations and scrubbing. Thormann trays can be engineered for low pressure drops and can be provided with coils for auxiliary heating or cooling for endothermic or exothermic applications.

Sieve Trays. Sieve trays in one form or another have been in use at least as long as bubble cap trays and like bubble caps, are still used in some applications today. They are used in a somewhat wider variety of applications than bubble caps because of their simplicity of construction and consequent modest cost. They consist of flat perforated plates to provide the active area between downcomers and, although more limited in range of throughput than bubble caps, can deliver good efficiencies with proper design (Figure 7.30). In addition, another benefit of sieve trays' mechanical simplicity is better fouling resistance than bubble caps because of fewer quiescent zones and surfaces for solids to deposit and build up on.

Unlike bubble caps which provide a positive liquid seal against liquid leakage and bypassing around trays, sieve trays rely on maintaining sufficient vapor velocity up through the tray perforations to maintain liquid level above the tray. If the vapor velocity drops low enough, some of the tray liquid begins weeping down through the holes, bypassing the trays and reducing mass transfer efficiency. If the vapor velocity drops low enough, virtually all of the liquid flows down through the holes in the tray rather than through the downcomers, a condition called dumping which results in a severe loss of efficiency and very little separation power. In general, sieve trays with smaller diameter holes deliver higher tray efficiencies than those with larger holes, but trays with smaller holes also are more susceptible to fouling and exhibit higher pressure drops.

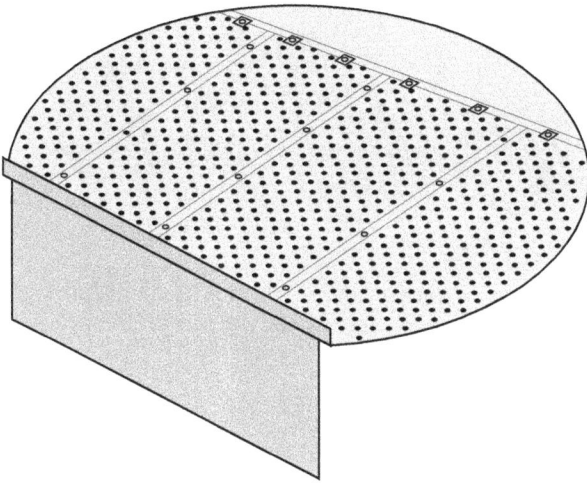

Figure 7.30: Sieve Tray.

Valve Trays. Valve trays were developed in the 1950s and came into wide industrial practice in the 1960s and 1970s, displacing many applications of bubble cap trays and sieve trays. Valve trays can have either movable or fixed valves and consist of a flat perforated plate like sieve trays but with a movable or fixed valve device inserted or formed in each hole. One-piece movable valves are generally round or rectangular plates with legs inserted in each hole in the tray deck (Figure 7.31). The legs have retaining tabs at the bottom that fit under the tray deck to prevent the valves from blowing completely out of the hole but leave them free to move up and down so as to provide a variable opening for vapor flow. As a result, as the operating rate changes the valves can open or close, thus changing the effective tray open area for vapor flow and reducing the amount of weeping that occurs at low operating rates.

As an alternative to one-piece movable valves, caged valves can be provided which have a stationary metal frame or cage affixed to the deck above each vapor orifice (Figure 7.32). Inside of each cage are one or two movable disks. As vapor rate increases with a single-disk caged valve, the disk rises in the cage to allow greater open area for vapor flow similar to the one-piece movable valve. But caged valves potentially allow for wider operating ranges than with one-piece movable valves because of greater lifts possible within a stationary cage and better, stable sealing that can be accomplished to prevent weeping between the caged valve disk and the tray deck. In addition, caged valves can have two disks, a lower, lighter, flat disk which starts lifting at low vapor rates and an upper, heavier disk with integral legs that only begins opening when the lower disk reaches it. In this manner, a two-disk caged valve can be engineered to handle very low liquid rates as well as a wide range of vapor rates.

Figure 7.31: Examples of Movable Valves: left) Koch-Glitsch FLEXITRAY® Valve Tray; right) Koch-Glitsch Type A Full-Size, One-Piece Valve (movable). Taken from ref. [113]. Reprinted with permission from Koch-Glitsch, LP.

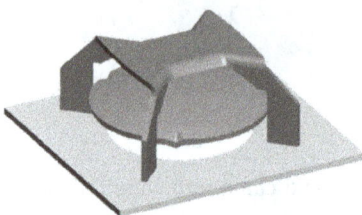

Figure 7.32: Sulzer RC1 Caged Valve with Single Disk (movable). Taken from ref. [114]. Reprinted with permission from Sulzer, Ltd.

In the case of fixed valves (Figs. 7.33 and 7.34), the perforations are often punched in the tray deck in such a way that the metal removed from each perforation is left attached to the deck at two or more points to form a covering plate held slightly above the tray deck. In other fixed valve tray designs, an oversized valve plate and legs formed from separate plate stock fully overlaps the hole opening and is attached to the tray deck after the holes are punched rather than formed as part of the process of punching the holes in the deck. Whether determined during the hole punching process or afterward by the length of the legs on the separate valve plates, the fixed height of the valve plate above the tray deck is set based on the application vapor flow requirements and vapor-liquid physical properties. Even though fixed valves do not move to reduce the open area during low vapor flow conditions, the horizontal deflection of the vapor flow by the fixed valve plates serves to minimize weeping under low flow conditions.

In general, for both movable and fixed valves, the smaller the size of the valve, the greater the vapor capacity and sometimes the better the turndown capability and/ or efficiency [115]. However, smaller valve units usually are more susceptible to fouling issues as well as more expensive because of the higher cost of fabricating tray decks with a larger number of smaller valve units per unit area of deck. But fouling

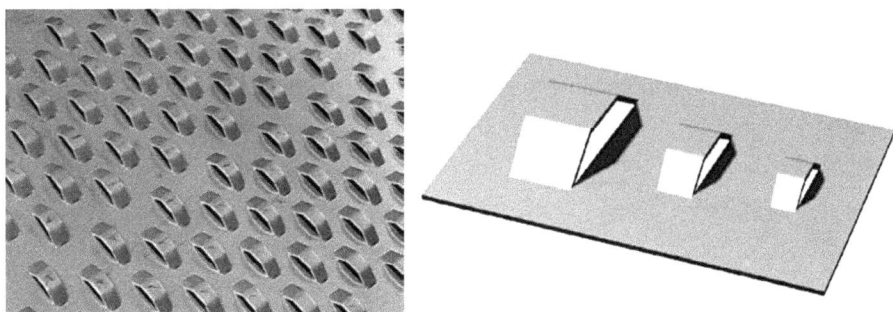

Figure 7.33: Examples of Fixed Valves for Valve Trays: left) Sulzer MMVG™; right) Sulzer V-Grid. Taken from ref. [114]. Reprinted with permission from Sulzer, Ltd.

Figure 7.34: left) Koch-Glitsch VG-10 Full-size Valve; middle) Koch-Glitsch VG-0 MINIVALVE®; and right) Koch-Glitsch PROVALVE®. Taken from ref. [113]. Reprinted with permission from Koch-Glitsch, LP.

issues may be less of a factor with fixed valves than with movable valves because of their lack of moving parts.

As a result of the ability of valve trays with either movable or fixed valves to reduce weeping or dumping during low flow operation, they usually have a much broader range of operation than sieve trays, and although somewhat more expensive than sieve trays, are usually considerably less expensive than bubble cap trays. With considerably wider operating range and better performance than sieve trays at considerably lower capital and installation cost than bubble cap trays, valve trays have become the preferred tray type for most mainstream trayed distillation applications.

Trays without downcomers

Dual Flow Trays. Among trays without downcomers, Dual Flow trays (Figure 7.35) are the most widely used, although in much fewer applications than trays with downcomers. Consisting of flat plates covering the entire cross-sectional area of the column and perforated with larger holes than sieve plates to allow vapor to pass up and liquid to flow down through the same holes, Dual Flow trays in early forms were used as early as the nineteenth century. However, their widest use came after extensive research and development work in the 1950s and 1960s by Fractionation Research, Inc. (FRI). FRI measured performance data on multiple distillation systems, providing information on pressure drop, efficiency, and flooding to greatly improve the reliability of new application designs.

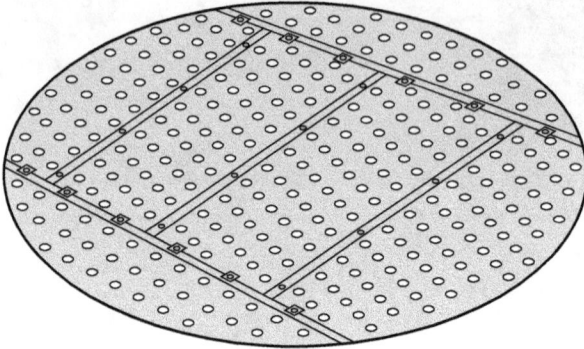

Figure 7.35: Dual Flow Tray.

Because vapor and liquid flow through the same holes on Dual Flow trays, a balance between those flows is necessary to maintain sufficient liquid froth on the tray to provide adequate interfacial area and retention time for good mass transfer. As a result, the range of vapor and liquid flows, or throughputs, that will provide good mass transfer performance of Dual Flow trays is invariably much narrower than that of trays with downcomers such as valve trays. But because of Dual Flow trays' simplicity of construction, lack of unagitated dead zones, and the significant turbulence and scouring action that occurs with opposing vapor and liquid flows through the same holes, Dual Flow trays are much more resistant to fouling than other trays with downcomers. Accordingly, Dual Flow trays find many applications today in services where solids buildup would foul trays with downcomers.

Turbogrid and Ripple Trays. Other types of trays without downcomers include the Turbogrid tray developed by Shell Development Company and the Ripple Tray™ introduced by Stone & Webster Engineering Corporation in the 1950s and now manufactured by TechnipFMC. As trays without downcomers, they are similar to the Dual Flow tray, but have mechanical differences from the flat perforated Dual Flow tray. The Turbogrid tray has slats and slots instead of circular holes, and the Ripple Tray has circular holes but in a plate with multiple sinusoidal corrugations on its surface. As these are both proprietary tray designs, there is little open literature on design and performance correlations, so direct discussions with the manufacturers are recommended for new applications or retrofits.

7.5 Process Control Considerations

In earlier sections we touched on several common approaches to distillation process control. For batch rectification, a common approach involves manipulating reflux to control an overheads temperature (see Section 7.3.2: Batch Distillation with Reflux). For a stripping column, a common approach involves maintaining a constant ratio of boilup rate to feed rate (see Section 7.3.3: Continuous Stripping and Absorption). These and other control schemes are covered in many books and articles on the subject, such as refs. [116, 117, 38, 118]. And James Fair has provided a brief review of four commonly used control schemes [119].

In continuous fractional distillation, there are a number of operating parameters and control loops that need to be defined and specified in order to allow stable column operation. Parameters such as feed and reflux flow rates, liquid levels in the reflux drum and column bottoms vessel, and column pressure must be maintained at constant values if the column is to develop a stable composition profile and deliver reliable separation performance. The process characteristic that is most often correlated with the column composition profile and thus separation performance is the temperature profile along the length of the column. For this reason, a key control strategy often employed involves controlling some aspect of the temperature profile by manipulating the column's material balance split (see Section 6.5.1: Material Balance Split and Separation Power). Temperature measurement is robust and easy to implement, so it is often used to infer composition instead of using a more complex online analytical instrument.

As a result, many distillation control strategies involve manipulation of one or more process variables to maintain a representative temperature at a set-point, which may be located anywhere within the column. Intuitively, one might think that the column top temperature best reflects the distillate composition and the column bottom temperature best reflects the bottoms composition, and that is true to an extent. But the issue is that when very low levels of impurities are specified for the distillate and/or bottoms streams, the top and bottom temperatures may not vary appreciably even with order of magnitude changes in impurities. As a result, choosing the top or bottom temperature for control can make it very difficult to maintain the distillate and bottoms streams within specifications, so they are often not the best choice for the representative temperature location.

Since a representative temperature measurement is used to infer the state of the column temperature profile and thus the composition profile, the required temperature sensor should be located so that the measured temperature is most sensitive to changes in composition, a temperature generally called the most-responsive temperature (MRT). One or more process variables may be manipulated to control the MRT, including:

– Feed rate
– Reflux rate or reflux ratio

- Boilup rate or boilup ratio
- Boilup to feed ratio
- Material balance split as determined by distillate-to-feed or bottoms-to-feed ratios

The location of the MRT may be determined by modeling the composition and temperature profiles that exist within the specific distillation column for the system of interest. One approach is to start with a base computer simulation of the column that comfortably overachieves both the distillate and bottoms specifications, for instance with twice the minimum required separation power. Then, two more simulations are performed at that same separation power but with higher and lower distillate (or bottoms) rates to perturb the material balance split higher and lower in the column. A convenient way of setting the magnitude of the perturbations in the material balance split is to adjust the split higher in one simulation so that the required distillate specification is exactly met while overachieving the bottoms specification. Then in the other simulation, adjust the split lower so that the required bottoms specification Is exactly met while overachieving the distillate specification. This results in hotter and cooler temperature profiles which can be compared to determine which location in the column provides the greatest temperature sensitivity to composition changes.

The stage-by-stage temperatures in the two perturbation simulations can be inspected to determine where the greatest difference in stage temperatures occurs. However, a convenient approach is to calculate the differences in stage temperatures between the high and low material balance split cases and the base simulation case and display them graphically on a plot of stage number versus temperature deviation from the base simulation case. The stage where the maximum difference in the temperature deviations from the base simulation occurs is then identified as the MRT location. We should note that in many cases, rather than a single stage location, a range of stages is identified as the most likely MRT location. This is important since frequently it may not be possible to locate a temperature sensor in exactly the location dictated by the simulations.

Example 7.10. Perform computer simulations of the methanol-isopropanol distillation covered in Example 7.1 and determine the stage where the MRT is located. The production specifications are that 95 mol% methanol is to be recovered from a flashing feed of 55 mol% methanol/45 mol% isopropanol with a residual methanol concentration of 5 mol% in the column bottoms. Match the 14.3 required theoretical stages determined by the McCabe-Thiele analysis in Example 7.1 as closely as possible.

In this example, the ChemSep™ distillation simulator [120] (by Harry Kooijman and Ross Taylor) was used to calculate the column performance and perturb the material balance split to determine how the column temperature profile changes and the MRT location. In the simulation, 15 column stages were specified where Stage 1 is the total condenser, leaving 14 stages for the actual separation with the flashing feed entering on Stage 10. Built-in ChemSep physical property and binary interaction parameters (Van Laar) were used for convenience even though these resulted in somewhat lower relative volatilities and higher reflux ratios than were used in the Example 7.1 McCabe-Thiele analysis. The first simulation was set up to determine the stage-by-stage temperature profile for the base case with twice the minimum separation power needed to exactly achieve the distillate and bottoms purity specifications. Splitting the excess separation power equally between the distillate and bottoms concentrations resulted in the following specifications:

$$x_{dist} = \frac{0.05}{\sqrt{2}} = 0.03535 \text{ mol frac isopropanol}$$

$$x_{btm} = \frac{0.05}{\sqrt{2}} = 0.03535 \text{ mol frac methanol}$$

With these revised specifications, the base simulation resulted in a required external reflux ratio of 4.955 and the temperature profile shown in Table 7.7.

Table 7.7: Stage Temperature Profile for Example 7.10 Methanol-Isopropanol Distillation Base Case.

Stage	Temp, °C	Press, kPa	Stage	Temp, °C	Press, kPa
1	66.71	101.325	9	71.72	101.325
2	66.88	101.325	10	73.36	101.325
3	67.1	101.325	11	75.43	101.325
4	67.41	101.325	12	77.67	101.325
5	67.83	101.325	13	79.64	101.325
6	68.42	101.325	14	81.08	101.325
7	69.23	101.325	15	82.01	101.325
8	70.34	101.325			

Now, holding the reflux ratio (or alternatively, the boilup ratio, reflux-to-feed ratio, or other parameter) constant to maintain the separation power relatively constant, we performed two additional simulations, one to meet the original distillate specification of 0.05 mol fraction isopropanol in the distillate and the other to meet the original bottoms specification of 0.05 mol fraction methanol in the bottoms. The stage-by-stage temperature profiles for these two perturbation simulations were then compared to the base simulation temperature profile, and the stage-by-stage deviations from that base profile were calculated. As mentioned earlier, a good way of visualizing these results is with a plot of the temperature deviations for both perturbation simulations versus stage number which can illustrate the stage location where the maximum temperature difference between the two simulations occurs. This stage location in the simulated column is usually very close to the location in the actual column where the MRT is likely to occur.

The plot of temperature deviations versus stage number for this example is shown in Figure 7.36 where we can see that the maximum difference between the temperature deviations from the base simulation occurs on Stage 11 where the temperature difference is 3.09 °C. But we can also see that the temperature differences on Stages 10 and 12 are also quite generous and could be used for control, or even possibly any location from about Stages 8 to 13. In practice, the best location for the temperature sensor might be in the packing immediately above the feed point if a packed column, or in the vapor space or downcomer above the feed tray if a trayed column. This would provide some isolation of the control temperature sensor from swings in the feed temperature which could cause excessive variability in column control.

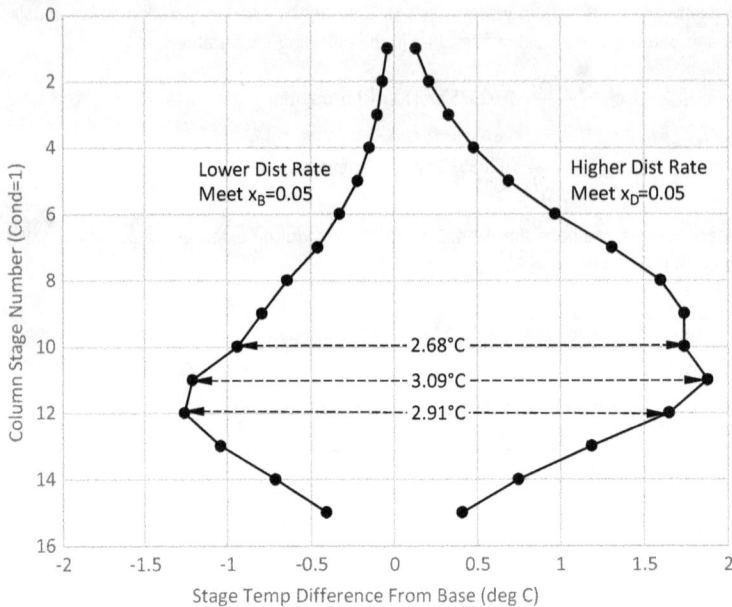

Figure 7.36: Column Stage vs. Temperature Difference Plot for Locating MRT.

For distillations that do not exhibit significant change in temperature along the length of the column (as in many steam stripping applications, for example), then other strategies must be used, such as maintaining a constant ratio of boilup rate to feed rate or a constant material balance split. In special cases, online instruments may be used to directly measure composition and provide input for open loop or even fully automatic closed loop control. Or, frequent manual sampling of distillate and bottoms may be used to check that the operation is performing as intended. Examples of online instruments include online sampling-valves feeding pulse-injection gas chromatography and liquid chromatography instruments, as well as various kinds of spectroscopy instruments. These instruments can respond quickly and they allow for sophisticated model-based control schemes, but they also are expensive to maintain and if they ever go out of calibration (or, more realistically, when they go out of calibration) they can give misleading readings that can cause poor control or process upsets.

Up to now we have focused on several process control strategies commonly associated with distillation operations. Taking a more general view, process control strategies may be categorized into three main types [121, 122]:

– Continuous process control involving steady-state operation
– Batch process control involving sequential (cyclical) and non-steady-state operation
– Discrete operation control typically involving on/off or open/closed states of operation

These may be applied to any separation process or chemical operation, and often a single operation involves application of two or more control loops working in concert. With this in mind, the implementation of a given process control scheme, whatever it may be and however complex it may be, includes the following considerations:

- Control strategy
- Control theory
- Instrumentation
- Oversight of the operation

The first consideration has to do with choosing the process variable or variables to manipulate in order to achieve the desired responses and stable control. This is where knowledge of process fundamentals is essential. The second consideration involves choosing and optimizing a suitable control algorithm; that is, deciding on the mathematical relationships between measured inputs and calculated outputs. This is the subject of general control theory. It is an area of expertise for electrical engineers and industrial engineers as well as chemical engineering specialists. Theory may be used to optimize the desired response and minimize deviation from the set-point, as in minimizing overshoot, off-set, lag time, or cycling about a desired set point. Many books on control theory discuss these aspects in detail. Common control algorithms often involve traditional proportional-integral-derivative (PID) feed-back control. Other more complex approaches also are used, some involving modeling of the distillation process with online adjustment of model parameters, feed-forward control, and non-steady-state control (batch process control). Distributed control schemes also may be used, allowing for integration of a network of multiple unit operations. The third category covers how to implement the control scheme in an economical and reliable manner. This includes the choice of hardware such as the choice of instruments and sensors, as well as the choice of the type of signal to be used and methods for signal conditioning (analog and digital, hard-wired and wireless).

These aspects often are the focus of discussions of process control, but they are not the full story. Another critical aspect involves oversight of the overall operation. The control room is the center for information gathering, data analysis, and decision-making for most-effective process control. Human operators (engineers, technicians, and operator specialists) must be fully informed in order to adjust and guide the operation beyond what the automated controller is doing. This is true for most any operation large or small, such as large-scale operations employing specialized computers and multiple unit operations, networks of on-board Programmable Logic Controllers (PLC controllers), distributed control networks, and PC computers employed in the laboratory for data acquisition and control of experiments. The idea is to provide operators with information needed to understand the operating history and current state of the process – for situational awareness to borrow a military term – so they can use this information to make informed changes, usually small incremental changes in setpoints and control constants that improve the operation. And if needed, the operators can immediately respond to any upsets that may occur. A computer can be programmed to analyze such data, place the process in a safe holding condition if needed, and provide operators with recommended changes to set-points or indicate when the process is ready to move to the next step in the operation. With this approach, the operator has oversight and can quickly make key decisions.

Another objective of process control is to minimize variability in the overall operation [123, 124]. This generally refers to variability in the purity or composition of the desired product. The idea is to minimize both under-purification, which produces product that customers do not want, and over-purification, which adds extra cost that customers are not willing to pay for. In practice, mixing under-purified product with over-purified product is sometimes done out of necessity to meet product specifications. But a situation like this is an excellent opportunity to reduce operating costs as well as improve product consistency. Measuring process variability in terms of the standard deviation or another statistical metric allows the operator to characterize the current state of the process so that the effect of an adjustment can be quantified. An approach described by Lanny Robbins involves calculation of the Process Capability Index (Cpk) for process measurements collected over time, using specific rules to inform operators in making small adjustments to the control scheme [123]. The idea is to make a small change in the process operation, a change that is suggested by a set of rules (derived from a process model or from experience), and then allow plenty of time for the change to manifest itself in the operation. Then, once the quantitative effect of the earlier change is known, the rules can be refined before making another small change. This learning approach allows for continual improvement and refinement of the control scheme – all the while avoiding big changes that might cause upsets.

So, the control room is the place where all process information comes together so operators can make well-informed and timely decisions. Key information includes flow rates, pressures, pressure drops, temperatures, controller set-point values, results of chemical analysis (either results from grab samples analyzed in the quality control lab or from online analyzers, or both), as well as calculated performance measures such as the material balance, product recovery, cycle time, production rate, capacity factor (or percent of design capacity), material balance split, separation power, and Cpk. These are plotted as a function of time in run charts and compared with desired values so operators can quickly gauge trends and the current status of the process and then steer the process where it needs to go. This can be done by the automated control scheme to some extent, but often the operator is given the responsibility of making key control decisions at key points in a process. This approach ensures safety in handling situations in which a sensor goes bad, a key separator enters a flooding condition, or a slowly developing problematic trend in performance occurs that needs to be addressed before getting out of hand. For example, a slow decline in separation power and/or an increase in pressure drop over time might serve as an early warning of a fouling issue that needs attention. Often plots of key data and online calculations (performance trend lines) are continually displayed at the operator's station along with online video of key equipment operations.

This information-intensive approach can facilitate making changes that improve the operation in terms of better product purity and yield, higher overall production rate, reduced energy consumption, or whatever the goal may be.

7.6 Energy Utilization

A discussed earlier, it is estimated that distillation consumes as much as 40% of the energy required for a typical chemical manufacturing plant. This dominance in the use of energy derives not only from inherent energy consumption (which can be high, but this depends on the application and the overall process scheme), but also from the fact that distillation is by far the most popular separation method, with roughly 40,000 units in operation in the U.S. alone. These are very large numbers, so even small improvements in energy utilization can add up to significant savings. In discussing energy, it is important to make a distinction between energy consumption and energy efficiency. The former refers to how much energy is expended in producing a unit of product. The latter is a measure of the opportunity to recover that energy; that is, how much can be recovered from the process outlet streams and how much is lost work or lost heat that cannot be recovered (see Section 5.8: Energy Consumption and Energy Efficiency).

Understanding the maximum energy efficiency for a given application allows the designer to gauge how far a given design is from the theoretical limit, and from this, gauge the opportunity for improvement in terms of the potential for reduction in overall energy consumption for better energy utilization. The designer can then determine if the potential gain in efficiency is worth the added complexity and cost that may be required. In a typical steam power plant (another thermal process), the maximum theoretical energy efficiency may be as high as 55%, with actual efficiencies of 40% or 45% depending on operating temperatures and the type of process [125, 126]. Typical internal combustion engines have maximum theoretical energy efficiencies near 50%, with actual efficiencies around 35% or somewhat higher depending on the specific design [125]. Just like these other thermal processes, some distillation processes can have maximum energy efficiencies as high as 50% or even as high as 70% [127]. However, due to the great variety of distillation applications, each application requires its own assessment, and the energy efficiency of a given process may be less than 10%, especially when high reflux ratios are required.

The actual energy efficiency of a distillation process may be calculated from the amount of energy contained in the outlet streams relative to the amount of energy input to the process. For the simple case involving saturated liquid streams and total condensation in the condenser, the overall energy balance around a distillation column yields:

$$\varepsilon_{actual} = \frac{\text{Energy recovered in distillate and bottoms}}{\text{Energy input in feed and reboiler}} \tag{7.63}$$

or,

$$\varepsilon_{actual} = \frac{\dot{M}_D H_D + \dot{M}_B H_B}{\dot{M}_F H_F + Q_{reb}} \tag{7.64}$$

where ε_{actual} = actual energy efficiency of the process

$\dot{M}$ = mass flow rate (of Distillate, Bottoms, and Feed streams)

H = specific enthalpy of liquid (for Distillate, Bottoms, and Feed streams)

Q_{reb} = energy input through the reboiler (reboiler duty)

Note that taking advantage of this efficiency requires that the energy contained in the distillate and bottoms streams is utilized through cross-exchange with other unit operations or in other ways. This simple calculation neglects the energy removed in condensing the overhead vapor because this energy generally is not recovered in a useful form unless it is recovered using extra methods, as discussed below. Additional terms may be added to account for extra energy recovery or input.

The calculation of maximum theoretical energy efficiency involves analysis of the process as a type of heat engine coupled with its surroundings, which in this case are the hot and cold heat transfer fluids at temperatures T_{hot} and T_{cold} (assuming zero heat lost to the environment). The exergy analysis accounts for useful (recoverable) work and lost work expended during operation between the extremes of distillate and bottoms temperatures (T_D and T_B). The maximum thermodynamic efficiency of a typical stand-alone, adiabatic distillation column in doing the work of separating is approximated by [127]:

$$\varepsilon_{max} = \left(\frac{x_{FA}\ln x_{FA} + (1 - x_{FA})\ln(1 - x_{FA})}{\ln\alpha \times \left(\frac{1}{\alpha - 1} + x_{FA}\right)} \right) \left(\frac{\left(1 - \frac{T_D}{T_B}\right)Q_{reb}}{\left(1 - \frac{T_{cold}}{T_{hot}}\right)Q_{reb}} \right) \tag{7.65}$$

where ε_{max} = efficiency of distillation column including condenser, reboiler, and utilities

x_{FA} = mole fraction of the light key component in the feed

α = relative volatility of light key to heavy key

T_D = distillate temperature (near boiling point of light key)

T_B = bottoms temperature (near boiling point of heavy key)

T_{hot} = reboiler heat transfer fluid temperature

T_{cold} = overheads condenser coolant temperature

A similar analysis is given in ref. [128]. Equation (7.65) is derived for a binary separation with low to moderate relative volatility for a process with sharp separation of light and heavy key components. In that case, distillation is most energy efficient when the light key is present at concentrations of 20 to 50 mole% in the feed, depending on relative volatility. At higher concentrations, efficiency begins to decrease, perhaps because too much of the feed must be taken overhead. For high relative volatilities around 100 or higher, the process is most efficient when the product is present in dilute concentration in the feed and most of the feed exits in the bottoms stream [127]. However, unlike energy efficiency, energy consumption per unit of product is likely to be less sensitive to changes in feed composition, as more of the light product will be processed as its concentration in the feed increases. Internal energy consumption may be approximated by $Q_{reb}/\dot{M}_D$ for a distillate product and by $Q_{reb}/\dot{M}_B$ for a bottoms product. This measure of energy consumption is a direct function of Q_{reb}, but energy efficiency in eq. (7.65) is not. And the overall energy consumption may be reduced to some extent by recovering energy from the separator outlet streams in an integrated process scheme. So, in conducting an analysis of energy efficiency, it is important to also calculate and compare overall energy consumption per unit of product, because a change in one is not always commensurate with a change in the other.

Other expressions may be derived for other types of distillation and evaporation processes [127, 128]. The difference between actual and maximum efficiencies can then be used to gauge the opportunity for improvement. Or, the calculation may be used to compare distillation with alternative separation methods. Keep in mind that distillation is not necessarily more energy intensive compared to other separation methods, especially when all requirements of the overall process are taken into account. This depends on the specific application [19]. An exergy analysis, as illustrated by eq. (7.65), provides a convenient way of quickly screening various conceptual design options for a given application, in terms of their relative energy efficiency – one that is easily programmed for rapid computer evaluation. But as mentioned above, energy consumption per unit of product should also be compared, and this depends on the overall process scheme.

A stand-alone fractional distillation design normally is optimized for energy utilization by paying close attention to the mass transfer efficiency of various packing or tray designs, making sure to design for a sufficient number of stages or transfer units and sufficient column diameter with low pressure drop to allow operation with a reasonably low reflux ratio to minimize Q_{reb}. Retrofitting an existing column with more modern, higher mass transfer efficiency packing or trays that provide a greater number of theoretical stages/transfer units is a common way of reducing energy consumption or improving separation performance (or both). The design should also carefully match the feed stage location with composition profiles occurring within the column (obtained from simulations) to avoid retrograde separation. Furthermore, adjustment of operating pressure may yield operating temperatures that improve the relative volatility driving force, and the key temperature ratios in eq. (7.65) may be favorably altered, as well (although this normally requires a clean-sheet design). The choice of process control strategies also can have a significant impact on energy efficiency. As discussed in the previous section, working to achieve sharp separation without over-separation can provide significant energy savings [123, 124]. This is an excellent approach to energy savings, sometimes under-appreciated, that may require little or no capital investment.

Energy utilization also may be improved by applying various external-to-the-column heat integration strategies. Options include integration of condenser and reboiler in an external heat pump loop as mentioned above (called mechanical vapor recompression), and cross-exchanging incoming feed with the bottoms, effectively reducing Q_{reb}. The application of a heat pump becomes more attractive for larger differences in overhead and bottoms temperatures, say more than 20 Celsius degrees difference.

Steam and air stripping are relatively simple distillation processes that can deliver high energy efficiency and low energy consumption through cross-exchange of energy. In this special case, energy utilization is maximized when feeds are dilute, relative volatility is high, a tall distillation column is used to provide a large number of transfer units, the column is sufficiently well insulated, and the bottoms stream is cross-exchanged with the incoming feed. In this special case, only a small fraction of the feed is taken overhead, most of the feed goes out with the bottoms, a portion of the heat content of the overheads is recovered by mixing hot condensate with the feed, and much of the energy expended by the process is recovered by cross-exchanging the bottoms with the incoming feed (see Section 7.3.3: Continuous Stripping and Absorption).

Energy utilization may also be improved by integrating distillation heat exchangers with those of nearby processes in a heat transfer network (commonly called pinch technology). Placing a distillation unit within an integrated production site can yield a significant reduction in energy consumption compared to stand-alone operation. This is a common practice at large integrated petrochemical facilities. Opportunities include integration with hot reactor effluent streams, for example. For a new design, designers also should consider replacing distillation with an alternative separation method. This approach has been successfully applied in petrochemical facilities, replacing distillation with melt crystallization, for example, to purify select aromatics and isocyanates (see Section 11.6).

Another well-known approach to increasing energy utilization is to combine distillation with another separation method in a hybrid process scheme. The idea is to take some of the load off of the distillation column (preferably corresponding to the part of the McCabe-Thiel diagram where the separation is most difficult), and have that part of the separation done by another unit that is better suited for that specific part of the overall separation. A simple example involves the use of a membrane module to preconcentrate the feed to a distillation unit – increasing desired solute concentration and reducing the mass and volume of the stream entering the distillation (so there is less to distill). Other examples are distillation + PSA and distillation + extraction. Hybrid schemes are discussed in more detail in Chapter 14: Inventive Engineering: Process Simplification, Intensification, and Hybrid Processing.

Distillation efficiency may further be improved by performing heat exchange at multiple locations within the column itself, rather than just at the external reboiler and condenser, an approach called diabatic distillation. A well-known example is the HIDiC column [129]. The application of column-internal diabatic heat integration has the potential for significant improvement in energy efficiency; however, it has not been widely adopted, apparently due to the requirement for very significant increases in column complexity and cost. Additional improvements in energy consumption may be obtained by applying the principle of thermally coupled columns, such as the Petlyuk column and dividing-wall column. For review of these concepts, see refs. [130, 131].

References

[1] J. R. Fair, "Historical Development of Distillation Equipment," in *AIChE Symposium Series*, American Institute of Chemical Engineers, New York, 1983.

[2] X. Li, T. Zhao, Y. Wang, Y. Wang and Z. Zhu, "Operational Design and Improvement of Conventional Batch Distillation and Middle-Vessel Batch Distillation," *Brazilian Journal of Chemical Engineering*, vol. 35, no. 2, (available on-line), 2018.

[3] J. Cahill, "Reducing Distillation Column Energy Usage," Emerson Automation Experts Blog, 2 April 2010. [Online]. Available: https://www.emersonautomationexperts.com/2010/industry/down stream-hydrocarbons/reducing_distil/#:~:text=Did%20you%20know%20that%20there,the%20chemi cal%20and%20refining%20industry. [Accessed 23 October 2022].

[4] Asphalt Production and Oil Refining, Pavement Tools Consortium, [Online]. Available: https://pavementinteractive.org/reference-desk/materials/asphalt/asphalt-production-and-oil-refining/. [Accessed 22 Sep 2022].

[5] J. Ford, "Distillation Columns: Product Composition Control – Process Identification Models," *Control Engineering*, 12 Nov 2013.

[6] T. C. Frank, B. S. Holden and A. F. Seibert, "Section 15: Liquid-Liquid Extraction and Other Liquid-Liquid Operations and Equipment," in *Perry's Chemical Engineers' Handbook*, D. W. Green and M. Z. Southard, Eds., 9th ed., McGraw-Hill, New York, 2019.

[7] A. A. Kiss and C. S. Bîldea, "Revive Your Columns with Cyclic Distillation," *Chemical Engineering Progress*, vol. 111, no. 12, pp. 21–27, 2015.

[8] M. Baldea and T. F. Edgar, "Dynamic Process Intensification," *Current Opinion in Chemical Engineering*, vol. 22, pp. 48–53, 2018.

[9] J. Pirie, "The Manufacture of Hydrocyanic Acid by the Andrussow Process," *Platinum Metals Review*, vol. 2, no. 1, pp. 7–11, 1958.

[10] J. G. Kunesh, H. Z. Kister, M. J. Lockett and J. R. Fair, "Distillation: Still Towering Over Other Options," *Chemical Engineering Progress*, vol. 91, no. 10, pp. 43–54, Oct 1995.

[11] A. Górak and H. Schoenmakers, *Distillation: Operations and Applications*, Elsevier, Amsterdam, 2014.

[12] J. G. Stichlmair and J. R. Fair, Distillation: Principles and Practice, Wiley-VCH, Hoboken, New Jersey, 1998.

[13] J. G. Stichlmair, H. Klein and S. Rehfeldt, *Distillation: Principles and Practice*, AIChE-Wiley, Hoboken, New Jersey, 2021.

[14] J. Pendergast, D. Jewell, D. Vickery and J. Bravo, "Save Energy in Distillation," *Chemical Processing*, 27 March 2018.

[15] A. A. Kiss, "Distillation Technology – Still Young and Full of Breakthrough Opportunities," *Journal of Chemical Technology & Biotechnology*, vol. 89, no. 4, pp. 479–498, 2014.
[16] A. A. Kiss and R. Smith, "Rethinking Energy Use in Distillation Processes for a More Sustainable Chemical Industry," *Energy*, vol. 203, article no. 117788, 2022.
[17] E. Worrell, D. Phylipsen, D. Einstein and N. Martin, "Energy Use and Energy Intensity of the U.S. Chemical Industry (LBNL-44314)," in *Energy Analysis Department, Environmental Energy Technologies Division*, Ernest Orlando Lawrence Berkeley National Laboratory, University of California, Berkeley, California, 2000.
[18] S. K. Ritter, "Putting Distillation out of Business in the Chemical Industry," *Chemical & Engineering News*, vol. 95, no. 25, pp. 19, 19 June 2017.
[19] J. A. C. Velasko, M. Tawarmalani and R. Agrawal, "Systematic Analysis Reveals Thermal Separations Are Not Necessarily Most Energy Intensive," *Joule*, vol. 5, no. 2, pp. 330–343, 2021.
[20] A. B. Hinchliffe and K. E. Porter, "A Comparison of Membrane Separation and Distillation," *Chemical Engineering Research and Design*, vol. 78, no. 2, pp. 255–268, 2000.
[21] A. S. Stillwell and M. E. Webber, "Predicting the Specific Energy Consumption of Reverse Osmosis Desalination," *Water*, vol. 8, no. 12, article no. 601, 2016.
[22] C. Staudt-Bickel and W. J. Koros, "Olefin/Paraffin Gas Separations with 6FDA-Based Polyimide Membranes," *Journal of Membrane Science*, vol. 170, no. 2, pp. 205–214, 2000.
[23] R. Abhishek, S. R. Venna, G. Rogers, L. Tang, T. C. Fitzgibbons, J. Liu, H. McCurry, D. J. Vickery, D. Flick and B. Fish, "Abhishek Roy, Surendar R. Venna, Gerard Rogers, Li Tang, Thomas C. Fitzgibbons, Junqiang Liu, Hali McCurry, David J. Vickery, Derrick Flick, and Barry Fish," *Membranes for Olefin-Paraffin Separation: An Industrial Perspective, PNAS*, vol. 118, no. 37, e2022194, 9 pages, 2021.
[24] L. Rayleigh, "On the Distillation of Binary Mixtures," *The London, Edinburgh, and Dublin Philosophical Magazine and Journal of Science (6th Series)*, vol. 4, no. 23, pp. 521–537, 1902.
[25] C. Kendall and E. A. Caldwell, "Chapter 2, Fundamentals of Isotope Geochemistry," in *Isotope Tracers in Catchment Hydrology*, C. Kendall and J. J. McDonnell, Eds., Elsevier, Amsterdam, 1998, pp. 51–86.
[26] V. Dohnal and I. Horáková, "A New Variant of the Rayleigh Distillation Method for the Determination of Limiting Activity Coefficients," *Fluid Phase Equilibria*, vol. 68, pp. 173–185, 1991.
[27] P. E. Minton, *Handbook of Evaporation Technology*, Noyes Publications, West Wood, New Jersey, 1986.
[28] R. L. Shilling, P. M. Bernhagen, W. E. Murphy, P. S. Hrnjak and D. Johnson, "Section 11, Heat-Transfer Equipment," in *Perry's Chemical Engineers' Handbook*, D. W. Green and M. Z. Southard, Eds., 9th ed., McGraw-Hill, New York, 2019, pp. 11–1 to 11–98.
[29] R. E. Treybal, *Mass Transfer Operations*, 3rd ed., McGraw-Hill, New York, 1980, pp. 369–370.
[30] M. Doherty, Z. T. Fidkowski, M. F. Malone and R. Taylor, "Section 13, Distillation," in *Perry's Chemical Engineers' Handbook*, D. W. Green and M. Z. Southard, Eds., 9th ed., McGraw-Hill, New York, 2018.
[31] P. C. Wankat, *Separation Process Engineering*, 3rd ed, (Includes Mass Transfer Analysis), Prentice Hall, Upper Saddle River, New Jersey, 2012, pp. 331–336.
[32] J. Cvengroš and J. Lutišan, "Evaporator with Wiped Film as Reboiler of the Vacuum Rectifying Column," *Separation and Purification Technology*, vol. 15, pp. 95–100, 1999.
[33] J. D. Fonseca, A. Suaza, R. F. Cortes, I. D. Gil, G. Rodríguez and A. Orjuela, "Rapid Feasibility Assessment for the Use of Wiped-film Evaporation in the Purification of Thermally Labile Products," *Chemical Engineering Research and Design*, vol. 146, pp. 141–153, 2019.
[34] J. Lopez-Toledo, "Heat and Mass Transfer Characteristics of a Wiped-Film Evaporator, PhD Dissertation," University of Texas, Austin, Texas, 2006.
[35] W. L. F. Armarego, *Purification of Laboratory Chemicals*, 8th ed., Elsevier, Amsterdam, 2017.
[36] K. A. Berglund, *Artisan Distilling: A Guide for Small Distillers*, Department of Chemical Engineering and Materials Science, Michigan State University, East Lansing, Michigan, 2004.

[37] W. L. Luyben, "Chapter 16, Pressure-Compensated Temperature Control in Distillation Columns," in *Distillation Design and Control Using AspenTM Simulation*, 2nd ed., Wiley, Hoboken, New Jersey, 2013.

[38] U. M. Diwekar, *Batch Distillation: Simulation, Optimal Design, and Control*, 2nd ed., CRC Press, Boca Raton, Florida, 2016.

[39] M. Mujtaba, *Batch Distillation: Design and Operation, Series on Chemical Engineering*, vol. 3, R. T. Yang, Ed., Imperial College Press, London, 2004.

[40] E. Sørensen and S. Skogestad, "Comparison of Regular and Inverted Batch Distillation," *Chemical Engineering Science*, vol. 51, no. 22, pp. 4949–4962, 1996.

[41] W. L. Luyben, "Multicomponent Batch Distillation. 1. Ternary Systems with Slop Recycle," *Industrial & Engineering Chemistry Research*, vol. 27, pp. 642, 1988.

[42] S. Zhao, P. Bai, X. Guo, K. Tang and G. Li, "Time Requirements in Open and Closed Batch Distillation Arrangements for Separation of a Binary Mixture," *Polish Journal of Chemical Technology*, vol. 16, no. 4, pp. 66–74, 2014.

[43] M. Barolo, "Batch Distillation Optimization Made Easy," *Chemical Engineering Education*, pp. 280–285, 1988.

[44] M. J. Gentilcore, "Quickly Estimate Batch Distillation Time," *Chemical Processing*, 7 July 2014.

[45] M. J. Gentilcore, "Estimate Heating and Cooling Times for Batch Reactors," *Chemical Engineering Progress*, vol. 96, no. 3, pp. 41–45, 2000.

[46] J. R. Ortiz-Del Castillo, G. Guerrero-Medina, J. Lopez-Toledo and J. A. Rocha, "Design of Steam-Stripping Columns for Removal of Volatile Organic Compounds from Water Using Random and Structured Packings," *Industrial & Engineering Chemistry Research*, vol. 39, no. 3, pp. 731–739, 2000.

[47] J. L. Bravo, "Design Steam Strippers For Water Treatment," *Chemical Engineering Progress*, vol. 90, no. 12, pp. 56–63, 1994.

[48] Y. L. Hwang, G. E. Keller II and J. D. Olson, "Steam Stripping for Removal of Organic Pollutants from Water. 1. Stripping Effectiveness and Stripper Design," *Industrial & Engineering Chemistry Research*, vol. 31, no. 7, pp. 1753–1759, 1992.

[49] Y. L. Hwang, J. D. Olson and G. E. Keller II, "Steam Stripping for Removal of Organic Pollutants from Water. 2. Vapor-Liquid Equilibrium Data," *Industrial & Engineering Chemistry Research*, vol. 31, no. 7, pp. 1759–1768, 1992.

[50] L. Kohl, "Chapter 6, Absorption and Stripping," in *Handbook of Separation Process Technology*, R. W. Rousseau, Ed., Wiley, New York, 1987.

[51] L. A. Robbins, "Method of Removing Contaminants from Water", U.S. Patent 4,236,973, 2 Dec 1980.

[52] J. M. Prausnitz, R. N. Lichtenthaler and E. G. D. Azevedo, *Molecular Thermodynamics of Fluid-Phase Equilibria*, 3rd ed., Prentice Hall PTR, Upper Saddle River, New Jersey, 1999.

[53] C. Frayne, *Boiler Water Treatment. Principles and Practice*, vol. 1, Chemical Publishing Company, Palm Springs, California, 2002.

[54] G. Matisoff and C. Narquis, "Computer Modeling of Scale Formation During Treatment of Ground Water in Air Strippers," *Groundwater Monitoring Review*, vol. 11, no. 2, pp. 137–144, 1991.

[55] P. Aggarwal, "Discussion and Reply," *Groundwater Monitoring Review*, vol. 11, no. 4, pp. 165, 1991.

[56] H. Y. Yaws and X. Pan, "Henry's Law Constants for 362 Organic Compounds in Water," *Chemical Engineering Magazine*, pp. 179–185, Nov 1991.

[57] H. L. Clever and R. Batlino, "The Solubility of Gases in Liquids", in *Solutions and Solubilities. Techniques in Chemistry, vol. VIII, part I*, M. R. J. Dack Ed., Wiley, Hoboken, New Jersey, 1975, pp. 429–433.

[58] M. J. Lazzaroni, D. Bush, C. A. Eckert, T. C. Frank, S. K. Gupta and J. D. Olson, "Revision of MOSCED Parameters and Extension to Solid Solubility Calculations (plus Supplemental Information. Database of Published Limiting Activity Coefficients)," *Industrial & Engineering Chemistry Research*, vol. 44, no. 11, pp. 4075–4083, 2005.

[59] C. Tsonopoulos and J. M. J. M. Prausnitz, "Activity Coefficients of Aromatic Solutes in Dilute Aqueous Solutions," *Industrial and Engineering Chemistry Fundamentals*, vol. 10, no. 4, pp. 593–600, 1971.

[60] J. Setschenow, "Über die Konstitution der Salzlösungen auf Grund ihres Verhaltens zu Kohlensäure," *Zeitschrift für Physikalische Chemie*, vol. 4, pp. 117–125, 1889.

[61] R. Sander, "Compilation of Henry's Law Constants (Version 4.0) for Water as Solvent," *Atmospheric Chemistry and Physics*, vol. 15, pp. 4399–4981, 2015.

[62] H. Z. Kister, *Distillation Design*, McGraw-Hill, New York, 1992, pp. 365.

[63] W. L. McCabe and E. W. Thiele, "Graphical Design of Fractionating Columns," *Industrial & Engineering Chemistry*, vol. 17, no. 6, pp. 605–611, 1925.

[64] W. H. Rodebush, "A Simple Graphical Method of Calculating the Number of Plates Required for a Distillation Column," *Industrial & Engineering Chemistry*, vol. 14, no. 11, pp. 1036–1037, 1922.

[65] M. Ponchon, "Étude Graphique de la Distillation Fractionée Industrielle, Parties 1 et 2," *Technology Moderne*, Vol. 13, no. 1 and no. 2, pp. 20–24, and 55–58, 1921.

[66] R. Savarit, "Definition of Distillation, Simple Discontinuous Distillation, Theory and Operation of Distillation Column, and Exhausting and Concentrating Columns for Liquid and Gaseous Mixtures and Graphical Methods for Their Determination," Arts et Mètiers, vol. 3, pp. 65, 142, 178, 241, 266, and 307, 1922.

[67] J. G. Dunlop, "Vapor-Liquid Equilibrium Data," Master's Thesis, 1948, obtained from ref. 68.

[68] "Vapor-Liquid Equilibrium Data, Methanol/2-Propanol," [Online]. Available: http://www.ddbst.com/en/EED/VLE/VLE%202-Propanol%3BMethanol.php. [Accessed 9 Oct 2022].

[69] J. Gmehling et al., "DDBST," Dortmund Data Bank Software & Separation Technology GmbH, [Online]. Available: http://www.ddbst.com/. [Accessed 10 July 2022].

[70] Lee, et al, "A Graphical Method for Designing Reactive Distillation Columns I. The Ponchon-Savarit Method," *Proceedings: Mathematical, Physical and Engineering Sciences*, vol. 456, no. 2000, pp. 1953–1964, 2000.

[71] M. R. Fenske, "Fractionation of Straight-Run Pennsylvania Gasoline," *Industrial & Engineering Chemistry*, vol. 24, no. 5, pp. 482–485, 1932.

[72] A. J. V. Underwood, "Fractional Distillation of Multi-Component Mixtures – Calculation of Minimum Reflux Ratio," *Journal of the Institute of Petroleum*, vol. 32, no. 274, pp. 614–626, 1946.

[73] A. J. V. Underwood, "Fractional Distillation of Multicomponent Mixtures," *Chemical Engineering Progress*, vol. 44, no. 8, pp. 603–614, 1948.

[74] E. R. Gilliland, "Multicomponent Rectification Estimation of the Number of Theoretical Plates as a Function of the Reflux Ratio," *Industrial & Engineering Chemistry*, vol. 32, no. 9, pp. 1220–1223, 1940.

[75] C. J. King, *Separation Processes*, McGraw-Hill, New York, 1971, pp. 460.

[76] R. H. Perry, C. H. Chilton and S. D. Kirkpatrick, *Perry's Chemical Engineers' Handbook*, 4th ed., McGraw-Hill, New York, 1963.

[77] W. L. McCabe, J. C. Smith and P. Harriott, *Unit Operations of Chemical Engineering*, 7th ed., McGraw-Hill, New York, 2005.

[78] Y. K. Molokanov, N. I. Mazurina, G. A. Nikiforov and T. P. Korablina, "Approximation Method for Calculating Basic Parameters of Multicomponent Fractionation," *International Chemical Engineering*, vol. 12, no. 2, pp. 209, 1972.

[79] A. Kremser, "Theoretical Analysis of Absorption Process," *National Petroleum News*, vol. 22, no. 21, pp. 43–49, Jan 1930.

[80] M. Souders and G. G. Brown, "Fundamental design of absorbing and stripping columns for complex vapors," *Industrial & Engineering Chemistry*, vol. 24, pp. 519–522, 1932.

[81] P. Dalager, "Vapor-Liquid Equilibriums of Binary Systems of Water with Methanol and Ethanol at Extreme Dilution of the Alcohols," *Journal of Chemical & Engineering Data*, Vol. 14, no. 3, pp. 298–301, 1969.

[82] H. S. Hu, "Determination of Vapour-Liquid and Vapour-Liquid-Liquid Equilibrium of the Chloroform-Water and Trichloroethylene-Water Binary Mixtures," *Fluid Phase Equilibria*, vol. 289, pp. 80–89, 2010.

[83] G. M. Ramapriya, A. Selvarajah, L. E. J. Cucaita, J. Huff, M. Tawarmalani and R. Agrawal, "Short-Cut Methods versus Rigorous Methods for Performance-Evaluation of Distillation Configurations," *Industrial & Engineering Chemistry Research*, vol. 57, no. 22, pp. 7726–7731, 2018.

[84] J. L. Falconer, J. W. Medlin, J. deGrazia, N. Hendren and M. L. Medlin, "LearnChemE.com – Excel Spreadsheets – other," Department of Chemical and Biological Engineering at the University of Colorado Boulder, [Online]. Available: https://learncheme.com/student-resources/excel-files/. [Accessed 10 Oct 2022].

[85] CheSheets.com, "Binary Distillation McCabe Thiele Diagram," [Online]. Available: https://www.chesheets.com/mccabe_theile.html. [Accessed 10 Oct 2022].

[86] CheCalc.com, "Distillation – Packed Tower Sizing," [Online]. Available: https://www.checalc.com/calc/packcol.html. [Accessed 10 Oct 2022].

[87] M. Kumar, "cheresources.com – Distillation Theoretical Stages Calculator," [Online]. Available: https://www.cheresources.com/invision/files/file/124-distillation-theoretical-stages-calculator/. [Accessed 10 Oct 2022].

[88] Engineering ToolBox, "Miscellaneous – Concentration Units Converter," 2017. [Online]. Available: https://www.engineeringtoolbox.com/molarity-molality-mole_fraction-weight_percent-grams_solute_liter_solution-conversion-formula-calculator-d_1998.html. [Accessed 10 Oct 2022].

[89] Wikipedia contributors, "List of chemical process simulators," Wikipedia, The Free Encyclopedia, [Online]. Available: https://en.wikipedia.org/w/index.php?title=List_of_chemical_process_simulators&oldid=1109213691. [Accessed 10 Jan 2022].

[90] K. E. McIntush, D. Mamrosh, K. Fisher and C. A. M. Beitler, "Caustic Scrubber Designs for Refinery Fuel Gas, Sour Water Stripper Gas, and Other Refinery Applications," Brimstone Sulfur Symposium (Vail, Colorado), September 2012. [Online]. Available: https://brimstone-sts.com/wp-content/uploads/2015/11/09V12-McIntush-Caustic-Scrubber-Designs-for-Refinery-Fuel-Gas.pdf. [Accessed 24 January 2023].

[91] H. Z. Kister, P. M. Mathias, D. E. Steinmeyer, W. R. Penney, V. S. Monical and J. R. Fair, "Section 14, Equipment for Distillation, Gas Absorption, Phase Dispersion, and Phase Separation," in *Perry's Chemical Engineers' Handbook*, D. W. Green and M. Z. Southard, Eds., 9th ed., McGraw-Hill, New York, 2019.

[92] R. Pohorecki and W. Moniuk, "Kinetics of Reaction Between Carbon Dioxide and Hydroxyl Ions in Aqueous Electrolyte Solutions," *Chemical Engineering Science*, vol. 43, no. 7, pp. 1677–1684, 1988.

[93] R. R. Bottoms, "Process for separating acidic gases". US Patent 1,783,901, 2 Dec 1930.

[94] Wikimedia Commons contributors, "'File:AmineTreating.svg', Wikimedia Commons, the free media repository," License: https://creativecommons.org/licenses/by-sa/4.0/legalcode, 3 Oct 2022. [Online]. Available: https://commons.wikimedia.org/w/index.php?title=File:AmineTreating.svg&oldid=479333964. [Accessed 10 Oct 2022].

[95] J. Price and D. Burns, "Clean Amine Solvents Economically and Online," *Hydrocarbon Processing*, vol. 74, no. 8, Aug 1995.

[96] K. McCarley, "Finding the Capacity of a Distillation Column," *Chemical Engineering Progress*, vol. 117, no. 12, pp. 23–28, Dec 2021.

[97] Mach Engineering LLC, "Evolution in Random Tower Packings," [Online]. Available: https://www.machengineering.com/evolution-in-random-tower-packings/. [Accessed 10 Oct 2022].

[98] Wikipedia contributors, "Random column packing," Wikipedia, The Free Encyclopedia, 13 Dec 2021. [Online]. Available: https://en.wikipedia.org/w/index.php?title=Random_column_packing&oldid=1060031983. [Accessed 10 Oct 2022].

[99] Wikipedia contributors, "Raschig ring," Wikipedia, The Free Encyclopedia, 7 Oct 2021. [Online]. Available: https://en.wikipedia.org/w/index.php?title=Raschig_ring&oldid=1048686794. [Accessed 10 Oct 2022].

[100] E. Berl, "Filling Body for Reaction and Washing Towers". US Patent 1,796,501, 17 Mar 1931.

[101] R. F. Strigle Jr and K. E. Porter, "Tower Packing". US Patent 4,303,599, 1 Dec 1981.

[102] Sulzer Ltd, "Conventional Random Packings – Nutter Ring™," [Online]. Available: https://www.sulzer.com/en/shared/products/conventional-random-packings. [Accessed 11 Oct 2022].

[103] Koch-Glitsch LP, "Metal Random Packing – CASCADE MINI-RINGS® Random Packing," [Online]. Available: https://koch-glitsch.com/products/packing-and-internals?productcategory=packing-and-internals&categoryname=Metal-Random-Packing. [Accessed 11 Oct 2022].

[104] Raschig GmbH, "Plastic Random Packings – Hacketten® / Jaeger Tri-Packs®," [Online]. Available: https://stofftrenntechnik.raschig.de/plastic-random-packings/. [Accessed 11 Oct 2022].

[105] Koch-Glitsch LP, "Metal Random Packing – INTALOX® ULTRA Random Packing," [Online]. Available: https://koch-glitsch.com/products/packing-and-internals?productcategory=packing-and-internals&categoryname=Metal-Random-Packing. [Accessed 11 Oct 2022].

[106] Sulzer Ltd., "High performance random packings – NeXRing™," [Online]. Available: https://www.sulzer.com/en/shared/products/high-performance-random-packings. [Accessed 11 Oct 2022].

[107] Raschig GmbH, "Metal Random Packings – Raschig Super-Ring®," [Online]. Available: https://masstransfer.raschig.de/products/metal-random-packings/. [Accessed 11 Oct 2022].

[108] Raschig GmbH, "Metal Random Packings – Raschig Super-Ring® Plus," [Online]. Available: https://masstransfer.raschig.de/products/metal-random-packings/. [Accessed 11 Oct 2022].

[109] Koch-Glitsch LP, "Plastic Random Packing – INTALOX® SNOWFLAKE® High Performance Packing," [Online]. Available: https://koch-glitsch.com/products/packing-and-internals?productcategory=packing-and-internals&categoryname=Plastic-Random-Packing. [Accessed 6 Jun 2022].

[110] M. Schultes and D. R. Summers, "100 Years of Distillation with Trays and Packings and Beyond," *12th International Conference on Distillation & Absorption 2022*, Toulouse, France, 18–21 September, 2022.

[111] Sulzer Ltd., "Sulzer Structured Packings / Gauze Packings (Sulzer BX Wire Gauze Packing)," [Online]. Available: https://www.sulzer.com/en/shared/products/gauze-packings. [Accessed 28 November 2022].

[112] Sulzer Ltd, "Structured Packings, Energy-Efficient, Innovative, and Profitable", Brochure E10686, 2020.

[113] Koch-Glitsch LP, "FLEXITRAY® Valve Trays" Bulletin FTCVT-02, 2015.

[114] Sulzer, Ltd., *Leading Tray Technology, Reliability and Performance, Sales Brochure E10503*, 2018.

[115] M. J. Lockett, *Distillation Tray Fundamentals*, Cambridge University Press, Cambridge, England, 1986.

[116] C. L. Smith, *Control of Batch Processes*, Wiley-AIChE, Hoboken, New Jersey, 2014.

[117] C. L. Smith, *Distillation Control: An Engineering Perspective*, Wiley, Hoboken, New Jersey, 2012.

[118] F. G. Shinskey, *Distillation Control: For Productivity and Energy Conservation*, 2nd Ed., McGraw-Hill, New York, 1984.

[119] J. R. Fair, "Section 5.11, Distillation Column Control," in *Handbook of Separation Process Technology*, R. W. Rousseau, Ed., Wiley, New York, 1987, pp. 328–331.

[120] H. Kooijman and R. Taylor, "ChemSep™ Modeling Separation Processes," [Online]. Available: http://www.chemsep.org/. [Accessed 11 Oct 2022].

[121] J. Alford and G. Hida, "Discrete Systems in Process Control," *Chemical Engineering Progress*, vol. 118, no. 6, pp. 57–63, 2022.

[122] B. R. Mehta and Y. J. Reddy, *Industrial Process Automation Systems Design and Implementation*, Butterworth-Heinemann, Oxford, England, 2015.

[123] L. A. Robbins, *Distillation Control, Optimization, and Tuning: Fundamentals and Strategies*, CRC Press, Boca Raton, Florida, 2011.

[124] D. C. White, "Optimize Energy Use in Distillation," *Chemical Engineering Progress*, vol. 108, no. 3, pp. 35–41, 2012.

[125] Y. A. Çengel and M. A. Boles, *Thermodynamics: An Engineering Approach*, McGraw-Hill Education, New York, 2015.

[126] X. Liu and R. Bansal, *Thermal Power Plants*, CRC Press, Boca Raton, Florida, 2016.

[127] M. Blahušiak, A. A. Kiss, S. R. A. Kersten and B. Schuur, "Quick Assessment of Binary Distillation Efficiency Using a Heat Engine Perspective," *Energy*, vol. 116, pp. 20–31, 2016.

[128] J. L. Humphrey and G. E. Keller II, Separation Process Technology, McGraw-Hill, New York, 1997, pp. 301–303.

[129] M. Marin-Gallego, B. Mizzi, D. Rouzineau, C. Gourdon and M. Myer, "Concentric Heat Integrated Distillation Column (HIDiC): A New Specific Packing Design, Characterization and Pre-Industrial Pilot Unit Validation," *Chemical Engineering and Processing – Process Intensification*, vol. 171, article no. 108643, 2022.

[130] R. Smith, M. Jobson, L. Chen and S. Farrokhpanah, "Heat Integrated Distillation System Design," *Chemical Engineering Transactions*, vol. 21, pp. 19–24, 2010.

[131] M. Jobson, "Chapter 6, Energy Considerations in Distillation," in *Distillation: Fundamentals and Principles*, Academic Press, Cambridge, Massachusetts, 2014, pp. 225–270.

8 Enhanced Distillation: Taking Advantage of Specific Molecular Interactions

Thermal energy is the motive force or separating agent used to make a distillation process function. In a typical process, heat is input to the bottoms reboiler and removed from the overheads condenser. Excluding distillations at very high pressures or other conditions where vapor phase non-idealities come into play, the relative volatility driving force is a function of pure component vapor pressures (the ideal contribution) and any specific molecular interactions that may be present, as reflected in the liquid-phase activity coefficients (the non-ideal contribution):

$$\alpha_{i,j} = \frac{p_i^{sat}}{p_j^{sat}} \quad \text{for an ideal mixture} \tag{8.1}$$

$$\alpha_{i,j} = \frac{\gamma_i \, p_i^{sat}}{\gamma_j \, p_j^{sat}} \quad \text{for a real mixture} \tag{8.2}$$

When the relative volatility of components is insufficient for standard distillation, enhanced distillation methods may be used. These methods enhance the non-ideal contribution to relative volatility by altering the activity coefficients. This may be accomplished by adding a material separating agent to the feed, or it might be accomplished simply by changing the distillation operating conditions.

Azeotropic distillation involves addition of a light entrainer. It is commonly practiced in either a batch or a continuous process scheme (Figures 8.1 and 8.2). The entrainer normally enters the distillation column as reflux to the top of the column. It exits in the overheads carrying (or entraining) one or more components with it, normally in the form of a heterogeneous azeotrope. This can increase the relative volatility of key components, sometimes breaking a homogeneous azeotrope present in the feed, to facilitate production of a pure bottoms product. Extractive distillation, on the other hand, employs a relatively heavy solvent (compared to feed components) that enters the column at a point somewhere above the feed point but below the top (Figure 8.3). The solvent exits the column at the bottom in the form of a heavy liquid solution (not an azeotrope), selectively carrying a portion of the feed components with it. This can facilitate production of a pure overheads product or a bottoms product that is later separated in a second column. In both azeotropic and extractive processes, the function of the material separating agent is to selectively interact with a key feed component, altering its activity and relative volatility. With pressure change distillation, no material separation agent is added. Key activities and relative volatilities are altered simply by changing the operating temperature – which is accomplished by changing the operating pressure. In this special case, the feed liquid must already be highly non-ideal to be sufficiently sensitive to change in temperature, and this often

https://doi.org/10.1515/9783110695052-013

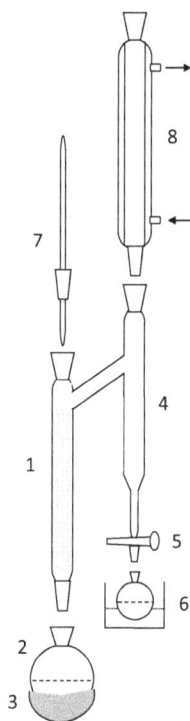

Figure 8.1: Laboratory Batch Azeotropic Distillation Setup (Dean-Stark Apparatus).
Shown for an entrainer with liquid density less than that of water, such as toluene: 1) distillation column; 2) still pot; 3) heating mantle; 4) condensate collection tube; 5) drain valve; 6) receiver with cooler; 7) thermometer; 8) vented condenser.

involves the presence of a homogeneous azeotrope in the feed, although not necessarily. In practice, an enhanced distillation process may involve both the addition of a material separating agent and change in operating temperature/pressure to optimize relative volatility.

The principles of enhanced distillation have been known for a long time. Steam distillation, which employs water as the entrainer, is said to date back to the tenth century when it was used to distill essential oils. In 1902, Sidney Young, Professor of Chemistry at University College (Bristol) in England, studied batch distillation of alcohol + water azeotropes and the impact of adding benzene or n-hexane [1]. In 1915, Konrad Kubierschky introduced a continuous azeotropic distillation process employing a heterogeneous azeotropic entrainer to break a homogeneous azeotrope [2]. And in 1920, Ernest W. Dean and David D. Stark introduced the batchwise Dean-Stark distillation apparatus for removing water from petroleum samples in the laboratory [3, 4]. These innovations were followed by various industrial-scale applications including removal of water from ethanol by using various alternatives to benzene (in 1928) [5], removal of water from acetic acid by using benzene (in 1929) [6], and removal of water from acetic acid by using propyl acetate or propylene chloride (in 1936) [7]. The general concept of pressure change distillation was introduced in 1928 by Warren K. Lewis, Professor of Chemical Engineering at Massachusetts Institute of Technology, focusing on removal of water from ethanol [8]. The concept

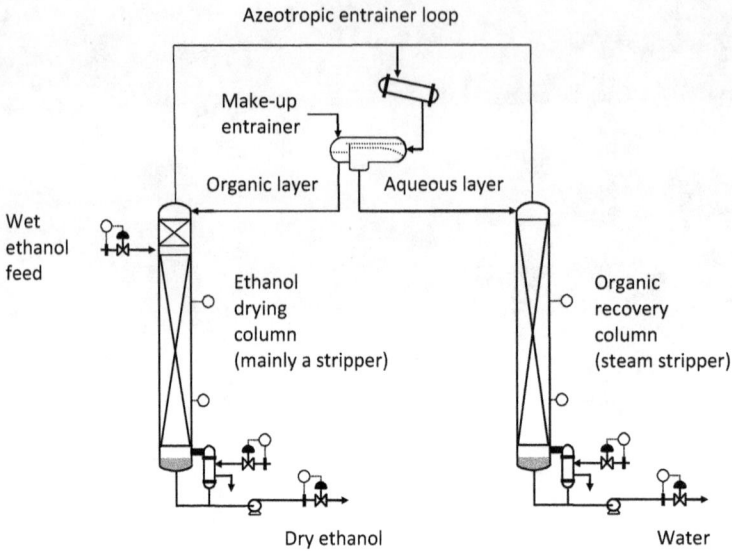

Figure 8.2: Continuously Fed Dual-Column Azeotropic Distillation Process. Shown for the system ethanol + water + benzene (for illustration).

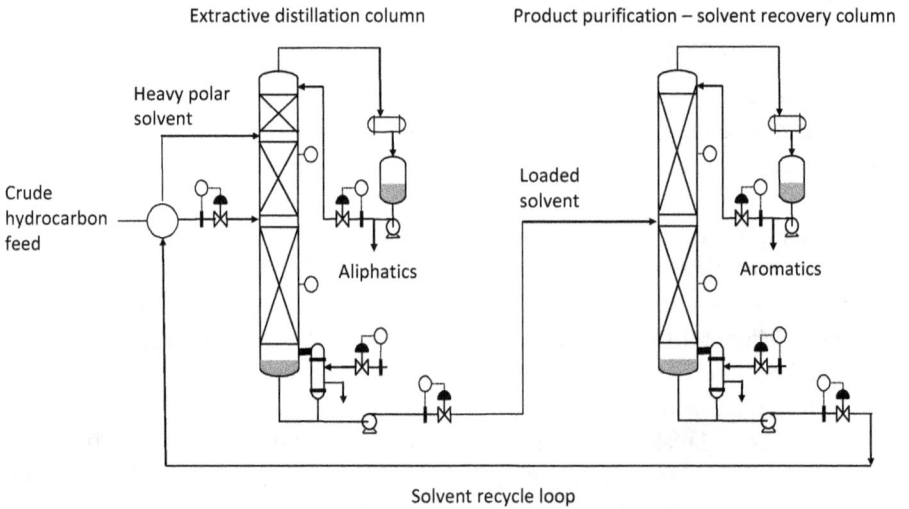

Figure 8.3: Continuously Fed Extractive Distillation Process. Shown for separating aromatics from aliphatics by using a heavy polar solvent, the classic application.

of extractive distillation came later. It appears it was first described by Clarence L. Dunn and Robert B. McConaughy around 1940. They developed it as an improvement over the liquid-liquid extraction + distillation process introduced by Lazăr Edeleanu in 1908 to extract aromatics from aliphatics using liquid SO_2 as solvent and later using phenol [9]. Dunn and McConaughy combined the extraction and distillation steps into a single fractionation column, using alkyl phenol as the heavy solvent. They called it vapor phase extraction in the presence of a relatively high-boiling selective solvent – or extractive distillation.

Many enhanced distillations have subsequently been introduced. In the following discussion, we focus on some of the better-known applications in explaining the basic process concepts. The use of a membrane to enhance the selectivity of liquid to vapor mass transfer may also be considered a form of enhanced distillation. For that discussion, see Section 12.5: Pervaporation.

8.1 Heterogeneous Azeotropic Distillation and Steam Distillation

As mentioned above, azeotropic distillation normally refers to the use of an azeotrope-forming additive called an entrainer to improve the separation of close-boiling feed components, by selectively forming an azeotrope with one or more of the feed components. An entrainer that forms a heterogeneous azeotrope may be preferred, because the azeotrope acts as a pseudo light component that is easily distilled overhead, and once there, the azeotrope is easily separated simply by condensing it, allowing it to phase split, and then decanting the individual condensate layers [10]. The entrainer layer can then be returned to the distillation column to continue selective removal of the specific feed component. This effectively increases the relative volatility of close-boiling feed components and makes the distillation easier to perform, allowing use of shorter columns but not necessarily consuming less energy. Both batch and continuous processes may be used. The basics of azeotropes are reviewed in the Section 5.6: Phase Diagrams: Equilibrium Curves, Azeotropes, and Eutectics.

Batchwise heterogeneous azeotropic distillation is often used in organic chemistry laboratories to remove water from an organic reaction mixture. Or it may be used during the course of running a reaction to remove a light byproduct (typically water or an alcohol) to drive conversion of an equilibrium-limited reaction such as an esterification or transesterification (a reactive distillation). A typical apparatus is shown in Figure 8.1. The laboratory technique is often called Dean-Stark distillation. As previously mentioned, Ernest Dean and David Stark introduced it in 1920 to quantify the amount of water present in petroleum samples. Toluene or cyclohexane are commonly used as entrainers for removing water because they form a heterogeneous azeotrope with water, mutual solubility with water is low, and the organic condensate layer is less dense than water, so it simply overflows the top of the overheads condensate collector to automatically return back to the column. But other apparatus designs allow use of entrainers that are denser than water, such as chloroform (trichlorome-thane). A scaled-up version of this laboratory technique is sometimes used for drying reaction mixtures in the manufacturing plant. The control of this type of process is simple and robust. The entrainer is added to the pot, the pot heater is turned on at a constant rate, and cooling water is turned on for the overheads condenser. This causes the entrainer to be boiled overhead and out of the pot, carrying water with it. The overhead vapors are then condensed and collected in a condensate collection tube or decanter, where two liquid layers form and separate. In the case of a less-dense entrainer such as toluene, the top layer (the entrainer layer) overflows the collection tube/decanter back to the top of the column, and the lower layer (the aqueous layer) accumulates in the tube/decanter where it can be drained periodically. This is continued until water stops accumulating. No other control is needed [4].

A continuously fed heterogeneous azeotropic distillation process may be used to break a homogeneous azeotrope present in the feed. The classic process, introduced by Konrad Kubierschky in 1915, involves drying ethanol by using an organic entrainer such as benzene. A typical process employing two stripping columns is shown in Figure 8.2. As in the batchwise process, the entrainer forms a heterogeneous azeotrope with water, and in so doing is able to break the homogeneous azeotrope present in the feed (ethanol + water). This is possible because the entrainer forms a ternary azeotrope with water and with ethanol, one that is lower boiling than either binary azeotrope or the individual components. In thinking about how this process works, it is convenient and informative to think of the various azeotropes as if they were real components with reduced boiling points (pseudo components), as discussed by James Fair [10]. The ordering of normal boiling points is: ethanol + benzene + water ternary azeotrope (64.9 °C); ethanol + benzene binary azeotrope (68.2 °C); benzene + water binary azeotrope (69.3 °C); ethanol + water binary azeotrope (78.2 °C); ethanol (78.4 °C); benzene (80.1 °C); and water (100 °C).

In Figure 8.2, wet ethanol is fed to the first column (the ethanol drying column) where the chosen entrainer (benzene) enhances the removal of water to the point where dry ethanol exits the column as bottoms product. Virtually all of the benzene and water as well as some ethanol leave the top of the drying column at close to the ternary heterogeneous azeotrope composition. There, the overheads are condensed and enter the overheads decanter. The condenser/decanter essentially breaks the ethanol + benzene + water heterogeneous azeotrope by phase separating the mostly benzene + ethanol organic liquid from the mostly water + ethanol aqueous liquid. The organic layer is sent back to the top of the drying column to recover the ethanol and recycle the benzene entrainer. Make-up entrainer is added to the decanter as needed. Meanwhile, the aqueous layer from the decanter is sent to the top of the second column (the organic recovery column) which serves to recover small amounts of benzene and ethanol that saturate the rejected water from the overheads decanter. The purified water exiting the bottom of the organic recovery column, a steam stripper, is nearly devoid of benzene and ethanol.

Each column operates to produce overheads that approach the ethanol + benzene + water heterogeneous ternary azeotropic composition. In effect, the shared overheads condenser and decanter serve the function of rectification, so the columns themselves normally do not need rectification sections, although a short rectification section may be used in the ethanol drying column to better approach the ternary azeotrope composition overhead. Taller rectification sections should be used for feeds containing high amounts of water, say greater than 25% water. Alternatively, it is sometimes convenient at larger scales to provide a simple pre-fractionator prior to the ethanol drying column to remove much of the water from the wet ethanol feed under favorable conditions and deliver concentrated ethanol (still well below the ethanol + water azeotrope composition) to the ethanol drying column to reduce its size and energy consumption.

With this arrangement, the organic entrainer always remains within an internal overhead loop, serving the function of breaking the ethanol + water azeotrope present in the feed. Process control may be as simple as maintaining constant steam to feed ratios for each column, maintaining sufficient cooling to the condenser, and automatically routing liquids from the decanter back to the appropriate column. The solute activity coefficients, phase diagrams, and stripping design calculations discussed in earlier chapters all apply to the design of such a process. And, as stated earlier, other entrainers such as toluene, cyclohexane, or even trichloromethane may be used instead of benzene. Similar process schemes employing somewhat different decanting and recycle schemes may be used depending on the choice of entrainer and feed composition. But the basic concepts are the same.

Historically, some applications of heterogeneous azeotropic distillation have been difficult to operate due to formation of two liquid phases in the top of the distillation column, but this may be mitigated with the use of trays designed to handle those phases placed in the top of the column (typically two or three trays), with additional trays or packing in the rest of the column. The use of such trays can improve operation by preventing a phenomenon involving delayed boiling of the more volatile liquid as liquid droplets occasionally migrate down the column and suddenly boil, causing surges in operating pressure and disruption of the process or damage to column internals. Another potential problem involves difficulty obtaining well-clarified liquid layers in the decanter. This can happen periodically when surface-active impurities contaminate the feed and are volatilized or entrained into the overheads, which then accumulate in the decanter and interfere with phase separation. Careful attention to the decanter's design, providing sufficient liquid residence time, low volumetric flow rates at the main coalescing interface, and the use of specific types of internals that promote coalescence, can help to avoid these situations [11].

Specialized applications of azeotropic distillation include:
- Formic acid from acetic acid using isopropyl acetate as the entrainer (forming a heterogeneous formic acid + isopropyl acetate azeotrope) [12],

– 2-Butanol from *t*-pentanol using ethyl acetoacetate or nitroethane as entrainers (forming heterogeneous azeotropes with 2-butanol) [13],
– *o*-Xylene from *m*-xylene or *p*-xylene using an aliphatic ester as entrainer [14]. Typical examples of effective entrainers are: for *m*-xylene from *p*-xylene, propyl butyrate or methyl valerate; for *m*-xylene from *o*-xylene, hexyl formate or methyl valerate. These entrainers form heterogeneous azeotropes with *m*-xylene but homogeneous azeotropes with *o*-xylene.

Steam distillation of high-boiling organic oils is another kind of azeotropic distillation. Steam forms a heterogeneous azeotrope with the oils because they are only partially or sparingly miscible with water. In adding steam to the distillation of the oil, the steam serves as an entrainer that lowers the partial pressures of organics in the vapor (compared to boiling the organics by themselves), and by doing so, it lowers the distillation temperature. This is the same phenomenon described earlier in Section 5.6.3: Heterogeneous Azeotropes, in discussing the use of eq. (5.51) to determine the boiling point of a heterogeneous azeotrope. With steam distillation, the total distillation pressure is the sum of the vapor pressures in the oil phase plus the vapor pressure exerted by water at boiling conditions, and this lowers the boiling point of the mixture. Steam distillation is particularly useful whenever there is a need to minimize the distillation temperature to avoid degradation of the organic components (although hydrolysis of the oil may be a concern depending on the kind of oil being processed). Steam and air stripping of sparingly miscible and partially miscible organics out of water are related processes, because they also take advantage of the properties of a heterogeneous azeotrope to enhance performance. This is readily apparent in Figure 8.2 which shows the classic heterogeneous azeotropic distillation process comprised of a feed stripper plus a steam stripper. See Section 7.3.3: Continuous Stripping and Absorption. As a side note, the reduction in operating temperature that occurs on addition of water to the distillation of high-boiling oils is analogous to the reduction in operating temperature that occurs with the addition of solvent to a eutectic crystallization process (see Section 5.6.3: Heterogeneous Azeotropes and Section 5.6.5: Solid-Liquid Eutectic Behavior).

In general, heterogeneous azeotropic distillation is an easy-to-control and robust process that has been used for many years, but energy consumption can be high. This depends on the composition of the heterogeneous azeotrope formed by the added entrainer. To reduce energy consumption, the entrainer should have the following properties:

– The azeotrope formed by the entrainer should contain a significant amount of the component to be removed overhead. This may involve formation of a binary azeotrope or a ternary azeotrope depending on the feed composition.
– The enthalpy of vaporization should not be too high, so in this respect organic entrainers are preferred over the use of water.

- The mutual solubility of the two liquid phases in the overheads condenser should be low for efficient phase splitting and decantation of the overheads; that is, for efficient rectification.
- The entrainer must also meet other important criteria concerning its use in a chemical facility, such as having properties that allow for safe storage, safety in use, and effective control of emissions.

Many other azeotropic process schemes are possible and may offer advantages. For example, the use of a homogeneous azeotropic entrainer avoids any difficulties with the presence of two liquid phases in the distillation column and any difficulty with liquid-liquid phase separation. It might also allow a reduction in energy consumption depending on the composition of the azeotrope formed by the entrainer and the feed composition. On the other hand, a homogeneous azeotropic entrainer cannot be re-covered for recycle using simple stripping-mode distillation plus a decanter. Process analysis and process design typically involve construction of residue curve maps and in-depth process simulation.

A residue curve map (RCM) is a plot of the composition of liquid remaining in the pot of a batchwise single-stage distillation, starting with a specific feed composition and continuing from there as vapor is removed from the pot [15–17]. An example is shown in Figure 8.4. Plot lines are constructed for a range of feed compositions originating in different regions of the diagram. The resulting RCM facilitates distillation design by highlighting composition boundaries that border specific regions of separation feasibility for standard distillation. Such boundaries make certain separations infeasible, but as we have been discussing, sometimes they can be crossed by condensing and decanting a het-erogeneous azeotrope or by changing the operating conditions (changing pressure and thus temperature). Constructing an RCM is helpful whenever analyzing distillation op-tions for highly non-ideal multi-component mixtures of any kind. The fundamentals un-derlying such phase diagrams were described by Franciscus A. H. Schreinemakers in 1901 [18–20] and by Friedrich Wilhelm Ostwald in 1902 [21]. Ostwald is also known for the Ost-wald-Miers diagram used in analyzing crystallization (see Section 6.7.4: Crystallization Rates: Nucleation, Growth, and the Population Balance). But it wasn't until the 1980s and 1990s that analysis of RCMs became standard practice in industry, following on the work of Michael Doherty, Michael Malone, and others [22, 23]. Many commercially available distillation simulation software programs now include RCM construction and analysis capabilities.

Discussion of the many possible types of RCM diagrams and their application in process design is beyond the scope of our discussion here. Many helpful discussions are available elsewhere [22–28]. Interestingly, RCM diagrams may be used to augment a McCabe-Thiele graphical design, as discussed by Raymond Rooks [29]. Keep in mind that quantifying the results of these methods requires an accurate vapor-liquid equi-librium (VLE) model of a highly non-ideal, multi-component system. Validation of the

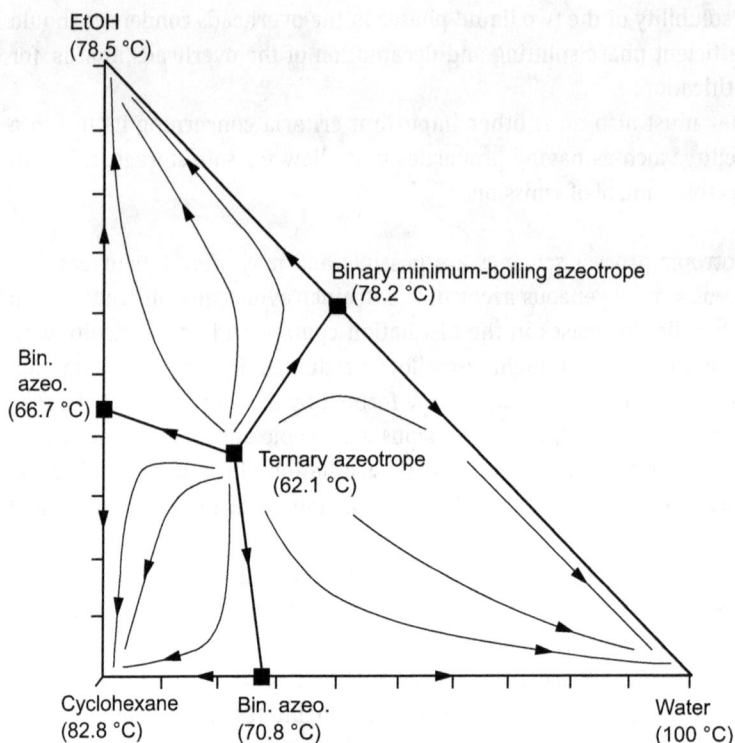

Figure 8.4: Example Residue Curve Map for the System Ethanol + Cyclohexane + Water, showing three operating regions isolated by azeotropic boundaries. Taken from ref. [15]. Reprinted with permission from *Ind. Eng. Chem. Res.*, copyright 1994, American Chemical Society.

VLE model for the system of interest is critical; otherwise, the results of the analysis and the design methods can be misleading.

Azeotropic distillation can be an effective and useful process scheme; however, we recommend also looking at alternative processes that may be considerably more energy efficient. For example, the removal of water from alcohols may best be accomplished today by first using standard distillation to generate a near-azeotrope overheads product (containing 90+% ethanol, for example), followed by further purification (breaking the azeotrope) by using pressure swing adsorption (PSA) to remove the remaining water down to ppm levels if needed, or by using pervaporation membranes to remove the remaining water. For more discussion, see Chapter 14: Inventive Engineering: Process Simplification, Intensification, and Hybrid Processing.

8.2 Extractive Distillation

The term extractive distillation refers to the addition of a heavy, miscible solvent to a distillation column to improve the separation of close-boiling components by selectively carrying (or extracting) a portion of the feed into the bottoms stream, as shown in Figure 8.3. The solvent is chosen so that it selectively interacts with a key feed component (or class of components), either by decreasing its activity coefficient in the mixture through a relatively attractive interaction (pulling it down into the solvent) or, less often, by increasing its activity coefficient (pushing it up into the overheads). Doing so increases the relative volatility of key feed components and makes the distillation easier to perform, allowing use of shorter columns and less reflux, and it may break an azeotrope present in the feed. The addition of a heavy solvent may in some cases actually reverse the relative volatility of key components. In any case, the added solvent itself normally does not form an azeotrope with any of the feed components (unlike an azeotropic entrainer). A variety of solvent screening methods may be used to identify candidate solvents having the desired influence on the activity of specific feed components. A well-known experimental method involves injecting a pulse of feed onto a packed chromatography column that has been pre-wetted with a candidate solvent and then measuring the solvent's effect on feed-component retention times [30, 31]. Activity coefficient models also may be used to estimate activity coefficients of feed components dissolved in the candidate solvent. For feed components that are pure-component solids, estimation and ranking of solid solubility in various solvents can also highlight candidate solvents (see Section 5.7: Solvent Selection Methods).

Many extractive distillations are developed for applications that also may be practiced by using liquid-liquid extraction followed by distillation. Extractive distillation simply combines extraction with distillation in a single unit operation. This is why the added separation agent is often called a heavy solvent. In an extractive distillation process the heavy solvent is added directly to a primary distillation column, and the solvent-rich bottoms from that column are then routed to an auxiliary distillation column (or stripper) for product recovery and solvent recycle. On the other hand, in a process involving liquid-liquid extraction followed by distillation, the liquid-liquid extraction column is coupled with two or more distillation columns used to treat the raffinate and extract streams for product recovery and solvent recycle. Both types of processes may be used to accomplish a desired separation, but the driving forces are different. With extractive distillation, the solvent added to the distillation column impacts the main separation through both its vapor pressure and its influence on liquid-phase activity (as in eq. 8.2). With extraction followed by distillation, the solvent in the main extractor impacts the main separation only through its influence on activity. Because of this difference in driving forces, one process scheme may have advantages over the other depending on the application. Other differences have to do with the potential for capital cost reduction when using extractive distillation versus liquid-liquid extraction followed by distillation, due to a reduction in the number of required columns. Other options include the use of salt as a heavy material separating agent [32]. For more discussion, see refs. [10, 33–35].

Extractive distillation typically uses much less energy compared to azeotropic distillation, because the added solvent goes out with the bottoms stream. It is not distilled overhead. Heterogeneous azeotropic distillation typically requires more energy because it involves repeated boiling of the entrainer overhead with recycle back to the top of the column in an internal loop. The amount of heavy solvent added to an extractive distillation process needs to be sufficient for the solvent to dissolve the feed (to ensure miscibility) and for the solvent to have a significant influence on feed component activities. It must act as the main solvent for the feed components. Usually this requires at least a 1:1 solvent to feed ratio. The solvent does not need to be completely miscible with the feed over the entire composition range, as long as the extractive distillation operation can be conducted at conditions that ensure liquid-liquid miscibility.

Various extractive distillation applications (both commercial operations and those with commercial potential) include:
- Separation of aromatics from aliphatics. This is the classic commercial application. It was originally developed using alkyl phenols as the heavy solvent [9] and later using heavy glycols such as diethylene glycol, triethylene glycol, and tetraethylene glycol and their mixtures. Later still, specialized polar, aprotic solvents were introduced including N-formyl morpholine, N-methyl pyrrolidone, and sulfolane, which gave better selectivity and lower energy consumption overall. Here, extractive distillation is an alternative to processes involving extraction followed by distillation [11, 36, 37]. Both types of processes are practiced today.
- Vinyl acetate from ethyl acetate using certain glycols or glycol ethers [38].
- Separation of hydrocarbon mixtures by using bio-derived solvents [39].
- Cyclopentane and cyclohexane recovery from naphtha or natural gas liquids using N-methyl pyrrolidone (NMP) [40].
- Acrylic acid from water using sulfolane, dimethylformamide, or dimethylacetamide [41].
- Acetone from methanol using glycerol, ethylene glycol, or dimethylsulfoxide [42].

8.3 Pressure Change Distillation

Pressure change distillation (PCD) is a dual-column, steady-state distillation process used to overcome or *break* a homogeneous azeotrope already present in the feed [43, 44]. Typically, one column is operated near atmospheric pressure, the other at a significantly higher pressure, and the two columns are connected by an internal recycle loop. Two pure products are produced as bottoms products, one from each of the two columns. The process is illustrated in Figure 8.5 with data for the binary system tetrahydrofuran (THF) + water. In some respects, a dual-column PCD process is analogous to the dual-column process shown in Figure 8.2 for heterogeneous azeotropic distillation, but instead of using a common decanter to separate the overheads, the two columns are operated at different pressures to do so.

A PCD process takes advantage of a change in azeotropic composition that accompanies the change in operating pressure; that is, a change in the location of the azeotropic point on the phase diagram (as in Figure 5.4). In this way, the process is able to move the feed point for a given column from one side of the azeotropic point to the other. In other words, the process is able to cross the azeotropic boundary. This sounds quite elegant, because it eliminates any need for an added entrainer or solvent. Historically, commercial applications have included the minimum-boiling azeotropes THF + water, acetonitrile + water, and methanol + acetone. Additional systems that may be separated using PCD are given in ref. [44]. PCD has also been applied to separate the maximum-boiling azeotrope HCl + water, in which case purified products are produced overhead. However, in practice (especially nowadays), PCD is not employed as often as alternative methods because in many cases the magnitude of the change in azeotropic composition is not sufficiently large to provide a sufficiently economical process. With too small a change, the volume of the recycle loop becomes quite large, and this greatly increases energy consumption and the required column diameters.

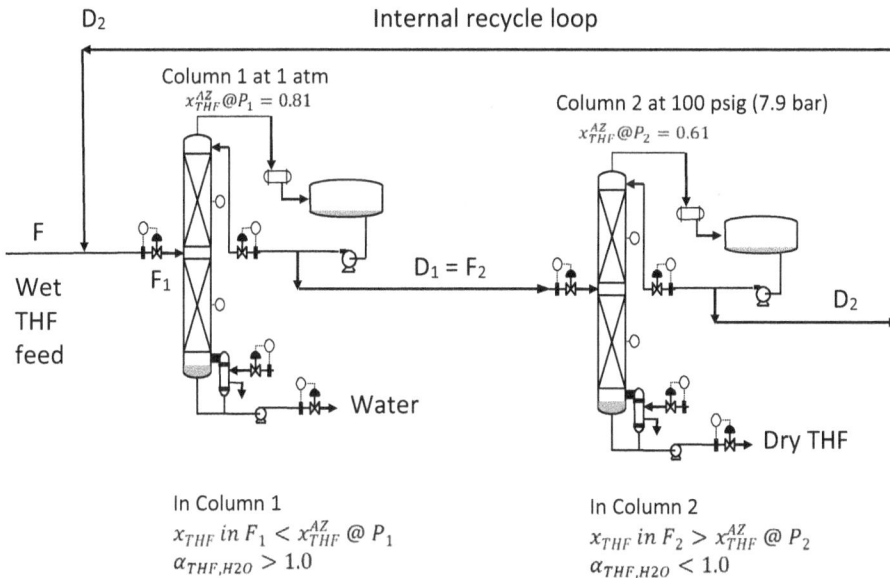

Figure 8.5: Continuously Fed Dual-Column Pressure Change Distillation Process. Shown for separation of tetrahydrofuran (THF) and water [45].

The enhancement in distillation performance that is achieved by this type of process is fundamentally due to the change in vapor-liquid equilibrium (VLE) that occurs with change in operating temperature, not pressure. So, from a thermodynamic point of view, this process might best be called temperature change distillation. But in practice, it is commonly known by the names pressure change distillation, pressure swing distillation, dual pressure distillation, or pressure sensitive distillation, because the required change in the operating temperature range is accomplished simply by changing the operating pressure. Today, we prefer the name pressure change distillation because it seems well suited to a steady-state process scheme. The name pressure swing distillation is also commonly used, but to us it suggests more of a cyclical process scheme like that of pressure swing adsorption.

Evaluating a PCD process involves determining azeotropic composition as a function of temperature, and from this, calculating the size of the required internal recycle loop and the impact this has on economic feasibility. For an initial evaluation and to illustrate the work process, consider the following approximate method used to evaluate separation of a feed containing components i and j. Given azeotropic composition at one temperature and pressure (typically atmospheric pressure data from a database), azeotropic composition may be extrapolated as a function of temperature by using Joffe's method [44]. For details and definitions of the symbols and model constants A, B, and C used below, see Section 5.6.2: Homogeneous Azeotropes. The superscript AZ denotes an azeotropic condition. First, calculate the activity coefficients corresponding to the given azeotropic data point:

$$\gamma_i^{AZ} = \frac{P_T}{P_i^{sat}}, \text{ for } T^{AZ} \text{ and } x_i^{AZ} @ P_1 \tag{8.3}$$

Then, calculate azeotropic composition for a new value of T^{AZ} via Joffe's method using eqs. (5.45) to (5.50). The corresponding pressure at the new temperature is then given by:

$$P_T = \exp\left(\frac{BA^2 \left(x_i^{AZ}\right)^2}{RT^{AZ}C^2}\right) P_2^{sat} \tag{8.4}$$

Calculations can be made over a range of T^{AZ} values, determining the corresponding values of P_T. Referring to Figure 8.5, the internal recycle for the proposed PCD process can then be estimated for a given change in operating pressure from P_1 to a new pressure P_2:

$$\frac{D_2}{F} \approx (1 - x_{1F})\left(\frac{x_i^{AZ} @ P_1}{\Delta x^{AZ}}\right), \text{ between operating pressures } P_1 \text{ and } P_2 \tag{8.5}$$

where D_2 is the flow rate of the stream recycled back to the first column and F is the feed rate to the overall process. The results of this analysis can be compared to well-known commercial PCD processes to gauge feasibility. A typical recycle load for a commercial process is on the order of 25%. Of course, the required column height and diameter will also need to be calculated for each of the two distillation columns, using methods discussed in Chapter 7, to determine their feasibility. Here we have outlined a general procedure using an approximate VLE model for illustration. An accurate VLE model will be needed for accurate design calculations.

As mentioned above, it is not common for the change in azeotropic composition to be sufficiently large to make a full PCD process competitive with other methods. However, it is common to adjust the pressure of a single distillation column to alter the temperature range to obtain more favorable VLE even if the azeotrope composition may not change very much. This may allow positioning the feed point on the desired side of the azeotropic point, in order to recover some of the desired product as a bottoms product (for a minimum-boiling azeotrope) or simply to allow for higher purity and recovery of a desired product due to more favorable VLE. This approach is sometimes used in distilling ethanol from water, for example. Of course, compared to the full dual-column process with recycle, the recovery of the desired component will be somewhat reduced when a single column is used. Or, the distillation might be done in steps using a single column (either batch or continuous), such that multiple steps or campaigns are conducted at different pressures. This might be particularly attractive at small scale using single-column batch distillation in which the residual pot contents from distillation at one pressure are stored and accumulated from multiple batches for later distillation at another pressure. But in any case, whether this approach is attractive still depends on the size of the change in azeotropic composition between steps. As we have emphasized in earlier discussions, it is more common nowadays to augment a single-column distillation process by adding another separation method instead of adding another distillation column or additional distillation steps; that is, by using a hybrid process scheme. See Chapter 14: Inventive Engineering: Process Simplification, Intensification, and Hybrid Processing.

References

[1]   S. Young and E. C. Fortey, "The Properties of Mixtures of the Lower Alcohols with Benzene and with Benzene and Wate," *Journal of the Chemical Society*, vol. 81, pp. 739–752, 1902.
[2]   K. Kubierschky, "Verfahren zur Gewinnung von Hochprozentigen, beziehungsweise Absoluten Alkohol-Wasser Gemischen in Unterbrochenem Betriebe," German Patent 287,897, 1915.
[3]   E. W. Dean and D. D. Stark, "A Convenient Method for the Determination of Water in Petroleum and Other Organic Emulsions," *Industrial & Engineering Chemistry*, vol. 12, no. 5, pp. 486–490, 1920.
[4]   P. Laszlo, "On the Origins of a Tool for Chemists: The Dean-Stark Apparatus," *Bulletin for the History of Chemistry*, vol. 38, no. 1, pp. 67–72, 2013.
[5]   D. B. Keyes, "Process of Obtaining Absolute Alcohol," US Patent 1,676,735, 10 Jul 1928.
[6]   A. H. Maude, "Process for Manufacturing Glacial Acetic Acid," US Patent 1,722,532, 30 Jul 1929.
[7]   D. F. Othmer, "Process for Concentrating Acetic Acid," US Patent 2,028,800, 28 Jan 1936.
[8]   W. K. Lewis, "Dehydrating Alcohol and the Like," U.S. Patent 1,676,700, 10 July 1928.

[9]   C. L. Dunn and R. B. McConaughy, "Recovery of Pure Aromatics," U.S. Patent 2,288,126, 30 June 1942.

[10] J. R. Fair, "Section 5.5, Special Distillations," in *Handbook of Separation Process Technology*, R. R. Rousseau, Ed., Wiley, New York, 1987, pp. 261–274.

[11] T. C. Frank, B. S. Holden and A. F. Seibert, "Section 15: Liquid-Liquid Extraction and Other Liquid-Liquid Operations and Equipment," in *Perry's Chemical Engineers' Handbook*, 9th ed., D. W. Green and M. Z. Southard, Eds., McGraw-Hill, New York, 2019.

[12] L. Berg, "Separation of Formic Acid from Acetic Acid by Azeotropic Distillation," US Patent 5,633,402, 27 May 1997.

[13] L. Berg, "Separation of 2-butanol from *tert*-Amyl Alcohol by Azeotropic Distillation," US Patent 5,338,410, 16 Aug 1994.

[14] L. Berg, "Separation of *m*-Xylene from *p*-Xylene or *o*-Xylene by Azeotropic Distillation," US Patent 5,039,380, 13 Aug 1991.

[15] G.-J. A. F. Fien and Y. A. Liu, "Heuristic Synthesis and Shortcut Design of Separation Processes: A Review," *Industrial & Engineering Chemistry Research*, vol. 33, no. 11, pp. 2505–2522, 1994.

[16] V. Kiva, E. Hilmen and S. Skogestad, "Azeotropic Phase Equilibrium Diagrams: A Survey," *Chemical Engineering Science*, vol. 58, pp. 1903–1953, 2003.

[17] N. Shcherbakova, V. Gerbaud and I. Rodriguez-Donis, "On the Riemannian Structure of the Residue Curves Maps," *Chemical Engineering Research and Design*, vol. 99, pp. 87–96, 2015.

[18] F. A. H. Schreinemakers, "Dampfdrucke Ternaer Gemische. Theoretischer Teil: Erste Abhandlung (In German)," *Zeitschrift fuer Physikalische Chemie*, vol. 36, no. 3, pp. 257–289, 1901.

[19] F. A. H. Schreinemakers, "Dampfdrucke Ternaer Gemische. Theoretischer Teil: Zweite Abhandlung (In German)," *Zeitschrift fuer Physikalische Chemie*, vol. 36, no. 4, pp. 413–449, 1901.

[20] F. A. H. Schreinemakers, "Dampfdrucke Ternaer Gemische. Theoretischer Teil: Dritte Abhandlung (In German)," *Zeitschrift fuer Physikalische Chemie*, vol. 36, no. 6, pp. 710–740, 1901.

[21] W. Ostwald, *Lehrbuch der Allgemeinen Chemie*, Verwandtschaftslehre Erste Teil., Verlag von Wilhelm Engelmann,, Leipzig, Germany, 1902.

[22] M. F. Doherty and M. F. Malone, *Conceptual Design of Distillation Systems*, McGraw-Hill, New York, 2001.

[23] H. N. Pham and M. F. Doherty, "Design and Synthesis of Heterogeneous Azeotropic Distillations – II. Residue Curve Maps," *Chemical Engineering Science*, vol. 45, no. 7, pp. 1837–1843, 1990.

[24] M. Doherty, Z. T. Fidkowski, M. F. Malone and R. Taylor, "Section 13, Distillation," in *Perry's Chemical Engineers' Handbook*, 9th ed., D. W. Green and M. Z. Southard, Eds., McGraw-Hill, New York, 2018.

[25] J. R. Knight and M. F. Doherty, "Optimal Design and Synthesis of Homogeneous Azeotropic Distillation Sequences," *Industrial & Engineering Chemistry Research*, vol. 28, no. 5, pp. 564–572, 1989.

[26] E. R. Foucher, M. F. Doherty and M. F. Malone, "Automatic Screening of Entrainers for Homogeneous Azeotropic Distillation," *Industrial & Engineering Chemistry Research*, vol. 30, no. 4, pp. 760–772, 1991.

[27] L. Laroche, N. Bekiaris, H. W. Andersen and M. Morari, "The Curious Behavior of Homogeneous Azeotropic Distillation – Implications for Entrainer Selection," *AIChE Journal*, vol. 38, no. 9, pp. 1309–1328, 2004.

[28] S. D. Barnicki, "Put Your Column on the Map," *Chemical Processing, vol. 67*, no. *9*, pp. 39, September 2004.

[29] R. E. Rooks, "Separation Technology: Draw Insights on Distillation," *Chemical Processing*, vol. 69, no. 5, 2006.

[30] A. Vega, F. Díez, R. Esteban and J. Coca, "Solvent Selection for Cyclohexane–Cyclohexene–Benzene Separation by Extractive Distillation Using Non-Steady-State Gas Chromatography," *Industrial & Engineering Chemistry Research*, vol. 36, no. 3, pp. 803–807, 1997.

[31] L. Jiménez and J. Costa-López, "Solvent Selection for a Reactive and Extractive Distillation Process by Headspace Gas Chromatography," *Separation Science and Technology*, vol. 38, no. 1, pp. 21–37, 2003.

[32] W. F. Furter, "Extractive Distillation by Salt Effect," *Chemical Engineering Communications*, vol. 116, no. 1, pp. 35–40, 1991.

[33] J. P. Knapp and M. F. Doherty, "Minimum Entrainer Flows for Extractive Distillation: A Bifurcation Theoretic Approach," *AIChE Journal*, vol. 40, no. 2, pp. 243–268, 2004.

[34] L. Berg and A.-I. Yeh, "The Unusual Behavior of Extractive Distillation – Reversing the Volatility of the Acetone-Isopropyl Ether System," *AIChE Journal*, vol. 31, no. 3, pp. 504–506, 1985.

[35] V. Gerbaud, I. Rodriguez-Donis, L. Hegely, P. Lang, F. Denes and X. You, "Review of Extractive Distillation. Process Design, Operation, Optimization and Control," *Chemical Engineering Research and Design*, vol. 141, pp. 229–271, 2019.

[36] L. H. Horsley, V. S. Morello and N. Poffenberger, "Aromatic Hydrocarbons by Solvent Extraction with a Solvent of Diethylene Glycol and Water," US Patent 2,770,664, 13 Nov 1956.

[37] A. H. Zahed, Y. Yorulmaz and M. A. Gashghari, "Extraction of BTX Hydrocarbons from Saudi Arabian Refinery Platformates," *JKAU: Engineering Science*, vol. 4, pp. 101–114, 1992.

[38] L. Berg and M. W. Paffhausen, "Separation of Vinyl Acetate from Ethyl Acetate by Etractive Distillation," U.S. Patent 4,897,161, 30 January 1990.

[39] T. Brouwer and B. Schuur, "Bio-based Solvents as Entrainers for Extractive Distillation in Aromatic/ Aliphatic and Olefin/Paraffin Separation," *Green Chemistry*, vol. 22, pp. 5369–5375, 2020.

[40] R. K. Joshi, "Separation of Cyclopentane from Close Boiling Paraffins Using Selective Solvents in Extractive Distillation Step," *Jetir*, vol. 5, no. 5, pp. 486–490, 2018.

[41] L. Berg, "Dehydration of Acrylic Acid by Extractive Distillation," U.S. Patent 5,154,800, 13 October 1992.

[42] L. Berg and K. J. Warren, "Separation of Acetone from Methanol by Extractive Distillation," U.S. Patent 4,584,063, 22 April 1986.

[43] J. P. Knapp and M. F. Doherty, "A New Pressure-Swing Distillation Process for Separating Homogeneous Azeotropic Mixtures," *Industrial & Engineering Chemistry Research*, vol. 31, no. 1, pp. 346–357, 1992.

[44] T. C. Frank, "Break Azeotropes with Pressure-Sensitive Distillation," *Chemical Engineering Progress*, vol. 93, no. 4, pp. 52–63, 1997.

[45] Y. Tanabe, J. Toriya, M. Sato and K. Shiraga, "Process for Preparing Tetrahydrofuran," U.S. Patent 4,093,633, 6 June 1978.

9 Liquid-Liquid Extraction: Employing Molecular Interactions Alone

We have already co-authored a review of liquid-liquid extraction technology in Section 15 of *Perry's Chemical Engineers' Handbook* [1]. We refer readers to that book as well as refs. [2, 3] and other publications listed in the Bibliography for Chapter 9. Here, we offer some additional comments. Liquid-liquid extraction is a sub-topic of solvent extraction in general, which also includes liquid-solid extraction or leaching. Leaching technology involves many of the same process concepts and methods of analysis, but mass transfer rates often are slower and specialized equipment is required to handle solids. Discussions of leaching technology are available elsewhere [4].

As discussed earlier, distillation is the workhorse method for separating liquid feeds – except when component volatilities are too close, components are not sufficiently volatile, or the required process is too energy-intensive. It is in these special cases that liquid-liquid extraction, the transfer of solute between two liquid phases instead of liquid and vapor, may be an attractive alternative (among other separation methods). With liquid-liquid extraction, an immiscible or partially miscible solvent is added to the feed to transfer select components from the feed phase to the solvent phase. The spent feed is called the raffinate. The loaded solvent is called the extract. Additional complications with this approach include the need to store and handle an added solvent plus the need for extra operations for product isolation (from the solvent), product purification, and solvent recovery and recycle – operations that often involve distillation. In fact, in many cases liquid-liquid extraction may be viewed as a way of enabling or enhancing the use of distillation. In these cases, the desired product is first recovered and concentrated using extraction to generate a new feed phase from which the product can more easily be isolated and purified by using distillation. The ultimate example of this is the process involving addition of a solvent directly to a distillation column, as described in Section 8.2: Extractive Distillation.

Furthermore, in many ways, liquid-liquid extraction is analogous to distillation. The mass transfer design equations look much the same because both extraction and distillation involve diffusion of solute from one fluid phase to another. In fact, standard liquid-liquid extraction design equations involving dilute feeds are exactly the same as the equations used for distillation stripper design. Both involve straight equilibrium and operating lines, and the solute transfer factor (the extraction factor) in liquid-liquid extraction ($\varepsilon = m \times S/F$) is completely analogous to the stripping factor in distillation ($S = m \times V/L$). So both the Kremser-Souders-Brown theoretical-stage equation and the Colburn transfer unit equation, described in Section 6.6.1: Theoretical Stage Relationships and in Section 6.7.1: Mass Transfer Units and Mass Transfer Coefficients, may be used for extraction. And, McCabe-Thiele-type graphical calculations also may be used including calculations to evaluate the use of reflux. But unlike distillation, the thermodynamic driving force for liquid-liquid extraction is entirely due to

https://doi.org/10.1515/9783110695052-014

non-ideal interactions of the feed components, as reflected in differences in their activity coefficients in liquid solution. In distillation, the driving force depends on pure-component vapor pressures as well as activity coefficients (see Section 5.5.1: Vapor-Liquid Equilibria and Section 5.5.2: Liquid-Liquid Equilibria). So, unlike liquid-liquid extraction, distillation may be effective even when handling a completely ideal feed mixture so long as pure-component vapor pressures are sufficiently different (and the desired product is sufficiently volatile). And of course, the hydraulics of the two processes are very different due to different densities and viscosities of vapor versus liquid, so mass transfer rates and equipment capacities can be considerably different. Another key differentiator is the strong influence of interfacial tension on extractor performance. Although vapor-liquid surface tension is indeed a factor in distillation performance, it is not generally a focus of design. But in the design of a liquid-liquid extractor, one of the first things to do is to measure liquid-liquid interfacial tension because this has a tremendous impact on liquid droplet dispersions, and for that reason, on the choice of extractor type and its design. These factors are reviewed in the references cited above.

Because distillation and liquid-liquid extraction are analogous in the ways just described, similar types of process schemes may be used. Standard extraction is analogous to a top-fed stripping-mode distillation column, and fractional extraction is analogous is some respects to a middle-fed distillation column. A fractional extraction is a middle-fed process with two distinct sections for purification of product and for obtaining high product recovery (washing and stripping sections), and these may be compared to a middle-fed distillation with analogous rectification and stripping sections. Differences have to do with whether reflux is employed (not common in extraction operations) and the types of equipment that are most effective for generating high interfacial area. With extraction, the implementation of a fractional extraction scheme often involves a cascade of multiple extractors, whereas in distillation a single column more often than not will suffice.

Over the years, industrial liquid-liquid extraction has developed into two general schools of practice:

– Chemical-reaction-driven extraction (or reaction-enhanced extraction) involving significant change in the chemical form of the solute species in solution as it transfers from the feed phase into the product phase. This category is a key technology area of hydrometallurgy [5].

– Solvation-driven extraction involving transfer of solute from the feed phase to the solvent phase with little or no change in the solute's general chemical form. Applications may involve simple change from ionic to non-ionic form, but without change in the main functional groups or chemical moiety comprising the product solute [1]. This category is a key technology area of the chemical process industries.

Because chemical reaction and complexation are key to chemistry-driven extraction, they tend to be the focus of extraction courses taught in industrial chemistry departments.

Hydrometallurgy is a highly successful area of commercial practice. A well-known application involves recovery of copper metal from acidic leachate solution using hydroxyoxime extractant dissolved in kerosene. On the other hand, solvation-driven extraction tends to be the focus of extraction courses taught in chemical engineering and process technology departments and practiced by the petrochemical and specialty chemical industries. Well-known applications include the extraction of aromatic components out of aliphatic feeds by using heavy, polar aprotic solvents, and the extraction of penicillin out of aqueous fermentation broth by using amyl acetate solvent. Textbooks tend to focus on one or the other of the two schools of extraction. The distinction between these two areas of practice is somewhat arbitrary, of course, as both involve the same basic concepts of phase equilibrium and diffusion-based mass transfer. However, the various commercial applications, specific jargon used by practitioners, and popular types of equipment often are different. Practitioners should become familiar with each approach because opportunities for improvement may become evident by comparing the best practices of each. For example, ideas for improvement may come from combining concepts of fractional extraction as practiced in hydrometallurgical applications with the use of efficient equipment popularly employed in petrochemical operations.

Note that some of the terms used by the two schools have different meanings, so this can be a source of confusion. For example, hydrometallurgy involves first extracting metal ions from an acidic leachate solution into an organic phase containing a dissolved extractant. The extractant forms a chemical complex with the metal ions which enables their transfer away from the aqueous feed into the organic phase. The term stripping in this context refers to transfer of the metal ions from the loaded organic solvent phase (the extractant phase) into a clean aqueous phase (the electrolyte phase) where the metal can be recovered via electrolysis (electrowinning) or another method. In the chemical process industries, stripping is a more general term that refers to the extraction of product from a feed of any kind (aqueous or organic) into an immiscible or partially miscible solvent phase (organic or aqueous). A stripping operation such as this normally does not yield a pure product solute as co-solutes are extracted along with the desired solute. A fractional extraction process including both stripping and washing sections is needed for solute purification [1].

Because hydrometallurgy depends on formation of an extractant-solute complex in solution, the mass transfer rate may be controlled by relatively slow reaction kinetics, so in many applications mixer-settler-type equipment is used to provide long contacting times on the order of 10 min per stage, and processes often involve many mixer-settler vessels (stages) in series. It is not uncommon for these vessels to be rectangular in shape for space efficiency. Also, unlike petrochemical processes, hydrometallurgy commonly utilizes fractional extraction probably because the partition ratios and extraction factors employed in hydrometallurgy generally are not far from unity. In petrochemical applications, on the other hand, liquid-liquid extraction tends to be used for recovery of a desired product from a crude feed (stripping-mode extraction), and then another separation method is used to purify the recovered product. Petrochemical applications also tend to utilize column-type extractors, and if mixing is used, it is with efficient mixing designs in more optimal vessels (no tanks with square corners).

In any case, a liquid-liquid extraction operation generally is designed to generate an unstable dispersion that can easily be settled, coalesced, and decanted after extraction. The design process involves determining the interfacial tension between the two liquid phases, as this not only determines droplet size but it also controls the ease with which droplets form and settle. So knowing interfacial tension helps in deciding which phase should be dispersed and what type of contacting equipment to use. Options include static column extractors with either packing or sieve trays (for lower interfacial tension systems), mechanically agitated columns (generally for higher interfacial tension systems), mixer-settlers, and centrifugal extractors. In addition to whether interfacial tension between phases is high or low, the choice of equipment depends on requirements for residence time (either long contact time due to slow kinetics or short contact time due to rapid degradation of solute at harsh extraction conditions), and the number of stages or transfer units needed for high recovery and/or purity (depending on the partition ratio and extraction factor). See ref. [1] for detailed discussion of these and other factors.

References

[1]   T. C. Frank, B. S. Holden and A. F. Seibert, "Section 15: Liquid-Liquid Extraction and Other Liquid-Liquid Operations and Equipment," in *Perry's Chemical Engineers' Handbook*, 9th ed., D. W. Green and M. Z. Southard, Eds., McGraw-Hill, New York, 2019.
[2]   T. C. Lo, M. H. I. Baird and C. Hanson, *Handbook of Solvent Extraction*, Wiley-Interscience, New York (reprinted by Krieger, Malabar, Florida, 1991), 1983.
[3]   J. C. Godfrey and M. J. Slater, *Liquid-Liquid Extraction Equipment*, Wiley, New York, 1994.
[4]   J.-P. Duroudier, *Liquid-Liquid and Solid-Liquid Extractors*, Elsevier, Amsterdam, 2016.
[5]   M. Nicol, *Hydrometallurgy: Theory, Vol. 1, and Practice, Vol. 2*, Elsevier, Amsterdam, 2022.

10 Adsorption-Based Processes: Adding an Active Solid Surface

Adsorption and desorption are ongoing dynamic processes found everywhere in nature. The biological functioning of plant and animal cells involves highly selective binding of biomolecules within cells and tissue. And adsorption/desorption of water, minerals, and organic matter within soil, driven by changing weather conditions and the action of soil microbes, greatly affects the bioavailability of nutrients and the growth of plants. Adsorption/desorption also is involved in the formation of surface chemistries (formation of specific molecular groups at solid surfaces) and the interaction of moisture with these surfaces, phenomena that determine many of the characteristics we associate with naturally occurring solids. Think of water beading up on the hydrophobic surface of wax-covered plant leaves and contrast that with water wetting the surface of hydrophilic clay particles, making clay-rich soils moist and slippery in humid conditions.

Over the centuries, people have learned to exploit adsorption phenomena for many practical purposes. Milestones in the development of adsorption as a separation process are listed in Section 3.2: Pioneers. Examples include the use of naturally occurring clay and zeolite minerals to remove color and moisture from oily liquids, and the use of charcoal to remove color and improve the taste of drinking water. The use of charcoal to purify water is said to date back as far as 3750 BC in ancient Egypt and Mesopotamia [1]. People have also developed non-separation applications involving adsorption and surface reaction. These include various chemical treatments to passivate metals to minimize corrosion such as bluing and parkerizing of steel [2] and to minimize wear of pressure-bearing metal surfaces as in zinc dialkyldithiophosphate treatment [3]. The deliberate modification of surface chemistry has also been used to improve the surface properties of naturally occurring adsorbents, as in the making of activated carbon [1]. Similarly, adsorption and surface chemistry are critical in the manufacture and functioning of solid catalysts used to promote desired reactions. In fact, molecular adsorption may be viewed as a kind of surface reaction, so methods used to model solid-catalyzed reactor performance may also be applied in adsorber design.

The large-scale industrial application of adsorption for chemical separation appears to have its beginnings in the late eighteenth century when solids began to be used in large volumes to improve liquid products – by removing unwanted colors or off-putting odors and flavors [4]. One example is the use of charcoal in the English sugar refining industry to remove color from raw sugar solutions (around 1794) [5]. Another is the famous Lincoln County Process for making Tennessee whiskey, which includes a charcoal filtration (or steeping) step prior to aging the whiskey in new charred oak barrels. Introduced around 1825 or so, this processing step was developed to remove grainy or harsh odors and flavors, possibly following on previous practices

https://doi.org/10.1515/9783110695052-015

by others in the craft of making whiskey. Today we recognize sugar decolorization and whiskey steeping as examples of the use of adsorption to remove organic colorant compounds in the first case [6] and to remove complex ketones, esters, and organic acids in the second [7]. But at the time these processing steps were first developed, the underlying mechanisms were not known. That understanding came much later. The concept of molecular *adsorption* as distinguished from *absorption* was not introduced until 1881 when Emil du Bois-Reymond suggested the term to Heinrich Kayser who then used it in his writings [8–10].

The capability of an adsorption process depends first and foremost on the properties of the adsorbent material. Today, adsorption is widely used in industrial processing due to the wide variety of available adsorbents, including naturally occurring solids, modified or *activated* versions of naturally occurring solids, and synthetically derived solids. These materials are used to process gases and liquids of all kinds – aqueous, organic, and inorganic – and for removal or recovery of both small (volatile) and large (non-volatile) molecules. Most applications involve purification of the feed by preferential adsorption of impurity components present in small concentrations; however, adsorption is also used to recover desired products present in larger concentrations and for bulk separations. Either thermodynamic or kinetic factors may control adsorption performance, and normally both are involved to some extent. The fundamentals of thermodynamic equilibrium are discussed in Section 5.5.4: Adsorption Isotherms. Adsorption kinetics are discussed in Section 6.7.3: Adsorption Dynamics: Breakthrough Curves.

10.1 Adsorbent Materials

Adsorption involves the selective accumulation of molecules onto the surface of highly porous solid particles in the form of spheres, granules, or pellets. Solid surface area can be very high, typically between 100 and 1200 m^2/g. The great majority of the surface area exists within the internal pore structure and not at the exterior of the particles, so one might say that feed molecules are adsorbed *into* the porous particles. However, a clear distinction is made between the term *adsorption*, which refers to attachment of molecules at the solid surface existing within the pores, and *absorption*, which refers to dissolution and spreading of solute within a bulk phase.

Table 10.1 lists the characteristics of two main categories of adsorption: physical adsorption and chemical adsorption. Physical adsorption has been further classified into five major types [12]. In this book, we focus on Type I physical adsorption where adsorption capacity increases linearly with solute concentration (or partial pressure) at low solute concentrations. Then at higher concentrations, capacity begins to plateau toward a constant saturation amount. This is the most common type of physical adsorption behavior. Chemical adsorption, on the other hand, depends on the specific reactivity of the adsorbing molecules and the solid surface. In many cases, chemical adsorption capacity

Table 10.1: Comparing Physical Adsorption with Chemical Adsorption. Adapted from ref. [11], Copyright 2020. Reprinted with permission from Elsevier.

Adsorption category	Physical adsorption	Chemical adsorption
	Van der Waals interaction	**Chemical bond**
Selectivity	Non-selective adsorption (within a given class of compounds)	Selective adsorption
Adsorption layer	Single or multiple layers	Single layer, titration of active sites
Adsorption heat	Low	High*
Adsorption rate	Fast	Slow
Stability	Unstable, easily reversed at elevated temperature or at reduced pressure.	Stable, not easily reversed (a kind of titration reaction)

* May require elevated temperature to overcome significant activation energy.

can be related to the number density of active sites at the solid surface, and so chemical adsorption may be considered a type of non-reversible titration.

Adsorbents generally are classified according to the bulk material from which they are made. The most widely used materials include carbon, alumina, silica, aluminosilicate, and synthetic polymers. Box 10.1 lists two main categories of these materials: naturally derived and synthetic, plus some comments about their preparation or activation. Box 10.2 lists examples for some of the more widely used adsorbents and typical applications. Further information is widely available in the literature. For example, see refs. [13, 14].

Fundamentally, the surface chemistry and inner structure or porosity of a solid material determines its behavior as an adsorbent, and this can vary greatly even for the same bulk material. Adsorbent properties can also vary significantly according to the presence or absence of impurities. For example, coal-derived carbon typically contains trace metals that can catalyze unwanted trace reactions such as hydrolysis or oxidative decomposition of organic substrates. So it is important to be aware of the presence of impurities and their potential impact. The composition of the feed phase can also affect adsorption behavior. That is, adsorption capacity depends on the affinity a solute molecule has for the feed phase relative to its affinity for the adsorbent's solid surface. For example, if a solute molecule is made to form stronger interactions with a liquid feed mixture by using a better feed solvent or by ionizing the solute in solution, then the capacity for adsorption of solute from out of the liquid and onto the solid surface likely will be reduced. For more discussion of adsorption capacity, see Section 5.5.4: Adsorption Isotherms.

Many commercial adsorbents are manufactured to meet standard specifications for particle shape and size (particle size distribution), internal surface area and porosity (pore size distribution), and certain properties associated with specific surface chemistries. Naturally, different kinds of adsorbents will have different standard specifications and different applications. And adsorbent manufacturers often provide specification sheets listing key properties and typical applications. Examples are listed in Box 10.3. Interested readers can search the internet for more information. Specification sheets are useful guides to identifying candidate adsorbents, but in most cases the user will need to generate their own design data by measuring adsorption capacity and by running dynamic breakthrough tests using representative feed. For more information, see Section 5.5.4: Adsorption Isotherms, Section 6.7.3: Adsorption Dynamics: Breakthrough Curves, and Section 10.4: Fixed-Bed Design Methods.

Box 10.1: Typical Adsorbent Materials

- Naturally Derived Raw Materials (modified to make adsorbents)
  - Coal, with high levels of trace metals and ash.
  - Woody plants such as coconut shell, with lower levels of trace metals and ash.
  - Petroleum, with very low levels of trace metals and ash.
  - Natural zeolite mineral. Microporous, volcanic hydrated aluminosilicate minerals of alkaline and alkaline-earth metals with negatively charged solid surfaces within pores of well-defined structure (size). The dimensions of the zeolite framework varies depending on the metal and conditions of formation.
  - Clay mineral. Hydrated alumina/silica structures, often containing substituted potassium and calcium, with Bronsted and Lewis acid sites.
  - Silica mineral.
  - Sodium bicarbonate.
  - Calcium chloride (adsorbent and absorbent)
- Activation Treatments (to increase active surface area and modify surface chemistry)
  - Various treatments may be used to alter the surface chemistry, increase porosity and internal surface area, and obtain a desired particle size distribution.
  - Activated carbon (from coal or plant materials). Made by processing material with hot gases (usually air or steam) to greatly increase surface area and alter surface chemistry from partial oxidation or hydrolysis, increasing porosity and surface area and forming oxide or hydroxide surface groups. In general treatments with various hot gases can form various surface functional groups such as $C = O$, C-O, $-CO_3$, and C-H, and O-H.
  - Activated alumina. Made from aluminum hydroxide by heating in a furnace or kiln to remove bound moisture (dehydration or dehydroxylation) to form a partially dehydrated alumina with surface hydroxyl groups.
  - Partial pyrolysis of carbon materials. Affects pore structure, surface area, and surface properties.
  - Chemical treatment. Using an acid, base, oxidant, or another specific reagent designed to obtain desired surface groups by reaction with the surface. Used to obtain specific surface properties and molecular surface structures, or for regeneration of the adsorbent.
- Synthetic Materials
  - Silica gel (partially dehydrated form of hydrated silica).
  - Synthetic zeolites. Aluminosilicates with metal cations. These are called molecular sieves due to highly uniform pore structures. Different synthesis conditions and cations alter pore size, and the resulting material may be weakly acidic or basic. Zeolites that are commonly used for separations include: 3A with 3 Å (0.3 nm) pore size and 13x with 10 Å (1 nm) pore size.
  - Polymeric beads (styrene divinylbenzene, acrylonitrile, acrylic, . . .).

- Pyrolyzed polymeric beads (carbon molecular sieve).
- Ion exchange media (cationic or anionic styrene divinylbenzene).
- Supported thiourea on polymeric beads, for adsorption of trace metals.
- Affinity resins, supported ligands for specific protein adsorption.

Box 10.2: Popular Adsorbents and Typical Uses

- Activated carbon (granular, pelletized, powder). Removal of organic vapors from air and other gases. Removal of organics from water. Odor control. Taste control. Food processing. Medical uses. Respirators.
- Activated alumina. Removal of moisture from air and other gases. Removal of moisture and contaminants from oil. Removal of fluoride, arsenic, and trace metals from water.
- Silica gel. Removal of moisture from air and other gases. Removal of gases and vapors from air. Food processing.
- Fuller's Earth. Clay material. Used to remove color from fats and oils and other liquids.
- Sodium bicarbonate powder. Removal of acidic vapors. Odor control.
- Zeolite 3A molecular sieve: weakly acidic, 3 Å (0.3 nm) pore size. Removal of moisture from gases and liquids. Separation of gases.
- Carbon molecular sieve beads (for example, from pyrolysis of polyvinylidene chloride beads). Separation of gases.
- Thiourea supported on polymeric resins. Removal of trace metals from liquids.
- Affinity ligand resins for protein separations. Chemically treated to form specific molecular groups (ligands) at the solid surface. Highly selective recovery of proteins.
- Calcium chloride beads (adsorbent/absorbent). Removal of moisture from air and other gases, eventually transitioning from a hygroscopic solid bead to a deliquescent aqueous liquid. Also used to melt ice by formation of a low-freezing-point liquid brine.

Box 10.3: Typical Specifications for a Commercial Adsorbent

- Particle shape and size: such as 4 × 10 mesh granules, 3 mm (1/8 inch) square pellets, or 2.5–3.0 mm diameter spheres.
- Packed bulk density (lb or kg per unit volume).
- Crush strength (weight of bed, lb or kg per cross-sectional area, before material begins to crush).
- Hardness: Characterizes the particle's ability to resist dust formation over time.
- Pressure drop curve: For a given flow rate, the pressure drop per cross-sectional area per unit height of bed. Reported for a range of flow rates. Needed for sizing an adsorber bed.
- Total ash (%). For carbon, the fraction of non-carbon minerals in total mass of adsorbent.
- Trace metals. For example, coal-derived carbon contains various trace metals.
- Iodine No. Characterizes surface area and porosity of carbon.
- Surface area per unit wt. or unit volume, by Brunauer-Emmett-Teller (BET) N_2 adsorption method.
- Porosity. Characterization of micro-, meso-, and macro-sized pores. For example, information may be given in terms of the Barrett-Joyner-Halenda equation for pore sizes of 1.7–100 nm or mercury porosimetry results for pore sizes below 1.7 nm.
- Statement about general suitability: Example: May be used to adsorb polar organic compounds from aqueous solution. Or, may be suitable for adsorbing hydrocarbons under appropriate conditions.
- Specifications may include information about surface functional groups, but normally this kind of information is not specified.

10.2 General Process Features

In addition to the development of numerous adsorbent materials over the years, many adsorption process schemes have been developed for cost-effective use of these materials. Most processes employ a packed column called a fixed-bed adsorber. A typical example is shown in Figure 10.1. A similar design is described in ref. [16]. Most adsorption processes involve cyclical operation of two or more fixed-beds, with on-line (in-place) regeneration of the beds at either elevated temperature (for processes treating gases or liquids) or under reduced pressure (for processes treating gases or volatile liquids that have been evaporated into the gas phase). A typical process utilizes two or more fixed beds. While one bed is on-line, the other bed or beds are undergoing various regeneration steps. After a pre-determined time period, the on-line bed is switched off-line for regeneration, and another, newly regenerated bed is switched back on-line to take its place. In this way, the cyclical, multi-step operation is able to handle a continuous feed stream. In-place regeneration of adsorbent is desirable because it avoids any need to remove and replace adsorbent with new material, an expensive and time-consuming proposition. Still, some specialty processes that process very dilute feeds for extended on-line times are designed for off-site regeneration, especially those involving removal of low-volatility impurities requiring high-temperature thermal regeneration of the adsorbent or those involving irreversible chemical adsorption. An example involves adsorption of non-volatile impurities from wastewater using carbon adsorbents that require high-temperature furnaces for regeneration. Sometimes, especially for small volume operations, entire vessels filled with adsorbent particles are exchanged with fresh vessels containing newly regenerated adsorbent such as with the Calgon Carbon VENTSORB® process [17]. The used vessels are shipped off-site for regeneration at a regional facility. In some cases, the spent adsorbent is not regenerated at all, but instead is sent to landfill or to incineration for disposal. Specialized regeneration process schemes also may be used [17]. An example is the pulsed-bed adsorber. With this design, a portion of the adsorbent is periodically removed from the bottom inlet-end of a vertical bed, and fresh adsorbent is added to the top – in a "pulse" amount that is only a portion of the entire bed volume. This can be an efficient alternative to periodic removal of all the adsorbent, when removal of spent adsorbent is necessary. Also, chemical treatments sometimes are used for regeneration, such as acid and base treatment (or salt treatment) of ion exchange resins, and chemical treatment may be used for ligand regeneration of affinity adsorbents for protein separations. Whenever periodic regeneration of the adsorbent proves difficult, the use of liquid-liquid extraction to remove contaminants from liquids may be an attractive alternative [18]. For more discussion of adsorption processes, see Perry's Handbook [12] and the book by Chi Tien [19].

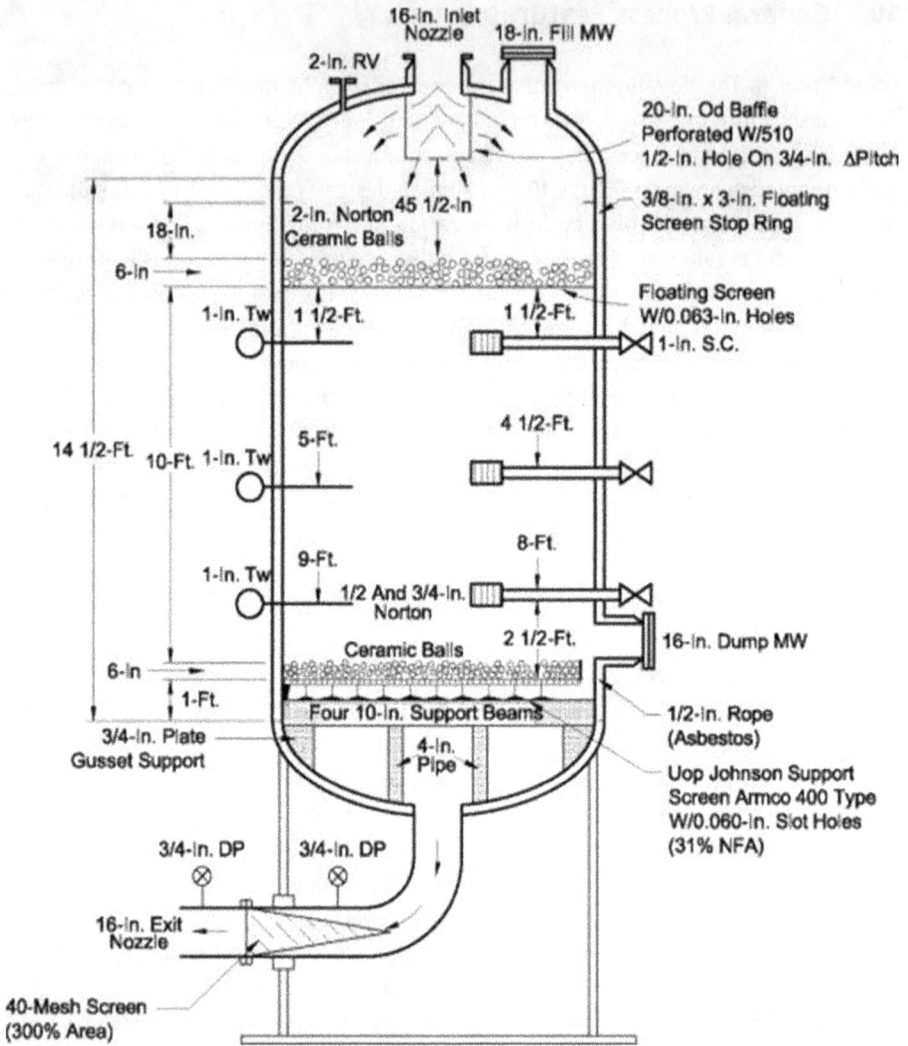

Figure 10.1: Typical Vertical Fixed-Bed Adsorber Internals. Shown with graded-ball type flow distributors at each end of the adsorbent bed. Taken from ref. [15]. Reprinted with permission from Gulf Professional Publishing, copyright 2014, Elsevier.

10.3 Continuous Versus Batch Operation

Continuous adsorption involves continuous flow of a gas or liquid through a stationary or fixed bed of solid adsorbent particles (or a porous monolith); or, in rare cases, through a moving bed of solid particles (more on this approach later). As discussed above, a continuously fed, fixed-bed process normally is designed as a cyclical process that handles a continuous feed stream by periodically switching two or more adsorbent beds on- and off-line, alternating between on-line processing and off-line regeneration steps. Batch liquid-solid adsorption consists of thoroughly mixing the solid particles with a batch of liquid using a specific liquid-to-solids ratio, allowing the mixture to equilibrate for some time (sometimes called steeping), and then separating the liquid from the solids by settling the solids or by filtering them from the liquid. Batch adsorption may also be used to treat gases by adding packets of solids to the volume of gas to be treated. For example, activated carbon and silica gel packets are often added to small packages of product to adsorb moisture and odors and extend shelf life. Although the use of batch adsorption has a number of commercially important applications, especially in packaging applications, batchwise operation generally is not the best approach to use for large-scale chemical separations. Here, continuously fed, fixed-bed adsorbers generally are best because they can operate at higher adsorption capacity. This is because batchwise contactors necessarily operate at the low end-point solute concentration within the product liquid or gas. Think of a perfectly mixed CSTR reactor which must operate at the low solute concentration in the outlet stream, not the higher concentration in the feed. On the other hand, continuously fed fixed-bed contactors operate at the solute feed concentration, a concentration higher up on the isotherm where adsorption capacity is greater. This argument assumes preferential equilibrium adsorption (physisorption) is the fundamental adsorption mechanism, not irreversible chemisorption. In some applications the adsorption mechanism is indeed reactive chemisorption, and in that case the adsorption capacity may be independent of the flow configuration, as it is essentially a titration of active sites. But in the majority of large-scale industrial applications, continuously fed fixed-bed design has an inherent advantage of higher capacity compared to batchwise design. For our purposes, then, we will focus on fixed-bed design. Batchwise design is covered elsewhere [12, 20].

With a continuously fed, fixed-bed process, the fixed bed develops a mass transfer zone (MTZ) that moves along the length of the bed as adsorption proceeds in time. Adsorption occurs at the leading edge of the MTZ where solute concentration is highest (equal to the feed concentration). The MTZ forms an adsorbed solute concentration profile that mirrors the effluent breakthrough curve, as shown in Figure 6.4. It is the moving section of the adsorbing profile, and it generally takes on a constant shape and length as it moves down the bed.

From a mechanistic point of view, a typical adsorption process involves three overall steps: 1) contact of feed fluid with stationary solid adsorbent particles held within a fixed bed; 2) diffusion of solute from the fluid bulk phase, through a boundary layer at the fluid-solid interface, and into the pores of the stationary adsorbent; and 3) molecular interaction with a specific surface adsorption site (molecular adhesion involving surface chemical reaction or weaker Van der Waals physical interactions). The surface reaction may be essentially irreversible (when strong chemical bonds are formed); or, more commonly, it may be reversible with change in physical conditions (temperature, pressure, or chemical composition of the bulk fluid). The latter is the basis for in-place regeneration of the adsorbent in a multi-step cyclical process scheme. In applications involving in-place regeneration, the process may be considered a "time-averaged" steady-state process, or a cyclic steady-state process. In other words, when looking at the process over multiple cycles, the solute material balance may conveniently be modelled using standard steady-state, countercurrent design equations (for example, see Section 10.6: Pressure Swing Adsorption).

10.4 Fixed-Bed Design Methods

A fixed-bed adsorber is simply a column filled with stationary adsorbent particles. It may be oriented in horizontal or vertical fashion. Feed, either liquid or gas, is introduced at one end of the column, a portion of the feed is retained by the bed, and the remaining feed exits the bed. This continues until the bed becomes saturated with adsorbed components (or nearly so), at which point these components begin to break through into the exit stream. A useful concept for characterizing the capacity of such a bed is the concept of bed volumes to breakthrough (BVTB). This refers to the number of empty bed volumes of feed liquid or gas that can pass through the bed before the bed becomes saturated and adsorbate begins to breakthrough into the effluent stream. It is a normalized term that is independent of the actual bed size. For background, see Section 6.7.3: Adsorption Dynamics: Breakthrough Curves.

There are many decisions to be made in designing an adsorber that affect BVTB and the shape of the breakthrough curve. Besides the choice of adsorbent and its particle size, perhaps the most important of these is the face velocity (or superficial velocity) of the feed. This refers to the volumetric flow rate of entering feed divided by the vessel's cross-sectional area. Too fast gives too little time for solute to diffuse into adsorbent. Too slow gives poor productivity. So, an intermediate velocity must be chosen for optimal design. With this in mind, the designer must decide whether to specify a short and fat bed or a tall and skinny one (for the same volumetric feed rate and equal volumes of adsorbent) and what the adsorbent particle size should be. Small particles increase the sharpness of the breakthrough curve but at the expense of greater pressure drop through the bed. Another consideration is the particle shape and size distribution. Use of small monospheres with a narrow size distribution can sharpen the breakthrough curve significantly, but the required adsorbent can be costly. Yet another consideration is the uniformity of the feed flow entering the front of the bed. The design of the feed distributor to obtain uniform flow is important in maximizing the capacity of the bed; that is, in extending the operation of the bed before breakthrough occurs.

The adsorber design also depends on the level of breakthrough that is acceptable. In other words, where on the leading edge of the breakthrough curve does the concentration exceed the maximum allowable value? Is it 1 ppm or 0.1 wt%? Other key design decisions have to do with how long the bed needs to remain on-line before being switched off-line to be regenerated, as well as how the bed is to be regenerated.

10.4.1 General approach to design

Analyzing the on-line performance of an adsorber often is separated into two parts: 1) calculation of "idealized" performance which is determined solely by the equilibrium adsorption capacity; and 2) calculation of a correction term that takes into account mass transfer resistance and deviation from plug flow – the spreading of the breakthrough curve around $BVTB_{50}$. (This is the terminology introduced in Section 6.7.3: Adsorption Dynamics: Breakthrough Curves.) This high-level view of the design problem often is the focus of textbook discussions, but it is not the whole story.

For illustration, let's first consider the design of a simplified fixed-bed adsorption process used to remove trace contaminants from either a liquid or gaseous feed. In this case, we can neglect thermal effects and focus on the general aspects of adsorber design and specification of an overall process scheme. We will assume "self-sharpening" Type I adsorption which is a common type and focus on adsorption of a single solute. With multiple solutes, the same basic approach to design may be used, but it is important to appreciate the fact that the presence of a more strongly adsorbing component can greatly reduce the capacity for adsorption of a more weakly adsorbing component. Simply taking single-solute isotherms and using them together to calculate multi-solute adsorption capacity is not valid. Multicomponent adsorption capacity must be determined in testing with multiple solute feeds. For more discussion, see Section 5.5.4: Adsorption Isotherms.

With these simplifications, the design of a single-trace-component adsorption process involves the following considerations:

- Working adsorption capacity (dynamic, reversible, physical adsorption).
- Face velocity and required bed dimensions.
- Hydraulic considerations such as feed distribution and pressure drop.
- Adsorbent regeneration method.
- Process configuration. Number of beds and piping diagram.
- Cyclical operation control scheme.
- Valve selection.
- Waste management.

Boxes 10.4 to 10.6 describe these considerations in more detail along with typical design goals and general recommendations. The design work process includes screening candidate adsorbents and measuring breakthrough curves. It also involves deciding on the time period that an adsorber needs to be on-line as well as deciding on the allowable maximum concentration of adsorbate in the effluent at breakthrough. The required on-line time period often is determined in large part by the time required to regenerate a bed off-line and return it to operation. For regeneration at elevated temperatures, this normally requires many hours. For regeneration at reduced pressure, it may only require minutes. Based on these requirements, work can proceed to determine the adsorber dimensions and a suitable cyclical process scheme including regeneration.

Box 10.4: Fixed-Bed Design Considerations

–   Equilibrium or Kinetic (Dynamic) Capacity. Selecting an adsorbent with adequate working (regenerable) adsorption capacity and service life (with good availability at an affordable price). This normally is determined by the equilibrium adsorption capacity as a function of temperature (for liquids and gases) and pressure (for gases). The working capacity is the reversible capacity after repeated cycles of regeneration, not necessarily the initial capacity or that reported by the supplier for fresh adsorbent. A dynamic capacity or kinetic capacity is also possible. This involves elution of different solutes at different rates through a packed adsorber column, determined in large part by different rates of diffusion through the adsorbent's internal pore structure.
–   Mass Transfer Resistance. This refers to quantifying the shape of the solute breakthrough curve – as a function of adsorbent particle size and flow rate through the bed at full scale operation. The $BVTB_{50}$ value represents adsorber capacity with zero mass transfer resistance.
–   Face Velocity and Required Bed Dimensions. Determining the bed diameter and height needed to deliver the required on-line time. For a given production rate, smaller bed diameters result in higher face velocities, longer mass transfer zones, and therefore larger bed heights. In all cases, the adsorbate concentration in the effluent must be kept below a maximum allowed breakthrough value.
–   Hydraulic Considerations. This involves designing for uniform feed distribution while avoiding excessive pressure drop through the vessel.
–   Adsorbent Regeneration method. Selection of an appropriate adsorbent regeneration scheme (preferably in-place regeneration).
–   Process Configuration. Specifying the number of vessels for the overall process scheme including regeneration and the required interconnecting piping and valves. See Box 10.5.
–   Cyclical Operation and Step Times. Specifying the operating cycle (required on-line time, cycle steps and timing, valve switching schemes).
–   Valve Selection. The choice of the type of valve and its specifications often is critical to achieving long-term reliable performance.
–   Waste Management. In many cases, regeneration of the bed produces a waste stream that must be properly managed in some way (involving recycle or disposal).

Box 10.5: Typical Design Goals

–   Choose a suitable adsorbent. Identification and specification of a suitable adsorbent material, including selection of the general type of adsorbent, its physical properties, and particle size. Selection involves finding the best compromise of factors such as price and availability, pressure drop, adsorption capacity, and options for in-place regeneration.
–   Specify adsorber vessel dimensions, diameter and length. The bed diameter is determined by the maximum effective feed velocity entering the bed (called the face velocity or superficial velocity). If it is too fast, there is too little time for solute to diffuse into the adsorbent internal pore structure. Too slow gives poor productivity (production rate per volume of adsorber vessel). In our experience,

optimum face velocities are roughly 1 ft/sec for gases (0.3 m/sec) and roughly 1 gal/min per ft^2 for liquids (80 liters/min per m^2). These are rough guidelines to use in outlining an initial application. Refinements come with further design calculations and experimentation, as needed.

–   Generate performance test data. Determining the optimum face velocity generally requires access to performance test data. The required adsorber length is determined by the designer's selection of the required online time for the adsorber, which is often determined largely by the time required for off-line regeneration. Use of the methods described above requires generation of experimental data or at least access to some experimental data including some measure of adsorption capacity and some measure of mass transfer resistance (or spreading) which causes the breakthrough curve to spread about a mean breakthrough time. The choice of method will depend on the availability of particular databases and test methods and the preference of the designer.

–   Specify adsorber bed internals. These include the adsorbent (packing) support and feed flow distributor. A feed distributor is needed to insure uniform distribution of feed onto the adsorbent bed. Otherwise, mass transfer performance suffers due to initial spreading of the mass transfer zone. See the discussion given in the text.

–   Specify a suitable regeneration scheme. An in-bed regeneration scheme is highly desirable as the most cost-effective method compared to unloading and reloading new adsorbent. A very long on-line time will be needed to justify replacing adsorbent instead of in-place regeneration (at least several months and often several years capacity for removal of trace components). Options include in-place thermal regeneration, removal of the spent adsorbent for thermal regeneration at another facility, displacement of adsorbates with strong solvents (liquid or gaseous desorbents), or in-place (off-line) reduction in pressure to induce desorption.

Box 10.6: Typical Work Process for Adsorber Design

–   Adsorbent Screening. First, screen adsorbents by measuring equilibrium adsorption capacity in batch tests. After all, working capacity is mostly determined by the equilibrium capacity (the material balance), and less so by the kinetic limitations (which reduce working capacity). See Sections 5.5.4: Adsorption Isotherms and 6.7.3: Adsorption Dynamics: Continuous Breakthrough Curves.

–   Pulse Tests. Run pulse injection tests using a scaled-down test column to assess mass transfer resistance from peak shape. Use exactly the same adsorbent to be used at full scale with the same form (granular or pelletized, etc.), the same pore size characteristics, and the same particle size distribution (4 x 12 mesh, for example). Use adsorbent that has been exposed to solute and regenerated several times using the envisioned regeneration method (in order to assess the practical working capacity of the adsorbent). Run tests at three different face velocities over a typical range expected for commercial operation. In our experience, this typically is 0.5 up to 2 ft/sec for gaseous feeds (up to 0.6 m/sec), and 0.5 up to 2 gal/min per ft^2 for liquid feeds (up to 80 liters/min per m^2). Other face velocities may be appropriate depending on the application. Particle size must be chosen to allow good mass transfer performance with reasonably low pressure drop.

–   Maximum Face Velocity. From among the top candidate adsorbents, select an adsorbent that exhibits good mass transfer characteristics in pulse tests (that is, sharp peak performance) and run a breakthrough test over the range of face velocities. Choose the maximum face velocity that still gives sufficiently sharp peaks. Design the breakthrough test so that the mass transfer zone stabilizes and the test measures the full breakthrough curve. The test does not need to provide the full on-line time envisioned for the full-scale application, just establishment of a stable breakthrough curve and constant pattern.

–   Maximum Allowed Breakthrough. Characterize the breakthrough curve spreading using a suitable model framework (see Section 6.7.3). In doing so, decide what the maximum solute outlet concentration can be.

Normally, this is specified as some percentage of the solute feed concentration; for example, commonly 1% of the feed concentration, or a minimum 100-fold reduction from the feed concentration. This will affect the length of the breakthrough curve to be considered.

– Required On-Line Time. For a given application, choose the on-line time that the design must meet. This requires consideration of the total cycle time requirements including required regeneration time and other operational requirements (manpower availability, vessel size limitations, . . .).

– Theoretical Capacity. Determine the step change limit ($BVTB_{50}$) determined by the equilibrium capacity from the material balance and measured adsorption capacity.

– Actual Capacity. Using the chosen design model as framework, calculate spreading about $BVTB_{50}$ to establish the MTZ bed length needed to meet the on-line time. The actual column length will be longer than the column length at $BVTB_{50}$.

– Pressure Drop. Consider how much pressure drop the overall process can handle as the driving force for flow through the beds. This can affect the choice of particle size and well as adsorber diameter and length.

– Bed Geometry. Discuss short and fat versus tall and skinny as design choices. Which is best (in general)? Tall and skinny may be favored to maximize productivity while maintaining acceptable mass transfer performance using large particles of adsorbent. Short and fat columns may be favored in some cases to minimize the volume of very expensive small-particle media.

– Bed Dimensions. Specify the adsorber dimensions (diameter and length) and the vessel internals (feed distributor design and any packing supports). The total bed length must include the length of the initial distributor section as well as the saturation zone in the middle and the mass transfer zone at the end.

– In-Place Regeneration. Based on test work for regeneration of the adsorbent, determine whether in-place regeneration is practical; that is, does it give sufficient working capacity? This information is then used to develop a suitable regeneration scheme including the required number of adsorber vessels and the required piping diagram. Economic considerations normally favor minimizing the number of vessels as much as possible.

10.4.2 Adsorber dimensions, internals, and piping diagram

As we've already emphasized, mass transfer efficiency is a strong function of face velocity. This is reflected in the well-known van Deemter effect in chromatography, where peak broadening is minimized (mass transfer efficiency is at a maximum) at some moderate range of velocities. Below this optimal range the diffusion path length within the fluid phase approaches the same length as the mass transfer zone, so broadening increases. But this effect occurs only at very low velocities, normally far too low to be of concern in designing a commercial adsorption process. At velocities above the optimal range, however, broadening begins to increase again, now because solute no longer has sufficient time to diffuse into the adsorbent's pores to access all of the internal surface area. So, in practical terms, as face velocity is increased in an attempt to increase adsorber productivity, a point is reached where the fluid phase begins to move past the adsorbent particles faster than it can equilibrate with the particles, and broadening of the mass transfer zone begins to increase with face velocity.

So, in designing an adsorber, it is important to determine the maximum face velocity that gives good productivity while maintaining sharp breakthrough performance. Then the column diameter is set by the required cross-sectional area needed to achieve the specified throughput capacity (face velocity × $(\pi D^2/4)$ = volumetric flow rate). Normally the optimum face velocity is determined experimentally using a miniplant column containing the same adsorbent particles envisioned for the commercial process. The miniplant column diameter should be at least 8 times the particle diameter. See Box 10.6 for recommended test conditions.

Once the optimal face velocity and corresponding adsorber diameter for a given throughput capacity have been determined, the focus shifts to specifying the total adsorber bed length (or height). The required length of an adsorber depends on the required period of time it is to remain on-line. Once this is chosen, the total adsorber length may be conveniently broken down into three separate zones: 1) a feed distribution zone; 2) a saturated zone; and 3) a mass transfer zone:

$$L_{total} = L_{dist} + L_{sat} + L_{MTZ} \tag{10.1}$$

The design process involves specifying the lengths of each zone. The feed distribution length is that needed to contain the feed distributor and uniformly distribute the feed. This depends on the specific design but normally can be less than 0.3 m or so. Of course, additional length generally is also needed at the ends of the adsorber vessel for the vessel heads, piping connections or nozzles, and access ports.

Neglecting the relatively small amount of solute adsorbed within the distributor and the MTZ zones, the saturated zone length may be calculated directly from the material balance:

$$L_{SAT} \approx \frac{C_{in} u_s \Delta t}{q_i(at\ C_{in})\rho_{bed}} \tag{10.2}$$

where C_{in} = solute concentration at the inlet to the bed (grams solute per unit volume)
u_s = liquid velocity at the inlet to the bed (superficial velocity or face velocity)
Δt = on-line time
q_i = adsorption working capacity (grams solute adsorbed per gram of adsorbent)
ρ_{bed} = adsorbent bulk density (grams adsorbent per unit volume)

In most applications, the adsorber is designed so that the saturation zone is much longer than the other zones. In that case, a very rough initial estimate of total adsorber length may be obtained simply by estimating the degree of adsorbent saturation. For example, if we estimate that L_{SAT} represents something on the order of 75% of total required bed length, then the total adsorber length is simply $L_{SAT}/0.75$. This approach can be used for early estimation purposes, but not for the final design of an adsorber.

As discussed above, the mass transfer zone length is a strong function of operating conditions, especially face velocity. It is also affected by the adsorbent particle shape, size, and porosity. The MTZ may be characterized and modeled using a number of methods that rely either on statistical residence time distribution models, kinetic rate expressions, or phenomenological models – normally assuming self-sharpening type I adsorption behavior. These methods are reviewed in Section 6.7.3: Adsorption Dynamics: Breakthrough Curves. See Figure 6.4. It is commonly observed that the length of the MTZ is constant (or nearly so) and independent of the length of the saturated zone. This differs from pulse injection chromatography involving movement of solute peaks through a column. In GC and LC chromatography, peak shape tends to narrow with added column length; whereas, for a continuously fed adsorber, the s-shaped breakthrough curve (the MTZ) normally takes on a constant shape. In that case, it appears that the length of the saturated zone in front of the MTZ does not matter because once the liquid velocity profile is established, solute passes through the saturated zone as if it were completely inert; the adsorption sites are saturated at the prevailing operating conditions. In other words, the solute undergoing adsorption within the MTZ has no connection to the length of the saturation zone. On the other hand, in chromatography, an injected peak travels through a column that is nowhere saturated; so chromatographers can normally improve resolution by increasing column length.

Flow Distributors. Uniform distribution of feed gas or liquid to the front of the adsorbent bed is critical for getting the best performance out of the adsorbent, particularly in adsorbers with short bed lengths. In our experience, it seems that many existing adsorber designs may be improved in this regard. This is the same issue that designers faced in the 1980s when it became clear that better liquid distributors were needed to improve the performance of packed distillation columns. In the case of adsorber design, attention paid to ensuring uniform distribution of feed can significantly extend the on-line time of the adsorber before breakthrough. Typical adsorber feed distributors are of several types:

- Graded-ball distributor. Spherical media, all of the same size or of graduated sizes, placed in layers on top of a bed support screen, as in Figure 10.1.
- Lateral style distributor. These are pipe distributors made of perforated pipe or cylindrical wedge-wire screen, often placed in a parallel pipe configuration as in Figure 10.2. Wedge wire is a type of wire screen made with wires of triangular cross-section that resists becoming clogged with dust particles.
- Multiple nozzle distributor. Typically these are lateral style distributors constructed using multiple, individual nozzles (Figure 10.2). The nozzles are spaced uniformly on parallel lateral pipes for uniform coverage across the entire adsorber cross-section. This may require 10 nozzles per m^2 of cross-sectional area (1 nozzle per ft^2) as a general rule-of-thumb. Nozzles may employ wedge-wire screens or other designs that resist plugging.

- Highly specialized designs used for chromatography applications. These are custom distributors designed to minimize extra-column volume and backmixing of liquid within piping.

Figure 10.2: Lateral Style Distributor. Laterals may be constructed from cylindrical screens, or individual nozzles may be uniformly placed on laterals. Taken from ref. [21]. Reprinted with permission, courtesy of Johnson Screens, Aqseptence Group.

With the graded-ball type of distributor, the idea is to place spherical media in layers on the adsorbent-bed support screen. Ceramic spheres of uniform (graded) sizes typically are used to provide uniform open area and pressure drop characteristics. It is important to keep the layer thickness of each sphere size uniform across the cross-sectional area of the bed so as not to create zones of differing pressure drops. The resulting pressure drop across the multiple layers of ceramic spheres should be sufficiently greater than the velocity head loss for flow in the feed piping to ensure uniform flow through the layers. A pre-distributor or diffuser is used so the full force of the incoming flow is not directed at the center of the bed, as shown in Figure 10.1.

Lateral style flow distributors must be designed with care to ensure that no particular section of piping handles more than its share of the flow. As with a graded-ball distributor, this requires designing the feed piping for low velocity head loss, so the major pressure drop is through the openings in the lateral pipes. But this is not always possible or practical. This issue may be addressed with the design of distributors with multiple nozzles, by specifying a sufficient number of nozzles needed for uniform coverage and controlling the pressure drop at each nozzle so that no single nozzle is favored. In some cases, as before, the feed piping can be designed so that the velocity head loss for flow in the distribution piping is significantly less than the pressure drop across the distribution points. In that case, all of the nozzles can be of the same size with the same pressure drop characteristics. Another approach is to specifically size each nozzle so that each has a unique pressure drop to give uniform distribution. This latter method of achieving uniformity is more often used in cases where the feed loading in volumetric flow rate per cross-sectional area is so high that the required diameters of distributor headers and laterals to keep velocity heads low would be so large as to be impractical. Although a useful and sometimes necessary approach in special cases, designing distributors with individually specified distribution points with high header and lateral velocities is complex and makes fabrication difficult and time-consuming. Whether the distributor has to handle a wide range of flow or not will also affect the required design. For example, large distribution-point open areas specified to handle high flow rates may result in maldistribution at much lower flow rates, so special designs are needed to allow for significant turndown in the feed rate. Additional discussion of flow distributors is given elsewhere [22].

Piping Diagram and Number of Beds. Specifying the piping diagram involves deciding how many adsorber vessels to employ, how they are to be interconnected, and when specific valves are to be switched to cycle each bed through the various process steps (the timing of the operating cycle). It also involves deciding where purge streams are to be sent for recycle, treatment, or disposal. The number of adsorber vessels often is chosen to accommodate in-place regeneration and to minimize the potential for breakthrough into the product stream. For thermal regeneration, this involves use of at least two adsorbers and normally they are operated in a lead-lag configuration. Designers normally try to minimize the number of beds to just two to minimize overall cost. In the two-bed example, while one bed is on-line, the other bed is off-line being regenerated. The term "lead-lag" refers to first placing a newly regenerated bed back on-line downstream of the on-line bed, so that any incipient breakthrough will be caught by the newly regenerated bed. With this configuration, sometime later the newly regenerated bed is switched on-line as the main adsorber, and the other bed is switched off-line for regeneration. As a result, the cycle time must include the time for regeneration plus some additional time for switching beds and allowing for some cushion in case of premature breakthrough. A two-bed system often is used for dilute feeds that allow for long on-line times. A "lead-trim" process scheme utilizes three beds to minimize bed size. The lead bed is left on-line long enough to become nearly or completely saturated, allowing for significant breakthrough into the second bed (the trim bed), while a third bed is being regenerated and ready to be put back on-line. When the lead bed is saturated, it is taken off-line for regeneration, the trim bed becomes the lead bed, and the newly regenerated bed becomes the trim bed. This scheme is sometimes used for thermally regenerated adsorption beds. It allows for minimum bed size as most of each bed's capacity is utilized. These are examples of just two process schemes among many possibilities. Additional discussions of specific types of processes are given later in this Chapter.

Thermal Effects. In describing adsorber design models, we have neglected the energy balance, because in many cases the heat of adsorption and corresponding temperature effects do not significantly impact the breakthrough curve. This is because in many cases the rate of energy generation (from the heat of adsorption released within the adsorbent particles) is less than the capacity for energy removal by transfer of energy into the fluid flowing through the bed. That is, the heat of adsorption is constantly being removed by the flowing fluid and the resulting temperature rise is not significant. So, the effect of these thermal factors generally can be neglected with little error. However, this is not always the case, especially in applications involving adsorption of large amounts of solute from bulk mixtures. In these cases, thermal effects must be taken into account. Furthermore, the fluid velocity within the adsorber may decrease dramatically with distance down the bed as significant amounts of solute leave the fluid phase and enter the adsorbent. This can also have a significant effect on mass transfer performance, particularly for pressure-swing adsorption. Discussion of thermal effects and operation at high solute concentrations are given elsewhere [23, 24]. Of course, thermal effects are key to thermal regeneration, which is a separate step in the overall process of temperature swing adsorption. Each step can be designed separately, and normally thermal regeneration is developed and tested using experimental methods.

Thermal effects are critically important when thinking about the potential for a runaway fire situation: when the heat generation rate due to adsorption is greater than the capacity for heat removal by heat transfer into the flowing fluid. This can be a particular problem when starting up a new bed of activated carbon designed to remove organic contaminants from air. The initial adsorption rate for the new carbon may be so great as to generate excessive heat of adsorption and high temperatures, high enough in some cases to cause the carbon to reach its autoignition temperature and begin burning in air. Measures used to minimize the potential for fire in such situations include maintaining a high gas flow rate to continually remove heat from the bed, use of water or steam spray to keep the bed moist, and addition of nitrogen during start-up so that the oxygen level is kept below that needed for combustion (normally less than 16% oxygen by volume) [25].

Practical Considerations. So far, we have been discussing idealized applications with constant feed composition and flow rate. But for many applications, the feed is actually quite variable in both flow rate and composition during operation. For example, an activated carbon bed used to recover organic vapors from the vent of a solvent storage tank will see large changes in the vent-gas flow rate and solute concentration. This will depend on the time of day, whether sunny or cloudy, the resulting temperature of the tank, and whether the tank is being filled or emptied. The tank has to expel vent gases due to an increase in the volume (or expansion) of the liquid and/or the gas contents. Furthermore, the presence of humidity and any impurities or co-solutes can alter the adsorption capacity. For these reasons, the design of an adsorber normally cannot be exact, and the designer must work to specify an appropriate design that is likely to work for all anticipated scenarios. To mitigate this uncertainty in design, it is sometimes useful to add inferential sensors to monitor the state of the adsorbent bed. In some cases (not all), temperature sensors can detect the position of the MTZ and warn when it is getting too close to the bed outlet. This is commonly seen with pressure-swing adsorption processes. It is also useful in some cases (again, not all) to include a relatively small guard bed to catch solute in case of breakthrough from the main adsorber.

10.5 Temperature Swing Adsorption

Adsorption capacity is a strong function of temperature. If temperature is raised above a certain critical value, capacity decreases dramatically. This is the basis for a common temperature swing adsorption process scheme involving adsorption at relatively low temperature followed by in-place thermal regeneration of the adsorbent at an elevated temperature – without removing adsorbent from the adsorber vessel. Regeneration may require only moderate elevation of temperature, as when desorbing volatile molecules such as adsorbed water from alumina or light hydrocarbons from activated carbon. On the other hand, sometimes very high temperatures are required to overcome strong adsorbate-adsorbent interactions, as in the desorption of water from certain zeolite adsorbents. In cases involving non-volatile adsorbates such as the adsorption of heavy hydrocarbon oils out of wastewater onto activated carbon, regeneration of the adsorbent may require furnace temperatures to burn off or decompose (pyrolyze) the adsorbed components. In that case, the used adsorbent normally must be physically removed from the adsorber for treatment in a separate furnace.

As discussed in Section 10.1: Adsorbent Materials, a variety of materials may be used as adsorbents, and adsorption normally involves physical adsorption or adhesion of molecules to a solid surface by formation of van der Waals type bonds. For light hydrocarbons adsorbed onto activated carbon, adsorption within the pores involves the same forces as in vapor condensation. In other applications, physical adsorption also involves specific molecular interactions between the adsorbate molecules and the solid surface. This is the case for adsorption of arsenic onto modified activated alumina [26]. So, physical adsorption involves formation of a condensed phase of adsorbate molecules close together within the pores of the adsorbent. In order to desorb these molecules, the bonds they form with the surface and with each other must be broken. When dealing with adsorbent in contact with gas or vapor, breaking these bonds to desorb molecules is analogous to boiling a liquid to vaporize the molecules. The temperature must be greater than some critical temperature (analogous to the boiling point), and this temperature depends on both adsorbate and adsorbent properties. (It also depends on pressure, the subject of the next section.) For liquid-solid adsorption involving removal of volatiles such as water from organic liquids, regeneration may be accomplished by first draining the adsorber of the bulk liquid and then proceeding with evaporation of residual liquid and desorption of contaminants into the vapor phase, using a sweep of inert gas or vapor to purge the vessel. As mentioned above, whenever the adsorbed molecules are not volatile, thermal decomposition at some extreme temperature may be required. This is commonly the case when regenerating activated carbon used to purify wastewater contaminated with heavy organics.

In practicing regeneration at an elevated temperature, there are several practical issues that must be addressed:

- How to quickly heat a large bed of adsorbent for good productivity.
- How to efficiently flush the desorbing molecules from the bed to minimize the volume of gas or vapor and maximize the concentration of desorbed components.
- How to quickly cool the bed after regeneration so it is ready to go back on-line.
- What to do with the components that exit the bed during regeneration.

The ideal thermal regeneration method allows for rapid transfer of heat to and from the adsorbent particles and promotes formation of a sharp desorbate concentration profile exiting the bed. Heating a fixed bed of adsorbent may be accomplished by embedding heat-transfer coils within the bed. But this makes for an expensive vessel design and uneven heating, especially at large industrial scale. Most often, fixed beds of adsorbent are heated by flowing hot steam, hot gas, or hot condensable vapor through the bed.

When feasible, the use of steam for thermal regeneration perhaps comes closest to the ideal. Superheated steam, because of its high heat capacity and direct contact with the adsorbent particles, can rapidly heat the bed and flush desorbed components out of the bed. It also allows for complete condensation of the exiting vapors. If the contaminants are partially miscible in water, this allows for separation of condensate into an organic layer and an aqueous layer (as with steam stripping). In an industrial setting, the organics can be further treated for recycle back to the upstream manufacturing process or efficiently disposed of via incineration or some other method. The aqueous layer may be treated in a relatively small steam stripper or a steam stripper that serves as a utility for the entire plant, or sent directly to a waste-water treatment plant. In cases not involving phase splitting, the condensate may be sent directly to waste-water treatment or to a stripper. Steam is readily available as a utility at most industrial sites (and has been since the industrial revolution), so the infrastructure for its generation, distribution, use, and disposal of condensate is well developed. A typical example involves use of steam to regenerate activated carbon to remove volatile organics from vent gases. This often provides an economical process option. A steam-regenerated carbon bed process commonly is a two-bed process used to recover thermally stable and hydrolysis-resistant organics from air and nitrogen vents. An example process diagram is shown in Figure 10.3. In this particular example, the adsorber vessel is horizontal, giving a relatively shallow bed with large cross-sectional area. With proper design, the face velocity may be kept low for good mass transfer performance in a shallow bed, and regeneration using steam may be sufficiently frequent as to accommodate the fact that the bed is shallow with limited capacity.

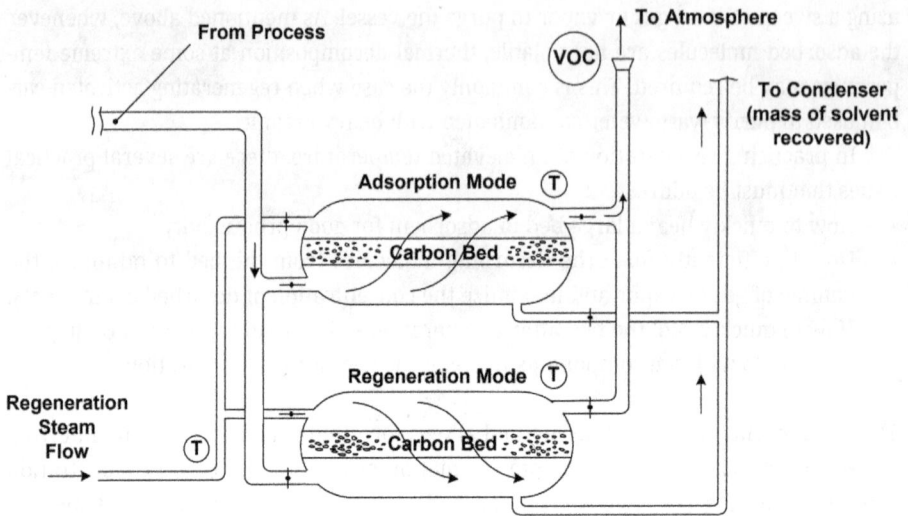

Figure 10.3: Example Adsorption Process Utilizing Steam-Regenerated Carbon. The required flow distributors are not shown. Taken from ref. [27]. Reprinted from a U.S. federal government publication in the public domain.

Sometimes an adsorption bed is used as a concentrator designed to reduce the volume of a contaminant-laden gas stream – to increase the concentration of contaminants and reduce the volume of the stream to facilitate economical treatment. This typically involves using a hot gas to regenerate an adsorbent bed. The temperature of the hot gas must be sufficiently high so that the volume of hot gas can be significantly less than the volume of feed to the adsorber. Such units are available in many styles, from conventional fixed-bed designs to various types of rotating carousels. Often these units are used to concentrate VOCs in large building ventilation streams prior to some final treatment such as catalytic incineration. An example is shown in Figure 10.4. This particular example involves using this concept to recover and concentrate CO_2 out of a dilute flue gas so it can more easily be captured for sequestration using aqueous amine absorption or some other method.

Thermal regeneration is used for all kinds of adsorption processes involving either liquid-solid or gas-solid adsorption. As discussed in the previous section, lead-lag and lead-trim are common process configurations. A typical TSA process cycle involves operation of two fixed beds of adsorbent in lead-lag fashion, as illustrated in Figure 10.5. Each bed is operated in the following cycle:

– On-line adsorption (lead position)
– Off-line draining of the bed (for liquid-solid adsorption applications)
– Off-line heating to regeneration temperature (often combined with the next step)
– Off-line regeneration with purge
– Off-line cooling to on-line temperature

- On-line adsorption (lag position, potentially combined with the cooling step)
- The cycle is repeated

DESORPTION / ADSORPTION

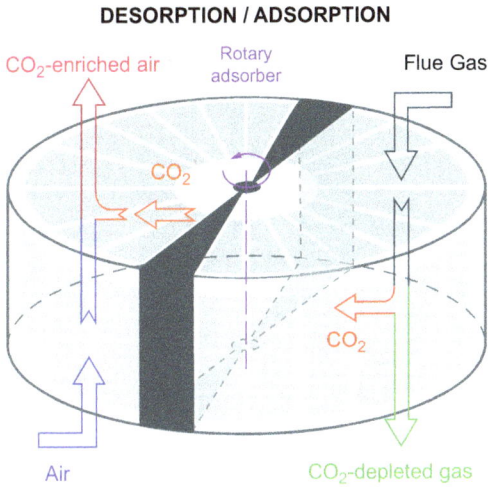

Figure 10.4: Rotary-Bed Concentrator. Taken from ref. [28]. Reprinted with permission from *Frontiers in Energy Research, Section on Carbon Capture, Utilization, and Storage*, Copyright 2020, Frontiers Media SA.

While one bed is on-line removing contaminants from the feed, the other bed is off-line undergoing heat-up, regeneration, and cool-down. The beds must be sized at least large enough to handle the feed load during the time a bed is off-line, which often is many hours. Often these steps can be combined; for example, by using steam or hot gases to simultaneously heat and regenerate a bed. That is, to begin regeneration by purging an initially cool bed with hot steam or hot gas and begin collecting the de-sorbing gases immediately. Similarly, the cool-down step may take place while the bed is placed back on-line in the lag position (when handling a gaseous feed). Then, the clean feed gases leaving the lead bed may be used to cool down the bed in the lag position before it is switched to the lead position (and to dry the bed of any conden-sate in cases involving the use of steam or another hot vapor). Or, for liquid-solid ap-plications, atmospheric air (or industrial nitrogen if oxygen must be excluded) might be used to cool and dry the bed before filling it with liquid feed and returning it to service. As an alternative, a three-bed system may be used in a lead-trim cycle in which each bed is cycled through regeneration to trim to lead positions to maximize the utilization of bed capacity and minimize bed size [30].

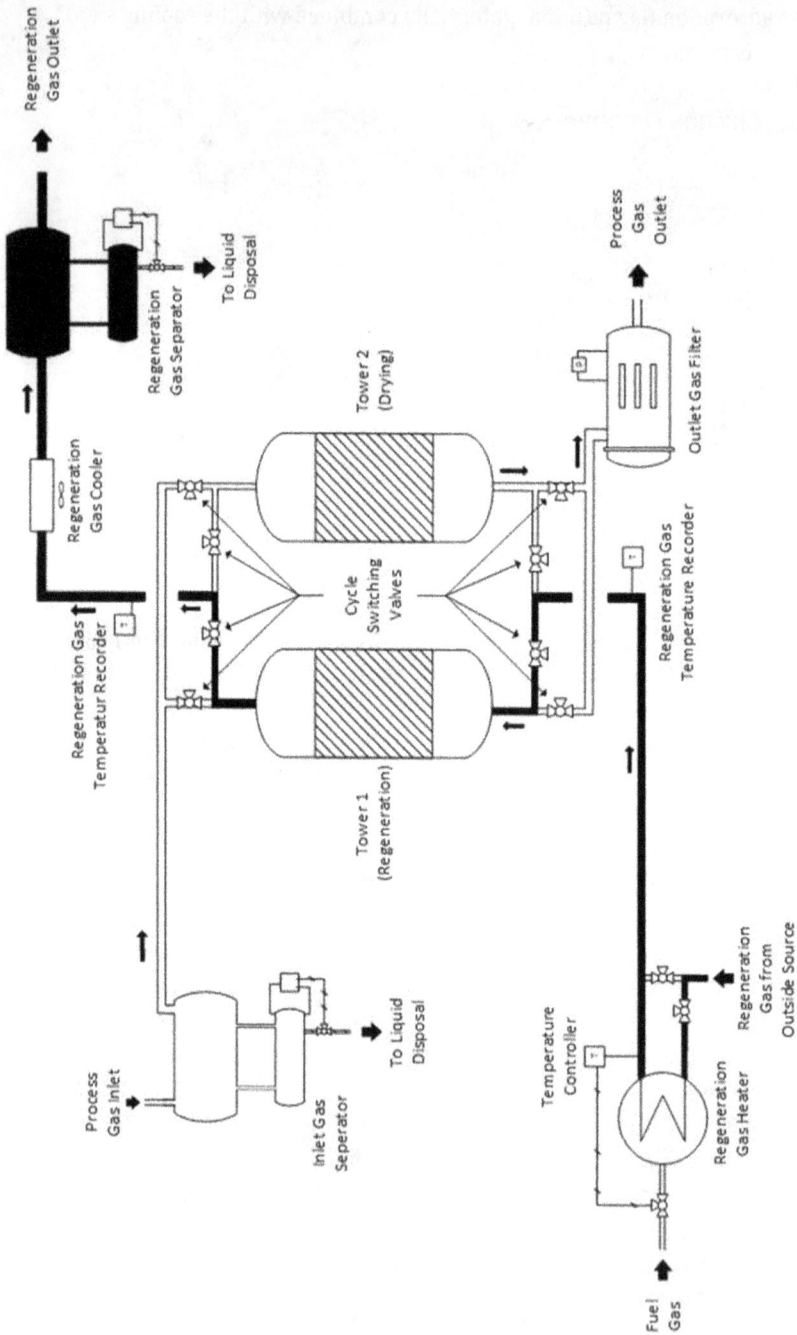

Figure 10.5: Example Temperature Swing Adsorption Process Scheme Showing Hot-Gas Regeneration Mode. Taken from ref. [29]. Reprinted with permission from *ChemBioEng Reviews*, copyright 2019, Wiley-VCH.

10.6 Pressure Swing Adsorption

As discussed above, thermal regeneration of fixed adsorbent beds can be very effective. It is the oldest of regeneration methods often used for applications involving adsorption of low-volatility compounds out of liquid feeds. With this type of application, thermal regeneration of the adsorbent may be necessary. But it takes considerable time to heat up an adsorbent bed and cool it down, and the adsorbent bed needs to be sized to accommodate the required down time. Similar considerations apply when using thermally regenerated adsorption beds to process gaseous feeds, as in removing organic vapors or water vapor from air or nitrogen. Although thermal regeneration may be an economical option in certain cases, as in the use of steam to regenerate carbon beds designed to remove organic contaminants from air (as previously discussed), there is another regeneration method to consider that may offer advantages.

This other method involves reducing the pressure in the bed to induce volatile contaminants to desorb out of the adsorbent and into the gas phase, as illustrated in Figures 10.6 and 10.7. A clean sweep-gas can then be used to carry the desorbed components out of the bed (at the reduced pressure). The step times for adsorption and regeneration are kept low so that the adsorbent bed retains the heat of adsorption released during adsorption to aid in desorption during regeneration. This approach, called pressure swing adsorption (PSA), has proved to be very effective for separating gases. Strictly speaking, it is not used to separate liquids. However, PSA may be used to separate a volatile liquid mixture if the liquid is first vaporized and then fed to the PSA unit in the vapor phase. Such a process is called Hot-Gas PSA [31]. In that case, the feed vapor must be superheated by 10 to 20 Celsius degrees to avoid any

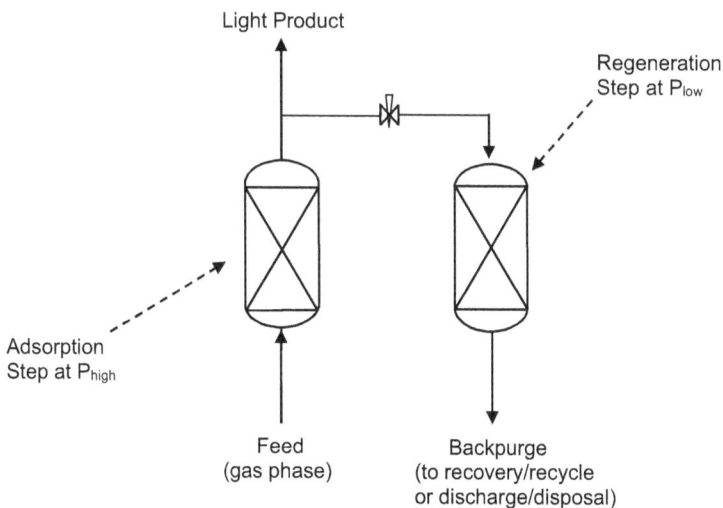

Figure 10.6: Basic Pressure Swing Adsorption Process Concept.

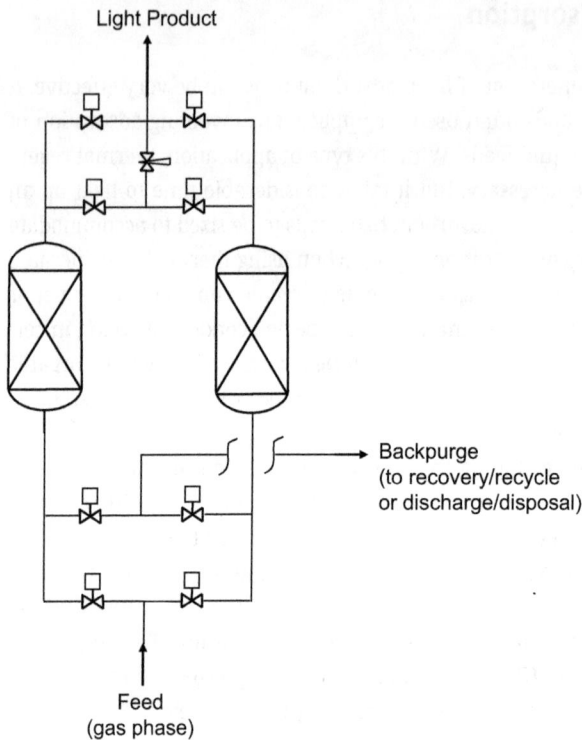

Figure 10.7: Basic Pressure Swing Adsorption Process Flow Diagram. This dual-bed process normally is operated by using a Skarstrom 4-step cycle.

condensation within the beds, because condensation will quickly plug the pores of adsorbent particles, greatly reducing performance. A number of commercial applications for both gases and volatile (evaporated) liquids are listed in Box 10.7. Production rates range from very small units such as personal PSA oxygen concentrators (about 300 standard ft^3/day (SCFD) equivalent to 8.5 standard m^3/day (SCMD)) on up to large industrial-scale units used for production of oxygen or nitrogen from air or for production of purified hydrogen (up to 100 million SCFD or about 3 million SCMD). Scaling PSA to larger and larger sizes becomes limited by practical limits on the size and cost of the required valves, so PSA units are not as large as the largest distillation units.

Box 10.7: Examples of Commercial PSA Processes

- Removing water vapor from air or nitrogen (Skarstrom Heatless Dryer)
- Gasoline engine tank vent emissions control (Skarstrom application in the auto industry)
- Recovery of solvent vapors from process vents and storage tank vents, especially for corrosive and reactive components that cause problems for TSA

- Air separation: enriched oxygen production (for pilots and for medical uses, using a zeolite molecular sieve) and enriched nitrogen (for blanketing purposes, using a carbon molecular sieve)
- Removing contaminants (such as methane, ethane, CO_2, and H_2S) from hydrogen gas
- Recovery of hydrogen in the steam-methane reformer process
- Recovery of hydrogen and ammonia from reactor off-gases
- Generation of concentrated oxygen out of air using a zeolite molecular sieve
- Generation of concentrated nitrogen out of air using a carbon molecular sieve
- Generation of both concentrated oxygen and concentrated nitrogen in a single unit
- Noble gas (He, Ar, Xe) purification
- Normal isoparaffin separation
- Removal of CO_2 from methane (methane upgrading) and other gaseous hydrocarbons, as in upgrading landfill off-gases or crude natural gas
- Removing water from vaporized, near-azeotropic ethanol and other alcohols (breaking the azeotrope)
- Removing water from temperature-sensitive compounds such as monomers
- Removing organic vapors from air or nitrogen vent streams, as in gasoline storage tank vents
- Removing chlorinated organics from air or nitrogen, instead of using steam regeneration which can induce hydrolysis to form HCl and severe corrosion
- Removal of temperature-sensitive monomer from air or nitrogen vents, avoiding fouling due to polymerization

Although desorption of adsorbed components by reduction in pressure was described in the literature as early as the 1870s (by John Hunter), it appears that this phenomenon was not exploited for chemical separation until the 1950s. Charles Skarstrom described the basic four-step PSA process in 1957, and this basic process scheme is often referred to as a Skarstrom type of process (see Section 3.2: Pioneers). Since that time, various versions of PSA have proved commercially successful in many applications – because compared to TSA, a PSA process allows the use of much smaller adsorption beds with much shorter cycle times. The process trades the need for heating and cooling systems (unless the feed has to be vaporized) with the need to manipulate the operating pressure and supply a sweep gas. This normally requires use of some sort of compressor or vacuum pump, which is costly, so designers try to minimize the size of this equipment or eliminate it altogether if the required pressures are already available.

For this reason, in many commercial PSA applications, atmospheric pressure (or near atmospheric pressure) is used either for regeneration or for adsorption. For instance, a typical PSA process is designed to operate at some high (elevated) pressure during adsorption and near atmospheric pressure during regeneration; or, the process is designed to operate near atmospheric pressure during adsorption, and regeneration is accomplished under vacuum. The latter process scheme is sometimes referred to as vacuum swing adsorption or VSA. These pressures may be adjusted, of course, to obtain the desired purity and recovery, as needed. When using elevated pressure for the adsorption step (above atmospheric), sometimes the feed comes from an upstream process already at the required pressure. This is sometimes the case for purification of hydrogen produced in an ethylene plant, for example. When dealing with a volatile liquid feed that needs to be vaporized (in Hot-Gas PSA), the required pressure may be obtained via thermal compression, avoiding the need for a mechanical compressor. This refers to boiling the feed in a confined vessel to generate pressurized vapor (its vapor pressure at the boiling temperature). This is the same principle used by a stove-top pressure cooker to generate elevated steam temperature and pressure. As mentioned above, the resulting pressurized vapor should then be further heated to gain 10 to 20 Celsius degrees of superheat.

The adsorption mechanisms underlying PSA can involve: 1) differences in the equilibrium adsorption capacity of components, allowing for separation according to the relative strength of thermodynamic interactions with the adsorbent's internal surface area; or 2) differences in the dynamic mobility of components, allowing separation according to differences in the relative amount of time components spend within the maze of the adsorbent's internal pores, resulting in differences in their elution times. Depending on the adsorption mechanism, bed temperatures normally will cycle up and down somewhat about the feed temperature (forming a temperature profile) due to heating and cooling effects from the physical adsorption and desorption of components. These temperature effects may be used to monitor a bed's status, and if sufficiently large, to enhance the efficient functioning of the beds. The cycle timing is adjusted so that a bed is switched off-line for regeneration before the temperature profile can leave the bed. The heat released during adsorption is then available to help drive desorption.

Pressure swing adsorption normally is a continuously fed process that separates the feed mixture into two outlet fractions: a stream enriched in weakly adsorbed or fast eluting components (called lights) and a stream enriched in relatively strongly adsorbing or slower eluting components (called heavies), as illustrated in Figures 10.6 and 10.7. By analogy with other fractionation processes, the stream rich in lights is sometimes called the raffinate, and the stream rich in heavies is called the extract or the loaded backpurge. In most cases, the raffinate is the desired product. Performance is highly dependent on the properties of the adsorbent because this determines the adsorption mechanism and the order in which components pass through the bed of adsorbent. Adsorbents that allow production of a desired product as the light component (the raffinate) are much preferred, because this allows for a much simpler process scheme.

The simple and elegant two-bed Skarstrom PSA process scheme produces a light product as illustrated in Figure 10.7. Cyclical operation of the two adsorbent beds allows continuous processing of the feed. While one bed is on-line processing the feed, the other bed is undergoing regeneration, a cyclical steady-state operating scheme. In the Skarstrom operating cycle, each bed is switched through four main steps in succession:

1. Adsorption. Continuous addition of feed gas for adsorption of heavier components at the higher pressure, P_{high}.
2. De-pressurization. Reducing pressure in the bed to P_{low}.

3. Regeneration (desorption of heavy components). Continuous addition of clean sweep gas at P_{low} in a reverse (countercurrent) direction. Sweep gas is taken from the product stream. It is called backpurge or light reflux.
4. Re-pressurization. Increasing pressure to P_{high} using pressurized feed gas, and back to step 1.

Sometimes this basic 4-step 2-bed operating scheme is modified by adding additional steps to improve PSA performance, such as the addition of pressure equalization steps and rinse steps, or re-pressurization with product gas instead of feed. A PSA process design can become highly complex, with many additional steps (and many additional beds) in the cycle, all added to achieve greater purity, higher recovery, or higher productivity, or to consume less energy. While these additions are important for specific purposes, we will focus our attention here on the 4-step 2-bed Skarstrom process to illustrate the main PSA concepts. In general, the process designer must choose among many design options to strike an appropriate balance between lower-cost simplicity or higher-cost complexity for somewhat better performance. The appropriate balance can vary greatly depending on the chemical system and the specific requirements of a given application.

Note that the direction of backpurge during the regeneration step is countercurrent or opposite the direction of feed gas during the adsorption step. This ensures that the product-discharge end of the bed is the cleanest end after the regeneration step is finished, to minimize breakthrough of contaminants when the cleaned bed is switched back online. Although steps 1 and 3 are the longest steps in the cycle, they typically last only about 5 to 10 min. Faster cycling of the beds is sometimes used for small units such as personal oxygen concentrators. The duration of steps 1 and 3 is kept low to ensure that the heat of adsorption released during step 1 remains in the bed where it can aid in regeneration during step 3. This may be determined by measuring the cyclical temperature profile that is maintained within each bed during operation, making sure that the maximum bed temperature obtained during adsorption does not reach the bed outlet before switching to the next step.

Steps 2 and 4 are shorter transition steps. The timing for steps 2 and 4 must be adjusted to obtain a rapid but sufficiently gradual change in pressure to avoid any movement of particles within the bed. Movement can cause attrition of the adsorbent particles and formation of dust that contaminates the product and may plug valves and the bed itself. Typical valve types are butterfly valves or other types of valves with large open area. Valves used in PSA service must be designed for very long service life (many hundreds of thousands of open/close cycles).

The basic Skarstrom process scheme is possible only when the faster-eluting light component is the desired product. In this special case, a small portion of the clean product gas can be used as sweep gas to purge the bed of relatively heavy (slower eluting) components during step 3. This backpurge sweep gas is analogous to reflux in distillation because, just as in distillation, a portion of the purified light product is recycled back to the main column or bed to improve the column's ability to purify the light product. In each case, this comes at the cost of a reduction in the product production rate (reduced productivity). In the case of distillation, the amount of reflux needed to produce pure product (the reflux ratio) is determined by thermodynamics (relative volatility) and the number of theoretical stages or transfer units achieved by the column (column height). In the case of pressure swing adsorption, the amount of reflux depends on the ratio of pressures used by the process as well as the properties of the adsorbent (relative adsorption capacities or elution dynamics). But unlike distillation, in PSA the reflux or backpurge used by the process may or may not be recovered and recycled back to the PSA process and into the product. So, the use of backpurge not only reduces PSA process productivity, it may also reduce product recovery depending on PSA design choices (as we discuss below).

As a first-pass design rule for the PSA configuration shown in Figure 10.7, the pressure ratio P_{high}/P_{low} may be specified such that the actual (not standard) volumetric flow rate of backpurge gas at low pressure during regeneration is somewhat greater than the actual volumetric flow rate of the product gas stream produced at the higher pressure during adsorption:

$$V_{actual} \text{ at } P_{low} > V_{actual} \text{ at } P_{high} \tag{10.3}$$

This is one of the design rules established by Charles Skarstrom for design of the Heatless Dryer PSA process for removal of water vapor from air [32]. In other words, the ratio V_{regen}/V_{ads} expressed in terms of actual volumetric flow rates is the transfer factor for pressure swing adsorption (called the desorption factor), and it should be greater than unity. This rule reflects the fact that at cyclical steady-state, the material balance between adsorption and regeneration must close; that is, during the regeneration step the backpurge gas must carry the same amount of contaminants out of the bed as were carried into the bed during the adsorption step. By setting the desorption factor greater than unity and confining our analysis to a reasonably dilute feed, equal adsorption and regeneration step times, equal temperatures, and ideal gases, we obtain an expression for the required pressure ratio:

$$\frac{P_{high}}{P_{low}} > \frac{V_{regen}}{V_{ads}} \left(\frac{1 - y_{contaminants}}{F_{backpurge}} \right) \tag{10.4}$$

where
P_{high}	= adsorption pressure (absolute)
P_{low}	= regeneration pressure (absolute)
V_{ads}	= feed volumetric flow rate during adsorption (actual at P_{high})
V_{regen}	= backpurge volumetric flow rate during regen (actual at P_{low})
$y_{contaminants}$	= contaminant concentration in feed (mole fraction)
$F_{backpurge}$	= portion of feed used as backpurge (mole fraction)

Equation (10.4) is derived from the ideal gas law, by equating n = PV/RT during adsorption with PV/RT during regeneration. In other words, if V_{regen}/V_{ads} is to be greater than 1.0, then what does the pressure ratio have to be given the amount of clean product gas and the amount of backpurge taken from the product? Now, assuming V_{regen}/V_{ads} should be kept constant at a value of 1.2 or so for reliable performance (usually a good practice), and using eqs. (10.3) and (10.4) as rough guides, we can estimate the required pressure ratio P_{high}/P_{low}. We do so here to illustrate typical design trade-offs. For example, if the concentration of contaminants in the feed is 5 mole% and we want to use only about 10% of the product gas as backpurge (light reflux), then eq. (10.4) indicates that the pressure ratio should be about 10.8 or so to obtain a desorption factor of about 1.2. The use of only 10% of the product as backpurge may be considered an acceptable loss. If so, this amount does not need to be recovered, allowing for a simpler process. If a pressure ratio of 10.8 is considered too high, then it may be reduced if the designer is willing to use up more of the product as backpurge. If the designer chooses to use as much as 50%, then P_{high}/P_{low} may be dramatically reduced to about 2.2. The pressure range required by the design is then much reduced, but now the addition of backpurge recovery to the overall process likely will be needed (but maybe not when dealing with air, for example). On the other hand, if the designer wants to achieve very high recovery of light product without the need for backpurge recovery, then eq. (10.4) suggests that increasing the pressure ratio dramatically, up to 54 or so, will allow 98% product recovery with only a 2% loss of product as backpurge. But such a high pressure ratio may not be practical. High pressure ratios are particularly difficult to implement at elevated pressures above atmospheric due to practical limits to the absolute-pressure compression ratios achievable with typical compressor equipment per stage of compression (normally no higher than 3) and significant thermal effects (heating of the compressed gas). On the other hand, compression ratios of 50 or so may be obtained with vacuum pumps operating below atmospheric pressure. This is one reason why certain PSA processes are operated under vacuum for regeneration. Of course, our analysis here is qualitative in nature, using only rough estimates that also assume the adsorbent properties and bed dimensions are adequate. But our discussion highlights key factors that need to be considered in specifying a PSA design.

As a first step in evaluating a PSA application, the operating pressure range (P_{high}/P_{low}) may be estimated using eq. (10.4) as a starting point, to roughly assess how much backpurge will be needed and whether it will need to be recovered. Then, this preliminary assessment will need to be refined using knowledge of adsorption capacity as a function of pressure and flow rate and using process simulation to evaluate the potential impact of adding extra beds and extra steps in the PSA cycle. These options or the addition of extra processing operations for backpurge recovery may be needed if the overall process is to achieve both high product purity and high product recovery using a practical pressure ratio. Adsorption conditions, bed diameter, and bed height should be specified to ensure that the adsorption capacity will be sufficient to remove contaminants and achieve high purity during the adsorption step – using methods discussed earlier. Detailed discussions of PSA process simulation and design are available elsewhere [33–37]. The simulation will need to be tested and optimized in miniplant tests, as in ref. [37]. Additional design options include the use of mixed adsorption media and layered adsorbent beds for applications involving multiple kinds of contaminants.

If the contaminant-laden backpurge gas exiting step 3 is not recovered, it will have to be purged somewhere. If it is comprised of non-hazardous components such as oxygen and nitrogen, it can be purged directly to the atmosphere. This is how it is done with PSA processes designed to produce concentrated oxygen or concentrated nitrogen from air. In other applications where venting to the atmosphere is not appropriate, the backpurge gas may be sent back to another part of the overall process, such as a reactor upstream of the PSA unit (allowing recovery from there), or it may be sent directly to a thermal treatment unit for disposal (such as an incinerator or catalytic oxidizer). This option might be used when recovering mixed organic contaminants from a gaseous product, as in, for example, cleaning a gaseous vent stream for emissions control.

In many cases, however, the backpurge stream will need to be recovered and recycled. This might be accomplished simply by treating the backpurge using cold condensation if the components are condensable, producing a liquid condensate that can be stored as a liquid and then conveniently recycled or disposed of as needed. This condensate liquid then constitutes the purge fraction leaving the PSA process. The treated (post-condenser) backpurge is then mixed with the incoming feed to the PSA unit. This is how the Dow Sorbathene PSA process operated to recover organic vapors from air or nitrogen vents, and it worked quite well (Figure 10.8) [38]. In this application, the condenser is placed after the vacuum pump so condensation of the organic content can occur at a higher pressure, as most of the gas stream is not condensable. This option generates an internal loop that in some cases represents the greatest load to the online adsorption bed. That is, the concentration of contaminants in the "cleaned" and saturated backpurge gas may be far greater than the concentration of contaminants in the feed to the PSA unit itself. Lanny Robbins at the Dow Chemical Company envisioned and led the development of this process in the 1980s [38]. It may be used to recover a wide variety of organic vapors. It was originally developed to recover chlorinated organics from tank vents as an alternative to steam regenerated carbon beds. The use of steam was problematic because it caused hydrolysis of the chlorinated organic, resulting in formation of wet HCl and severe corrosion of the equipment. PSA avoided this problem.

Another approach to backpurge recovery involves coupling the PSA unit with another stand-alone separation unit to achieve acceptable overall product recovery (a hybrid process scheme). Typically, this involves use of distillation, absorption, or a membrane module in addition to PSA. A commercial example of this is the combination of Hot-Gas PSA and distillation to break an alcohol-water azeotrope and produce anhydrous ethanol or isopropanol (Figure 10.9) [31, 39]. In this application, distillation

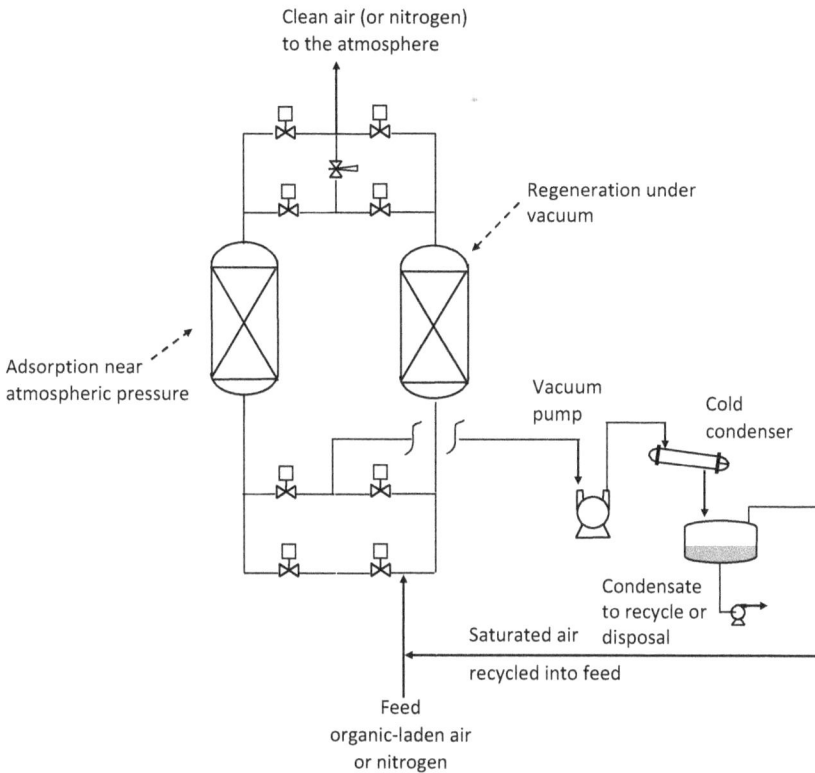

Figure 10.8: PSA Process for Recovering Organic Contaminants from Air or Nitrogen Vents.

is used to produce near-azeotropic alcohol as the overheads distillate. This enriched product is processed as vapor using PSA to adsorb the azeotropic water and produce dry alcohol effluent (breaking the azeotrope). Achieving less than 100 ppm water in the alcohol product is possible with careful design (specifying sufficiently tall adsorbent beds and eliminating sources of tramp water). The water-rich backpurge stream is sent back to the distillation unit to recover its alcohol content and purge the water.

PSA can involve use of either light reflux, as in the Skarstrom cycle described above, or heavy reflux. Light reflux is much preferred because it generally allows use of a simpler process. The use of heavy reflux is called for when the desired components are preferentially adsorbed or are the slow eluters (the heavy components), so the desired product ends up in the backpurge stream and has to be recovered from there. This is also important when designing a PSA process to isolate two products; that is, when the "purge" fraction is a desired co-product. This situation requires additional beds to produce pure products. Many commercially successful PSA applications are of the simpler Skarstrom type involving the use of two adsorbent beds and light reflux, as can be seen by inspecting the list of commercial applications in Box 10.7. Some units

Organic containing less than 100
ppm water (typical).
Less than 10 ppm is possible.

Desorption near
atmospheric pressure

Backpurge
containing water
and organic

Adsorption at
elevated pressure
using 3 A
molecular sieve
adsorbent
(typical).

Recycled back to
the distillation
process where
water is purged in
the bottoms.

The organic
components are
too large to enter
the pores, so only
water is adsorbed.

super heater

Organic + water feed to PSA
(vapor phase, near azeotropic
overheads from the partial
condenser on the distillation
column. Superheated to avoid
any condensation)

Distillation
column

Water purge

Wet organic
feed to the
overall process

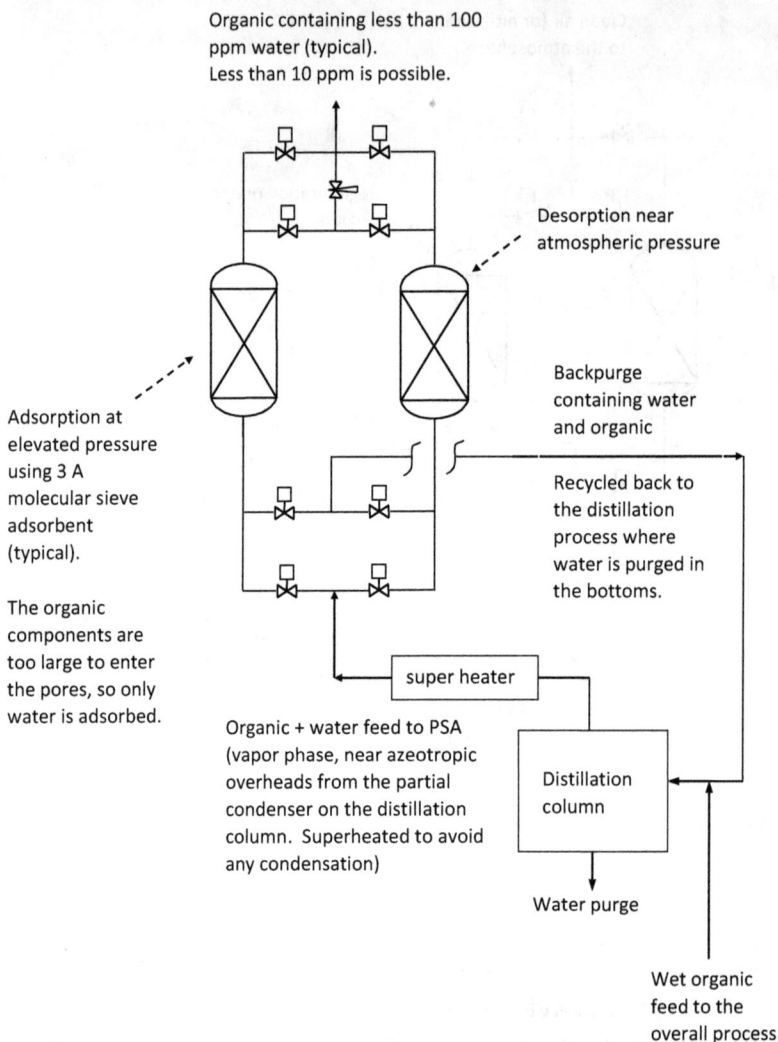

Figure 10.9: Hot Gas PSA Process for Drying Volatile Organics that Form Homogeneous Azeotropes
with Water.

can be even simpler, involving only a single bed and a two or three step operating
cycle. This is possible with a dynamic "pulsed pressure" type of process involving a
pressure increase or pulse (for adsorption) and pressure release or de-pressurization
(for regeneration) without ever operating at a steady pressure. This approach is used
for an ultra-compact-sized, personal oxygen concentrator [40]. Other PSA oxygen concentrator designs utilize as many as 4 beds to achieve specific performance goals [41].

So, a PSA process design can be fairly simple or it can become quite complex. The
required complexity is determined largely by the properties of the adsorbent and the

resulting adsorption mechanism, plus the severity of the specifications for product purity and recovery (and compact size). If adsorbent properties are favorable and design
goals are not too stringent, the PSA designer can minimize the number of beds and
the size and cost of the compressor or vacuum pump system. A good example of this
is the use of PSA to separate air as described above. PSA is used commercially to generate either concentrated oxygen or concentrated nitrogen from air, with product purities on the order of 92 to 95%. Higher purity normally is not needed for many
applications. Both options can be accomplished using two-bed systems with minimal
compressor requirements (or even a single-bed system in some ultra-small designs),
and the required backpurge can be vented directly to the atmosphere – so in each
case, a fairly simple PSA process design may be used. The main difference in how oxygen or nitrogen is produced as the product is in the choice of adsorbent. The adsorbent should preferentially adsorb or, in this case, slow the elution of the unwanted
component, allowing the desired component to pass through the bed, becoming more
concentrated in the process. To produce concentrated nitrogen from air, an effective
adsorbent is a carbon molecular sieve with faster oxygen diffusion into pores for
slower elution of oxygen and faster elution of nitrogen, so nitrogen behaves as the
light product [42]. To produce concentrated oxygen from air, an effective adsorbent is
a zeolite molecular sieve with faster nitrogen diffusion into pores for slower elution
of nitrogen and faster elution of oxygen, so in this case oxygen behaves as the light
product [43]. This example shows how important it can be, in setting out to design a
PSA system, to first understand and define the requirements for the optimal adsorbent. If such an adsorbent is not readily available, it may be best to focus on developing a better one before entering into detailed process design.

A good example of an application requiring a much more complex PSA process
design is the use of PSA to purify hydrogen for industrial uses. Here, the process must
deal with a range of contaminants of different types (typically CO, CO_2, water, and
light hydrocarbons), and the requirements for hydrogen purity and recovery can be
extreme. This is why PSA purification of hydrogen often utilizes extra steps in the
cycle and requires six or more adsorbent beds. These more complex cycles are discussed elsewhere [44, 45].

Whether a PSA process scheme is simple or complex, it can be analyzed without
having a dynamic process simulation model (in an initial analysis) by using the concepts of material balance split and separation power. This is because even though PSA
is a cyclical process, it is a binary separation for which the material balance must be
satisfied over a complete cycle of operation. For example, consider a PSA oxygen concentrator that generates an enriched oxygen product stream containing 92% oxygen
(by volume or mole percent). The feed air is comprised of 21% oxygen and 78% nitrogen. If we analyze the purge stream and find it to be comprised of 10.3% oxygen and
89.7% nitrogen, then we calculate separation power SP = (92/8)/(10.3/89.7) = 100. The
material balance then indicates that a material balance split of MBS = 0.15 is needed to
generate 92% oxygen. (MBS is the mass of the enriched oxygen product obtained per

mass of purge gas.) The resulting recovery of oxygen from the air is 57%. This amount may be satisfactory, since air is free and the purge stream can safely be purged to the atmosphere. But if we wanted higher recovery of oxygen from the air, we would either have to settle for less concentrated oxygen in the product gas, or we would need to design the PSA unit for higher separation power. A unit with SP = 500 can generate 92% oxygen using MBS = 0.26 to give oxygen recovery of 91%. The nitrogen-rich purge stream would contain 2.4% oxygen and 97.6% nitrogen (on average over the cycle), which is even more concentrated than the intended product. The challenge is to design a PSA unit that can achieve SP = 500. These numbers may be obtained using the equations described in Section 6.5.1: Material Balance Split and Separation Power (Magnified Separation Power).

10.7 True Countercurrent Moving-Bed Adsorption

Moving particles against a stream of gas or liquid in true countercurrent fashion is not often attempted. Most adsorbents eventually suffer from significant particle wear and attrition (dust formation). Compared to a fixed-bed design, a true moving-bed process is more mechanically complicated, and such a process requires a sufficiently rapid means for desorption and regeneration of the adsorbent particles. These issues have successfully been addressed in only a few cases that we know of, specifically for removing VOCs from air or nitrogen vents using specialized (often proprietary) adsorbent particles and specialized columns or fluidized beds specifically designed to move the particles. We are not aware of any systems designed to handle liquid feeds.

One example is the Polyad® system for control of VOC from very-high-volume air streams. It was jointly developed in the early 1980s by Chematur Engineering AB (formerly Nobel Chematur) with a few applications for VOC control from vent streams including one for Oakley Sunglasses [46]. It was a fully countercurrent continuous adsorption process with a vertical typically rectangular contactor vessel equipped with a cascade of slightly pitched perforated trays and downcomers similar to a trayed distillation column. In operation, a stream of beads of macro-porous polymer adsorbent was fed to the top of the contactor with the beads flowing down through the trays, fluidized by the up-flowing process gas stream. The countercurrent nature of the contact allowed the gas to be progressively cleaned and discharged at the top while the adsorbent beads were progressively loaded with contaminants and continuously discharged from the bottom of the contactor to be regenerated and returned to the contactor. As a continuous process, a moving-bed adsorber will reach a steady state with relatively constant gas discharge and adsorbent concentrations. A moving-bed adsorber has the ability to handle high volume gas streams with a modest vessel size and quantity of adsorbent compared to the bed sizes required for a fixed-bed process for the same service. The Polyad process was adopted by American Purification, Inc.

of Bakersfield, CA in the late 1990s with external microwave regeneration, but it appears this design is no longer commercially available.

Another true moving-bed adsorption design is currently in use for control of VOCs from industrial painting and printing operations – especially those involving very-high-volume air streams. It is called the fluidized-bed concentrator process [47, 48]. VOC-laden air is contacted with adsorbent in a column containing fluidized adsorbent particles. Regenerated particles added to the top of the column make their way to the bottom where they are taken into a regeneration loop. The specifics of this approach vary somewhat for different installations. In one design, regeneration is accomplished by heating a portion of the side-loop using microwave heating or by using steam heat [48]. The hot particles are flushed with a relatively small stream of N_2 purge gas (between 1/10 and 1/100 the size of the VOC laden feed stream). This yields a sufficiently large concentration of VOCs in the purge gas to allow condensing a majority of the VOC content in a cooled condenser. The cooled purge gas stream leaving the condenser is then recycled back to the contaminated inlet air loop. Alternatively, the concentrated VOCs may be treated in a thermal oxidizer. In this case, the concentrator process yields a VOC concentration high enough to support combustion with significant reduction in the use of supplementary natural gas.

10.8 Simulated Moving-Bed Chromatography

As discussed above, implementing a true countercurrent adsorption process is rarely attempted. Instead, process schemes have been devised to simulate a moving-bed process. One such process involves use of multiple vessels or fixed-bed columns in a batchwise, countercurrent process scheme. This is the arrangement often used in liquid-solid leaching operations called extraction batteries. Here, soluble components are extracted out of multiple batches of solid particles using a suitable solvent – as in processing coffee grounds with hot water to brew coffee. These fixed beds of solids are repeatedly filled with solvent, allowed to steep for a while, and then drained. Or, each bed may be treated with a continuous flow of solvent. In either case, the solvent leaving one bed is routed to the next bed in countercurrent fashion: Fresh solvent entering the process first sees spent solids on their way out of the process, and solute-rich solvent on its way out of the process sees fresh solids on their way into the process. This is a stripping type of countercurrent process scheme that is analogous to stripping-mode liquid-liquid extraction. Such a process produces a concentrated solvent phase called the extract and an extracted or depleted feed phase (the spent solids). It works well for liquid-solid leaching, yielding both high recovery of soluble solute out of the solids and high concentration of solute in the extract liquid. This stripping-mode approach is designed to efficiently transfer soluble components from the solids into the liquid phase, but not to separate them once they are there. Doing that requires a middle-fed, countercurrent process scheme. Such a process, designed to separate the components of a liquid by contact with solid particles, but without moving the solids, is called simulated moving-bed (SMB) chromatography [49].

Chromatography is a particular kind of adsorption process in which feed components are separated according to differences in how fast they move or elute through a bed of solid adsorbent particles. Their progress is slowed compared to bulk flow because of interactions with the solids – both thermodynamic and kinetic interactions (see Section 6.7.3: Adsorption Dynamics: Breakthrough Curves). Gas and liquid chromatography are widely used in analytical chemistry for separating the components of a given sample for component identification. Here, a carrier gas or liquid eluent (the solvent or mobile phase) constantly flushes a column (filled with solids or, more typically, some other form of stationary phase), and a sample is injected into the bulk flow as a pulse. Sample components then move through the column at different rates depending on the specific type of column, the composition of the gas or liquid mobile phase, and temperature [12]. The components exit the column in different component peaks which are used for identification. Pressure swing adsorption (PSA) is sometimes referred to as a kind of scaled-up gas chromatography, and SMB chromatography, a kind of scaled-up liquid chromatography. (For discussion of PSA, see Section 10.6: Pressure Swing Adsorption.) Both processes take advantage of differing elution rates to separate feed components, but they differ from the analytical methods in that the scaled-up versions are fed a constant flow of feed (normally) instead of pulse injection. This allows for more productive use of the adsorbent, but reduces the number of product streams that can be isolated (generally only two). As a result, SMB chromatography uses much less solvent compared to pulse injection, reducing dilution of product. Sometimes chromatography methods are indeed scaled up using pulse injection, especially when multiple products need to be isolated as in certain high-value biopharmaceutical applications (see Section 13.1: Bioseparations).

The classic continuously fed SMB chromatography process typically utilizes eight or more fixed beds to separate a feed stream into two fractions: a fraction enriched in the slower components (customarily called the extract) and a fraction enriched in the faster moving components (called the raffinate). Valves are switched to direct flows through four zones, forming an internal loop of flowing solvent as shown in Figure 10.10. These zones and their basic functions are (in sequence):

Zone I.   Addition of fresh solvent (eluent). Used to flush slow eluters forward into the extract outlet.

Zone II. Purification of extract. Used to move fast eluters forward away from the extract outlet.

Zone III. Addition of feed and purification of raffinate. Begins the separation and keeps slow eluters from moving too far forward into the raffinate outlet.

Zone IV. Retention of fast eluter. Keeps fast eluters from going too far forward into Zone I.

Analogous to fractional liquid-liquid extraction, Zones I and II effectively form the washing section, producing purified extract, and Zones III and IV effectively form the stripping section, producing purified raffinate. Note that feed is added in-between these sections, giving a middle-fed process scheme.

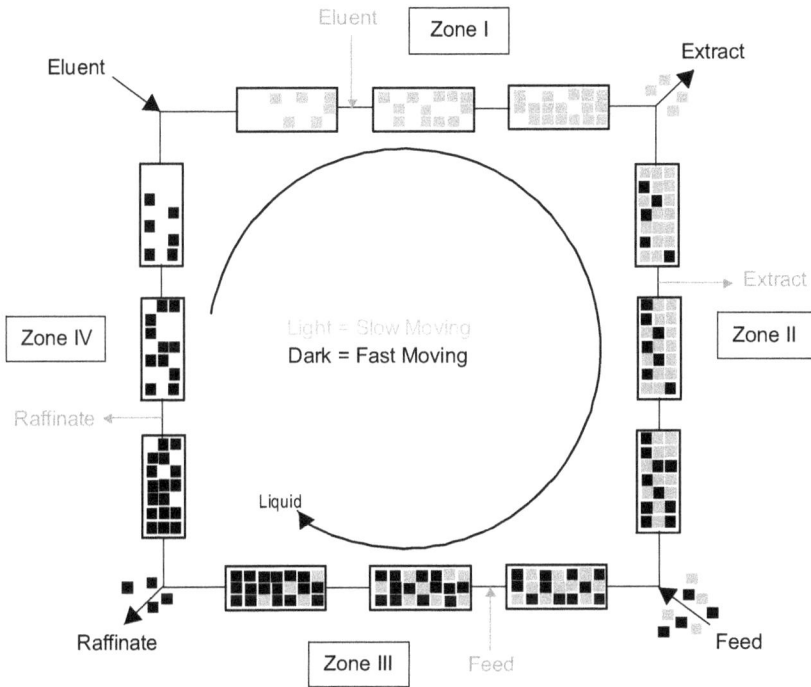

Figure 10.10: Four-Zone SMB Chromatography Operation. Taken from ref. [50]. Reprinted with permission from *AIChE J.*, copyright 2009, American Institute of Chemical Engineers.

At pre-determined intervals, valves are switched to redirect the various flows from one zone to the next. Most separation and all of the purification occurs within zones II and III, so typically more fixed beds (typically referred to as columns) are used within these zones. The use of 12 columns in a 2-5-4-1 configuration is typical: 2 columns for zone I; 5 columns for zone II; 4 columns for zone III; and 1 column for zone IV. In realizing such a design, these columns may be individual columns or they may be separate sections of a smaller number of columns. And liquid flows may be switched using multiple valves in a manifold arrangement, or a multi-port valve may be used. The design of the piping between columns should minimize unnecessary holdup of liquid to minimize axial dispersion and unwanted backmixing within the piping.

The SMB process is optimized by adjusting the intervals between valves being switched as well as the rates at which feed and eluent are continuously added to the loop and extract and raffinate are continuously removed from the loop. The goal is to coordinate the timing of valve switching with the movement of components within the loop so that slow-eluting components exit at the extract outlet and fast-eluting components exit at the raffinate outlet. The rate at which valves are switched gives an effective rate of progression for each zone in terms of the zone length per step-time period. This is adjusted to be in-between the movement of the fastest and slowest eluters within the loop in terms of solute elution-length per minute. As a result, fast eluters move in one direction in the loop and slow eluters effectively move in the opposite direction. A composition profile is established within the columns such that most of the adsorbent is utilized for separation at any given moment, much more than with a standard fixed-bed adsorption process in which a single adsorbent particle is utilized only a small fraction of the time as the adsorption front moves through the bed. The efficiency of SMB chromatography approaches that of a true moving-bed system as more fixed beds are used (usually 8 or 12, but sometimes as many as 20 or more). The classic 4-zone SMB process is constrained to a binary separation just like classic continuously fed fractional extraction and distillation, and just like those operations, it remains a popular commercial option. If additional cuts or fractions are needed beyond a binary separation, then standard pulse-injection fixed-bed processes may be used (as mentioned above), or various modified (and more complex) SMB process schemes that allow for additional product removal locations are available [49]. The latter are analogous to taking side-draw streams off a fractional extraction or distillation column.

The classic SMB process scheme was invented in the 1950s and patented in 1961 by Donald B. Broughton and Clarence G. Gerhold at Universal Oil Products (now Honeywell UOP) (see Section 3.2: Pioneers). The process was first used commercially to separate xylene isomers in the petrochemical industry, an important application that continues today. In the 1960s and 1970s, SMB chromatography was developed to enrich fructose in fructose-glucose mixtures produced by the corn refining industry via enzyme catalyzed conversion of corn starch to sugars, for production of corn syrup sweeteners. This has become one of the best-known and most widely used applications of SMB chromatography. More recent applications include purification of pharmaceuticals.

Since its inception, various modifications to the classic SMB process have been introduced for even greater efficiency or to isolate multiple fractions. These include:
- Asynchronous valve timing (Varicol® process scheme) [51].
- Non-linear feed rates: partial feed [52], Power Feed [53] and ModiCon [54].
- Potential for ternary separations (beyond binary) [55].
- Hybrid processes comprised of SMB + another separation unit (see Chapter 14: Inventive Engineering: Process Simplification, Intensification, and Hybrid Processing).

SMB processes have been mathematically modeled using a variety of approaches [49, 56]. Typical ones include:
- True countercurrent models (as in distillation).
- Standing wave models
- Linear driving-force models. See Section 6.7.3: Adsorption Dynamics: Breakthrough Curves.
- Detailed mechanistic models involving characterization of internal porosity and tortuosity

The commercial development of an SMB chromatography process typically involves the following steps:
- Use of pulse injection tests (similar to analytical chromatography) to screen potential adsorbents and operating conditions
- Analysis of successful peak separation to parameterize an initial SMB process model
- Use of these results to determine initial conditions for an SMB miniplant run
- Generation of SMB miniplant data in several runs to adjust operating conditions, and if feasible, to demonstrate desired performance
- Use of the SMB miniplant run data to refine and validate an improved SMB process simulation
- Use of the process simulation and fundamental design rules to specify the functional design of a commercial process.

In a pulse injection test (or pulse test for short), a feed sample is injected into a single adsorbent column just as in analytical chromatography, but the columns are larger (perhaps half an inch in diameter) and are filled with the same (or very similar) solid particles envisioned for the scaled-up SMB process. A pulse test measures retention time and peak shape for each component. Results showing satisfactory separation of the component peaks are then analyzed to determine model parameters for initial simulation of the SMB process and to estimate initial operating conditions for an SMB miniplant run. The detailed procedures depend on the specific process model. A recent study is described in the literature [57], and vendor software may be used to fit parameters. The SMB miniplant is then operated to further refine and adjust the operating conditions (flow rates and switching times). An SMB process is very sensitive to even small changes in these operating parameters, and the initial SMB model derived from pulse test data typically is not sufficiently accurate to predict these sensitivities. For this reason, a miniplant study is needed to adjust and demonstrate the required separation and to improve the SMB process simulation. These validated results can then be applied to reliably specify a large-scale design. Details of the miniplant approach are described elsewhere [50, 58].

Typical results of a pulse test are shown in Figure 10.11 for an example system with a feed containing two different proteins (denoted EHM and BSA) dissolved in water [50]. Note that baseline resolution of the two protein peaks is not achieved; that is, the two peaks overlap and are not completely separated. In fact, for most-effective SMB processing, baseline resolution actually is not desirable. In effect, in an SMB process scheme, purified components are isolated from the leading edge of the first peak and the trailing edge of the second. Baseline resolution is avoided to allow for operation at high solute concentrations for high process productivity, yielding product streams with maximum solute concentration. Figure 10.12 shows the concentration profile for a 12-column SMB process that corresponds to the pulse test in Figure 10.11.

Pulse Test Perloza MT 100 (100-250 µm)
Feed 10 mg/mL, 1.0 BV/hr

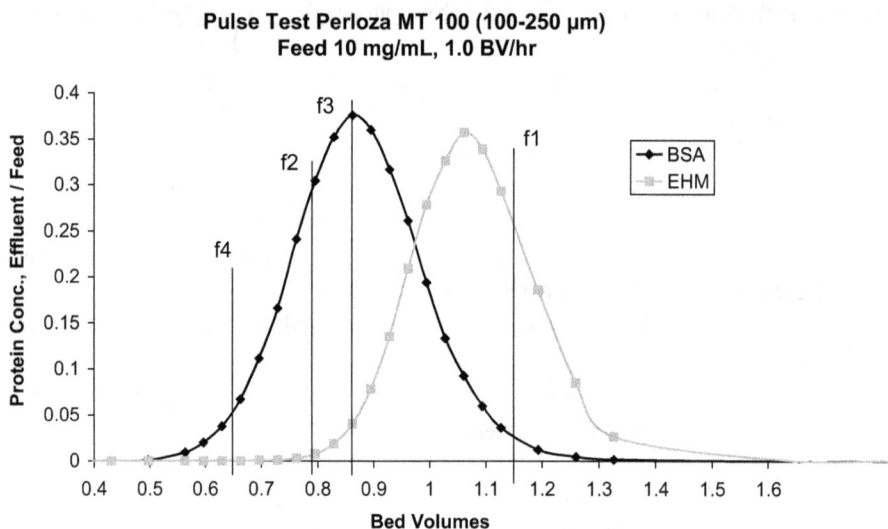

Figure 10.11: Pulse Test Results used to Generate Initial SMB Miniplant Run Parameters. Taken from ref. [50]. Reprinted with permission from *AIChE J.*, copyright 2009, American Institute of Chemical Engineers.

As with PSA, an existing SMB unit can be analyzed as to its overall separation performance without having a dynamic process simulation model, using the concepts of material balance split and separation power. This is because even though SMB is a complex cyclical process, it is a binary separation for which the material balance must be satisfied over a complete cycle of operation. For example, consider an SMB unit designed to concentrate fructose out of a fructose + glucose feed containing 30% fructose and 70% glucose (by weight on a solvent-free basis). The SMB unit generates an enriched fructose product stream containing 77% fructose and 23% glucose. Fructose is the more strongly adsorbed component enriched in the extract. If we analyze the raffinate purge stream and find it to be comprised of 6.4% fructose and 93.6% glucose, then we calculate separation power equal to SP = (77/23)/(6.4/93.6) ≈ 50. The material balance then indicates that a material balance split of MBS = 0.5 (5: 10 extract to raffinate) is needed to generate 77% fructose, and the recovery of fructose from the

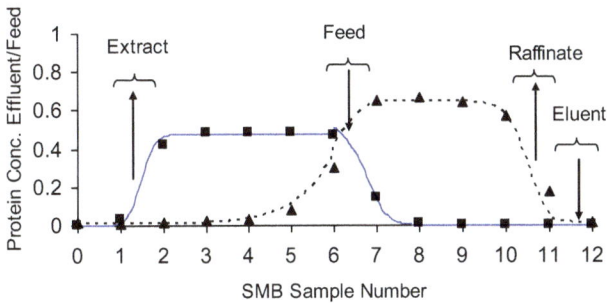

Figure 10.12: Miniplant Data from a 12-Column SMB Miniplant Run. With reference to Figure 10.11, the slow eluter (BSA) is in the extract, and the fast eluter (EHM) is in the raffinate. Note that the resulting concentration profiles reflect the fact that SMB chromatography is a middle-fed binary fractionation process. Taken from ref. [50]. Reprinted with permission from *AIChE J.*, copyright 2009, American Institute of Chemical Engineers.

feed is about 86%. In this application, high recovery of fructose is desired but not required because the raffinate can be recycled back to the enzymatic isomerization reactor. So if we wanted to obtain greater fructose purity in the extract at a higher concentration, we would need to design the SMB unit for higher separation power and use a smaller MBS. For example, an SMB unit with SP = 100 can generate 93% fructose using MBS = 0.30 to give fructose recovery of 71%. The raffinate purge stream would contain 11% fructose and 89% glucose on average. The lower recovery may be acceptable considering the greater enrichment of fructose in the extract. Then the challenge is to design an SMB unit that can achieve SP = 100 or even greater. This kind of analysis allows for a quick assessment of performance, helping to set targets for process improvement. In this example, the challenge is to design an SMB unit that can achieve SP = 100 at a reasonably high concentration of sugar dissolved in water.

References

[1]   F. Çeçen, "Water and Wastewater Treatment: Historical Perspective of Activated Carbon Adsorption and Its Integration with Biological Processes, Chapter 1," in *Activated Carbon for Water and Wastewater Treatment: Integration of Adsorption and Biological Treatment*, F. Çeçen and Ö. Aktaş, Eds., Wiley-VCH, Hoboken, New Jersey, 2011.

[2]   P. Marcus and V. Maurice, Eds., *Passivation of Metals and Semiconductors, and Properties of Thin Oxide Layers*, Elsevier, Amsterdam, 2006.

[3]   J. Zhang, J. P. Ewen, M. Ueda, J. S. S. Wong and H. A. Spikes, "Mechanochemistry of Zinc Dialkyldithiophosphate on Steel Surfaces under Elastohydrodynamic Lubrication Conditions," *ACS Applied Materials & Interfaces*, vol. 12, no. 5, pp. 6662–6676, 2020.

[4]   E. Robens, "Some Intriguing Items in the History of Adsorption," *Studies in Surface Science and Catalysis*, vol. 87, pp. 109–118, 1994.

[5]   J. R. West, "Some Industrial Aspects of Adsorption," *Journal of Chemical Education*, vol. 22, no. 8, pp. 398, 1945.

[6]   R. Riffer, "The Nature of Colorants in Sugarcane and Cane Sugar Manufacture, Chapter 13," in *Sugar Series*, Vol. 9, M. A. Clarke and M. A. Godshall, Eds., Elsevier, Amsterdam, 1988, pp. 186–207.

[7]   T. Kerley and J. P. Munafo Jr., "Changes in Tennessee Whiskey Odorants by the Lincoln County Process," *Journal of Agricultural and Food Chemistry*, vol. 68, no. 36, pp. 9759–9767, 2020.

[8]   H. Kayser, "Ueber die Verdichtung von Gasen an Oberflaechen in ihrer Abhaengigkeit von Druck und Temperatur," Analen der Physik und Chemie, vol. 248, no. 4, pp. 526–537, 1881.

[9]   H. Kayser, "IV. Ueber die Verdichtung von Gasen an Oberflaechen in ihrer Abhaengigkeit von Druck und Temperatur," Analen der Physik und Chemie, vol. 250, no. 11, pp. 450–468, 1881.

[10]   E. Robens, S. Amarasiri and A. Jayaweera, "Early History of Adsorption Measurements," *Adsorption Science & Technology*, vol. 32, no. 6, pp. 425–442, 2014.

[11]   H. X. K. Hu, *Chapter 8 – Physicochemical Technologies for HRPs and Risk Control, High-Risk Pollutants in Wastewater*, Elsevier, Amsterdam, 2020, pp. 169–207.

[12]   M. D. LeVan, G. Carta, J. A. Ritter and K. S. Walton, "Section 16, Adsorption and Ion Exchange," in *Perry's Chemical Engineers' Handbook*, 9th ed., D. W. Green and M. Z. Southard, Eds., McGraw-Hill, New York, 2018.

[13]   A. Dąbrowski, "Adsorption – From Theory to Practice," *Advances in Colloid and Interfacial Science*, vol. 93, no. 1–3, pp. 135–224, 2001.

[14]   M. J. B. Matos, A. S. Pina and A. C. A. Roque, "Rational Design of Affinity Ligands for Bioseparation," *Journal of Chromatography A*, vol. 1619 (article 460871), 2020.

[15]   M. I. Stewart, "Chapter 7, Dehydration," in *Surface Production Operations*, 3rd ed., Vol. 2, M. I. Stewart, Ed., Gulf Professional Publishing, Oxford, England, 2014, pp. 279–373.

[16]   R. H. M. Herold and S. Mokhatab, "Gas Processing News, Optimal Design and Operation of Molecular Sieve Gas Dehydration Units – Part 1," Gas Processing & LNG, Gulf Publishing Co., Gulf Energy Information, August 2017. [Online]. Available: http://gasprocessingnews.com/features/201708/optimal-design-and-operation-of-molecular-sieve-gas-dehydration-units. [Accessed 15 February 2023].

[17]   G. McKay, "Design Models for Adsorption Systems in Wastewater Treatment," *Journal of Chemical Technology & Biotechnology*, vol. 31, no. 1, pp. 717–731, 1980.

[18]   T. C. Frank, B. S. Holden and A. F. Seibert, "Section 15: Liquid-Liquid Extraction and Other Liquid-Liquid Operations and Equipment," in *Perry's Chemical Engineers' Handbook*, 9th ed., D. W. Green and M. Z. Southard, Eds., McGraw-Hill, New York, 2019.

[19]   C. Tien, *Introduction to Adsorption: Basics, Analysis, and Applications*, Elsevier, Amsterdam, 2019.

[20]   C. Tien, "Chapter 5: Batch Adsorption Models and Model Applications," in *Introduction to Adsorption: Basics, Analysis, and Applications*, Elsevier, Amsterdam, 2019, pp. 119–153.

[21]   Anonymous, "Industrial Screens, Sales Brochure, 20 Pages," Aqseptence Group, Inc., [Online]. Available: https://johnsonscreens.com/wp-content/uploads/2021/11/Johnson_Industrial_Screens.pdf. [Accessed 9 February 2023].

[22]   M. S. A. Baksh, B. T. Neu and D. H. Zadeh, "Flow Distributor for PSA Vessel," U.S. Patent 7,166,151, 23 January 2007.

[23]   H. Miyajima, T. Hirose and M. Goto, "Heat Effect and Performance in Pressure Swing Adsorption," in *Fundamentals of Adsorption, the Kluwer International Series in Engineering and Computer Science*, Vol. 356, M. D. LeVan, Ed., Springer, Boston, MA, 1996, pp. 643–650.

[24]   M. Boscherini, F. Miccio, E. Papa, V. Medri, E. Landi, F. Doghieri and M. Minelli, "The Relevance of Thermal Effects during CO_2 Adsorption and Regeneration in a Geopolymer-Zeolite Composite: Experimental and Modelling Insights," *Chemical Engineering Journal*, vol. 408 (article 127315), 2021.

[25]   S. Mannan, *Lees' Loss Prevention in the Process Industries*, 4th ed., Butterworth-Heinemann, Waltham, MA, 2012.

[26] T. S. Singh and K. K. Pant, "Kinetics and Mass Transfer Studies on the Adsorption of Arsenic onto Activated Alumina and Iron Oxide Impregnated Activated Alumina," *Water Quality Research Journal* of *Canada*, vol. 41, no. 2, pp. 147–156, 2006.

[27] Anonymous, "Carbon Adsorber Schematic," U.S. Environmental Protection Agency, [Online]. Available: https://www3.epa.gov/ttnchie1/mkb/documents/Carbon_Adsorber_Schematic.pdf. [Accessed 9 February 2023].

[28] L. Herraiz, E. Palfi and E. S. L. M. Fernandez, "Rotary Adsorption: Selective Recycling of CO_2 in Combined Cycle Gas Turbine Power Plants," *Frontiers in Energy Research, Section on Carbon Capture, Utilization, and Storage*, vol. 8, 2020 (Dec. 11), DOI 10.3389/fenrg.2020.482708.

[29] F. Berg, C. Pasel, T. Eckardt and D. Bathen, "Temperature Swing Adsorption in Natural Gas Processing: A Concise Overview," *ChemBioEng Reviews*, vol. 6, no. 3, pp. 59–71, 2019.

[30] J. L. Humphrey and G. E. Keller II, *Separation Process Technology*, McGraw-Hill, New York, 1997, pp. 191–192.

[31] Y. Liu, S. D. Feist, C. M. Jones and D. R. Armstrong, "Isopropyl Alcohol Dehydration by Hot Gas Pressure Swing Adsorption: Experiments, Simulations, and Implementation," *Industrial & Engineering Chemistry Research*, vol. 53, no. 20, pp. 8599–8607, 2014.

[32] C. W. Skarstrom, "Method and Apparatus for Fractionating Gaseous Mixtures by Adsorption," U.S. Patent 2,944,627, 12 July 1960.

[33] D. M. Ruthven, S. Farooq and K. S. Knaebel, *Pressure Swing Adsorption*, Wiley-VCH, Hoboken, New Jersey, 1996.

[34] M. D. Sees, T. Kirkes and -C.-C. Chen, "A Simple and Practical Process Modeling Methodology for Pressure Swing Adsorption," *Computers & Chemical Engineering*, vol. 147 (107235, ISSN 0098-1354), 2021.

[35] G. N. Nikolaidis, E. S. Kikkinides and M. C. Georgiadis, "Modelling and Simulation of Pressure Swing Adsorption (PSA) Processes for Post-combustion Carbon Dioxide (CO_2) Capture from Flue Gas," in *Computer Aided Chemical Engineering*, Vol. 37, K. V. Gernaey, J. K. Huusom and R. Gani, Eds., Elsevier, Amsterdam, 2015, pp. 287–292.

[36] V. Rama Rao, S. Farooq and W. Krantz, "Design of a Two-Step Pulsed Pressure Swing Adsorption-Based Oxygen Concentrator," *AIChE Journal*, vol. 56, no. 2, pp. 354–370, 2009.

[37] S. Krishnamurthy, V. Rama Rao, S. Guntuka, P. Sharratt, R. Haghpanah, A. Rajendran, M. Amanullah, I. A. Karimi and S. Farooq, "CO_2 Capture from Dry Flue Gas by Vacuum Swing Adsorption: A Pilot Plant Study," *AIChE Journal*, vol. 60, no. 5, pp. 1830–1842, 2014.

[38] D. J. Pezolt, S. J. Collick, H. A. Johnson and L. A. Robbins, "Pressure Swing Adsorption for VOC Recovery at Gasoline Loading Terminals," *Environmental Progress*, vol. 16, no. 1, pp. 16–19, 1997.

[39] J. A. Ritter, W. Fan and A. D. Ebner, "New Approach for Modeling Hybrid Pressure Swing Adsorption – Distillation Processes," *Industrial & Engineering Chemistry Research*, vol. 51, no. 27, pp. 9343–9355, 2012.

[40] V. Rama Rao, M. D. Urich and M. V. Kothare, "Experimental Design of a 'Snap-on' and Standalone Single-Bed Oxygen Concentrator for Medical Applications," *Adsorption*, vol. 27, pp. 619–628, 2021.

[41] A. Arora and M. M. F. Hasan, "Flexible Oxygen Concentrators for Medical Applications," *Scientific Reports*, vol. 11, 12 Jul 2021, article no. 14317.

[42] S. V. Patel and J. M. Patel, "Separation of High Purity Nitrogen from Air by Pressure Swing Adsorption on Carbon Molecular Sieves," *International Journal of Engineering Research and Tehnology (IJERT)*, vol. 3, no. 3, pp. 2278–0181, 2014.

[43] D. M. Ruthven and S. Farooq, "Air Separation by Pressure Swing Adsorption," *Gas Separation & Purification*, vol. 4, no. 3, pp. 141–148, 1990.

[44] M. Asgari, H. Anisi, H. Mohammadi and S. Sadighi, "Designing a Commercial Scale Pressure Swing Adsorber for Hydrogen Purification," *Petroleum & Coal*, vol. 56, no. 5, pp. 552–561, 2014.

[45] M. Luberti and H. Ahn, "Review of Polybed Pressure Swing Adsorption for Hydrogen Purification," *International Journal of Hydrogen Energy*, vol. 47, no. 20, pp. 10911–10933, 2022.

[46] L. Felix, R. Merritt and A. Williamson, "Evaluation of the POLYAD® FB Air Purification and Solvent Recovery Process for Styrene Removal," U.S. EPA Control Technology Center, Research Triangle Park, NC, EPA/600/R-93/212 (NTIS PB94130317), 1993. URL https://www.epa.gov/sites/default/files/2020-08/documents/solventrecoverystyreneremoval.pdf (retrieved 2022-11-14).

[47] "Fluidized Bed Concentrator (FBC)," Captis Aire LLC, [Online]. Available: https://www.captisaire.com/Technology. [Accessed 14 November 2022].

[48] D. W. Price and P. S. Schmidt, "VOC Recovery through Microwave Regeneration of Adsorbents: Comparative Economic Feasibility Studies," *Journal of the Air & Waste Management Association*, vol. 48, no. 12, pp. 1146–1155, 1988.

[49] A. E. Rodrigues, C. Pereira, M. Minceva, L. S. Pais, A. M. Ribeiro, A. Ribeiro, M. Silva, N. Graça and J. C. Santos, *Simulated Moving Bed Technology: Principles, Design, and Process Applications*, Butterworth-Heinemann (Elsevier), Oxford, 2015.

[50] S. D. Feist, Y. Hasabnis, B. W. Pynnonen and T. C. Frank, "SMB Chromatography Design Using Profile Advancement Factors, Miniplant Data, and Rate-Based Process Simulation," *AIChE J*, vol. 55, no. 11, pp. 2848–2860, 2009.

[51] O. Ludemann-Hombourger, R. M. Nicoud and M. Bailly, "The 'VARICOL' Process: A New Multicolumn Continuous Chromatographic Process," *Separation Science and Technology*, vol. 35, no. 12, pp. 1829–1862, 2000.

[52] Y. Zang and P. C. Wankat, "SMB Operation Strategy–Partial Feed," *Industrial & Engineering Chemistry Research*, vol. 41, no. 10, pp. 2504–2511, 2002.

[53] Z. Zhang, M. Mazzotti and M. Morbidelli, "Powerfeed Operation of Simulated Moving Bed Units: Changing Flow-Rates during the Switching Interval," *Journal of Chromatography A*, vol. 1006, no. 1–2, pp. 87–99, 2003.

[54] Z. Zhang, M. Mazzotti and M. Morbidelli, "Continuous Chromatographic Processes with a Small Number of Columns: Comparison of Simulated Moving Bed with Varicol, PowerFeed, and ModiCon," *Korean Journal of Chemical Engineering*, vol. 21, no. 2, pp. 454–464, 2004.

[55] J. W. Lee, "Expanding Simulated Moving Bed Chromatography into Ternary Separations in Analogy to Dividing Wall Column Distillation," *Industrial & Engineering Chemistry*, vol. 59, no. 20, pp. 9619–9628, 2020.

[56] M. Johannsen, "Chapter 9, Modeling of Simulated-Moving Bed Chromatography," in *Modeling of Process Intensification*, F. J. Keil, Ed., Wiley, New York, 2007, pp. 279–322.

[57] Z. Yuan, Y. Yamamoto, T. Yajima and Y. Kawajiri, "Estimation and Statistical Analysis of Model Parameters Using Sequential Monte Carlo for Phenol and p-Cresol Separation," *Journal of Chromatography A*, vol. 1688 (article 463703), 2023.

[58] B. W. Pynnonen, S. D. Feist, Y. Hasabnis, D. R. Albers and T. C. Frank, "Quickly Evaluate SMB Chromatography for Your Application," *Chemical Processing*, vol. 73, no. 5, pp. 29–33, May, 2010.

11 Crystallization from Solution and from the Melt: Taking Advantage of Eutectic Behavior

Crystallization processes occur everywhere in nature in spontaneous ways even though crystallization involves going from a disordered state of matter to a highly structured one. These processes are driven by deviation from solid-liquid equilibrium; that is, by the presence of supersaturation (see Section 5.2: Dynamic Equilibrium and Gibbs Free Energy and Section 6.7.4: Crystallization Rates: Nucleation, Growth, and the Population Balance). In nature, supersaturation may be generated by anything that upsets solid-liquid equilibrium, such as changing weather conditions caused by the rotation and tilt of the earth in its orbit around the sun, by geological processes such as groundwater leaching and slow mineral-forming reactions, and by the metabolic processes of living things (including ourselves) involving biochemical reactions driven by chemical energy stored in food. Familiar examples are listed in Box 11.1. This great variety in the ways crystallization can take place reflects its highly complex and varied nature.

Box 11.1: Crystallization in Nature

- Snowflakes in winter [1].
- Frost on car windows on cold clear mornings, even at air temperatures several degrees above freezing due to radiative cooling [2].
- Salt crystals left after evaporation of sea water [3].
- Sugar crystals forming in maple syrup or honey stored in the refrigerator [4, 5].
- Formation of stalagmites and stalactites in underground caves [6].
- The presence of silicate and carbonate crystals in rocks [7].
- Formation of crystalline gemstones [8].
- Crystalline structures seen in metal castings [9].
- Calcium phosphate crystals in bone and teeth (hydroxyapatite) [10].
- Formation of kidney stones (mainly calcium oxalate crystals) [11].

Over the centuries, people have found many ways to take advantage of crystallization, perhaps beginning with food processing [12, 13]. This includes the production of table salt and granulated sugar, the refining of various fats, waxes, oils, and plant extracts by partial solidification on cooling (called winterization), and baking and food preparation in general. Long before scientific studies of crystallization, chefs and bakers developed recipes to control the overall structure of solid foods such as chocolate, sugar candy, pasta, and ice cream. Many of these recipes work by controlling the crystallization of key ingredients such as sugars, fats, starches, proteins, and sometimes water (as in ice cream), to obtain desired crystal sizes and overall textures (mouth feel). Or, they may work by inhibiting unwanted crystallization in order to delay formation of white, powdery crystals on the surface of chocolate (called fat bloom), to

https://doi.org/10.1515/9783110695052-016

slow the formation of large water crystals in ice cream (freezer burn), or to delay crystallization of starches in bread (making it dry and stale). People have also learned to control crystallization when fashioning useful items out of metal (metalsmithing) in order to obtain specific metal compositions and structures that give desired properties such as surface hardness with greater flexibility and less brittleness [14].

In the mid to late nineteenth and early twentieth centuries, new methods of crystallization were introduced for large-scale production of various organic chemicals and industrial salts. Early examples include Norbert Rillieux's invention of the multiple-effect evaporator for crystallizing sugar (ca. 1843) (see Section 3.2: Pioneers). The Karl Josef Bayer process for producing purified alumina from bauxite ore involved the use of extraction into hot caustic soda followed by crystallization of alumina (ca. 1887) [15]. This marked the beginning of the hydrometallurgical industry. The production of ammonium sulfate crystals for use as fertilizer also began around this time, as a by-product from coal and coke gasification plants [16]. Other examples include the manufacture of crystalline aspirin powder around 1900 by the Bayer Company, a company founded in 1863 by Friedrich Bayer [17]. The production of aspirin marked the beginning of the pharmaceutical industry. Around the same time, the production of small silver halide crystals in making emulsions for photographic film marked the beginning of the large-scale photographic film industry [18]. Today, crystallization is a critical step in the manufacture of these and many other products including a wide variety of organic compounds (both bio-derived and synthetic) used in personal care products, medicines, agricultural pesticides, and other specialty products, plus numerous salts such as halides, hydroxides, sulfates, and citrates, used directly or as ingredients in product formulations.

11.1 General Characteristics and Approach to Design

Certain characteristics of crystallization set it apart from other chemical separation processes:

- The many ways that crystals may be formed (as suggested by the examples in Box 11.1).
- Potential for achieving high crystal purity in a single stage of crystallization (in many but not all cases).
- A highly path-dependent process. Final purity and crystal size depend on the path taken.
- Constrained molecular mobility in the solid phase. This allows for solid defects and non-equilibrium (or metastable) solid phases to persist.
- Part of a larger overall process scheme. The results of the crystallization operation can greatly impact downstream process operations (the particle technologies).

Crystallization is a complex process involving both thermodynamic drivers and kinetically controlled mechanisms. A great variety of nucleation and growth mechanisms are possible, yielding many different crystal structures, shapes, and structural defects, depending on the chemical system, the presence of impurities or additives, the process conditions that are employed, and the rate at which process conditions change. And the same compound may form several different crystal structures called polymorphs [19, 20]. Because of this complexity, crystallization behavior is difficult to predict, and crystallization process design relies heavily on experimental methods including operation of scaled-down crystallizer miniplants.

The potential for achieving high purity in a singe stage of crystallization is due to eutectic behavior, which is common but not always observed. We chose to highlight eutectic behavior in the title of this chapter because it sets crystallization apart from most other separation methods (when possible). Most solid-liquid systems exhibit some form of eutectic behavior, which derives from a solid's natural tendency to reject foreign components from its crystal lattice as it forms out of liquid solution. The term *eutectic* was coined in 1884 by British chemist Frederick Guthrie (1833–1886) from the Greek eutēktos meaning easily melted or thawed. Discussions of eutectic behavior often focus on this low-melting property. But here, we focus on the fundamental property of a eutectic crystal to reject foreign components from its crystal lattice.

Eutectic systems are discussed in Section 5.6: Phase Diagrams: Equilibrium Curves, Azeotropes, and Eutectics. A representative eutectic phase diagram for a binary mixture is shown in Figure 5.12. The melting temperature at the eutectic point is below the melting point of either pure component. It is the minimum melting temperature for that system (for a binary mixture). Pure crystals of one type form on one side of the eutectic point, pure crystals of the other type form on the other side, and both types of crystals can form at the eutectic point. And such a system typically retains eutectic behavior with the addition of solvent. Although crystal formation is never perfect – impurities or solvent molecules may be incorporated within the crystal as inclusions of trapped mother liquor or as structural defects – a eutectic crystallization process may still produce highly pure (if not perfect) crystals in a single stage of crystallization. However, in practice an additional stage of crystallization may be needed to obtain a particularly high purity (due to imperfections).

As mentioned above, not all crystallizations behave as eutectics. Some systems produce solid-solution crystals of variable composition that always require multiple stages of crystallization to achieve high purity [21]. Solid solution crystals melt at temperatures in-between those of the pure components. Differential scanning calorimetry may be used to determine whether a given solid mixture exhibits true eutectic behavior, solid solution behavior, or partial solid solution behavior [22]. In the latter case, purity is limited by some partial miscibility of components in the solid phase. Additional discussions regarding crystallization of eutectic systems and solid solutions are available elsewhere [23, 24]. Fortunately, eutectic behavior seems to dominate. But as stated above, this does not guarantee success. Obtaining crystals with minimal defects and of the desired size requires paying careful attention to how the crystallization is carried out (our next point).

The path taken also matters. As a solid crystal or particle forms, whether as a eutectic crystal, a solid solution crystal, or an amorphous particle, the results of its formation are locked into the solid structure and are not easily altered. So crystallization results are determined not only by the kind of chemical system, but also by the path taken; that is, by the specific processing steps, processing environment, and rate of change in going from feed conditions to final conditions. Fundamentally, the path determines the level of supersaturation present during crystallization, and this determines the relative rates of crystal nucleation and growth. This, in turn, determines the number of crystals produced and the number of defects within them, factors which greatly impact crystal size and purity. This is a reflection of the population balance concept, which accounts for the interdependence of simultaneous nucleation and growth rate phenomena and their competition for solute from the same source – as discussed in Section 6.7.4: Crystallization Rates: Nucleation, Growth, and the Population Balance. Normally, careful washing of the final crystals to minimize contamination with residual mother liquor also is required.

Furthermore, a crystallization operation must be viewed in the context of the larger process scheme that includes (in many cases) filtration, washing, and drying operations, and sometimes, additional operations such as particle classification, milling, conveying, and packaging. These many solids processing steps are generally referred to as particle technologies. The properties of the solids made in the crystallizer (especially regarding their shape and size distribution) can greatly impact the outcomes and productivities of these downstream operations. The overall process scheme normally must also include additional processing of the spent mother liquor to recover and recycle residual product left in solution after crystallization and recover solvent for recycle (or treat this stream for disposal). So, instead of focusing only on the crystallization operation itself, the overall processing scheme must be designed and optimized as a whole. For example, minimizing the time spent in the crystallizer in hopes of increasing productivity can actually reduce overall productivity by necessitating excessively long times spent in the downstream filter and dryer. This is because shortening the time allowed for crystallization often requires operating the crystallizer at relatively high levels of supersaturation, which can cause excessive nucleation and formation of very fine crystals. This can make filtration and drying difficult and time-consuming; or, to mitigate this effect, larger equipment and higher energy consumption may be required. As discussed in Section 6.7.4, nucleation tends to dominate at high supersaturation levels, so to maximize crystal growth, the supersaturation level must be maintained below a certain critical level (within the metastable zone) which varies depending on the particular chemical system of interest.

So, designing a crystallization process involves first determining the type of system one is dealing with (eutectic or otherwise) and then evaluating various ways of generating and controlling supersaturation to obtain the desired purity and crystal size (actually a distribution of crystal sizes). It also involves determining overall process requirements such as the required number of crystallization stages and any associated operations. Generally, the initial goal is to find a way of generating supersaturation that allows formation of a suitable crystal structure and shape in laboratory experiments, and then to devise a suitable overall process scheme that can yield the desired crystal purity and size with high product recovery. This normally involves further development at the miniplant scale (in liter or gallon size crystallizers), often using the crystals made there to test and develop various downstream operations such as filtration and drying (and any other particle technologies that may be required). It also involves testing ways to process the residual mother liquor to recover residual product and recycle solvent (using distillation or adsorption or some other means). Once satisfactory results are obtained there, the focus shifts to scaling up the process to replicate results, and this may require further development at the pilot-plant scale. The large-scale, crystallization process necessarily will differ in some way from the laboratory or miniplant process because it is not possible to scale-up a crystallizer while maintaining all physical relationships the same. Differences include such things as mixing and heat transfer capabilities (which tend to decline as scale increases) and this affects such things as recirculation (turnover) times in the crystallizer, temperature and concentration gradients, and solids concentration gradients. Scale-up often requires adjustments in the mixing design and allowance for additional residence time in the crystallizer to promote crystal growth.

11.2 Types of Processes and Equipment

As discussed above, crystallization is accomplished by generating and controlling the level of supersaturation. Depending on the specific chemical system of interest, supersaturation may be generated in several ways:
- Cooling the mother liquor solution or melt below the saturation temperature or freezing point.
- Evaporation of solvent from a saturated solution
- Addition of anti-solvent to decrease the solubility of solute in solution
- Addition of a reagent that reacts with the feed solute to form a product with lower solubility
- Combinations of the above, as in processes involving evaporation with cooling (evaporative cooling) and cooling followed by anti-solvent addition

A crystallization process often involves some combination of these methods to achieve high recovery of solute (high yield of crystals) in addition to the desired crystal properties. For example, a batchwise cooling crystallization may be followed by addition of anti-solvent to further reduce solubility and maximize solute recovery. Reactive crystallization may be initiated at an elevated temperature that favors nucleation of a desired structural form of the product, followed by cooling. Or, the addition of reagent may change the solubility characteristics of the liquid phase (the solvent) as well as alter the chemical form of the solute. Evaporative crystallization may involve both evaporation and heat exchange to counter evaporative cooling effects and maintain a desired temperature. And multiple crystallizers may be used to achieve these goals. This may involve using a single crystallizer for multiple crystallization steps, or the use of multiple crystallizers in series. A crystallization process may also involve heating the solution to begin operation at a desired point on the phase diagram, to nucleate a desired form of the solute compound, followed by evaporation or cooling.

So, crystallizer designs vary depending on how supersaturation is to be generated and controlled; that is, depending on the need for heat exchange, for metering liquids (or gases), for venting vapors, for dealing with a change in volume, and/or for mixing the vessel contents to achieve specific requirements. The latter may involve requirements for bulk circulation, heat transfer, dispersion of additives or reagents, suspension of solids, removal of fines, or washing of solids. For design purposes, crystallizers may conveniently be classified into various categories in terms of their basic function:

- Thermal crystallizers – designed to efficiently transfer thermal energy into or out of the mother liquor to change temperature or to evaporate solvent. These must have sufficient heat exchange capacity.
- Anti-solvent (or drowning-out) crystallizers – designed to meter and rapidly mix a miscible liquid or a soluble gas into the mother liquor to change the solubility characteristics of the liquid phase. These must handle a large change in volume as the crystallization proceeds.
- Reactive crystallizers – designed to meter and rapidly mix a reagent into the mother liquor to change the chemical form and solubility of a specific chemical component. These may require cooling and may need to handle significant change in volume.
- Melt crystallizers – designed to purify products via cooling crystallization without the use of an added solvent. This is a specialized kind of thermal crystallizer.
- Classifying crystallizers. These are special designs with special features for settling larger crystals from fine crystals within the body of the crystallizer or in a column attached to the crystallizer. Features are described by terms such as elutriation legs, settling/wash columns, and quiescent zones for clear liquor advance [25].
- Specialized mixing designs such as oscillating-flow tubular crystallizers for continuous crystallization from solution [26] and impinging-jet crystallizers for anti-solvent or reactive crystallization [27].
- Crystallizers with multiple features of these kinds.

Supersaturation must be characterized (or inferred) during operation and this becomes a key design variable, as well. Common ways of characterizing supersaturation at the commercial scale are outlined in Box 11.2.

Box 11.2: Methods for Characterizing Supersaturation within Commercial Crystallizers

- Cooling. In cooling crystallizations the supersaturation level almost always is characterized by the degrees of sub-cooling below the saturation temperature of the solution or the freezing point of the melt. An example is shown in Figure 6.5. Gradients in temperature and solute concentration exist throughout a crystallizer, so supersaturation will vary with location. The location of the temperature sensor then becomes a design variable that can affect the characterization of supersaturation during operation. Normally, the temperature sensor is placed within the bulk fluid away from the cooling surface, but there may be reasons to place it closer to the cooler (or have multiple sensors). In a batch cooling crystallization, the level of supersaturation is controlled by manipulating the rate at which the batch is cooled. In a continuous or fed-batch process, supersaturation is controlled by manipulating both feed temperature and bulk temperature. In any case, supersaturation as a function of location is greatly influenced by how the feed and bulk phases are mixed together.
- Evaporation of solvent. Here, supersaturation may conveniently be characterized by the temperature rise in the liquid at the heat-transfer surface or in the liquid exiting a flow-through heater, while monitoring (and controlling) the heat exchanger ΔT and the rate of heat input (the heat duty). All of this is influenced by the mixing design.
- Addition of anti-solvent to decrease the solubility of solute in solution. Here, supersaturation is characterized by the difference between actual solute concentration and the saturation concentration as a function of solvent composition. The level of supersaturation is a function of the rate of anti-solvent addition and the total amount added, so these measures are carefully controlled and monitored during crystallization, normally following a previously developed protocol. Again, the resulting supersaturation levels, which vary as a function of location within the crystallizer, are greatly influenced by the mixing design.
- Addition of a reagent that reacts with the feed solute to form a product with lower solubility. Supersaturation is the difference between the concentration of reaction product produced by reaction and its saturation concentration. As with anti-solvent addition, the rate of reagent addition and the total amount added serve to characterize the level of supersaturation during operation.
- The use of various on-line process analytical instruments may provide additional real-time information useful for control. Normally, on-line instruments of this kind are used in the laboratory and pilot plant to develop a suitable process scheme or protocol. They are less often used at the commercial scale due to their cost and maintenance requirements. A light-scattering turbidity probe is a robust instrument designed for use at large scale (see Section 11.3: Process Analytical Technology).

Furthermore, crystallizers are designed for either batchwise, fed-batch, or continuously fed modes of operation. Crystallization process schemes include the following (to be discussed later):

- Batch and fed-batch operation with solids (crystal) suspension. Vessels designed for suspending solids during batchwise operation often are similar in design to standard baffled stirred-tank vessels used for other batchwise operations such as batchwise reaction and liquid-liquid extraction.
- Batch operation with formation of crystals on solid surfaces (called layer crystallization). This is employed in batchwise melt crystallization.
- Continuously fed operation with solids suspension. Crystallizers designed for continuous operation with solids suspension include a great variety of vessel shapes

and custom features, for both crystallization from solution and for crystallization from the melt.

– Multi-stage processes with discrete stages performed by individual crystallizers. This may involve use of the same crystallizer for a second recrystallization step when a single stage of crystallization is inadequate, or it may involve use of multiple crystallizers in a middle-fed countercurrent process scheme used to achieve both high product purity and high product recovery. Or, it may involve operation of multiple crystallizers in series, each crystallizer taking a small step in supersaturation level.

– Specialized melt crystallizer designs that allow for true countercurrent contacting of liquid and suspended solids through partial melting and recrystallization along the length of the crystallizer.

In discussing the generation of supersaturation within a crystallizer, a distinction should be made between the bulk or average level of supersaturation within the crystallizer as a whole and local levels of supersaturation. The average level reflects the overall driving force for the process, as in the difference between the average solution temperature and the saturation temperature for a cooling crystallization. Local supersaturation levels can be much higher due to significant time required to uniformly mix fluids of different temperatures and compositions, as mixing is not instantaneous. In a thermal crystallization process, temperatures are more extreme next to the heat-transfer surface, and in the case of boiling, the generation of vapor bubbles may be more intense at the main liquid-vapor interface where liquid-head pressure is low (the main liquid level). In an anti-solvent process, concentrations of the anti-solvent are highest at the addition point. In a reactive crystallization, concentrations of the reagents are highest at the addition point, but the reaction products may or may not be highest there – depending on how fast the chemical kinetics are relative to mixing times.

In general, some mechanical mixing is needed to circulate the bulk fluid into and away from these intense regions – at a flow rate sufficient to avoid excessively high local supersaturation levels that lead to uncontrolled nucleation. The need for macro-, meso-, or micro-mixing will differ depending on how supersaturation is generated and its intensity at the point of generation. Most crystallization processes require good macro-mixing. Anti-solvent processes also require good meso-scale mixing at the addition point, and reactive crystallizations may also require good micro-scale mixing, especially those with very rapid kinetics and low solubility of reaction products (see Section 6.2: Fluid Flow and Mixing: Imagine the Best Flow Path through the Equipment, for more discussion).

For thermal crystallizers of any kind, careful attention should be made to provide sufficient heat exchange area to avoid excessive temperatures at the heat transfer surface. Designers may be tempted to use a smallish heat exchanger to reduce capital cost, but often this is done at the cost of poor crystallization performance because this results in higher temperature gradients and higher local levels of supersaturation. In our experience, as a general rule the maximum ΔT at the heat exchange surface should be less than 5 Celsius degrees, sometimes much less, but this should be determined in laboratory measurements of metastable zone width (see Section 6.7.4: Crystallization Rates: Nucleation, Growth, and the Population Balance). Cooling crystallization may be conducted in a jacketed stirred vessel or in a stirred vessel with an external forced-circulation cooling loop. Evaporative crystallizers often are designed with an external heating loop because use of a vessel jacket alone cannot provide sufficient heat transfer rates without excessive ΔTs. Liquid velocities in the loop may be kept around 7 ft/s or so (about 2 m/s) to allow for a self-cleaning, scouring action to minimize formation of deposits on the heat transfer surface. Higher velocities should be avoided to minimize crystal damage and erosion of the piping. In our experience, designing for a temperature rise of about 2 °C can give good performance, but again, this should be determined in tests for a given application. Submerged heat-transfer coils are sometimes added to stirred tanks to increase the heat transfer area, but this often leads to formation of deposits on the internals and the need for frequent (and costly) shutdown for cleaning. In the case of thermal crystallizers, the mixing system normally is designed for good circulation (turnover) using a high-flow, low-shear impeller (an axial-flow impeller) for good macro-mixing, and sufficient flow to avoid settling of crystals.

Anti-solvent methods involve changing the solubility characteristics of the solvent, either by addition of a solvent-miscible liquid or a soluble gas (yielding a gas-expanded liquid). Normally, the crystallizer must be designed to quickly disperse the anti-solvent liquid so that local supersaturation levels do not cause excessive nucleation. This usually involves specification of an addition pipe or dip tube with the outlet located near the suction side of an axial-flow impeller, plus appropriate baffling of the vessel. As mentioned above, mixing design may require meso-scale mixing for good dispersion, and the vessel must include sufficient added volume to handle the added anti-solvent while maintaining good mixing.

Reactive crystallization refers to reactions that form a stable stoichiometric compound with relatively low solubility in the mother liquor so it may be crystallized out of solution. In this case, the reagent addition rate and the mixing of reagent at the point of entry must be controlled to minimize high levels of localized supersaturation. Relative nucleation and growth rates vary depending on the mixing characteristics of the crystallizer, the inherent chemical reaction kinetics (relatively fast or slow compared to mass transfer), and the solubility of the reaction product. Nucleation will tend to be more dominant (more difficult to control) as reaction kinetics become faster and the solubility of the reaction product decreases. If possible, it is best to adjust the order of reagent addition to maximize product solubility at the start of a reactive crystallization. For acid-base reactions, the order of addition can have a dramatic impact on the ability to grow large product crystals. Adding acid to a pool of base results in production of reaction product under basic pH conditions. Adding base to a pool of acid results in production of reaction product under acidic conditions. Product solubility will be lower or higher depending on the product's pKa value relative to the pH conditions in the crystallizer, so the choice of how to add reagent to the crystallizer can be a critical one. Co-addition of reagents may also be an option, as when using impinging jet mixers. If supersaturation is high, sometimes the product will rapidly come out of solution as an oil (termed oiling out), failing to form an organized crystalline phase. In these cases, dilution of the reagent may help to control supersaturation levels, and programmed seeding techniques may help to reduce the supersaturation level and minimize nucleation.

Many reactive crystallizers are similar to anti-solvent crystallizers in terms of the types of crystallization vessels and the need to provide a means for adding liquids. Applications involving reactive reagents include pH-swing crystallization (with addition of acid or base), reversible adductive crystallization, hydrate or solvate crystallization (in which the solvent is a reactant), reactive crystallization involving addition of reagent to form a stoichiometric compound, and precipitation reactions, which generally refer to reactive crystallizations involving very high supersaturation levels. A pH swing crystallization involves crystallization of weak acids and bases by changing the pH of the aqueous solution to above or below the pKa of the solute of interest. A weak acid, present in its deprotonated, ionic form in the feed solution (as a soluble salt) is crystallized out of solution by adding a stronger acid to reduce the pH below the solute's pKa, typically by two pH units, to form the protonated, non-ionic form which is much less soluble. A weak base (such as an amine), present in its protonated ionic form (a soluble salt), is crystallized by adding a stronger base to increase the pH above the pKa.

Reversible adductive crystallization refers to processes involving formation of weakly bound adducts at crystallization conditions, such that the resulting adduct crystals can be isolated and then later decomposed into separate components of the adduct. A well-known example is the adduct formed by bisphenol-A and phenol [28]. Reaction of phenol with bisphenol-A in a phenolic reaction mixture allows selective crystallization of phenol-bisphenol-A adduct crystals, which later are thermally decomposed to yield bisphenol-A product (for use in the manufacture of epoxy and polycarbonate resins) and phenol reagent for recycle – thereby having isolated the product from the reaction mixture.

In processes involving crystallization of solvates or hydrates, the amount of solvate or hydrate incorporated within the compound may be varied by operating at a different temperature; that is, depending on the point on the phase diagram where the crystallization is operated. Some hydrates are stoichiometric with a precise composition and some are variable in composition. A well-known example is calcium chloride, which can crystallize out of aqueous solution as crystals with 1, 2, 4, or 6 molecules of water of hydration – depending on the temperature and composition of the liquid phase during crystallization (see the phase diagram in Figure 11.1). Another well-known example is calcium sulfate, which can crystallize as gypsum $CaSO_4 \cdot 2H_2O$, as the hemihydrate $CaSO_4 \cdot 0.5H_2O$, and as the anhydrous compound $CaSO_4$, depending on temperature and the concentration of ions in solution [29].

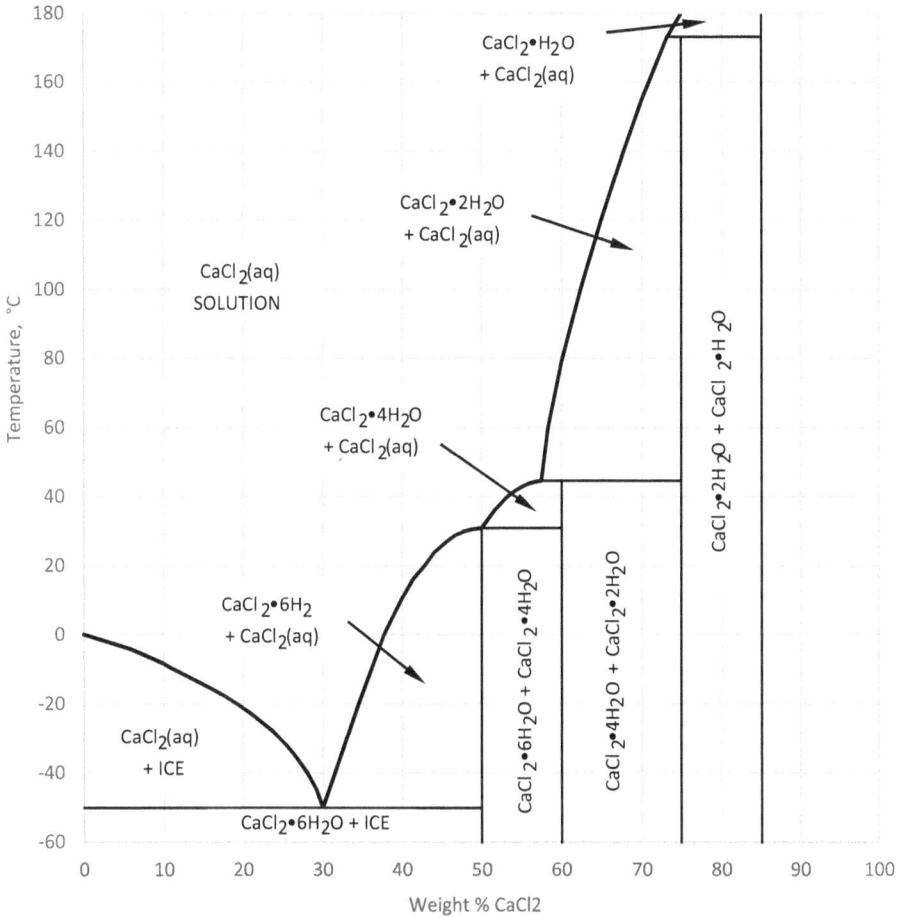

Figure 11.1: Calcium Chloride + Water Phase Diagram.

Precipitation reactions are extreme reactive crystallizations with very fast reaction kinetics and very low solubility of the resulting reaction product, so the level of supersaturation generated by the process is always very high, even if dilution or seeding techniques are used. In this case, nucleation tends to dominate growth, so such a process normally yields tiny particles or agglomerates of tiny particles, or the product may oil out of solution. High-shear micro-mixing may be called for in dealing with fast reactions that produce insoluble reaction products. But high-shear mixing may also cause crystal breakage, so a careful balance of mixing intensity may be needed for good results. Some authors also suggest that with such high supersaturation levels, nucleation proceeds without being influenced by other particles, so the main nucleation mechanisms are those of heterogeneous, primary nucleation and not secondary nucleation. This suggests that mixing shear generated by the mixing impeller or jet will not have a major effect on the nucleation rate, so high-shear Rushton impellors or imping-ing-jets may be used for their greater ability to mix at the micro-scale to help mitigate high supersaturation at the local level where the reagent is introduced to the crystallizer. So both macro-mixing (circulation) and meso-to-micro mixing (rapid dispersion at the local level) are needed. The order of reagent addition can also play a role. An example is the reaction of aqueous phosphoric acid with slaked lime (calcium hydroxide). The reaction involves rapid acid-base reaction kinetics with formation of calcium phosphate compounds [30]. Adding slaked lime to a pool of aqueous phosphoric acid avoids extreme precipitation conditions by operating under acidic conditions where the product compound (calcium phosphate dihydrate, $CaHPO_4.2H_2O$) is somewhat soluble. This helps to grow large crystals by avoiding high supersaturation levels. Operating the other way around by adding phosphoric aid to a pool of slaked lime produces hydroxyapatite, a calcium phosphate compound ($Ca_{10}(PO_4)_6(OH)_2$) that is much less soluble in aqueous base, and as a result tends to come out of solution as tiny particles.

No matter how supersaturation is generated, a crystallizer often operates without complete recovery of the product solute from the feed. This is because some amount of solute remains dissolved in the mother liquor after crystallization is complete. In some cases, this can represent a significant amount that is too large to discard and must be economically recovered in some way. And normally the solvent must be re-covered and purified for reuse in subsequent crystallizations. In this respect, crystalli-zation from solution is similar to liquid-liquid extraction in that additional steps normally are needed to process the left-over liquid (the raffinate in the case of liquid-liquid extraction and the spent mother liquor in the case of crystallization). These added steps are needed to recover residual solute, purge impurities, and recover and recycle the solvent. In principle, the same methods used to process raffinate may also be used to process spent mother liquor. Well-known examples involving extraction are described elsewhere [31]. These processing steps normally involve distillation which represents the major consumer of energy. That is, the extraction or crystalliza-tion itself is not energy intensive, but the associated processing steps can be. Alterna-tives to distillation, such as adsorption, chromatography, and the use of membranes, may be added to the process to help minimize energy consumption by lessening the requirements of the distillation step or, in some cases, by eliminating the need for dis-tillation altogether.

11.3 Process Analytical Technology

Many analytical tools have been developed to facilitate the design and operation of crystallization processes. These include real-time (*in-situ*) instruments referred to as Process Analytical Technology (PAT), as well as apparatus for rapid high-throughput ex-periments. A relatively simple and robust instrument is a fiber-optic, light-scattering probe used to monitor the turbidity of a crystal suspension. It is used to monitor and facilitate control of batch crystallizations [32] and may be helpful in monitoring continuous crystallizations, as well. Another instrument is the Focused Beam Reflectance Measurement (FBRM) probe that measures crystal chord length. Chord length is the dimension that a particle presents in front of the probe, which may be oriented in random ways. It is not a direct measure of crystal dimensions, but it can be interpreted to provide size information and to monitor and facilitate crystallization control [33]. Other instruments include various spectroscopy instruments including Attenuated Total Reflection – Fourier Transform Infrared (ATR-FTIR) spectroscopy and Raman spectroscopy [34–37]). ATR-FTIR may be used to measure the properties of the liquid solution such as key solute concentrations, and Raman spectroscopy may be used to measure properties of the solid particles (mostly), provided that solutes and crystals of interest can be detected by using these methods (not all molecules are responsive). Real-time imaging and video methods also may be used, and these generally are applicable to all types of processes (unless crystallization conditions are too harsh or corrosive). These methods are termed Particle Vision and Measurement (PVM) [38]. Video technology is getting better all the time, and software allows automatic image analysis to determine particle size distribution. High-throughput apparatus are used to screen solvents and crystallization operating conditions in very small-scale vials [39]. Detailed discussions of recent developments in PAT and PVM instruments and their applications are given elsewhere [40].

PAT tools may be used to measure solubility curves and characterize metastable zone width, to measure supersaturation levels in real time, and to characterize properties of the growing crystal mass as an aid to process control. Operating data may quickly be generated and useful operating conditions quickly identified by using directed or statistical process experiments in laboratory or miniplant apparatus fitted with PAT. Initial phases of process development often involve a search for crystallization conditions and protocols that can yield crystals with the desired size, habit, and purity plus good recovery. This includes screening for conditions that yield a desired crystalline structure or polymorph. A polymorph may be thermodynamically stable, or it may be produced as a metastable solid due to kinetic mechanisms. Screening a variety of solvents may also be important, as the choice of solvent affects the solubility curve and may form solvate compounds. In concert with experiments aimed at identifying a satisfactory crystallization protocol, solute solubility should be measured as a function of the process variable used to generate supersaturation. The solubility curve and metastable zone width should be measured before proceeding with process development, because this knowledge will guide selection of specific process conditions and protocols, and it will be needed to calculate the theoretical maximum recovery of product. The metastable zone is discussed in Section 6.7.4: Crystallization Rates: Nucleation, Growth, and the Population Balance.

Once a suitable crystallization protocol has been identified in the laboratory, generally a more limited suite of instruments is used to monitor and control the commercial process, with the goal of replicating the protocol developed using the larger suite of instruments in the lab. Besides measuring temperatures and any addition rates, this may involve measurement of crystal chord length using a focused beam reflectance type of probe, or it may involve a simpler turbidity probe which measures light transmission to gauge the size of crystals [32].

11.4 Batch Crystallization from Solution

A batch crystallization process is a highly flexible one that can be adjusted during operation to meet the requirements of a wide variety of applications. Batch operation allows for gradual change in supersaturation and adjustment of agitation intensity as needed. It generally requires large batch size at production scale for good productivity, perhaps up to 4000 gallons (15 m^3) or so. Designing for larger and larger vessel size becomes limited by the difficulty providing sufficient mixing of the vessel contents as scale increases. A typical goal of batch crystallization is to isolate large, easily filtered crystals with good purity, in high yield, and with few fine particles that can cause filtration problems and unwanted dust in the final dried product – and do so within a practical amount of time spent in the crystallizer and downstream filter and dryer (the total processing time). In specialized cases, however, the goal is more complicated, requiring production of specific crystal shapes (or crystal habit) with very narrow crystal size distribution, sometimes specifying production of very small uniformly sized crystals instead of large ones, and requiring very high purity which may require unusually low levels of solvent contamination from entrapment or inclusion of solvent within the crystals. More discussion of this type of process is available elsewhere [41].

Mixing Requirements. Designing a crystallizer vessel involves paying careful attention to mixing requirements. To produce large, easily filtered crystals, the mixing design must be capable of achieving sufficient distribution of the liquid slurry throughout the vessel to minimize gradients in liquid composition, solids concentration, and temperature. Completely uniform distribution is not possible in practice, but some approach to that ideal will be required. A typical 4000 gallon (15 m^3) batch crystallization vessel is equipped with internal baffles and an axial flow impeller for high volume circulation and rapid turnover of liquid at the main liquid-vapor interface (the main liquid level). Turnover rates vary depending on the application and the scale of operation. But in any case, the agitator must be able to adequately circulate the contents so that localized temperature and concentration gradients are not excessive and also to ensure that growing crystals are adequately suspended. Circulation typically is downflow at the center of the vessel and upflow at the vessel walls. This helps avoid settling of solids at the bottom of the vessel directly below the agitator shaft. Typical agitators employ axial mixing impellers designed for high-volume, low-shear mixing. These include various hydrofoil-shaped impellers and pitched-blade impellers. A draft tube surrounding the impeller is an option that can further facilitate circulation. Draft-tube design is often mentioned in the literature, but it appears that most batch crystallizers go without such internals in part because of the potential to foul the tube with deposits and because a draft tube can become ineffective a low liquid levels. A slotted draft tube may be used to overcome this issue [42]. Multiple impellers may be used, usually in a stacked config-uration with a single drive shaft. A useful option is another, smaller impeller located near the bottom drain to promote clean emptying of the vessel, leaving little or no sol-ids behind. This impeller is a retreat-style design that promotes wave action as the ves-sel is emptied, to clear and sweep solids into the drain [43]. The vessel may be outfitted with a variable speed drive to allow convenient adjustment of agitation speed during different phases of crystallization and for different applications, making for a flexible, multi-purpose design. The "crows foot" type of agitator commonly found in glass-lined vessels is not recommended for applications requiring high flow and low shear. For more detailed discussion, see Section 6.2.2: Mixing Design and ref. [42].

Nucleation and Seeding. A typical batch crystallization process scheme involves an ini-tial nucleation or seeding step followed by a crystal growth period. The goal is to first form a seed crop, either by *in-situ* nucleation or by adding seed crystals made in a sepa-rate process, and then grow those crystals with minimal nucleation of new crystals. Seeding procedures are described elsewhere [44]. Non-seeded nucleation may be in-duced at the start of a crystallization by sub-cooling the contents or by using another means of generating supersaturation. Different systems exhibit different nucleation be-haviors, with some sustaining high levels of supersaturation, say 20 or 30 Celsius de-grees of cooling, before the resulting supersaturation is released in an uncontrolled shower of nuclei. The level of supersaturation at which nucleation occurs may be re-duced by adding seeds, or it may be reduced by using various localized devices such as cold fingers and ultrasonic horns [45, 46]. If needed, the initial crop of seeds may be further modified in a subsequent processing step involving *in-situ* digestion of un-wanted very-fine crystals (fine nuclei). The modified seed crop is then grown up during the subsequent growth period. A specific digestion technique developed at Dow is illus-trated in Figure 11.2 [32]. Another approach that involves a seeded protocol with tem-perature conditioning of the seed crop is described by ref. [47]. This general approach is called temperature cycling. A suitable protocol can be developed in the lab by adjusting heating and cooling profiles, mixing intensity, and other process variables with the aid of PAT tools (see Section 11.3: Process Analytical Technology). Normally, the process can then be scaled up to the commercial scale by using the same general protocol but with adjustments as needed to obtain the desired results.

Template assisted crystallization is a new, promising method of inducing nucle-ation of crystals [48]. It may be used to control the nucleation rate or to obtain crystals of a desired form (or both). With this approach, the mother liquor is brought into con-tact with specially developed solid catalysts that provide active sites for crystal

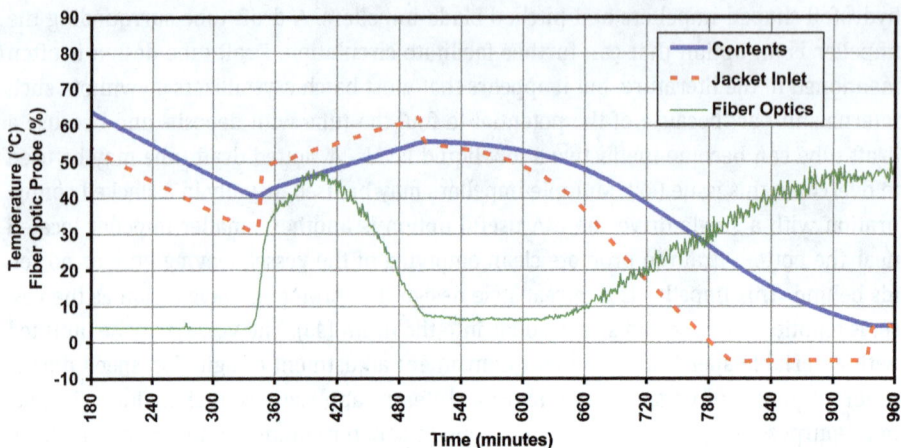

Figure 11.2: Use of a Fiber Optic Turbidity Probe to Monitor and Control an Unseeded Batch Crystallization involving Digestion. Taken from ref. [32]. Reprinted with permission from *Organic Process R&D*, copyright 2009, American Chemical Society.

nucleation. In this sense, the process is analogous to the well-known practice in chemistry labs of adding boiling-beads to a flask of super-heated liquid to catalyze nucleation of vapor bubbles and induce boiling. Potential applications include conditioning of hard water by inducing precipitation of calcium carbonate as stable microcrystals or amorphous micro-particles of calcite and aragonite. Water containing more than about 60 ppm of calcium carbonate is supersaturated and classified as hard water. Small inorganic particles such as TiO_2 supported on porous resin beads may be used as nucleation sites for crystals to form, for batch or continuous processing. Although published data about this concept is scarce, a search of the internet shows that several companies claim the technology is effective and the resulting microcrystals are so small and stable that they are easily flushed through the piping without forming scale or soap scum residues. The technology does not soften water by removing the calcium responsible for water hardness (as is done with ion exchange and reverse osmosis methods), but instead, it is claimed to transform the supersaturated calcium content into inert microcrystals. Other applications of template assisted crystallization currently under study include crystallization of carboxylic acids and various pharmaceuticals [49, 50].

Designing for Crystal Growth. A standard growth strategy for production of reasonably large, easily filtered crystals involves use of a parabolic operating protocol. This is illustrated in Figure 6.5 with reference to an Ostwald-Miers diagram showing regions of saturation, supersaturation, and undersaturation. As illustrated in the figure, once nucleation or seeding is accomplished, supersaturation is generated at a relatively low level at the start of the crystal growth period by using a low driving force (as in a small ΔT for cooling crystallization). The overall driving force is then gradually increased (by increasing the ΔT) in a parabolic profile until the batch is finished. This protocol is intended to allow for careful growth of crystal nuclei at the start of the batch (to avoid excessive nucleation of new crystals) and rapid growth of established crystals toward the end of the batch (for good process productivity). How this works may be explained with reference to Figure 6.6 showing how the various rate phenomena interact during crystallization. As the crystal mass grows, the total crystal surface area available for solute to join increases, and this increase in surface area works to modulate the local supersaturation level. So, as the crystal mass grows and the total crystal surface area increases, higher overall driving force may be used without incurring excessive nucleation of new crystals. This interpretation assumes that any increase in secondary nucleation due to high slurry densities (high crystal mass) is not significant. In any case, a parabolic operating strategy is often found to allow production of crystals that meet specifications while also achieving good productivity with reasonably short batch cycle time [32, 51, 52]. Figure 6.5 shows the path for an unseeded batch crystallization involving immediate growth of the resulting nuclei. The figure can easily be modified to include a digestion period following the nucleation event, to illustrate the path taken by the process described by Figure 11.2. This involves increasing the temperature after nucleation to temporarily re-enter the region of undersaturation in order to modify the crop of nuclei, followed by cooling to begin the growth period.

In practice, most batch crystallizations are not developed by directly measuring and quantifying the rates of nucleation and growth, because doing so unambiguously with well-characterized supersaturation levels within the appropriate mixing environment is quite a challenge and often not practical. Instead, information about regions of relative nucleation and growth is generated in the laboratory and used to guide process development (as in the Ostwald-Miers diagram, Figure 6.5). A suitable operating protocol is developed by using laboratory or miniplant-scale experiments (up to roughly 1 gallon or 4 liters in size). If a protocol can be found that yields good results at the small scale, then the designers job is to scale up the batch process to a much larger vessel (say 4000 gallons in size) to achieve the same results. Different protocols may be tried out in the laboratory until a suitable one is found. Whether a satisfactory protocol can successfully be scaled up primarily depends on how well the much larger vessel can mimic the laboratory vessel in terms of mixing performance and heat transfer capabilities – or whether adjustments in the protocol can be made to mitigate any performance deficiencies. Normally, both mixing and heat transfer capabilities decline on scale-up, and these differences must be taken into account by the designer.

For illustration, consider a typical cooling crystallization. The first step is to measure the solubility curve over a wide temperature range. The temperatures T_{start} and T_{end} are chosen to maximize solute recovery. We would like to be able to start with high solute concentration in solution and end with very little residual solute in solution. Maximum solute recovery is given by

$$\text{Solute recovery (\%)} = \frac{100}{1 - x_{end}^{sat}} \left(1 - \frac{x_{end}^{sat}}{x_{start}^{sat}} \right) \tag{11.1}$$

where x_{start}^{sat} = solute solubility at the start of crystallization

x_{end}^{sat} = solute solubility at the end of crystallization

and the productivity of the crystallization is given by

$$\text{Productivity (kg crystals/L/h)} = \frac{M_{To} x_{start}^{sat}}{V_{vessel} \Delta t} \left(\frac{Recovery(\%)}{100} \right) \tag{11.2}$$

where M_{To} = mass of the batch

V_{vessel} = vessel volume

Δt = total elapsed crystallization time

Neglecting the initial nucleation or seeding of crystal nuclei for now, the batch is started at some relatively high temperature T_{start}. A parabolic cooling profile may then be followed, such that

$$T(t) = T_{start} - \left[(T_{start} - T_{end}) \left(\frac{t}{\Delta t} \right)^2 \right], \quad T = T_{end} \text{ at } t = \Delta t \tag{11.3}$$

where T_{start} = temperature at the start of crystallization

T_{end} = temperature at the end of crystallization

t = elapsed crystallization time

In practice, such a profile may become constrained toward the end of the batch by the limited cooling capacity of the crystallization vessel, so the cooling rate reaches a maximum and the profile tends toward a linear one at the end. However, the idea is to start cooling slowly and then steadily increase the cooling rate. An example is shown in Figure 11.2 above. The cooling capacity of a jacketed stirred-tank crystallizer may be augmented by addition of a pump-around heat exchanger loop for additional heat exchange area, although this often requires a large circulation pump and may not be justified in terms of the added capital cost. The detailed design of heat transfer systems associated with stirred tanks is discussed elsewhere [53, 54]. The final design will depend on jacket type (such as dimpled or half-pipe), agitator design, jacket material of construction (carbon steel, stainless steel, glass coating, . . .), jacket area, vessel material of construction, jacket insulation, addition of an internal coil (optional), and so on.

Pump-around loops are costly, as they require large circulation pumps and associated piping to be effective at adding significant heat transfer capability. They also can affect the crystallization by causing significant attrition of the crystals, especially for needle shaped crystals. Low shear, high volume pumps are available that can reduce this effect, but normally not eliminate it. Examples include axial flow impeller pumps (an impeller in a pipe) and Discflo pumps. For smaller volume operations, options include progressive cavity pumps and diaphragm pumps.

The same design strategy may be applied using other ways of generating supersaturation. For a drowning-out crystallization process, supersaturation may be generated by slowly adding anti-solvent at the start with steady increase in the addition rate. For example, a parabolic profile for anti-solvent addition is given by:

$$\frac{M_{antisolvent}}{M_{total\ antisolvent}} = \left(\frac{t}{\Delta t}\right)^2 \qquad (11.4)$$

where $M_{antisolvent}$ = amount of antisolvent added at elapsed time t
$M_{total\ antisolvent}$ = total amount of anti-solvent added at $t = \Delta t$

Using this profile, at 50% of the total batch processing time, only a quarter of the anti-solvent has been added, and at 70% of the total time, 49% of the anti-solvent has been added. This means that most of the anti-solvent is added in the last 30% of the total batch processing time. The total volume of the batch will increase with processing time, and the potential for solute recovery is given by:

$$\text{Solute recovery (\%)} = \frac{100}{1 - x_{end}^{sat}}\left[1 - \frac{x_{end}^{sat}}{x_{start}^{sat}}\left(1 + \frac{M_{antisolvent}}{M_{To}}\right)\right] \qquad (11.5)$$

The productivity of the crystallization is calculated in the same way as in eq. (11.2).

For a batch evaporative crystallization, at any given time the temperature and liquid-phase concentration remain about the same throughout the batch, although the concentration of solids will increase with time, and gradients in temperature, liquid composition, and solids concentration will exist to some extent depending on how the batch is mixed, how it is heated, and where evaporation takes place. In any case, during evaporation the volume of the batch will decrease with time as solvent is evaporated overhead. Whenever a jacketed vessel is used, the heat transfer area covered by the batch will steadily decrease, slowing down the evaporation rate. Because of this, it is common practice to top-up a batch periodically by adding additional feed. This is repeated until the concentration of solids in the crystallizer has become sufficiently high, say 25% or so by weight, and the batch is ready for filtration. In some cases, an external heat exchanger loop will be used to heat the batch (as mentioned earlier), but the same topping off technique may still be useful.

Often the productivity and yield of a crystallization process can be further enhanced by using more than one way of generating supersaturation. It may be that the starting temperature is limited due to concerns about degradation of the active ingredient, or the solubility curve does not allow for low residual solute concentration at the endpoint T_{low}. Situations like this may be addressed by combining cooling with another method. Two common strategies are cooling – evaporation (evaporative cooling) and cooling – anti-solvent addition [55].

The descriptions and equations given above are intended to illustrate useful concepts and potential protocols. However, each application will be unique and the most-effective type of crystallization protocol will need to be determined by running crystallizations in the lab. Furthermore, the design of a commercial-scale batch crystallizer should allow some flexibility in operation to compensate for uncertainties in crystallization behavior, changes in the composition of the feed material from batch to batch, as well as uncertainty in scale-up. This may include flexibility in batch timing, flexibility in mixing intensity by using a variable speed impeller, and flexibility in having plenty of heat transfer capacity.

11.5 Continuous Crystallization from Solution

Potential advantages of continuous operation include the ability to use smaller crystallization vessels for a given production rate and the relative ease of controlling the process once the operation has reached steady state. Because a continuously operated vessel (or vessels) can be considerably smaller than a batch vessel for the same production rate, the mixing environment can be made more uniform throughout the (smaller) vessel and the mixing design may be easier to scale up with good results. Continuous operation also eliminates the need for repeated and unproductive start-up and shutdown steps (except when cleaning is needed). As a result, converting a batchwise operation to a continuous one is gaining interest as a potential productivity-improvement strategy in the pharmaceutical and specialty chemical industries [26, 56].

Historically, continuous crystallization has mostly been used for large-scale production of commodity chemicals, as in the production of various salts and sugars. Many different types of continuous crystallizers have been developed and commercialized. Specialized design types include the forced-circulation evaporative crystallizer, classified-suspension crystallizer (Oslo type), draft tube crystallizer, and vacuum crystallizer [16, 57]. As in the case of distillation, whether continuous or batchwise operation is best for a given application depends on many practical factors including the need to produce a single dedicated product or multiple products in campaigns, the scale of operation, and the availability of suitable equipment (whether significant investment in new capital is required). But batch versus continuous crystallization has another dimension as well – the fact that the ability to control supersaturation levels is different, and this can greatly affect the results. Batchwise operation allows for gradual change in supersaturation, so very slow change in temperature or another process variable, especially at the beginning of the growth period, may be used to obtain specific results. On the other hand, continuous operation normally requires a step change in the supersaturation level which is determined by the difference between inlet and outlet properties (primarily temperature and composition). Another difference involves formation of deposits on the vessels interior walls during crystallization. And the difference in supersaturation control may be a contributing factor. In continuous crystallization, formation of deposits (fouling) often requires periodic shut-down to clean the crystallizer, and this can be expensive, especially if fouling limits on-line time to only a week or so. Fouling is less of an issue for most batch crystallizations, because in effect the crystallizer is cleaned every time a fresh batch is added to the vessel (or it can more easily undergo cleaning prior to starting a batch), so fouling deposits do not have an opportunity to accumulate.

Continuous crystallization may be implemented in a variety of ways. When carried out in a single, well-mixed vessel, the crystallizer approaches the performance of an ideal design termed a Mixed Suspension Mixed Product Removal or MSMPR crystallizer. This type of continuous operation can be more challenging to develop compared to batchwise operation because it necessarily involves a step-change in liquid-phase solute concentration (and usually temperature as well) as feed enters the crystallizer vessel, with sudden generation of supersaturation at the feed inlet. Such a process necessarily operates at the average end-point or outlet concentration of solute in solution (and outlet temperature), just as in a Continuous Stirred Tank Reactor (CSTR). The concentration of solute in solution in the feed typically will be much higher than that in the outlet liquid. As a result, localized supersaturation levels at the feed point can be higher than those generated in optimized batch protocols. (Continuous evaporative crystallization is an exception in which the solute concentration remains relatively constant as solvent is evaporated to generate supersaturation.) Continuous crystallization processes may be designed to mitigate this step-change effect by operating a number of well-mixed vessels in series. The step change per vessel is then reduced and the residence time distribution within the overall process becomes narrower, approaching plug flow performance as more crystallizers are added in series (analogous to the use of multiple CSTRs). The use of multiple vessels in series is a time-honored approach to reducing the magnitude of the step change per vessel. A major issue, of course, is whether the capital cost of multiple vessels can be justified.

A newer design called the oscillatory-flow tubular reactor/crystallizer provides an alternative way to approach plug flow [26]. This design is derived from fundamental work done around 1990 on pulsating flow in tubes. For example, see refs. [58, 59]. With this design, feed is pumped through a long tube (or coil) that provides sufficient residence time for growth of crystals while also providing sufficient turbulence (due to the oscillation of flow back and forth within the tube) to insure that the crystals do not settle out of the mother liquor as they grow. The rate of forward progression may be adjusted to provide sufficiently long residence time.

But mixed-suspension type crystallizers remain the most common type. Important considerations in their design include: 1) designing a mixing system for sufficient circulation and solids suspension but without unnecessary mixing intensity that would cause excessive secondary nucleation; and 2) minimizing the potential to foul the equipment with deposits (called encrustation) [60]. Mixing designs are chosen to promote circulation and avoid high localized levels of supersaturation in order to promote crystal growth. This involves determining the level of macro-, meso-, or micromixing that is required and designing to provide just that level. This is discussed in

Section 11.2: Types of Processes and Equipment, in Section 6.2.2: Mixing Design, and in ref. [42]. The mixing design can also impact how quickly a crystallizer fouls because encrustation is driven by supersaturation. So the approach to good mixing design (minimizing localized high levels of supersaturation) can also help to minimize encrustation. But even so, encrustation normally occurs to some extent, and this requires periodic shut down to clean out the crystallizer and restart the process. One approach to minimizing the rate of encrustation involves operating at high solids concentrations (high slurry densities) to maximize the total amount of crystal surface area in the growing crystal mass. This may minimize encrustation by providing plenty of crystal area for solute to join as crystals grow, as an alternative to depositing onto vessel walls (see Figure 6.6 in Section 6.7.4: Crystallization Rates: Nucleation, Growth, and the Population Balance). Another approach involves designing the crystallization vessel with large internal spaces for fouling deposits to accumulate (called providing deposit inventory). With this approach, large deposits that form on vessel walls can break off and fall down into a lower part of the vessel where they can accumulate for a time without interfering with the on-going operation of the vessel, as discussed in ref. [61], and this may extend the operating time between cleanings.

A variety of other design considerations and options have been described in the literature for continuous crystallizer design, as summarized in Box 11.3. These options are not always cost-effective and may have unwanted effects on performance, so they should be carefully evaluated and tested before attempting large-scale implementation. Detailed discussions of specific continuous crystallizer vessel designs are given elsewhere [57, 62, 63].

Box 11.3: Continuous Crystallization Options

– Co-current, multiple-vessel processing in series. As mentioned in the text. Used to step down in solute concentration in the mother liquor by taking small, easy-to-control steps in supersaturation in a series of vessels. For example, in a cooling crystallization, instead of taking a large drop in temperature in a single vessel, two vessels are used in series, with temperature let down in smaller increments. This approach is used in many evaporative crystallization processes, which also involve multiple-effect evaporation in a series of vessels operated at progressively lower pressures (and thus temperatures) for energy conservation as well as for operation by taking small steps in supersaturation.

– Counter-current multiple stage processing. Used to improve crystal purity through countercurrent processing, especially for systems with solid-solution or partial solid-solution behavior. Example: Brodie countercurrent melt purification contactor. This approach is used primarily for crystallization from the melt (described below in Section 11.6: Crystallization from the Melt: Eliminating the Addition of Solvent), but in principle it may also be used for crystallization from solution by incorporating dissolution and recrystallization steps.

– Clarification of mother liquor within a specialized zone of the main vessel, allowing for withdrawal of clear liquor or liquor containing fines. This is used to enable clear liquor advance or fines destruction.

– Clear liquor advance (or clear liquor overflow) [25]. This is a long-known technique that in certain cases can help increase mean crystal size by increasing the population density of growing crystals. This can promote growth of existing crystals over nucleation of new ones, apparently by increasing solid surface area available for crystal growth – providing greater opportunity for solute to grow onto existing crystals and reducing the relative rate of crystal nucleation. The result is a reduction in the average supersaturation level, which promotes growth over nucleation. This technique may also reduce fouling by increasing the total crystal area within the crystallizer available for crystal growth as an alternative to deposit formation. After removal, the clear liquor may be processed (purified or concentrated) for recycle back into the feed stream to the crystallizer.

– Fines destruction [25]. Classification of crystal sizes within the crystallizer, allowing for discharge of a stream rich in fines. These are then treated outside the crystallizer to be dissolved and recycled into the feed. This may increase the mean size of product crystals.

– Retention of fines (classified product removal) [25]. Elutriation and classification of crystal sizes, allowing for retention of fines prior to discharge of product crystals. This may reduce the overall quantity of fines produced by the crystallizer by providing more residence time for small crystals to grow (but not necessarily). It may result in a narrower crystal size distribution but with a smaller mean particle size.

11.6 Crystallization from the Melt: Eliminating the Addition of Solvent

Up to this point we have focused on crystallization aided by the addition of a solvent. But this is not always needed. The term melt crystallization refers to crystallization without an added solvent. The melted feed *is* the solvent – as in crystallizing ice out of water containing dissolved impurities. Melt crystallization processes always employ cooling as the means to generating supersaturation. Different process types involve different ways of cooling the melt and different ways to form, purify, and collect the resulting crystal mass, either as layers of crystalline material grown on cold heat-exchange surfaces or as seed crystals formed on such surfaces and then scraped off and suspended in the melt.

The possibility of eliminating the addition of solvent presents a number of potential advantages. Foremost among these are: 1) no need to store, handle, and recycle a solvent; 2) no solvent-related safety hazards or emissions to deal with; 3) no need to filter crystals from solvent (even for suspended crystals, as we will see); and 4) the potential for ultra-high purification. Achieving high purity in a single stage of crystallization may be possible when dealing with eutectic systems. In that case, high product recovery requires starting with feed that is already fairly pure, say 90 +% or so. For further discussion of this point, see Section 5.6.5: Solid-Liquid Eutectic Behavior. Multi-stage processes also may be used, either to further increase the purity and recovery of eutectic products or to purify products that form solid solutions (or partial solid solutions) with high product recovery. A middle-fed, multi-stage melt crystallization process scheme is illustrated in Figure 11.3. Because crystallization from the melt avoids any possibility of contamination from inclusion of an added solvent into the forming crystal lattice, in special cases melt crystallization can yield purities of 99.99% and sometimes as high as 99.999%. This all depends on how crystals form

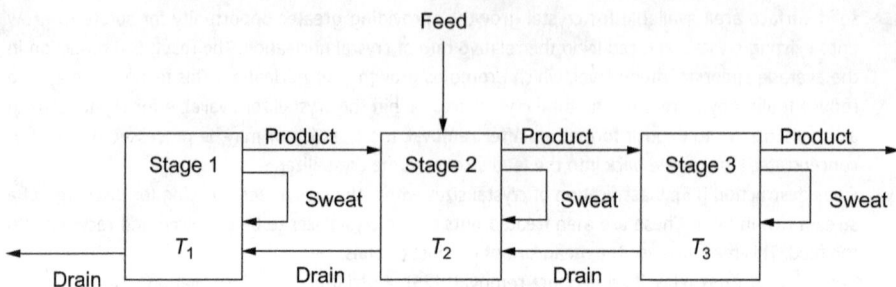

Figure 11.3: Multi-stage Countercurrent Melt Crystallization Process. Multi-stage operation of layer growth processes (temperatures $T_1 < T_2 < T_3$). Taken from ref. [65]. Reprinted with permission from *Encyclopedia of Separation Science*, copyright 2000, Elsevier.

within the process and how many stages of purification are used. Additional descriptions of multi-stage processes are given elsewhere [64, 65].

Melt crystallization can be practiced over a very wide temperature range from –50 °C to 200 °C or even higher. It may be feasible for compounds that melt within this range provided they do not degrade on heating to their melting point. One reason for adding a solvent is to lower the required crystallization temperature to avoid product degradation (if needed). A solvent may also be needed whenever supersaturation generation requires a method other than cooling, or whenever the final product must be delivered in a particular crystalline solid form. With melt crystallization, the purified product is melted and pumped away as a pure liquid. It is not delivered as a crystalline product. The melt may be sent to a relatively simple particle forming device such as a flaker, pelletizer, pastilles maker, or a prilling tower, but these options typically yield solid particles with poorly controlled crystallinity. So, when considering melt crystallization versus crystallization from solution, the tradeoffs to be considered include the potential for process simplification and ultra-high purification (by using melt crystallization) versus the potential to obtain a specific crystalline form of product and avoid thermal degradation when necessary (by using crystallization from solution). In some cases, an added solvent may provide an added ability to reject impurities by keeping certain impurities in solution, but the presence of the solvent itself makes obtaining ultra-high purity of the product unlikely due to difficulty preventing some inclusion of added solvent within the product crystals as defects.

Well-known industrial applications of melt crystallization mentioned in the literature include the processing of acrylic acid, benzoic acid, *p*-xylene, dichlorobenzene, isocyanates, phenol, caprolactam, bisphenol A, maleic anhydride, naphthalene, and various fatty acids. Melt crystallization has also been used for purifying low-melting inorganics such as phosphoric acid (as early as 1925 [66] and more recently [67]), and has been proposed for purification of aluminum nitrate nonahydrate [68]. Melt crystallization may also be a lower-energy-consuming alternative to distillation for products with close-boiling impurities, such as ethylene glycol contaminated with short-chain glycol impurities (1,2-butanediol and 1,2-propanediol) [69].

Compared to distillation, the use of melt crystallization alone requires considerably less energy. But in practice, melt crystallization often is used in combination with distillation to increase product recovery and purge impurities. Distillation is used to concentrate a crude feed up to 90% or 95% and purge the majority of impurities. Melt crystallization (single or multi-stage) is then used to further purify the product. An impurity-rich reject cut generated by the crystallization process is routed back to distillation (or sent off for disposal). A hybrid process scheme such as this can minimize the required number of crystallization stages while still requiring considerably less energy than the use of distillation alone.

Popular types of melt crystallizers include the following:
1. Batchwise static melt crystallizers (single stage)
2. Batchwise recirculating melt crystallizers (single stage)
3. Continuous suspended-crystal melt crystallizers (single stage or true countercurrent)

Single stage units may be operated as individual stages within a multi-stage process scheme. The first two types involve solid-layer growth processes. They operate by growing a layer of crystalline material at a cold heat-transfer surface. The first type involves formation of the layer out of a static batch of melted feed. The second type forces circulation and flow of the melt past the heat transfer surface to induce faster formation of the crystal layer for higher productivity. The third type involves continuous growth of crystals suspended in the melt, which may facilitate growth of highly pure crystals and offer the benefits of continuous steady-state operation. In this third type, seed crystals typically are grown in a layer on the surface of a cold heat exchanger and then scraped off using a rotating blade (a scraped freezer), but most crystal growth occurs within a dispersed or a partially settled suspension. All three types of crystallizers are commercially successful. All have their own particular set of attributes in terms of volumetric productivity, flexibility in processing multiple products, scale of operation, stability in operation, and overall cost for a given application. So the most cost-effective type varies depending on the specific needs of a given situation.

Batchwise Static Melt Crystallization. With this process, a batch of crude feed liquid, say 95% pure, is pumped to a non-stirred vessel containing internal heat-exchange surface. A typical unit is a box-shaped vessel containing rows of cooling fins. As the vessel is filled, the cooling surfaces become submerged in the melt. On cooling, crystals grow in a layer on the heat transfer surface until most of the batch has been frozen. This may take several hours or more (up to 10 or so). Typically, the solids that first begin to form are significantly higher in purity compared to the feed. Impurities are rejected from the growing crystal lattice and they become concentrated in the remaining melt (or liquor). When most of the batch has been frozen, the remaining liquor is drained. The crystals are then partially melted in several melt-back steps to further purify them (a processing step called sweating) and to collect various fractions. A static melt crystallization process was developed in France around 1950 [70]. This is the *Proabd®* MSC Melt Static Crystallization process currently offered by the Fives Group in France. Similar processes have since been commercialized by Sulzer Ltd. and other equipment suppliers.

The basic (typical) processing steps for a singe stage of crystallization are:

1. Filling. Pump melted feed liquid to a vessel containing internal heat-transfer surface.
2. Initial cooling. Cool the batch to the freeze point.
3. Nucleation. Hold the batch at 5 °C or so below the freeze point until nucleation of crystals occurs. At this point, the temperature of the batch will suddenly increase due to release of the heat of fusion.
4. Slow freezing. After initial nucleation, slowly freeze a large portion of the batch (the exact amount to be determined in testing).
5. Draining unfrozen liquor. Drain the residual mother liquor (rich in impurities) as a reject cut to be sent to waste treatment or recycled to distillation for further recovery of product and purging of impurities.
6. Sweating (melt-back). Slowly heat the batch to "sweat out" additional impurities. The term sweating refers to partial melting and annealing of the crystal mass. Impurities transfer to the melt (or sweated liquor) which then drains away from the remaining crystal layer. This is the ultra-purification step.
7. Draining sweated liquor. Drain the vessel of sweated liquor and collect it for further processing.
8. Taking multiple cuts. Steps 6 and 7 may be repeated to obtain several cuts, an initial purge cut and later cuts for recycle to other crystallization stages or to distillation.
9. Melting and pumping out product. Rapidly heat the remaining purified crystals to quickly melt them. Then drain and pump out the purified liquid product.

As described earlier, the overall process may involve carrying out multiple stages of crystallization (processing multiple batches in one or more crystallizers) with recycle of cuts from one stage to another in countercurrent fashion.

Batchwise Forced-Circulation Falling-Film Melt Crystallization. This process involves circulating a batch of mother liquor over a cooled heat transfer surface in a falling film, normally achieved by circulating melt inside the tubes of a vertical shell and tube heat exchanger. The falling film increases fluid motion and induces shear next to the crystallizing layer, allowing for faster crystal growth compared to a static batch design. This design may also facilitate the rejection of impurities from the growing layer. The processing steps are similar to those used with a static design, but the step times may be considerably shorter, so the volumetric throughput is higher per stage of processing. On the other hand, in some cases a static design may achieve higher crystal purity with fewer crystal defects precisely because it takes so much more time, and in that case, fewer stages of crystallization may be required. The pros and cons of each method for a given application can be determined in testing. The recirculated batch process scheme was invented by Kurt Saxer in Switzerland in the 1960s [71, 72] and later marketed by Sulzer Chemtech.

Continuous Single-Stage Suspension Melt Crystallization. With this process scheme, seed crystals typically are formed in some sort of scraped freezer [73]. The scraped seed crystals are then suspended in a stirred vessel called a ripening tank or growth vessel where they grow up in size. This results in what is essentially a single stage of crystallization. For a continuous feed, multiple scraped freezers may be used. While one is growing a layer of seed crystals, the other is scraping the seeds. This continuous process scheme may provide significantly higher volumetric throughput compared to batch process schemes and lower cost per unit of product, especially for high production volumes. However, continuous processing of crystal suspensions or slurries can be quite challenging, particularly concerning the continuous washing and rinsing of crystals before they are melted and discharged as purified melt (the final product). Competing wash-column designs are offered by TNO [74], Sulzer AG [75], and GEA Messo PT [76]. All involve countercurrent washing/rinsing of crystals using a portion of the purified melt (a kind of reflux). Other types of suspension crystallizers involving use of gravity wash columns also find commercial application, such as the TSK 4C process [64].

Continuous Countercurrent Melt Crystallization. Another type of melt crystallizer is a middle-fed, differential-style, multi-stage, countercurrent design called the Brodie purifier. This design was invented by John Alfred Brodie (1908–1996) of Union Carbide Australia and patented in 1972 [77]. The Brodie design uses a middle-fed, horizontal or slanted scraper/conveyor to move solids countercurrent to the flow of melt, achieving high recovery of product in one section and purification of product in the other. Product crystals exit the conveyor and enter a vertical wash column before being discharged as purified melt. Melt is heated or cooled within the various sections to promote countercurrent partial melting and recrystallization. Crystals slowly move within the scraper/conveyor in a partly settled suspension at the bottom of the conveyor. Various other configurations also may be used [77]. The Brodie purifier is a countercurrent devise, so in principle it can achieve more than one stage of crystallization in a single unit. Although not as widely practiced as other methods, the Brodie design has the potential for highly efficient operation. It may be particularly useful for processing products that form solid solutions.

Testing Services. Various laboratory and miniplant tests may be run to assess feasibility for a given application and type of crystallizer. A number of process R&D companies and suppliers of melt crystallization technology offer testing services. In small-scale tests of batch layer-growth processes, the product solute should form well-defined crystals at the wall of a cooled surface as cooling/freezing proceeds. The resulting crystal layer should consist of crystals with sizes and shapes that allow the remaining melt (or mother liquor) to easily drain away from the crystal layer. Feasibility is assessed by sampling various melted fractions to determine the purity of the deposited layer and the degree to which impurities partition into the mother liquor and drain away from the crystal mass. Test results often are sensitive to how fast the melt is frozen – slower freezing tends to form larger, better formed crystals that facilitate the separation, but a given chemical system can behave in unexpected ways, thus the need for testing. Sometimes only small, poorly shaped crystals form with a collective consistency of a wet slushy mass that does not allow for efficient draining of mother liquor and separation of impurities. In these cases, static batch processing is not promising. Freezing from a falling film may improve the results, but again, this needs to be determined in testing. If these results are not favorable, continuous suspension melt crystallization may still be feasible, because a suspension crystallization process does not rely on layer crystallization for crystal growth and purification. Suspended crystals are grown away from the cooling surface, and they can be washed of mother liquor residues in the wash column used to discharge purified product. The Brodie countercurrent crystallizer also requires certain properties and behavior of the liquid-solid slurry to perform as needed. Information about available testing services may be found by searching the internet.

Related crystallization processes include zone melting or zone refining, a process used to purify semiconductors and metals. A rod of material is purified by heating it in such a way as to create a zone of melted material that slowly moves down the length of the rod. This purifies the rod by causing impurities to diffuse into the melt and become concentrated within the moving zone of melt [78–80]. Another related process is called eutectic freeze crystallization [81–83]. With this process scheme, two distinctly different co-crystals are crystallized at the same time by operating the process at the eutectic point. For desalination of brine, this involves crystallizing separate water crystals and separate salt crystals at the same time, as a lower-energy alternative to evaporative crystallization. Ice crystals float and the salt crystals sink, so the two different types of crystals may be separated *in-situ* by density difference. Besides desalination, applications have been described for copper sulfate + water and monoammonium phosphate + water [84].

References

[1]   K. G. Libbrecht, *Snow Crystals: A Case Study in Spontaneous Structure Formation*, Princeton University Press, Princeton, New Jersey, 2021.

[2]   X. Du, Z. Yang, Z. Jin, Y. Zhu and Z. Zhou, "A Theoretical and Experimental Study of Typical Heterogeneous Ice Nucleation Process on Auto Windshield under Nocturnal Radiative Cooling and Subfreezing Conditions," *International Journal of Heat and Mass Transfer*, vol. 136, pp. 610–626, 2019.

[3]   N. Shahidzadeh-Bonn, S. Rafaï, D. Bonn and G. Wegdam, "Salt Crystallization during Evaporation: Impact of Interfacial Properties," *Langmuir*, vol. 24, no. 16, pp. 8599–8605, 2008.

[4]   C. Willits and C. Hill, "Maple Syrup Producers Manual," *USDA Agriculture Handbook No. 134*, United States Department of Agriculture, Washington, D.C., 1976.

[47] M. R. A. Bakar, Z. K. Nagy and C. D. Rielly, "Seeded Batch Cooling Crystallization with Temperature Cycling for the Control of Size Uniformity and Polymorphic Purity of Sulfathiazole Crystals," *Organic Process Research & Development*, vol. 13, no. 6, pp. 1343–1356, 2009.

[48] E. Bormashenko, Y. Bormashenko, R. Pogreb, O. Stanevsky, G. Whyman, T. Stein, M. H. Itzhaq and Z. Barkay, "Template-assisted Crystallization and Colloidal Self-Assembly with Use of the Polymer Micrometrically Scaled Honeycomb Template," *Colloids Surfaces A Physicochem Engineering Aspects*, vol. 290, no. 1–3, pp. 273–279, 2006.

[49] J. Urbanus, J. Laven, C. P. M. Roelands, J. H. T. Horst, D. Verdoes and P. J. Jansens, "Template Induced Crystallization: A Relation between Template Properties and Template Performance," *Crystal Growth & Design*, vol. 9, no. 6, pp. 2762–2769, 2009.

[50] V. Parambil, S. K. Poornachary, J. Y. Y. Heng and R. B. H. Tan, "Template Induced Nucleation for Controlling Crystal Polymorphism: From Molecular Mechanisms to Applications in Pharmaceutical Processing," *Crystal Engineering Commerce*, vol. 21, pp. 4122–4135, 2019.

[51] P. H. Karpiński and J. Bałdyga, "Chapter 12, Batch Crystallization," in *Handbook of Industrial Crystallization*, 3rd ed., A. S. Myerson, D. Erdemir and A. Y. Lee, Eds., Cambridge University Press, Cambridge, England, 2019, pp. 370–371 (Fig. 12.26).

[52] K. Pal and Z. K. Nagy, "Parabolic Temperature Profile, Paper 330e," in *AIChE Fall Meeting*, Pittsburgh (USA), 2018.

[53] D. Q. Kern, *Process Heat Transfer*, McGraw-Hill, New York, 1950.

[54] A. M. Flynn, T. Akashige and L. Theodore, *Kern's Process Heat Transfer*, 2nd Ed., Wiley, Hoboken, New Jersey, 2018.

[55] Z. K. Nagy, M. Fujiwara and R. D. Braatz, "Modelling and Control of Combined Cooling and Antisolvent Crystallization Processes," *Journal of Process Control*, vol. 18, no. 9, pp. 856–864, 2008.

[56] J. Orehek, D. Teslić and B. Likozar, "Continuous Crystallization Processes in Pharmaceutical Manufacturing: A Review," *Organic Process Research & Development*, vol. 25, no. 1, pp. 16–42, 2021.

[57] H. J. M. Kramer and R. Lakerveld, "Chapter 7, Selection and Design of Industrial Crystallizers," in *Handbook of Industrial Crystallization*, 3rd ed., A. S. Myerson, D. Erdemir and A. Y. Lee, Eds., Cambridge University Press, Cambridge, England, 2019, pp. 197–215.

[58] C. R. Brunold, J. C. B. Hunns, M. R. Mackley and J. W. Thompson, "Experimental Observations on Flow Patterns and Energy Losses for Oscillatory Flow in Ducts Containing Sharp Edges," *Chemical Engineering Science.*, vol. 44, no. 5, pp. 1227–1244, 1989.

[59] M. R. Mackley and X. Ni, "Mixing and Dispersion in a Baffled Tube for Steady Laminar and Pulsatile Flow," *Chemical Engineering Science*, vol. 46, no. 12, pp. 3139–3151, 1991.

[60] D. Acevedo, X. Yang, Y. C. Liu, T. F. O'Connor, A. Koswara, Z. K. Nagy, R. Madurawe and C. N. Cruz, "Encrustation in Continuous Pharmaceutical Crystallization Processes – A Review," *Organic Process Research & Development.*, vol. 23, no. 6, pp. 1134–1142, 2019.

[61] T. C. Frank, J. Moore, P. A. Larsen, P. D. Patil, R. M. Whittingslow and P. A. Gillis, "Process and Apparatus for Forced Circulation Evaporative Crystallization with Large Deposit Inventory." U.S. Patent 9,776,104, 3 October 2017.

[62] J. H. Wolf, "Chapter 16, Crystallization from Solutions and Melts," in *Chemical Process Equipment*, J. R. Couper, W. R. Penney, J. R. Fair and S. M. Walas, Eds., Gulf Professional Publishing, Houston, 2005, pp. 555–582.

[63] J. C. G. Moyers and R. W. Rousseau, "Chapter 11, Crystallization Operations," in *Handbook of Separation Process Technology*, Wiley-Interscience, Hoboken, New Jersey, 1987.

[64] N. Wynn, "Section 5.3, Melt Crystallization," in *Handbook of Separation Techniques for Chemical Engineers*, 3rd ed., P. A. Schweitzer, Ed., McGraw-Hill, New York, 1997, pp. 5–43 to 5–72.

[65] P. J. Jansens and M. Matsuoka, "Melt Crystallization," in *Encyclopedia of Separation Science*, Academic Press, Cambridge, Massachusetts, 2000, pp. 966–975.

[66]   W. H. Ross, R. M. R. M. Jones and C. B. Durgin, "The Purification of Phosphoric Acid by Crystallization," *Industrial & Engineering Chemistry.*, vol. 17, no. 10, pp. 1081–1083, 1925.

[67]   A. Chen, J. Zhu, B. Wu, K. Chen and L. Ji, "Continuous Melt Suspension Crystallization of Phosphoric Acid," *Journal of Crystallization Process and Technology*, vol. 2, pp. 111–116, 2012.

[68]   M. Kumashiro, Y. Izumi, T. Hoshino, Y. Fujita and I. Hirasawa, "Purification of Aluminum Nitrate Nonahydrate by Melt Crystallization," *Journal Chemical Engineering of Japan*, vol. 44, pp. 105–109, 2011.

[69]   T. Wang, X. Li and J. Dong, "Ethylene Glycol Purification by Melt Crystallization: Removal of Short-Chain Glycol Impurities," *Industrial & Engineering Chemistry Research.*, vol. 59, no. 18, pp. 8805–8812, 2020.

[70]   J. G. D. Molinari, "The Proabd Refiner," in *Fractional Solidification*, Vol. 1, M. Zief and W. R. Wilcox, Eds., Marcel Dekker, New York, 1967, pp. 393–400.

[71]   K. Saxer, "Fractional Crystallization Process." U.S. Patent 3,621,664, 23 November 1971.

[72]   K. Saxer, "A Large-Scale Industrial Process for Fractional Crystallization," *Chemische Produktion*, vol. 9, no. 11, pp. 38–39, 1980.

[73]   J. Ulrich and T. Stelzer, "Melt Crystallization," in *Handbook of Industrial Crystallization*, 3rd ed., A. S. Myerson, D. Erdemir and A. Y. Lee, Eds., Cambridge University Press, Cambridge, University, 2019, pp. 266–289.

[74]   Ultra-Pure Chemicals Production Using the TNO Hydraulic Wash Column (TNO HWC ®), TNO, [Online]. Available: https://www.tno.nl/media/1759/melt-crystallization2.pdf. [accessed 11 December Dec. 11, 2022].

[75]   "Suspension Crystallization Technology," Sulzer Chemtech, [Online]. Available: https://www.sulzer.com/-/media/files/products/process-techology/reaction_technology/brochures/suspension_crystallization_technology.pdf?la=en. [Accessed 11 December 2022].

[76]   GEA Suspension Melt Crystallization: The Efficient Purification Alternative, GEA Messo PT, [Online]. Available: https://www.gea.com/it/binaries/crystallization-melt-purification-chemical-gea_tcm39-34853.pdf. [Accessed 11 December 2020].

[77]   J. A. Brodie, "Solid-liquid Continuous Countercurrent Purifier Method and Apparatus." U.S. Patent 3,645,699, 29 February 1972.

[78]   W. G. Pfann, *Zone Melting*, 2nd ed., Wiley, Hoboken, New Jersey, 1966.

[79]   H. Schildknecht, *Zone Melting*, Verlag Chemie (now Wiley-VCH, Hoboken, New Jersey), 1966.

[80]   X. Zhang, S. Friedrich and B. Friedrich, "Production of High Purity Metals: A Review on Zone Refining Process," *Journal of Crystallization Process and Technology*, vol. 8, pp. 33–35, 2018.

[81]   A. J. Barduhn and A. Manudhane, "Temperatures Required for Eutectic Freeze Crystallization of Natural Waters," *Desalination*, vol. 28, no. 3, pp. 233–241, 1979.

[82]   F. Van Der Ham, G. Witkamp, J. De Graauw and G. Van Rosmalen, "Eutectic Freeze Crystallization. Simultaneous Formation and Separation of Two Solid Phases," *Journal of Crystal Growth*, vol. 198/199, pp. 744–749, 1999.

[83]   M. J. Fernández-Torres, D. G. Randall, R. Melamu and H. Von Blottnitz, "A Comparative Life Cycle Assessment of Eutectic Freeze Crystallization and Evaporative Crystallization for the Treatment of Saline Wastewater," *Desalination*, vol. 306, pp. 17–23, 2012.

[84]   K.-L. Wu, H.-Y. Wang and J. D. Ward, "Economic Comparison of Crystallization Technologies for Different Chemical Products," *Industrial & Engineering Chemistry Research*, vol. 57, no. 37, pp. 12444–12457, 2018.

12 Membrane-Based Separations: Adding a Semipermeable Barrier Between Phases

In devising a membrane-based separation process, designers can look to nature for examples and inspiration – as in the functioning of the human body (molecular transport in kidneys and lungs) and in the functioning of plants (transport of gases, water, and nutrients into and out of cells). The vision is very attractive: If we can mimic nature in devising materials and process schemes for selective transfer of desired components across a thin, semipermeable barrier, then we will have an elegantly simple process for mass transfer. With such a barrier, we can fix the interfacial area and precisely (and independently) control the flow rates of feed and product streams on each side, without doing the work of intermixing phases and later separating them back apart, as we do with other methods. But unlike nature's membranes, the use of man-made membranes at industrial scale typically requires use of significantly elevated feed pressure to obtain a sufficiently high production rate (with some exceptions). Nature's processes often involve biochemical transformations not present in industrial applications that help drive the process, and many are too slow to be economically attractive at industrial scale. And in industrial service, typical feeds are not clean; so tendencies to foul or plug the membrane surface can be severe and difficult to overcome. Even so, a number of membrane-based industrial separations have proved to be highly successful, and others show promise for the future. In this book, we focus on well-developed membrane-based processes typical of current large-scale industrial applications, mainly pressure-driven liquid separations such as desalination and ultrafiltration, and pressure-driven gas separations. For a brief discussion of enhanced (and promising) special cases, see Section 6.1.2: Drivers for Membrane-Based Separations.

The concept of using a semipermeable membrane for a specific separation purpose began with dialysis. In 1850, Thomas Graham demonstrated the use of a membrane (from an ox bladder) to remove excess urea from blood, and called the process dialysis [1]. Georg Haas performed the first human hemodialysis experiments in 1924, and in 1940, Willem Kolff and Hendrik Berk developed their high-surface-area, rotating-drum artificial kidney [2]. The same basic technology, with continual improvement over the years, is used today [3].

Membranes were commercialized for water purification (microfiltration) in the 1930s based on work by Richard Zsigmondy and Wilhelm Bachmann in the 1920s. In the 1950s, asymmetric (or composite) membranes were developed, comprised of a thin skin or barrier resin on top of a thicker porous support. This gave lower mass transfer resistance for good flux along with sufficient robustness to withstand elevated operating pressures, which made way for new applications for both gas and liquid separations. Industrial-scale separations began to appear in the 1960s, but it was not until the 1970s and early 1980s that the use of membranes in the chemical industry really began to take off, with the introduction of membrane-based processes for

https://doi.org/10.1515/9783110695052-017

desalination (reverse osmosis), recovery of hydrogen from nitrogen (gas permeation), and removal of water from alcohol (pervaporation) (see Section 3.2: Pioneers).

As mentioned earlier, the potential benefits that membranes offer are very attractive:

– Potential for reduction in energy consumption compared to distillation (in some but not all cases)
– Inherent isolation of feed and product phases (no intermixing of phases that would require phase separation after processing)
– Independent control of feed and product flow rates
– No need to add solvents or other separating agents (other than the membrane)
– No need for a regeneration step (except for periodic cleaning of the membrane)
– Straight-forward specification of commercial-scale operations using multiple membrane modules

The last point means that uncertainties in scale-up of the technology are minimal. Scale-up is achieved by packing as much membrane area into a single module as is practical. Then, large-scale production rates are achieved by specifying multiple commercial-size modules (called scale-out). But this also means that the process cannot benefit from the economy of scale offered by a single large unit.

12.1 General Features

Membrane-based processes generally are most effective when the permeate component is present in moderate or large concentration in the feed, because otherwise flux and selectivity tend to be low. It may also be desirable for the product to be produced as the retentate at the elevated feed pressure whenever the product stream is to be used under pressure. Membranes tend to be used for non-sharp separations per stage of contacting, with no more than three membrane stages employed overall. If needed, final purification or polishing operations may be performed by using another separation method such as adsorption. For more discussion, see Section 6.7.5: Membrane Permeance and Separation Factor.

Commercial Applications. Major industrial applications of membrane-based separations include:

– Microfiltration, ultrafiltration, and nanofiltration of aqueous liquids (water treatment)
– Reverse osmosis of aqueous salt solutions (desalination)
– Degassing of liquids
– Gas separations involving N_2, H_2, NH_3, CO_2, H_2O, or low molecular weight hydrocarbons
– Pervaporation of alcohol + water mixtures to remove water
– Nanofiltration of organic liquids.

<u>Membrane Materials</u>. Just as the properties of an adsorbent material are key to adsorption process design, the most important factor for a membrane-base separation is the membrane material and how it performs. Commercially available membrane materials include:
- Natural and synthetic rubber
- Composite polymers (an ultrathin discriminating barrier film supported on a thicker porous layer)
- Crosslinked composite polymers (for high thermal and chemical stability)
- Ceramic membranes (for even higher thermal and chemical stability)
- Microporous membranes (for microfiltration and ultrafiltration)

Commercial membrane materials include polyamide, polyimide, polyamide-imide, cellulose acetate, polysulfone, polyethersulfone, poly(phenyleneoxide), polycarbonate, aramid, and silicone rubber. Ceramic membranes may be made with alumina, silica, titania, silicon carbide, or other inorganic materials. Commercially available membranes can be found by searching the internet for specific membrane materials and separations.

<u>Membrane Modules</u>. Several types of membrane modules are available:
- Spiral-wound sheets
- Hollow-fiber bundles
- Tubular shell-and-tube or monolith
- Plate-and-frame

In principle, a given application may employ any one of these configurations. Membrane-based separations are scaled up by designing large individual modules, but there is an economical limit to the size of a single module. As mentioned earlier, the required large-scale production rate is then obtained by connecting multiple modules together within a large array of modules. For example, a single module designed for seawater desalination may be constructed by first connecting several individual spiral-wound elements in series within a single pressure vessel. The resulting module may be about 0.2 meters (8 inches) in diameter and up to about 6 meters (20 ft) in length, and contain about 60 m^3 of membrane area. The size of the module and the amount of area it contains will vary depending on the kind of module and the manufacturer's particular design details. Multiple modules may be assembled into a larger parallel array, and one or more of these arrays may be used for a given installation. Specific information may be obtained by contacting suppliers and by searching the internet.

Various pre-treatment processes are usually needed, such as pre-filtration of the feed to prevent plugging of the membrane material. Various designs are used to control flow to minimize fouling of membrane surfaces, provide efficient recovery of feed-side pressure, and minimize pressure losses. The choice of the style of membrane module depends on how the membrane and the module can most-effectively be manufactured (using flat sheets or hollow fibers), as well as how much area can be packed into a given device (m^2/m^3), and various operating factors such as the ability to withstand high feed pressure and resist fouling.

A spiral-wound module is made by winding sheets of various flat membrane materials. Typically, this involves a support web, a microporous inner layer, and a thin barrier layer, with spacers placed in-between the sheets and liquid collectors placed at the ends. Spiral-wound modules are designed to withstand high cross-membrane pressure drop, and most applications of reverse osmosis membranes use the spiral-wound type. In this application, the feed-side pressure may be on the order of 1000 psi. Many gas separations also use a spiral-wound design.

Hollow fiber bundles are made by gluing or thermally welding fine polymer fibers into a shell-and-tube configuration. They are used for gas separations, although spiral-wound modules also are used (as noted above). Hollow-fiber bundles are also used for ultrafiltration of water in wastewater oxygenation units because the movement of air bubbles on the shell-side of the fibers helps to minimize fouling. Ceramic membranes are offered in a larger (tubular) shell-and-tube style configuration, as a monolith structure, or sometimes as flat sheets.

The plate-and-frame configuration is normally used for testing membranes or for small-scale applications. It is generally not the most cost-effective design for large-scale operation unless the particular membrane is only available in flat sheets.

Test Methods. Membrane performance must be determined in laboratory and miniplant tests using representative multicomponent feeds. As mentioned above, the presence of multiple permeating species can affect the flux and selectivity for a given component. Polymeric membranes can absorb specific components, causing them to swell and alter their transport properties. So, test results obtained using single-component gases and liquids (to measure pure-component permeance) may be very different from results obtained using multicomponent mixtures. Also, temperatures above ambient should be evaluated, because operating at an elevated temperature may allow for significantly higher flux, without significant degradation in selectivity (and it may even allow some improvement in some cases) – up to some maximum temperature that varies for different membranes and different applications (different chemical systems). For most commercial polymeric membranes, it seems that the maximum operating temperature is less than 100 °C or so, but this should be determined in testing. Ceramic membranes and certain specialty polymeric membranes designed to achieve greater thermal and chemical stability may have advantages in this regard (assuming their transport properties are satisfactory). Examples include CoorsTek® and HybSi®ceramic membranes and Celazole® PBI polybenzimidazole membrane. Others may also be available (Search the internet).

Small-scale test apparatus generally are simple devices used to hold a small disc of flat membrane material, plus auxiliary equipment needed to generate feed pressure, control temperature, and collect retentate and permeate streams. The apparatus may be of the dead-end type or cross-flow type (or some other configuration of inter-est in process development). The apparatus should be designed to allow convenient sampling of all streams, either periodically or on-line, as well as quantitative mea-surement of flux using some type of volume flow sensor or weight-change sensor. Var-ious kinds of test apparatus are offered by vendors. Information can generally be found by searching the internet for membrane testing equipment, test apparatus, or test modules. Some of these apparatus allow testing of hollow fibers at small scale. Those membranes showing promise can then be tested in larger-scale membrane modules (spiral-wound or hollow-fiber bundles). Some vendors also offer testing serv-ices for specific applications.

12.2 Water Treatment

Water treatment is one of the most successful commercial application areas for mem-branes. Applications include desalination of seawater, forwarding of treated water in wastewater treatment plants, concentration of aqueous proteins, removal of harmful bacteria from drinking water and beverages, and processing of milk.

The commercial use of membranes for water treatment got its start in the 1960s. Commercial desalination membranes were introduced by Sidney Loeb and Srinivasa Sourirajan in 1960 [4]. These were cellulose-acetate membranes. In the late 1970s, com-posite membranes consisting of a thin polyamide barrier film on a polysulfone support were offered by Filmtec [5, 6]. Today, a great variety of membranes are available for a wide range of water treatment applications, each designed to retain molecules of differ-ent sizes (different molecular weight cut-off). These generally are classified as follows:
- Microfiltration, 0.1 to 10 μm (100 to 10,000 nm), retaining yeast, fungus, bacteria, and oil emulsions.
- Ultrafiltration, 0.01 to 0.1 μm (10 to 100 nm), retaining colloids, virus, proteins, and polysaccharides.
- Nanofiltration, 0.001 to 0.01 μm (1 to 10 nm), retaining sugars, small organic mole-cules, and multivalent ions.
- Reverse Osmosis, 0.0001 to 0.001 μm (0.1 to 1 nm), retaining inorganic ions.

Detailed information about these methods can be found in many handbooks and monographs, some of which are listed in the bibliography (for Chapter 12). Process schemes include cross-flow and dead-end filtration, as well as diafiltration in which small-molecule impurities are removed by permeation along with the solvent. Clean-ing methods include periodic backwashing with solvent and treatment with chemi-cal cleaners. Special pressure-exchanger designs allow for significant reduction in energy consumption for processes that require high feed pressures to overcome high osmotic pressure for product permeation, such as desalination. These work by using the hydraulic pressure of the retentate stream leaving a membrane module to help drive the feed pump for the same or another module [7]. Desalination of brackish water (salty to the taste) requires pressures of 250 to 400 psi (1700 to 2800 kPa), depending on feed composition and water recovery (the percentage of water recovered from the feed into the permeate). Desalination of seawater (containing about 3.5 wt% salt) requires pressures from 800 to 1000 psi (5500 to 6900 kPa) for high water recovery. See Section 6.7.5: Membrane Permeance and Separation Factor for discussion of osmotic pressure.

Manufacturers of membranes and membrane modules for water treatment include Pall Corp., 3M, FilmTec, Sanborn Technologies, Dupont Water Solutions, Keen-Sen Technology Company, Jay Water Management, Koch Membranes, Suez Water Technologies (Osmonics), Hydranautics, and many others. Search the internet for information about these and other manufacturers/suppliers of membrane modules and complete units.

12.3 Organic Solvent Nanofiltration

Following the success of membranes used for water treatment, the application of membranes for concentration of large molecules in organic solvents emerged in the early 2000s, especially for the recovery and reuse of expensive homogeneous organocatalysts [8–10]. For example, a Matrimid polyamide-imide membrane has been examined for its suitability as a membrane material for recovery of homogeneous catalyst for a 1-dodecene hydroformylation reaction catalyzed by a rhodium-triphenylphosphine transition metal catalyst [11].

Called Organic Solvent Nanofiltration (OSN), this new field of applications borrows many of the same process concepts and types of membranes used for water treatment. Like their aqueous feed counterparts, membranes used for organic solvent nanofiltration are rated according to their molecular size cut-off or molecular weight cut-off (MWCO), often expressed in Dalton units, Da. But the switch from an aqueous to an organic solvent environment presents some new challenges:
–   Swelling and degradation of polymeric membranes over time, and how this can limit membrane performance and service life
–   This can affect how sharp the molecular weight cut-off can be
–   And the ability to achieve high retention of large molecules and adequate flux of solvent

Post-fabrication cross-linking of the polymer matrix improves the robustness of the membrane. An early example is polyimide crosslinked with amines to form polyamide-imides [8]. Supported ceramic membranes also have been proposed to address this issue. Organic catalysts are typically very large (> 400 Da) compared to solvent molecules; so fairly sharp separations are possible. A number of membranes with specific pore structures and cut-off values are offered commercially. At present, these include polymeric OSN membranes with tradenames Duramem and Puramem (Evonik), Matrimid (Huntsman), Starmem (W. R. Grace), SelRO (Koch Membrane Systems), and SolSep (SolSep BV). Membranes developed for water treatment, such as Desal-5 and Desal-5-DK, manufactured by Osmonics, may also be effective for certain OSN applications [9, 10].

In principle, because of the large difference in molecular size between the solvent and the product, selectivity may be reasonably high. In that case, a single stage of membrane contacting may be all that is needed to recover solvents and concentrate large organic molecules. A number of processing schemes may be employed. For example, an OSN membrane that selectively permeates solvent and small-molecule impurities may be used to wash away impurities and concentrate the product (a diafiltration process). This may involve permeation of solvent containing impurities, with periodic addition of replacement solvent until satisfactory purity is achieved. This process scheme may allow product purification while maintaining reasonably high product retention. Further permeation of solvent may be used for product concentration. An OSN membrane may also be used to remove solvent and replace it with another solvent over several cycles of washing with replacement solvent (purification and solvent exchange). Of course, such units may be configured in a multistage countercurrent process scheme, if needed.

12.4 Gas Separations

One of the first commercial gas separation membranes was the Permea membrane developed by Monsanto for H_2/N_2 separation, introduced in 1979 [12]. Today, many other membrane-based gas separations are commercially successful. Most involve moderately sharp separations, without the need to produce very high purity gases. These applications include dehumidification (water from air), air separation, hydrogen from ammonia, and CO_2 from natural gas. Typical membrane selectivities are reported to vary from $\alpha > 200$ for H_2O/air with a polyimide/ polysulfone hollow-fiber module; $\alpha > 4$ to 6 for O_2/N_2 separation with a polyimide/poly(phenyleneoxide) hollow-fiber module; $\alpha \sim 100$ for H_2/N_2 separation with a polysulfone/polyimide hollow-fiber module; and $\alpha \sim 15$ to 40 for CO_2/CH_4 with a cellulose acetate/polyimide spiral- or hollow-fiber module [13]. Selectivity is defined as the ratio of permeabilities for the two key feed components to be separated (see Section 6.7.5: Membrane Permeance and Separation Factor). With the processing of gases, it is important to ensure that the feed gas is clean, with no entrained oils or particulates that could foul the membrane. Sometimes, it is necessary to prefilter the feed to ensure this is the case.

The use of membranes for air separation is well developed and provides an instructive example of how membranes may be used for gas separation. Various process designs are outlined in Figure 12.1. Detailed discussions of process variables and design tradeoffs are given elsewhere [14, 15].

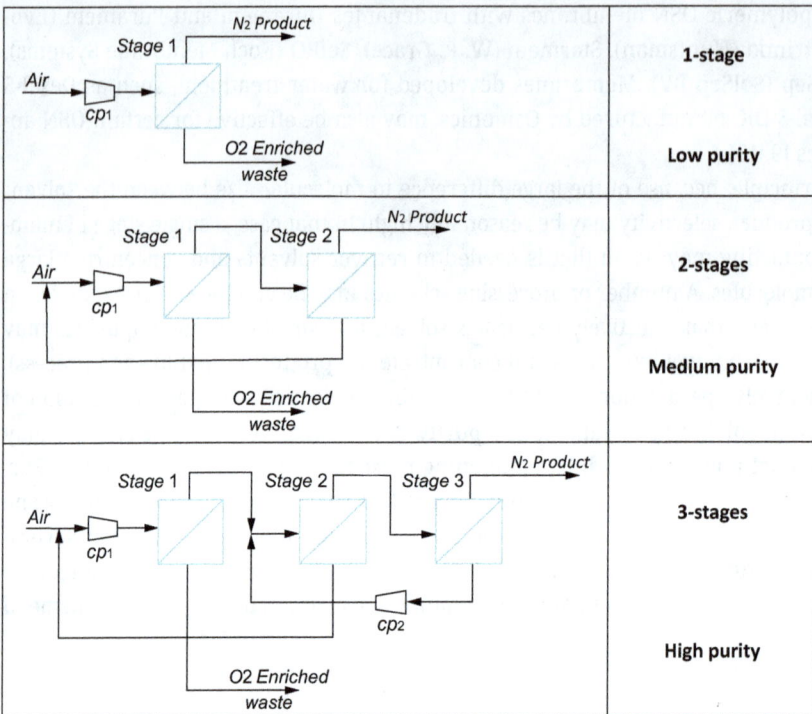

Figure 12.1: Process Configurations for Membrane-Based Air Separation. Taken from ref. [14]. Reprinted with permission from *Chem. Eng. Sci.*, copyright 2019, Elsevier.

Various multistage membrane cascades may be used to minimize the required number of compressors [16–18]. One concept used for design is to try and match (as closely as possible) the composition and pressure of a stream to be recycled with those at the point of recycle, to avoid degrading the separation and letting down pressure in such a way that it does not provide useful pressure somewhere else in the process. An example of pressure conservation is illustrated in Figure 12.2.

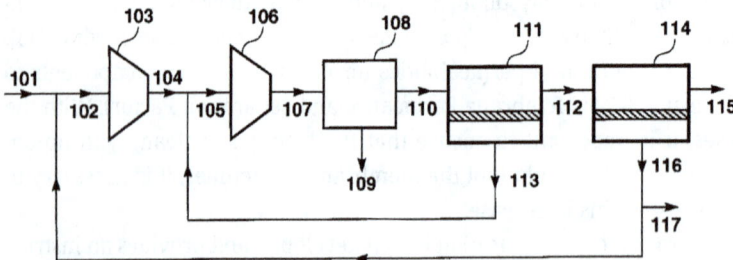

Figure 12.2: Pressure Matching Process Scheme for Membrane-Based Gas Separation. Taken from ref. [18]. Reprinted from a U.S. patent publication in the public domain.

Commercial processes generally involve rough-cut separations, not purifications. The issue has to do with difficulty in making a polymeric membrane that has both high selectivity and high permeability or flux. High selectivity tends to require the use of a more rigid polymer, but rigidity tends to reduce flux. Furthermore, polymeric membranes tend to absorb one or more of the gas phase components, causing the membrane to swell to some extent, and this can degrade performance. Current and future trends in the development of polymeric and ceramic membranes are discussed elsewhere [19–23]. A promising new application involves the development of molecular sieving membranes and processes for separating olefin from paraffin hydrocarbons (propylene/propane and ethylene/ethane) [24, 25]. Another area of research involves multistage separation of CO_2/N_2 for post-combustion capture of CO_2 from exhaust gases [26].

12.5 Pervaporation

Pervaporation is a membrane technology used to separate liquid mixtures containing volatile components. It involves processing liquid on the feed or retentate side of the membrane, with selective evaporation of solute into a vapor or gas phase on the permeate side. The permeation mechanism involves liquid-to-vapor phase change, so the feed side of the membrane must provide the necessary heat of vaporization. Pervaporation as a concept was introduced by Philip Kober in 1917 [27], but it seems it was not until the 1960s that it began to be developed [28]. And it was not until the early 1980s that it was first commercialized at large scale. At that time, the company, Gesellschaft für Trenntechnik (GFT at Hameln, Germany, now a part of Sulzer Ltd.), commercialized pervaporation for drying alcohols using a crosslinked membrane comprised of polyvinylalcohol supported on polyacrylonitrile [29].

Instead of requiring a high-pressure pump to drive sufficient flux, a permeation process requires only moderate pressure on the feed side of the membrane and a cold condenser plus vacuum pump on the permeate side. A slight stream of sweep gas may be added to the permeate stream to enhance the pervaporation rate by further reducing the partial pressure of the permeating component. Commercial operations handle feed mixtures containing polar + relatively non-polar components, especially water + organics. Depending on the type of membrane used (whether hydrophilic or hydrophobic), relatively polar or non-polar components selectively permeate the membrane and evaporate on the permeate side. The membrane serves to enhance the separation factor compared to what would be achieved by directly vaporizing the feed. For this reason, pervaporation may be considered a kind of enhanced distillation (see Section 6.7.5: Membrane Permeance and Separation Factor).

Figure 12.3 shows a three-stage pervaporation process for breaking the ethanol + water azeotrope in combination with distillation [30]. In this application, water preferentially permeates the hydrophilic membrane. The feed to each membrane module is heated to counter evaporative cooling and maintain a desired membrane temperature.

The permeate vapor is condensable; so the main condenser is placed in front of the vacuum pump and the exhaust gases are further cooled to remove residual condensables. Chilled coolant is used to obtain nearly complete condensation. Because the process permeates water, it is quite flexible, in that it can be used to dry a variety of volatile organics (as can distillation + pressure swing adsorption). Additional discussions of this type of process and other pervaporation processes are available elsewhere [31–34].

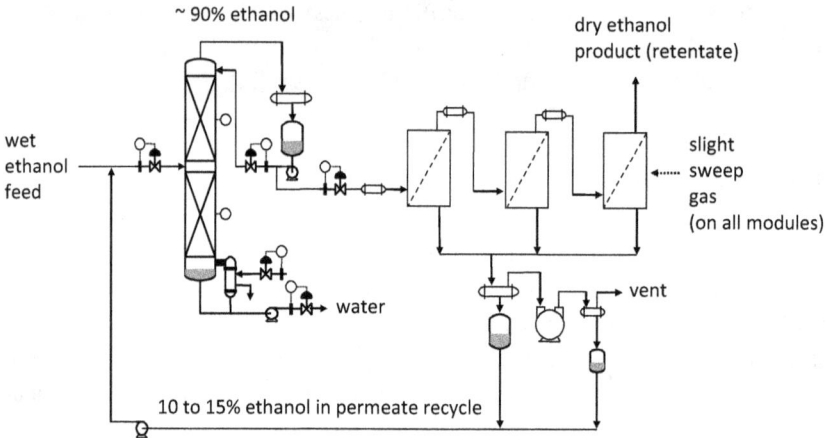

Figure 12.3: Distillation + Three-Stage Pervaporation Process Scheme. Optional heat exchangers for cross-exchange between streams are not shown. Adapted from Ref. [30], with permission.

References

[1]   G. Richet, "Early History of Uremia," *Kidney International*, vol. 33, no. 5, pp. 1013–1015, 1988.
[2]   D. N. Paskalev, "Georg Haas (1886–1971): The Forgotten Hemodialysis Pioneer," *Dialysis & Transplantation*, vol. 30, no. 12, pp. 828–832, 2001.
[3]   N. Said, W. J. Lau, Y.-C. Ho, S. K. Lim, M. N. Z. Abidin and A. F. Ismail, "A Review of Commercial Developments and Recent Laboratory Research of Dialyzers and Membranes for Hemodialysis Application," *Membranes*, vol. 11, no. 10, article no. 767, 2021.
[4]   S. Loeb, "The Loeb-Sourirajan Membrane: How It Came About," in *Synthetic Membranes (ACS Symposium Series)*, A. F. Turbak, Ed., American Chemical Society, Washington, D.C, 1981, pp. 1–9.
[5]   L. Wiles and E. Peirtsegaele, "Reverse Osmosis: A History and Explanation of the Technology and How It Became so Important for Desalination," in *79th International Water Conference*, www.eswp.com/water), 15 pages, Scottsdale, Arizona (USA), Nov. 4–8, 2018.
[6]   W. J. Lau and X. Feng, "60 Years of the Loeb-Sourirajan Membrane: From Fundamental Research to Industrial Application. Introduction to Volume 168, Special Issue.," *Chemical Engineering Research and Design*, vol. 168, p. 297, 2021.

[7]   A. A. Alanezi, A. Altaee and A. O. Sharif, "The Effect of Energy Recovery Device and Feed Flow Rate on the Energy Efficiency of Reverse Osmosis Process," *Chemical Engineering Research and Design*, vol. 158, pp. 12–13, 2020.

[8]   J. T. Scarpello, D. Nair, L. M. Freitas dos Santos, L. S. White and A. G. Livingston, "The Separation of Homogeneous Organometallic Catalysts Using Solvent Resistant Nanofiltration," *Journal of Membrane Science*, vol. 203, pp. 71–85, 2002.

[9]   P. Marchetti, M. F. Jimenez Soloman, G. Szekely and A. G. Livingston, "Molecular Separation with Organic Solvent Nanofiltration: A Critical Review," *Chemical Reviews*, vol. 114, no. 21, pp. 10735–10806, 2014.

[10]   P. Kisszékelyi, S. Nagy, Z. Fehér, P. Huszthy and J. Kupai, "Membrane-supported Recovery of Homogeneous Organocatalysts: A Review," *Chemistry*, vol. 2, no. 3, pp. 742–758, 2020.

[11]   D. J. Desrocher, Membranes for the Recovery of Homogeneous Catalyst, Atlanta: PhD dissertation, Georgia Institute of Technology, Atlanta, Georgia, 2004.

[12]   P. Bernardo and G. Clarizia, "30 Years of Membrane Technology for Gas Separation," *Chemical Engineering Transactions*, vol. 32, pp. 1999–2004, 2013.

[13]   C. Castel, R. Bounaceur and E. Favre, "Engineering of Membrane Gas Separation Processes: State of the Art and Prospects," *Journal of Membrane Science and Research*, vol. 6, no. 3, pp. 295–303, July 2020.

[14]   M. Bozorg, B. Addis, V. Piccialli, Á. A. Ramírez-Santos, C. Castel, I. Pinnau and E. Favre, "Polymeric Membrane Materials for Nitrogen Production from Air: A Process Synthesis Study," *Chemical Engineering Science*, vol. 207, pp. 1196–1213, 2019.

[15]   M. Micari and K. V. Agrawal, "Oxygen Enrichment of Air: Performance Guidelines for Membranes Based on Techno-Economic Assessment," *Journal of Membrane Science*, vol. 641, no. 1, p. 119883, 2022.

[16]   R. Agrawal and J. Xu, "Gas-Separation Membrane Cascades Utilizing Limited Numbers of Compressors," *AIChE Journal*, vol. 42, no. 80, pp. 2141–2154, 1996.

[17]   J. A. C. Velasco, R. T. Gooty, M. Tawarmalani and R. Agrawal, "Optimal Design of Membrane Cascades for Gaseous and Liquid Mixtures via MINLP," *Journal of Membrane Science*, vol. 636, article no. 119514, 2021.

[18]   D. Gottschlich and M. Ringer, "Membrane Gas Separation Process with Compressor Interstage Recycle," U.S. Patent 6,428,606, 6 August 2002.

[19]   R. W. Baker, "Future Directions of Membrane Gas Separation Technology," *Industrial & Engineering Chemistry Research*, vol. 41, no. 6, pp. 1393–1411, 2002.

[20]   D. F. Sanders, Z. P. Smith, R. Guo, L. M. Robeson, J. E. McGrath, D. R. Paul and B. D. Freeman, "Energy-efficient Polymeric Gas Separation Membranes for a Sustainable Future: A Review," *Polymer*, vol. 54, no. 18, pp. 4729–4761, 2013.

[21]   R. S. K. Valappil, N. Ghasem and M. Al-Marzouqi, "Current and Future Trends in Polymer Membrane-Based Gas Separation Technology: A Comprehensive Review," *Journal of Industrial and Engineering Chemistry*, vol. 98, pp. 103–129, 2021.

[22]   M.-D. Jia, B. Chen, R. D. Noble and J. L. Falconer, "Ceramic-Zeolite Composite Membranes and their Application for Separation of Vapor/Gas Mixtures," *Journal of Membrane Science*, vol. 90, no. 1–2, pp. 1–10, 1994.

[23]   V. Gitis and G. Rothenberg, *Ceramic Membranes: New Opportunities and Practical Applications*, Wiley-VCH, Hoboken, New Jersey, 2016.

[24]   C. Staudt-Bickel and W. J. Koros, "Olefin/Paraffin Gas Separations with 6FDA-Based Polyimide Membranes,," *Journal of Membrane Science*, vol. 170, no. 2, pp. 205–214, 2000.

[25]   R. Abhishek, S. R. Venna, G. Rogers, L. Tang, T. C. Fitzgibbons, J. Liu, H. McCurry, D. J. Vickery, D. Flick and B. Fish, *Membranes for Olefin-Paraffin Separation: An Industrial Perspective, Proceedings of the National Academy of Sciences (USA)*, vol. 118, no. 37, doi: 10.1073/pnas.2022194118, 9 pages, 2021.

[26] L. Zhao, E. Riensche, L. Blum and D. Stolten, "Multi-Stage Gas Separation Membrane Processes Used in Post-Combustion Capture: Energetic and Economic Analyses," *Journal of Membrane Science*, vol. 359, no. 1–2, pp. 160–172, 2010.

[27] P. A. Kober, "Pervaporation, Perstillation, and Percrystallization," *Journal of the American Chemical Society*, vol. 39, no. 5, pp. 944–948, 1917.

[28] R. C. Binning, R. J. Lee, J. F. Jennings and E. C. Martin, "Separation of Liquid Mixtures by Permeation," *Industrial & Engineering Chemistry*, vol. 53, no. 1, pp. 45–50, 1961.

[29] K. S. Lakshmy, D. Lal, A. Nair, A. Babu, H. Das, N. Govind, M. Dmitrenko, A. Kuzminova, A. Korniak, A. Penkova, A. Tharayil and S. Thomas, "Pervaporation as a Successful Tool in the Treatment of Industrial Liquid Mixtures," *Polymers*, vol. 14, no. 8, article no. 1604, 2022.

[30] W. Kujawski and Ł. Zieliński, "Bioethanol – One of the Renewable Energy Sources," *Environmental Protection Engineering*, vol. 32, no. 1, pp. 143–150, 2006.

[31] G. Liu, W. Wei and W. Jin, "Pervaporation Membranes for Biobutanol Production," *ACS Sustainable Chemical Engineering*, vol. 2, no. 4, pp. 546–560, 2014.

[32] F. Lipnizki, R. W. Field and P.-K. Ten, "Pervaporation-Based Hybrid Process: A Review of Process Design, Applications, and Economics," *Journal of Membrane Science*, vol. 153, no. 2, pp. 183–210, 1999.

[33] T. Zhu, Q. Xia, J. Zuo, S. Liu, X. Yu and Y. Wang, "Recent Advances of Thin Film Composite Membranes for Pervaporation Applications: A Comprehensive Review," *Advanced Membranes*, vol. 1, article no. 100008, 2021.

[34] A. J. Toth, B. Szilagyi, D. Fozer, E. Haaz, A. K. M. Selim, M. Szőri, B. Viskolcz and P. Mizsey, "Membrane Flash Index: Powerful and Perspicuous Help for Efficient Separation System Design," *ACS Omega*, vol. 5, no. 25, pp. 15136–15145, 2020.

13 Specialized Separations

Many separation processes have a special feature that sets them apart from the norm. This may be a novel solvent or adsorbent or membrane, or some new type of equipment for intermixing phases, or a new way of combining two or more separation processes together for better separation performance. However, in this section we focus on separation processes with specialized requirements or unique approaches to separation.

13.1 Bioseparations

Numerous chemical products are manufactured by harnessing a living organism. Biomanufacturing typically is carried out at industrial scale by growing cultures of bacteria or yeast or by growing specialized crops. The organism serves as the chemical "plant" in which the desired product is made, and the product may consist of one specific compound or a range of compounds. Normally, the desired product must be isolated from the biomass and later purified using various kinds of bioseparations. The resulting bio-derived product may be used directly as an ingredient in formulating a finished commercial product, or it may need to be modified in one or more subsequent reaction steps before it can be used. Biomanufactured products include fragrance and flavor compounds, beer and wine, refined alcoholic spirits, biofuels, vinegar, carboxylic acids, antibiotics, agricultural pesticides, amino acids, peptides, industrial enzymes, and various kinds of therapeutic proteins including biopharmaceuticals and antibodies. Detailed discussions of bioprocessing technology are available elsewhere [1–5].

With so many bio-derived products made using such a wide variety of organisms, the field of bioseparations has become very large and diverse. However, bioseparations may be classified according to two basic functions: 1) primary recovery of a desired product away from the biomass; and 2) concentration and purification of the product after recovery. In the course of doing either kind of separation, it may also be important to retain the activity of the bio-derived product necessary for its intended use. This is not always an issue, but normally it is critical for successful recovery of proteins, for example. Furthermore, bioseparations often involve feeds containing many unwanted components, and the desired component often is present in much lower concentration. For example, many microbial fermentations are single cell cultures that contain slimy biomass solids and numerous unwanted compounds in solution, in addition to one or more desired compounds produced by the microbe. Isolating a desired compound often involves primary recovery out of the cells and then out of the broth, followed by concentration and purification to yield the final product.

https://doi.org/10.1515/9783110695052-018

Methods used for primary recovery include cell disruption or lysis, cutting or grinding of plant material, filtration, centrifugation, washing, precipitation, sedimentation, solvent extraction, filter-cake or sediment pressing, flocculation, floatation, and skimming. These are rough-cut methods intended to recover the desired components in very high yield while rejecting a majority of the undesired components such as plant and cell debris. Other specialized methods include the use of pulsed adsorbers for periodic flushing of biomass solids from the adsorbent bed, use of evaporation and distillation for recovery of volatile products from biomass-containing feeds, and use of extraction, adsorption, and crystallization or precipitation for recovery of nonvolatile products.

Concentration and purification methods include specialized versions of the methods we have already discussed in Part III. Sometimes these methods must be carried out at mild conditions or with short contact times to avoid damage to the bio-derived product. Use of the Podbielniak centrifugal liquid-liquid extractor is a classic example of the need to minimize contact time in isolating a desired biomolecule such as penicillin from fermentation broth. (See Section 3.2, Pioneers). As another example, a number of specialized protein separations are specifically designed to maintain an aqueous environment to minimize loss of protein activity. These include methods with names such as aqueous two-phase extraction, reverse micellar extraction (involving molecular-scale dispersion of water in an organic liquid), ultrafiltration or nanofiltration of an aqueous feed, and affinity chromatography of an aqueous feed. Affinity chromatography involves the use of specialized adsorbents with highly-specific binding interactions for highly selective adsorption of select biomolecules. Some affinity media may even be non-porous materials, with essentially no diffusional resistance to mass transfer (which can be an issue for separation of large proteins) but with limited surface area.

Chromatography methods are widely used for purification of proteins and other biomolecules. These applications may benefit from short and fat column designs to minimize the volume of very expensive affinity adsorbents. Separation is optimized by using small, uniformly-sized (often spherical) adsorbent particles. Specific packing methods are needed to achieve highly uniform packing of the media particles. Specialized feed distributors provide extremely uniform distribution of feed with very little extra-column volume in the distribution zone. These and other design considerations are discussed elsewhere [6]. Separation via adsorption/chromatography may be further enhanced by using specialized multi-bed process configurations as in simulated moving bed chromatography, achieving good productivity with minimal use of mobile phase. Well known examples include fructose/glucose separation (See Section 10.8, Simulated Moving-Bed Chromatography). But other multi-bed configurations also are used, especially when the goal is to isolate multiple fractions beyond a binary separation. A number of so-called counter-current liquid-liquid chromatography methods are of particular interest for low-volume, high-value bioseparations [7].

13.2 Bio-designed Separations

A bio-designed separation is a type of intensified bioseparation operation in which the organism is deliberately modified (designed) so that the desired bio-product is secreted or produced within a specific part of the organism. This makes primary recovery and subsequent purification easier to accomplish using the kinds of bioseparation methods discussed above.

This approach is sometimes used to enhance microbial or plant-based production of chemicals. A typical microbial fermentation contains many unwanted contaminants such as cell debris plus unwanted proteins, residual sugars, and DNA and RNA fragments. A plant-based operation has to deal with plant debris plus proteins, starches, cellulose, lignin, and so on. Furthermore, such feeds often can be described as slimy, sticky, woody, mucky, or some other colorful descriptor indicating difficulty in handling and processing the biomass. Implementing a bio-designed approach involves modifying the organism in some way to induce it to produce the desired product in a more-easily recovered form. The required expertise falls within the field of biosynthesis or metabolic (genetic) engineering – the deliberate alteration of an organism to achieve desired features by using genetic modification or artificial selection. Close collaboration between biosynthesis specialists, process chemists, separation specialists, and others with specialized expertise is particularly important for commercial success – in process conceptualization, development, design, and commercial operation.

Bio-designed separation methods include:
- Modifying a single-cell microbe so that it secretes or releases a desired product outside of the cell.
- Modifying a microbe to express and concentrate a desired compound within a specific part of the organism. Then, after harvesting, various methods may be used to isolate that specific part away from most other parts of the organism.
- Modifying a plant to concentrate a desired compound within a specific part of the plant.

Many products are made via secretion from microbes during fermentation. This naturally occurs in the brewing of alcohol, and in special cases, other cell cultures may be developed to secrete a desired product. This avoids the need to lyse or severely disrupt the cell when it comes time to harvest the fermentation batch, minimizing the formation of cell debris and allowing cleaner isolation of the liquid containing the desired product.

Modifying a microbe to concentrate a desired product within a specific part of the cell can also avoid the need to lyse the cell or at least minimize the need to disrupt the cell. This minimizes release of undesired cell contents, allowing cleaner transfer of the desired compound into a series of downstream purification steps. An example involves modifying a gram-negative bacteria to express a desired protein within the periplasmic space between the outer cell wall and the inner periplasmic membrane.

Various methods such as heat treatment or treatment with surfactants or solvents may be used to disrupt the outer cell wall to release the desired compound, while maintaining the integrity of the inner membrane. The inner cell contents are kept away from the desired product, which can then be recovered and purified almost as if it were secreted [8, 9].

As another example, a corn or barley plant may be modified to concentrate a desired product such as a specific protein within the endosperm. The desired protein may then be extracted from the endosperm after isolating it from other parts of the plant in the harvesting and milling process [10].

13.3 Polarity Switching Separations

In certain kinds of separation processes involving liquids, manipulating a process variable may change or periodically switch the polarity of the feed liquid or solvent, and this technique may be used to enhance the desired separation. For example, a technology called Organic Aqueous Tunable Solvents (OATS) employs CO_2 pressurization to tune the properties of an organic liquid phase [11]. Under CO_2 pressure of 50 or 60 bar (elevated pressure but not supercritical pressure), an organic liquid generally will absorb a considerable amount of CO_2 to form a gas-expanded liquid phase with altered polarity, while an aqueous phase absorbs less CO_2 and is less affected. Another gas such as propane may also be used. This general phenomenon opens up many possibilities for enhancing and tuning separations involving organic liquids, and many studies of this kind have been described in the literature [12, 13].

This is a process intensification technique. It may be used to induce precipitation of a specific solute from the organic phase or alter the partition ratio in an organic + aqueous system to enhance liquid-liquid extraction. It may also be used to induce phase splitting, causing the system to transition from a single liquid to a two-liquid system. This may be particularly useful for improving reaction-extraction schemes used in specialty chemical manufacturing. In that case, a single-liquid-phase reaction may be carried out in a stirred tank. After reaction is completed, the vessel may be pressurized with CO_2 to induce phase splitting and enhance the transfer of components between the phases. Depending on the properties of the components, this may allow isolation of a desired reaction product or a homogeneous catalyst for recycle. After separating the liquids by decantation, the pressure may be vented and the cycle repeated for the next reaction batch [11].

Phase splitting and separation may also be enhanced simply by changing the temperature of the system, for systems that are particularly sensitive to change in temperature. Or, changing both temperature and pressure may provide greater tuning response to further optimize performance. An example of the use of temperature change to enhance performance involves recovery and concentration of aqueous proteins using glycol ether solvents [9].

13.4 Separations Employing a Supercritical Fluid

Various processes that employ supercritical fluids such as CO_2 or propane can be attractive options for product recovery and purification. Compared to liquid-liquid extraction with conventional solvents, supercritical fluid extraction may provide greater tunability for greater selectivity, by adjustment of pressure. When necessary, partition ratios may be enhanced by adding cosolvents such as methanol or tributylphosphate to the supercritical fluid [14]. And some products may be selectively crystallized out of supercritical fluids. Furthermore, supercritical fluids can minimize toxicity concerns, as most fluids are non-toxic and any residual fluid is easily removed from the final product. Recovery, purification, and recycle of the fluid is required, but normally doing so is fairly straightforward. CO_2, the most popular fluid, has the added advantage of being non-flammable. But the required processing equipment can be expensive. For supercritical CO_2, operating pressure is about 1100 psi (74 bar), and high-alloy construction may be needed to avoid corrosion whenever even small amounts of water are present due to formation of carbonic acid [15]. Discussions of supercritical processing are available elsewhere [16–19].

Many applications of supercritical fluids involve processing low-volume, high-value specialty products meant for human consumption, such as certain specialty pharmaceutical ingredients and vitamins derived from natural sources. Supercritical fluids may also be used to remove caffeine from green coffee beans and tea leaves in the production of decaffeinated coffee and tea, as an alternative to conventional solvents such as methylene chloride and ethyl acetate. Here, the flexibility of operation and ease with which residues can be removed from products may be important advantages compared to other methods. Various other industrial processes are operated at supercritical conditions, and change in pressure is used to facilitate desired separations. Examples include the processing of asphaltenes and other components present in crude oil by using supercritical or near-supercritical propane [16, 17].

13.5 Chiral Separations

Many organic compounds are mixtures of optical isomers called enantiomers or chiral compounds. Enantiomers are mirror reflections of one another with specific handedness. Simple examples include the molecules bromochlorofluoromethane (**CH**BrClF), propylene glycol (CH_3**CH**(OH)CH_2OH), and propylene glycol n-butyl ether ($CH_3CH_2CH_2$-CH_2-O-CH_2**CH**(OH)CH_3). Any organic molecule with four different atoms or molecular groups bonded to a single carbon atom forms an asymmetric carbon structure called a chiral center. The structure of the asymmetric carbon determines the handedness of the enantiomer. In these examples, the chiral center is highlighted in bold. Other, more complex examples include the flavoring agent menthol, a $C_{10}H_{20}O$ terpenoid alcohol with three chiral centers and four enantiomeric pairs [20]. Consult ref. [21] or other organic chemistry textbooks for discussions of chirality, which can be more complex. Chirality can also be exhibited by inorganic molecules.

Chirality does not affect a molecule's physical properties. And for many chiral products, its chiral purity is not important for its intended use. Racemic (mixed enantiomer) propylene glycol is used in antifreeze formulations, and racemic propylene glycol n-butyl ether is used in paint and cleaner formulations. But sometimes a chiral molecule has a desired biological activity associated with its use. Many flavor and fragrance compounds are chiral, such as menthol mentioned earlier, and taste and smell are greatly affected by chiral purity. Many drugs are chiral, such as ibuprofen (the active ingredient in Advil and Motrin, sold as the racemic mixture called the racemate) and atorvastatin (the active ingredient in Lipitor, sold in a pure chiral form). Typically, only one enantiomer is responsible for the compound's desired activity, and other enantiomers may contribute unwanted or even deleterious effects. In that case, the specific active enantiomer must be obtained from biological sources that produce it in its pure chiral form, or it must be synthesized and isolated in pure form. Small-molecule enantiomers called synthons can sometimes be used as reagents for synthesizing the larger, optically-pure active molecule. In other cases, it is necessary to separate the active compound from its racemic mixture. Because the physical properties of enantiomers are the same, separating a racemic mixture can be quite a challenge.

But there are certain differences between enantiomers that can be used as the basis for separating them. These include differences in reactivity, where only one enantiomer participates in a catalyzed reaction because a specific handedness is needed for interaction with the catalyst so the reaction can proceed. For example, enzyme-catalyzed transesterifications employing so-called activated acyl donors may be used to isolate desired enantiomers [22]. If conversion is sufficiently high, the racemic feed mixture can be converted to a mixture containing only the reaction product of one enantiomer, and the other enantiomer is left untouched. The mixture can then be separated by using standard physical separation methods such as distillation or crystallization. This approach requires the development of robust and highly specific catalysts, often enzymes, in order to produce highly pure enantiomers with high enantiomeric excess, or ee. Chiral separation processes of various kinds are described elsewhere [23–26].

A process that we helped to develop and demonstrate involves enantioselective transesterification of propylene glycol n-butyl ether with propylene glycol methyl ether acetate, both of which are chiral compounds [26]. The results of this study show how enantioselective reaction and subsequent distillation of the reaction mixture may be used to separate enantiomers in a continuous process scheme. In order to promote the transesterification reaction, a unique horizontal, compartmentalized reactive distillation apparatus was used, and the process was able to demonstrate chiral purities of slightly more than 90% ee for some key enantiomers. This is not high enough for some uses, but the work demonstrated the process concepts and various tradeoffs in operation. With a more selective catalyst, high %ee should be possible.

Other industrial processes that may be used to separate enantiomers include:

- Enantioselective adsorption/chromatography. This method generally involves use of a chiral adsorbent/stationary phase allowing for enantioselective interactions [27]. The result is a kind of reaction-enhanced (or at least attractive-interaction enhanced) chromatography.
- Enantioselective liquid-liquid extraction. This method generally employs an enantioselective extractant such as a chiral crown ether as a selective complexing agent [28], another kind of selective reaction-enhanced or interaction-enhanced separation method.
- Classic Pasteurian resolution. This involves enantioselective reaction with a specific chiral reagent to crystallize an optically-pure reaction product with low solubility in the reaction media [29].
- Enantioselective crystallization of eutectic enantiomers [29, 30]. This method is unique in that it does not require use of an added separating agent.

When enantiomers form a eutectic mixture, pure crystals of a specific enantiomer may be produced when the crystallization is seeded with that specific enantiomer and the process operates at or near the eutectic point. That is the strategy employed in carrying out a crystallization process called continuous preferential crystallization with dual coupled crystallizers [29–32]. The two crystallization vessels are interconnected by a mother-liquor recirculation loop. The feed to the process is racemic so the process operates near the eutectic point. Each vessel is seeded with crystals of a specific enantiomer to promote crystallization of that enantiomer in one vessel and the other enantiomer in the other vessel. The applicability of such a process is limited to systems that form eutectic enantiomers, which is a minority of racemic systems. It is more common for such systems to form racemate compounds or exhibit solid solution behavior [29]. In addition to separating eutectic racemates (called low-melting conglomerates), preferential crystallization may also be used to separate other kinds of eutectic mixtures comprised of compounds with similar structures. For example, Silva, Vieira, and Ottens [33] report separation of polyphenols naringenin and *trans* - resveratrol using this technique.

13.6 Environmental Protection

In previous sections we have emphasized the need for an approach to separation process design that includes concern for the Earth's environment. In addition, many of the separation technologies we have discussed may be used to remove contaminants from other operations to reduce and control their emissions and to remediate contamination of historical sites. Examples include:

- Stripping-mode distillation
    - Air and steam stripping of volatile and sparingly-soluble organic components from process water and wastewater.
    - Air and steam stripping for groundwater remediation.
    - See Section 7.3.3, Continuous Stripping and Absorption.
- Gas-liquid absorption
    - Use of reactive scrubbing solutions to remove contaminants such as HCl, ammonia, SO_2, and Cl_2 from air or N_2 vents or other industrial gas streams (reactive scrubbers). These are commonly used to protect against accidental releases (emergency scrubbers) [34].
    - Use of aqueous base or aqueous organic amines to remove CO_2, H_2S, and other acidic contaminants from hydrocarbon gases (gas sweetening) [35].
    - Modification of gas sweetening methods to remove CO_2 from air for carbon capture [36].
- Liquid-liquid extraction
    - Removal of contaminants from wastewater prior to steam stripping [37].
    - Removal of dissolved salts from sparingly water-soluble organic liquids prior to treatment for disposal (by incineration or otherwise).
- Activated carbon adsorption
    - Carbon bed treatment to remove highly toxic organics from wastewater prior to biotreatment (needed to avoid injury to microbes at the wastewater treatment plant).
    - Pressure swing adsorption to recover organic contaminants from storage tank and process equipment vents (typically air or nitrogen vents) [38].
    - Use of catalytic adsorption/oxidation to collect and destroy organic contaminants present in air and N_2 vents with minimal use of fuel for combustion (catalytic oxidation and incineration) [39].
- Crystallization or precipitation
    - Removal of salts from waste brine produced by acid-base neutralization operations at chemical manufacturing plants or as the byproduct of water desalination. The waste brine is treated to remove the majority of salt prior to its release to an inland river system or the ocean. Common salts include sodium chloride and sodium sulfate. Common processes involve multiple-effect evaporative crystallization, but other crystallization schemes also may be used [40].
    - Precipitation of phosphates and other dissolved solids prior to wastewater treatment [41].
- Use of membranes
    - Removal of salt from seawater or dilute brine (reverse osmosis). See Section 12.2, Water Treatment.
    - Removal of contaminants from vent gases [42].
    - Removal of contaminants from process water and wastewater [43].

For some of these applications, turn-key modular units are available from suppliers. Much information can be found by searching the internet. The use of separation technologies for environmental protection is also discussed in Chapter 4, Addressing Universal and Changing Needs, especially with regard to reduction of CO_2 emissions.

References

[1]   R. G. Harrison, "Bioseparation Basics," *Chemical Engineering Progress*, vol. 110, no. 10, pp. 36–42, 2014.

[2]   S. Ahuja, *Handbook of Bioseparations*, Academic Press, Cambridge, Massachusetts, 2011.

[3]   D. Forciniti, *Industrial Bioseparations: Principles and Practice*, Wiley-Blackwell, Hoboken, New Jersey, 2008.

[4]   P. A. Belter, E. L. Cussler and W.-S. Hu, *Bioseparations: Downstream Processing for Biotechnology*, Wiley-Interscience, Hoboken, New Jersey, 1988.

[5]   G. Subramanian, *Bioseparation and Bioprocessing: A Handbook*, vol. 2, set, 2nd ed., Wiley-VCH, Hoboken, New Jersey, 2007.

[6]   J. Mahar, T. C. Ransohoff, L. B. Klitgaard, S. Muller, A. Powell, T. Larkin and J. Hays, "Chromatography for the Bioprocessing Industry. Developed and Written by BPE Subject Matter Experts," *BPE Journal on System Design*. vol. 1, no. 1, pp. 1–8, 2015.

[7]   I. A. Sutherland, "Recent Progress on the Industrial Scale-up of Counter-Current Chromatography," *Journal of chromatography A*, vol. 1151, no. 1–2, pp. 6–13, 2007.

[8]   B. Balasundaram, S. Harrison and D. G. Bracewell, "Advances in Product Release Strategies and Impact on Bioprocess Design," *Trends in Biotechnology*, vol. 27, no. 8, pp. 477–485, 2009.

[9]   J. R. Allen, A. Y. Patkar, T. C. Frank, F. A. Donate, Y. Chun Chiu, J. E. Shields and M. E. Gustafson, "Use of Glycol Ethers for Selective Release of Periplasmic Proteins from Gram-Negative Bacteria," *Biotechnology Progress*, vol. 23, no. 5, pp. 1163–1170, 2007.

[10]   H. Horvath, J. Huang, O. Wong, E. Kohl, T. Okita, C. G. Kannangara and D. Von Wettstein, "The Production of Recombinant Proteins in Transgenic Barley Grains," *The Proceedings of the National Academy of Sciences*, vol. 97, no. 4, pp. 1914–1919, 2000.

[11]   P. Pollet, R. J. Hart, C. A. Eckert and C. L. Liotta, "Organic Aqueous Tunable Solvents (OATS): A Vehicle for Coupling Reactions and Separations," *Accounts of Chemical Research*, vol. 43, no. 9, pp. 1237–1245, 2010.

[12]   S. Abou-Shehada, J. H. Clark, G. Paggiola and J. Sherwood, "Tunable Solvents: Shades of Green," *Chemical Engineering and Processing: Process Intensification*, vol. 99, pp. 88–96, 2016.

[13]   C. A. Eckert, C. L. Liotta, D. Bush, J. S. Brown and J. P. Hallett, "Sustainable Reactions in Tunable Solvents," *Journal of Physical Chemistry*, vol. 108, no. 47, pp. 18108–18118, 2004.

[14]   J. F. Brennecke and C. A. Eckert, "Phase Equilibria for Supercritical Fluid Process Design," *AIChE Journal*, vol. 35, no. 9, pp. 1409–1427, 1989.

[15]   S. Sarrade, D. Féron, F. Rouillard, S. Perrin, R. Robin, J.-C. Ruiz and H.-A. Turc, "Overview on Corrosion in Supercritical Fluids," *The Journal of Supercritical Fluids*, vol. 120, no. Part 2, pp. 335–344, 2017.

[16]   M. McHugh and V. Krukonis, *Supercritical Fluid Extraction: Principles and Practice*, 2nd ed., Butterworth-Heinemann, Oxford, England, 1994, 1994.

[17]   J. L. Martinez, ed., *Supercritical Fluid Extraction of Nutraceuticals and Bioactive Compounds*, CRC Press, Boca Raton, Florida, 2008.

[18]   G. Brunner, "Applications of Supercritical Fluids," *Annual Review of Chemical and Biomolecular Engineering*, vol. 1, pp. 321–342, 2010.

[19] T. C. Frank, B. S. Holden and A. F. Seibert, "Section 15: Liquid-Liquid Extraction and Other Liquid-Liquid Operations and Equipment," in *Perry's Chemical Engineers' Handbook*, 9th ed., D. W. Green and M. Z. Southard, Eds., McGraw-Hill, New York, 2019.

[20] J. C. Leffingwell and D. Leffingwell, "Chiral Chemistry in Flavours and Fragrances," *Specialty Chemicals Magazine*, pp. 30–33, 2011.

[21] W. H. Brown, B. L. Iverson, E. Anslyn and C. S. Foote, "Stereoisomerism and Chirality, Chapter 3," in *Organic Chemistry*, 9th ed., Cengage Learning, Boston, 2022.

[22] R. Chenevert, N. Pelchat and P. Morin, "Lipase-mediated Enantioselective Acylation of Alcohols with Functionalized Vinyl Esters: Acyl Donor Tolerance and Applications," *Tetrahedron: Asymmetry*, vol. 20, pp. 1191–1196, 2009.

[23] A. N. Collins, G. Sheldrake and J. Crosby, *Chirality in Industry*, Vol. I and II, Wiley, Hoboken, New Jersey, 1995, 1998.

[24] H. Lorenz and A. Seidel-Morgenstern, "Processes To Separate Enantiomers," *Angewandte Chemie*, vol. 53, no. 53, pp. 1218–1250, 2014.

[25] O. Fellechner, M. Blatkiewicz and I. Smirnova, "Reactive Separations for in Situ Product Removal of Enzymatic Reactions: A Review," *Chemie Ingenieur Technik*, vol. 91, no. 11, pp. 1522–1543, 2019.

[26] P. H. Au-Yeung, S. M. Resnick, P. M. Witt, T. C. Frank, F. A. Donate and L. A. Robbins, "Horizontal Reactive Distillation for Multicomponent Chiral Resolution," *AIChE Journal*, vol. 59, no. 7, pp. 2603–2620, 2013.

[27] G. K. E. Scriba, Ed. *Chiral Separations: Methods and Protocols*, 2nd ed., Humana Press, Totowa, New Jersey, 2013.

[28] B. Schuur, B. J. V. Verkuijl, A. J. Minnaard,, J. G. De Vries, H. J. Heeres and B. L. Feringa, "Chiral Separation by Enantioselective Liquid-Liquid Extraction," *Organic & Biomolecular Chemistry*, vol. 9, pp. 36–51, 2011.

[29] C. Rougeot and J. E. Hein, "Application of Continuous Preferential Crystallization to Efficiently Access Enantiopure Chemicals," *Organic Process Research & Development*, vol. 19, no. 12, pp. 1809–1819, 2015.

[30] M. P. Elsner, G. Ziomek and A. Seidel-Morgenstern, "Simultaneous Preferential Crystallization in a Coupled, Batch Operation Mode – Part I: Theoretical Analysis and Optimization," *Chemical Engineering Science*, vol. 62, pp. 4760–4769, 2007.

[31] H. Lorenz, E. Temmel and A. Seidel-Morgenstern, "Continuous Enantioselective Crystallization of Chiral Compounds," in *The Handbook of Continuous Crystallization*, Vol.1, N. Yazdanpanah and Z. K. Nagy, Eds., Royal Society of Chemistry, Cambridge, England, 2020, pp. 422–468.

[32] E. Temmel, J. Gänsch, A. Seidel-Morgenstern and H. Lorenz, "Systematic Investigations on Continuous Fluidized Bed Crystallization for Chiral Separation," *Crystals*, vol. 10, article no. 394, 2020.

[33] V. Silva and Ottens, "Preferential Crystallization for the Purification of Similar Hydrophobic Polyphenols," *Journal Chemical Technology and Biotechnology*, vol. 93, no. 7, pp. 1997–2010, 2010.

[34] Ö. Yildirim, A. Kiss, N. Hüser, K. Leßmann and E. Kenig, "Reactive Absorption in Chemical Process Industry: A Review on Current Activities," *Chemical Engineering Journal*, vol. 213, pp. 371–391, 2012.

[35] A. Jamekhorshid, Z. K. Davani, A. Salehi and A. Khosravi, "Gas Sweetening Simulation and Its Optimization by Two Typical Amine Solutions: An Industrial Case Study in Persian Gulf Region," *Natural Gas Industry B*, vol. 8, no. 3, pp. 309–316, 2021.

[36] R. J. Perry and J. L. Davis, "CO_2 Capture Using Solutions of Alkanolamines and Aminosilicones," *Energy & Fuels*, vol. 26, no. 4, pp. 2512–2517, 2012.

[37] L. A. Robbins, "Liquid-liquid Extraction: A Pretreatment Process for Waste-Water. This Process Was Found to Be More Cost-Effective than Others for the Main Reduction of Toxic Pollutants before Biotreatment," *Chemical Engineering Progress*, vol. 76, no. 10, pp. 58–61, 1980.

[38] D. J. Pezolt, S. J. Collick, H. A. Johnson and L. A. Robbins, "Pressure Swing Adsorption for VOC Recovery at Gasoline Loading Terminals," *Environmental Progress*, vol. 16, no. 1, pp. 16–19, 1997.

[39] M. S. Kamal, S. A. Razzak and M. M. Hossain, "Catalytic Oxidation of Volatile Organic Compounds (Vocs). A Review," *Atmospheric Environment*, vol. 140, pp. 117–134, 2016.

[40] K.-L. Wu, H.-Y. Wang and J. D. Ward, "Economic Comparison of Crystallization Technologies for Different Chemical Products," *Industrial & Engineering Chemistry Research*, vol. 57, no. 37, pp. 12444–12457, 2018.

[41] D. Bian, S. Ai, J. Liu, Y. Zuo and X. Tian, "Treatment of Phosphorus Waste Water Using Crystallization Method," *Journal of Environmental Sciences*, vol. 23 (Supplement), pp. S106 –S109, 2011.

[42] Y. Kang, Z. Zhong and W. Xing, "Functional Membranes for Air Purification," in *Chapter 12, Advances in Functional Separation Membranes*, X. Li, J. Lin and S. Zhao, Eds., The Royal Society of Chemistry, London, 2021, pp. 279–315.

[43] J. E. Cevallos-Mendoza, C. Amorim, J. Rodríguez-Díaz and M. Montenegro, "Removal of Contaminants from Water by Membrane Filtration: A Review," *Membranes*, vol. 12, no. 6, article no. 570, 2022.

14 Inventive Engineering: Process Simplification, Intensification, and Hybrid Processing

Engineers and chemists involved in industrial chemical separation are problem solvers and change agents, like all engineers. This is true whether the task is to design a new process or to improve an existing one. Not only does an engineer need to exercise good judgment and skill in making design calculations, but an engineer must also be inventive in devising tailored solutions to the problems at hand. Inventiveness entails coming up with new ideas or new ways of combining existing ideas to yield a new and better result or accomplish a desired result in a new and better way. Inventiveness can be exhibited in the act of doing as well as in the final product. That is, inventiveness can be found in how one goes about specifying a design (coming up with a new and better design method), as well as in the features of a new design. Inventiveness can also be exhibited in identifying a new class of applications for existing technology. Box 14.1 lists some well-known examples of inventiveness from the world of chemical separations, many of which we have already discussed. The chemical engineering literature often refers to inventive advances such as these in terms of process simplification (making for a more elegant and cost-effective process), process intensification (maximizing driving forces and minimizing resistances), and hybrid processing (combining two or more operations into a single process scheme).

Box 14.1: Examples of Inventiveness in Chemical Separations

- Graphical method for specifying a distillation column assuming constant molar overflow and negligible heat effects. Warren McCabe and Ernest Thiele (1925). Easy-to-use, visual, and instructive.
- Combining extraction and distillation into a single unit operation called vapor-phase extraction with a heavy solvent (or extractive distillation) by Clarence L. Dunn and Robert B. McConaughy of Shell Dev. Co. (ca. 1940).
- Combining two distillation operations within a single distillation column (the dividing-wall distillation column). Invented by Richard Wright, Standard Oil Co. (1946), and developed/commercialized by BASF (1980s). Saves both capital and operating costs.
- Podbielniak centrifugal extractor, or "Pod," used for short-contact time extraction; for example, for recovery of penicillin out of fermentation broth. Walter Podbielniak (1940s). Allows efficient recovery of product at acidic conditions with little or no degradation.
- Extraction of aromatics from aliphatics using a high-boiling polar solvent followed by stripping-mode distillation with reflux of the top product back to the extraction column. The Dow Chemical Company (1940s).
- Extractive distillation of aromatics from aliphatics (1940s).
- Simulated moving-bed chromatography process scheme. Universal Oil Products (1950s).
- Large, white granular salt particles of uniform size with no additives, consisting of agglomerated salt crystals, used for topping pretzels (white, coarse food-grade salt). A product by process (undetermined date).
- Synthetic zeolite materials with well-defined pore sizes. Union Carbide (1950s). Enables sorting of molecules at the molecular scale for improved reaction and adsorption.

https://doi.org/10.1515/9783110695052-019

- Reciprocating-perforated-plate extraction column (Karr column). Andrew Karr (1960s). Provides uniform agitation for uniform fluid shear and uniform dispersed droplet size.
- Population balance method for crystallization process analysis and design. Alan Randolph and Maurice Larsen (1962). Improves reliability in the scale-up of continuous crystallization processes.
- Pressure-swing adsorption four-step Skarstrom process scheme for separating gases. Charles Skarstrom (1960s).
- Combined (hybrid) process scheme involving distillation + PSA. Hot-gas PSA for separation of liquids (by first vaporizing the liquid feed) (1980s).
- High-capacity distillation trays. A new design approach that uses the energy from the pressure drop across each tray to enhance vapor-liquid phase separation and achieve throughputs greater than traditional tray designs (1990s).
- Oscillatory-flow tubular reactor/crystallizer. A concept developed in the 1990s and commercialized in the 2000s.

New inventive ideas are not calculated or predicted. They come from pondering over a puzzling situation or problem for a time. The mind somehow offers up an idea. Exactly how this works is a mystery, but it can be made more likely by taking appropriate steps to be prepared – by being familiar with a variety of separation technologies (like those discussed in Chapters 7 to 13), by being mindful of the fundamentals (as discussed in Part II), and by giving yourself plenty of time to ponder the situation and open your mind to new possibilities. In this chapter, we outline a general thought process that we have found to be helpful. More specific methods include process synthesis [1–3] and process optimization [4–6], as well as various methods aimed directly at facilitating inventiveness [7]. One such method, called TRIZ or the "Theory of Inventive Problem Solving," facilitates ideation by recognizing patterns and analogies [8]. Another method, called Mind Mapping, fosters creativity by organizing and developing the thought process itself [9]. For discussions of these and other methods, see the publications listed in the Bibliography for Chapter 14. Of course, the most effective approach is likely to be some combination of methods.

To give our discussion a proper perspective, keep in mind that although creativity and invention are admired by many in the profession, in actual practice it may not matter much whether a proposed design is considered truly inventive or not. Inventiveness itself is not the measure of success. Instead, the work is judged first and foremost by how well it meets the need; that is, by whether it is cost-effective and meets the requirements for a balanced and robust design. If a standard design will do the job with only small refinements, then designers normally will focus on getting the job done in a timely manner using the standard approach without spending extra resources to explore a completely new method. This is because exploring and then developing a new approach can cause considerable delays and might add significantly to the overall cost and the technical risk. Of course, some inventiveness or innovation is still needed in making adjustments and refinements. The need for continuous improvement year after year often can be met by continuous refinement of standard designs, and this is an important way in which many designers pursue *best change*, as discussed in Part I.

Naturally, opinions differ as to whether standard designs can achieve what is needed, and this point is often debated. It is when standard designs cannot meet requirements that designers must look for alternatives, and this is when the ability to exercise inventiveness is especially prized.

The general thought process that we recommend is a practical one that aims to strike a balance between simplicity, intensity, and complexity, one that is most appropriate for the given application and its specific often-unique situation. We have found that a good way to begin is to first ask the question: What is the simplest process scheme I can imagine that might be used to produce the desired product? Then, evaluate its functionality and cost. Consider adding additional process complexity to address any deficiencies. Add complexity only if the potential benefit outweighs the extra cost and difficulty. The old adage, "Keep it simple," has a lot of merit. After all, a simple process that meets requirements is an elegant one. The complexity of sophisticated or elaborate schemes may impress, but they should be pursued only if they offer some additional value.

Consider the historical example of how the production of enriched ethanol from ethanol + water mixtures has evolved over time. The first and perhaps the simplest process scheme used to isolate ethanol out of water is atmospheric-pressure alembic or pot distillation, employed since ancient times. This process is indeed simple and quite easy to practice, but it is limited as to how concentrated it can make the ethanol product. A single pot might yield an overhead product of 40 to 50% ethanol (80 to 100 proof), depending on the composition of the feed and how much is taken overhead. The overhead product can be further enriched by using the pot to distill it again. But doing so is time-consuming and tedious and the recovery of ethanol from the feed is limited. Around 1800 or so, people began adding columns with reflux to achieve higher-purity ethanol. The added complexity of the new equipment was justified because it allowed for a more-concentrated overhead product in a single distillation and this made for a simpler and more productive overall operation. But, of course, the ethanol concentration that could be reached using this method was limited by the ethanol-water azeotrope (around 95% at atmospheric pressure). Around 1915, people began using an entrainer such as benzene to break the azeotrope. This added considerable complexity to the overall process scheme, but this was justified at the time by the greater separation power that could be achieved. Around 1928, people wanted a simpler and safer process, so they looked into operating the distillation under vacuum to avoid the azeotrope. This works, but relative volatility remains close to unity, so a very tall (and larger diameter) distillation column is required. Later it was also suggested that adding salt to the distillation liquid would break the azeotrope. This also works, but the salt is not volatile, so it has to be added and recycled back to the column. These options involving the use of vacuum or salt never really caught on it seems because they were too unwieldy for most applications. Sometime in the 1980s, people realized that a simpler and safer process could be devised simply by combining standard distillation employing a single short column along with another separation process in a hybrid scheme. As a result, today many ethanol drying processes involve the use of standard distillation to generate near-azeotropic ethanol overhead, followed by the use of pressure swing adsorption or pervaporation membranes to remove additional water and produce relatively dry or anhydrous ethanol. Both hybrid process schemes have proved to be popular and cost-effective in commercial practice.

This historical example serves to illustrate the benefits of process simplicity, the potential value of process intensification (making a single unit operation more effective), and the utility of hybrid processing (combining relatively simple operations together to improve overall effectiveness). This approach involves asking the following questions:

– Can it be simple? Can a single, uncomplicated unit operation achieve the desired separation? Is it less mechanically or operationally complex than other options? Does it involve fewer moving parts or operate at lower speeds?

– What difficulties or insufficiencies are encountered, and why? For example, does the proposed process lack the needed separation power due to thermodynamic or kinetic limitations, or does it consume too much energy? Can performance be improved by intensifying the process – by addressing key limitations of the single unit operation? Various potential intensification techniques are listed in Box 14.2. Also see ref. [10].

– At what conditions does the single unit operation have the most difficulty? In particular, at what concentration range is the unit operation least suitable for the job at hand? What additional unit operation would be better able to handle the job within that concentration range? Consider combining unit operations such that each is employed within the concentration range for which they are best suited.

– Is it possible to combine operations within a single column or vessel, as in reactive distillation or dividing-wall distillation?

– Would pre-treating the feed be helpful? Consider adding an up-front step to pre-concentrate the feed. Doing so may greatly reduce the volume and energy intensity of the downstream process.

Box 14.2: Techniques for Process Intensification

– Use of flow patterns and process configurations that increase separation power
  – Countercurrent contacting and use of reflux. See Section 7.1
  – Middle-fed fractional processing, for both continuous and batch operation. See Section 7.1
  – Fed-batch processing. For example, see Section 7.3.2
  – Multi-step or multi-zone cyclical processing (as in PSA and SMB processes)
    See Sections 10.6 and 10.8.
– Deciding which phase should be dispersed and how it should be dispersed
  – Which way of dispersing one phase into another, among many possibilities, gives best performance?
  – This can greatly affect the amount of interfacial area available for mass transfer.
  – See Sections 6.2, 6.3, and 6.4.
– Changing/manipulating operating conditions to alter thermodynamic drivers
  – Temperature change (for a single unit, or coupled units operating at different temperatures), in distillation, extraction, adsorption, crystallization, use of membranes, . . .

- – Pressure change (which may affect temperature)
  - – Temperature cycling in batch crystallization
  - – For examples, see Sections 5.6, 8.3, 10.6, and 11.4.
- – Addition of separating agents
  - – Taking advantage of specific molecular interactions
  - – Addition of entrainers, solvents, anti-solvents, and template compounds
  - – Use of specialized adsorbents and modifiers
  - – See Sections 8.1, 8.2, 10.1, 11.2, and 11.4.
- – Employing specialized mixing design
  - – High volume, low shear impellers for rapid, gentle circulation. See Section 6.2.2.
  - – Reciprocating-plate agitation for uniform shear, as in the Karr column extractor [11]
  - – Improved draining of slurries in batch vessels, as with the tickler impeller [12].
- – Maximizing effective interfacial area
  - – Operation near the flood point. See Section 6.3.
  - – High-efficiency internals such as packings, trays, and distributors. See Sections 7.4. and 10.4
  - – Mechanical agitation (gentle or intense). For LLE, see ref. [11].
  - – Use of a membrane to control interfacial area. See Section 12.1.
- – Application of extra forces
  - – Centrifugal force. For LLE, see ref. [11].
  - – Electric field. For LLE, see ref. [11].
  - – Ultrasound. For its use to facilitate nucleation in crystallization, see Section 11.4.
  - – Microwave heating. For its use to rapidly regenerate adsorbent beads, see Section 10.7.
  - – Use of rotating packed bed contactor to reduce equipment size and footprint in distillation [13]
- – Heat integration and heat pumps [14]
  - – Cross-exchange with other unit operations (pinch technology)
  - – Multiple-effect evaporation
  - – Mechanical vapor recompression [15]
  - – Heating and cooling performed within a packed or trayed column. For example, see ref. [16].
  - – Petlyuk configurations, including dividing-wall distillation [17]
- – Dynamic cycling of fluid flow rates
  - – Pulsed liquid-liquid sieve-plate column [11]
  - – Cyclic distillation [18]
  - – Oscillatory-flow tubular crystallizer [19]
  - – Applications in general [20]
- – Switching liquid polarity. See Section 13.3.
  - – Change in pH of aqueous liquid solution
  - – Gas pressurization to form a gas-expanded liquid.
- – Use of a supercritical fluid. See Section 13.4.
- – Improvement in process control strategies
  - – For sharper, more efficient separation with less variability without over-separation
  - – Applies to all processes, including continuous, batch, fed-batch, and cyclical
  - – For discussion of distillation control considerations, see Section 7.5.
- – Reaction-enhanced separation
  - – Use of reversible acid-base or other reactions. Examples include use of aqueous amines to increase partition ratios for removal of acid gases from hydrocarbons, and use of specialty amines dissolved in organic solvent to facilitate recovery of carboxylic acids from aqueous solution. See Section 7.3.7 and Chapter 9.

- Use of reaction to pre-treat a feed prior to separation processing, to convert difficult-to-separate impurities into easy-to-separate impurities. For example, pre-oxidation of sulfur-containing impurities can facilitate their extraction from fuel oil [21].
- Bio-designed separation
  - Facilitates primary recovery and purification of a product produced within a living organism
  - See Section 13.2
- Sequencing of multiple unit operations (of the same kind)
  - For efficient separation of multiple products (separation trains)
  - See refs. [1, 22] for crystallization and ref. [2] for distillation.
- Hybrid processing (which may be considered a kind of intensification)
  - Combining two or more operations of different kinds
  - Combining multiple separations within a single unit (as in dividing-wall distillation)

So, the thought process we recommend is to first look at a single unit operation, and if needed, consider the use of intensification techniques to maximize performance. As listed in Box 14.2, there are many ways to intensify a single unit operation. Some options add complexity but they may (or may not) be justified by significant improvement in performance. If the added complexity gets to be too unwieldy (a judgment call), then consider combining two relatively simple unit operations into a hybrid process scheme.

Hybrid processing is nothing new. It has been practiced for many, many decades, long before the term hybrid process became popular. And some long-standing processes that might very well be termed hybrid processes are not generally recognized as such. Liquid-liquid extraction, for example, may be considered a kind of hybrid process, because in most cases extra operations are needed to process extract and raffinate streams, and these operations often are closely integrated for recovery and final purification of product and for recovery and recycle of solvent. This usually involves close integration of extraction + distillation. The use of extractive distillation, which involves adding a heavy extraction solvent directly to a distillation column, certainly is a kind of hybrid processing. Other examples include distillation + melt crystallization, distillation + pressure swing adsorption, and distillation + pervaporation. Examples are outlined in Table 14.1 along with typical applications. Many other potential hybrid processes are mentioned in the literature, and no doubt there are many opportunities to apply hybrid schemes in new ways.

Normally, a hybrid separation process is not simply two or more separation operations conducted in series (a sequential series of operations). Instead, the various methods are integrated and closely coupled such that material from one method is recycled to the inlet of another. As discussed earlier, the idea is to take some of the load off of the main separator. This load is the more difficult portion of the separation for the main separator to manage – a portion that the added method is better suited for. As a result, the hybrid scheme is better equipped to handle the entire concentration range required of the given application, allowing for improvement in overall process performance and/or a reduction in the total processing cost. In other words, a hybrid process

Table 14.1: Example Applications of Hybrid Processing.

Process	Typical application
Distillation + PSA [23, 24]	Alcohol drying
Distillation + pervaporation [25]	Alcohol drying
Melt crystallization + distillation [26]	Purification of acrylic acid, dichlorobenzene, isocyanates, and other organics
Distillation + extraction (extractive distillation) [27]	Acetic acid recycle and purification, aromatics from aliphatics
Crystallization + extraction (extractive crystallization) [28]	Crystallization of salts from aqueous solution
Adsorption + crystallization [29]	Recycle and purification of active ingredients
Eutectic Freeze Crystallization [30, 31] (crystallization + in-situ separation of crystal types by density difference)	Separation of water from brine
Membrane-assisted crystallization (crystallization + membrane) [32]	Pre-concentration of feeds, recovery of organics from wastewater
PSA + membrane [33]	Hydrogen from CO_2
Membrane + combustion [34]	Nitrogen from air
Membrane + membrane (different types) [35]	Phenol from wastewater, wastewater treatment

can sustain greater separation driving force and efficiency over a wider concentration range compared to what can be achieved using either single unit operation alone. In some cases, hybrid processing is accomplished within a single vessel or unit operation, leading to further efficiencies. But this is not always possible or necessary.

In conceptualizing a hybrid process scheme, we recommend the general approach outlined in Box 14.3. It does not involve the use of simulation software. Most process simulation programs do not and cannot generate the process flow sheet *apriori*. In most cases, the overall process scheme has to be conceived of beforehand and then entered into the program by the designer. Only then can a potential process be evaluated further by using the program to calculate the mass and energy balances and to evaluate potential control schemes. For example, this is the general approach that we along with others at Dow used to devise a reactive distillation process for separating chiral glycol ethers. Our description of the process, given elsewhere [36], may serve as a useful case study.

Box 14.3: Hybrid Process Considerations

- Specify the requirements of the given application in terms of crude feed composition and the required product purity and recovery.
- Collect physical property and phase equilibrium data (or generate data and estimates as needed) to scope out the phase diagram for the system of interest. See Section 5.6.
- Using the phase diagram as a map of potential process paths, list possible separation methods that may be feasible.
- Make a judgment as to the concentration range best suited for each of these methods.
- Put the methods together in such a way that the feed to each method is within its most effective operating range. Try several configurations and choose the top two or three options that seem to give the best fit.
- Identify opportunities for internal recycle streams and purge streams.
- If any of the methods produce a reject stream that is in the most-effective concentration range of another method, then consider sending that reject stream to the inlet of the other method as internal recycle.
- If any of the methods can produce a stream that is highly concentrated in one or more impurity components, then consider options for purging that stream from the process at that point. The overall process must have at least one outlet for purging impurities, and sometimes multiple outlets for purging impurities of different kinds (lights and heavies, for example).
- Try several configurations and choose the options that seem to give the best overall fit.
- Further evaluate the top options via process modeling, and if attractive, further develop the process as needed to demonstrate capabilities.
- In selecting equipment, consider the pros and cons of different ways of dispersing one phase in another, and effective ways of contacting phases. Identify equipment with potential to provide sufficient residence time (or short contact time if needed) and sufficient theoretical stages or transfer units.
- Keep the number of separation methods to two or three at most (normally). A single unit operation can be too costly because the unit is not well suited to handle the entire composition range (the reason for considering a hybrid process), but adding too many units in a hybrid scheme can also be too costly.
- Consider combining multiple methods into a single vessel or shell. This may provide additional benefits such as greater separation power or reduced energy consumption for further reduction in cost (as in some forms of reactive distillation). But this is not always feasible.

For illustration, let's take a closer look at producing dry ethanol from an ethanol + water feed mixture and the steps involved in conceptualizing a hybrid process scheme. Consider questions that might arise in devising a process scheme to recover ethanol from fermentation broth. The broth contains roughly 85% water, 10% ethanol, 5% undesired organic lights and heavies plus slimy biomass solids. The goal is to recover at least 98% of the ethanol present in the feed broth. The ethanol product should contain less than 0.5% water (5000 ppm by weight) and meet certain other specifications regarding other impurities such as methanol (which we won't worry about for our purposes here).

We begin by considering various potential separation methods. The top options for primary recovery of ethanol from the broth and subsequent removal of water from the ethanol seem to be the possible use of distillation, membranes, and adsorption. We reject the use of liquid-liquid extraction (LLE), enhanced distillation, or crystallization as being less suitable. LLE and enhanced distillation are technically feasible, but they are rejected because they require adding a solvent (or entrainer) with the need for solvent recovery and recycle, which involves considerable added processing complexity and extra unit operations. Crystallization also is technically feasible; that is, it is possible to increase the concentration of ethanol by repeated formation and removal of water crystals in multiple crystallization stages. In fact this is a well-known way of making certain kinds of fortified wine and beer. But for our purposes here, this possibility is rejected because of considerable complexity involved in crystallizing and filtering ice crystals (requiring solids handling and extra unit operations beyond crystallization). Furthermore, it would be difficult to achieve sufficiently high recovery and purity of the ethanol product because of residual solubility of water in the ethanol mother liquor in equilibrium with ice and because of entrainment of trapped ethanol within the resulting ice crystals.

Now that we have pared down the list of options to distillation, adsorption, and the use of membranes, the next step is to identify operating ranges best suited for each of these methods. Distillation is most effective at processing the beer broth to generate near azeotropic overheads. Any need to pre-filter the beer is reduced or eliminated because distillation is able to handle some amount of solids with the proper column internals. Distillation becomes ineffective at high ethanol concentrations near the ethanol + water azeotrope because relative volatility approaches unity. In order to deal with the solids without requiring that they be filtered from the broth, a possible distillation scheme might involve repeated, topped-off batch rectifications for primary recovery of ethanol out of the solids-laden broth (with rejection of unwanted light impurities as first lights cut). This might be used to produce near-azeotropic ethanol in the overheads, or the accumulated overheads from a simpler batch rectification operation might be further concentrated using a continuous fractional distillation column to produce a near-azeotropic overheads distillate. Both options could be used to obtain high recovery of ethanol and reject unwanted lights and heavies. Another approach might be to centrifuge the broth to remove the majority of biomass and then feed the partially clarified broth directly to a continuous fractionation column. All of these options seem feasible and should be evaluated and compared. But distillations like this can only go so far.

Pervaporation involving preferential permeation of water is well-suited for removing water from near-azeotropic overheads obtained by distillation. We would need a membrane that can permeate sufficient water in three or fewer stages of membrane modules. An example is shown in Figure 12.3. A typical pervaporation process like this is capable of producing ethanol containing water in concentrations down to 0.5% or perhaps as low as 0.1% (1000 ppm water). Even lower levels may be possible, but at the expense of more stages and/or additional co-permeation and recycle of ethanol.

Adsorption can be effective at removing water from ethanol at both low and high ethanol concentrations. So, in principle, adsorption can be used alone to remove water from fermentation broth and produce anhydrous ethanol. But most likely the broth would need to be pre-filtered, and because the total amount of water to be removed is so large, the adsorption beds would need to be large with frequent regeneration. These difficulties are avoided by employing adsorption only at high ethanol concentrations near the ethanol + water azeotrope, where the amount of water that needs to be removed is much lower. Furthermore, compared to pervaporation, adsorption may be better suited to drying the ethanol to very low water concentrations as low as 100 ppm water or even lower if needed. This is because adsorbents such as 3A molecular sieves are highly selective for water even at very low concentrations of water in the feed.

In a distillation + pervaporation hybrid scheme, distillation is used to distill the beer to produce near-azeotropic overheads (as described above), which may involve a single fractionation column or it may involve use of a beer stripping column that then feeds a fractionation column. Either way, pervaporation is then used to permeate water to break the azeotrope and produce a retentate product containing dry ethanol. The permeate stream may be recycled back to the distillation operation to recover residual ethanol and reject water in the distillation bottoms. The feed to the membrane modules does not need to be pre-filtered, because it comes from the distillation overheads, so broth solids have already been excluded. This avoids problems in using a membrane in direct contact with bio-mass solids. For additional discussion of distillation + pervaporation, see Section 12.5: Pervaporation and Figure 12.3. The membrane employed in this process is a commercially available pervaporation membrane made from polyvinylalcohol and designed to permeate water. In this case, the selectivity of the membrane does not need to be extremely high. The process can tolerate some co-permeation of ethanol along with water without affecting the dryness of the ethanol product. However, further improvements in membrane selectivity and flux are desirable because this would allow the membrane modules to handle a higher water load; that is, the distillation column could then be operated to produce overheads containing more water (greater than 10%), reducing the required distillation reflux ratio for further reduction in energy consumption by the overall process.

A distillation + pressure swing adsorption (PSA) hybrid scheme is similar, except that adsorption may be used to produce even drier ethanol if needed. But this may not be needed, and pervaporation may have other advantages for a given installation. In any case, distillation + PSA often is an effective process scheme for such applications and should be evaluated as well as pervaporation. The backpurge from the PSA unit is recycled back to the distillation column to recover residual ethanol and purge water. For additional discussion, see Section 10.6: Pressure Swing Adsorption and Figure 10.9.

With these considerations, we now have two types of hybrid process schemes to consider, each with its own set of options and pros and cons. Both schemes are better able to handle the entire concentration range compared to the use of any one method alone. Now the job is to decide which scheme with what options will be best for the specific situation at hand. As it turns out, both types of hybrid process schemes have found commercial application [37].

As another example, consider the design of a process involving distillation + melt crystallization. Distillation is used to concentrate the component that is to be crystallized to a high concentration, say 98% or so, and melt crystallization is then used to produce highly pure product. A reject cut from the crystallizer that is concentrated in impurities can be purged periodically or recycled back to the distillation column to be further concentrated and purged. Another cut from the crystallizer (a recycle cut) with a composition near that of the initial feed can be recycled directly to the next crystallization batch. Or, multiple stages of melt crystallization can be carried out to further increase the recovery of product and purge a reject cut more concentrated in the impurities, a process scheme involving middle-fed, countercurrent crystallization with both purification and recovery sections (analogous to rectification and stripping). This kind of multi-stage crystallization process is described in Section 11.6: Crystallization from the Melt: Eliminating the Addition of Solvent (Figure 11.3). Another hybrid approach is associated with crystallization from solution, where the crystallizer is used to obtain the desired crystals and another separation method such as distillation, extraction, or SMB chromatography is used to recover residual product from the spent mother liquor and purge impurities from the solvent. Such a hybrid scheme can allow use of the most effective crystallization method needed to obtain the desired crystal structure (whether by cooling or another way of generating supersaturation) while also achieving high recovery of the desired product and necessary purging of impurities. And so the thinking goes in combining separation operations in general.

With that, we conclude our survey of industrial chemical separation. But the story continues, as engineers and chemists continue their work to change and optimize the various technologies – seeking better ways to address current challenges and opportunities and to generate additional value. Our best wishes to all.

References

[1] C. Wibowo and K. M. Ng, "Unified Approach for Synthesizing Crystallization-Based Separation Processes," *AIChE Journal*, vol. 46, no. 7, pp. 1400–1421, 2000.
[2] Z. Jiang and R. Agrawal, "Process Intensification in Multicomponent Distillation: A Review of Recent Advancements," *Chemical Engineering Research and Design*, vol. 147, pp. 122–145, 2019.
[3] A. Lucia, A. Amale and R. Taylor, "Energy Efficient Hybrid Separation Processes," *Industrial & Engineering Chemistry Research*, vol. 45, no. 25, pp. 8319–8328, 2006.
[4] R. W. Pike, *Essentials of Optimization for Chemical Engineering*, Primedia eLaunch LLC, Dallas, Texas, 2019.
[5] I. E. Grossmann, *Advanced Optimization for Process Systems Engineering*, Cambridge University Press, Cambridge, England, 2021.
[6] S. Terrazas-Moreno, I. E. Grossmann, J. M. Wassick and S. J. Bury, "Optimal Design of Reliable Integrated Chemical Production Sites," *Computers & Chemical Engineering*, vol. 34, no. 12, pp. 1919–1936, 2010.
[7] T. Arciszewski, *Inventive Engineering: Knowledge and Skills for Creative Engineers*, CRC Press, Boca Raton, Florida, 2016.

[8]   P. Livotov and V. Petrov, *TRIZ: Innovation and Inventive Problem Solving*. Handbook, TriS Europe – Innovation Knowledge Company, Freiburg im Breisgau, Germany, 2015.

[9]   K. Knight, *Mind Mapping: Improve Memory, Concentration, Communication, Organization, Creativity, and Time Management (Mental Performance)*, CreateSpace Independent Publishing Platform, Scotts Valley, California, 2012.

[10] F. J. Keil, "Process Intensification," *Reviews in Chemical Engineering*, vol. 34, no. 2, pp. 135–200, 2018.

[11] T. C. Frank, B. S. Holden and A. F. Seibert, "Section 15: Liquid-Liquid Extraction and Other Liquid-Liquid Operations and Equipment," in *Perry's Chemical Engineers' Handbook*, 9th ed., D. W. Green and M. Z. Southard, Eds., McGraw-Hill, New York, 2019.

[12] K. K. Kar, R. F. Cope, S. Sandor and A. Pennington, "Tickler for Slurry Reactors and Tanks". *U.S. Patent* 6,955,461, 18 October 2005.

[13] A. Mondal, A. Pramanik, A. Bhowal and S. Datta, "Distillation Studies in Rotating Packed Bed with Split Packing," *Chemical Engineering Research and Design*, vol. 90, no. 4, pp. 453–457, 2012.

[14] R. Smith, M. Jobson, L. Chen and S. Farrokhpanah, "Heat Integrated Distillation System Design," *Chemical Engineering Transactions*, vol. 21, pp. 19–24, 2010.

[15] B. W. Hackett, "The Essentials of Continuous Evaporation," *Chemical Engineering Progress*, vol. 114, no. 5, pp. 24–28, 2018.

[16] M. Marin-Gallego, B. Mizzi, D. Rouzineau, C. Gourdon and M. Myer, "Concentric Heat Integrated Distillation Column (Hidic): A New Specific Packing Design, Characterization and Pre-Industrial Pilot Unit Validation," *Chemical Engineering and Processing – Process Intensification*, vol. 171, no. 108643, 2022.

[17] M. Jobson, "Chapter 6, Energy Considerations in Distillation," in *Distillation: Fundamentals and Principles*, Academic Press, Cambridge, Massachusetts, 2014, pp. 225–270.

[18] A. A. Kiss and C. S. Bîldea, "Revive Your Columns with Cyclic Distillation," *Chemical Engineering Progress*, vol. 111, no. 12, pp. 21–27, 2015.

[19] T. McGlone, N. E. B. Briggs, C. A. Clark, C. J. Brown, J. Sefcik and A. J. Florence, "Oscillatory Flow Reactors for Continuous Manufacturing and Crystallization," *Organic Process Research & Development*, vol. 19, no. 9, pp. 1186–1202, 2015.

[20] M. Baldea and T. F. Edgar, "Dynamic Process Intensification," *Current Opinion in Chemical Engineering*, vol. 22, pp. 48–53, 2018.

[21] Y. Shiraishi and T. Hirai, "Desulfurization of Vacuum Gas Oil Based on Chemical Oxidation Followed by Liquid-Liquid Extraction," *Energy Fuels*, vol. 18, no. 1, pp. 37–40, 2004.

[22] C. Wibowo and K. M. Ng, *Conceptual Design of Crystallization Processes*, De Gruyter, Berlin/Boston, 2021.

[23] Y. Liu, S. D. Feist, C. M. Jones and D. R. Armstrong, "Isopropyl Alcohol Dehydration by Hot Gas Pressure Swing Adsorption: Experiments, Simulations, and Implementation," *Industrial & Engineering Chemistry Research*, vol. 53, no. 20, pp. 8599–8607, 2014.

[24] J. A. Ritter, W. Fan and A. D. Ebner, "New Approach for Modeling Hybrid Pressure Swing Adsorption – Distillation Processes," *Industrial & Engineering Chemistry Research*, vol. 51, no. 27, pp. 9343–9355, 2012.

[25] F. Lipnizki, R. W. Field and P.-K. Ten, "Pervaporation-based Hybrid Process: A Review of Process Design, Applications, and Economics," Journal of Membrane Science, vol. 153, no. 2, pp. 183–210, 1999.

[26] C. Bravo-Bravo, J. G. Segovia-Hernández, S. Hernández, F. I. Gómez-Castro, C. Gutiérrez-Antonio and A. Briones-Ramírez, "Hybrid Distillation/Melt Crystallization Process Using Thermally Coupled Arrangements: Optimization with Evolutive Algorithms," *Chemical Engineering Processing*, vol. 67, pp. 25–38, 2013.

[27] L. Petrescu and C.-M. Cormos, "Classical and Process Intensification Methods for Acetic Acid Concentration: Technical and Environmental Assessment," *Energies*, vol. 15 (8119) 23 pages, 2022.

[28] D. A. Weingaertner, S. Lynn and D. N. Hanson, "Extractive Crystallization of Salts from Concentrated Aqueous Solution," *Industrial & Engineering Chemistry Research*, vol. 30, no. 3, pp. 490–501, 1991.

[29] H. Qu, K. B. Christensen, X. C. Fretté, F. Tian, J. Rantanen and L. P. Christensen, "Chromatography-crystallization Hybrid Process for Artemisinin Purification from Artemisia Annua," *Chemical Engineering & Technology*, vol. 33, no. 5, pp. 791–796, 2010.

[30] A. J. Barduhn and A. Manudhane, "Temperatures Required for Eutectic Freeze Crystallization of Natural Waters," *Desalination*, vol. 28, no. 3, pp. 233–241, 1979.

[31] M. J. Fernández-Torres, D. G. Randall, R. Melamu and H. Von Blottnitz, "A Comparative Life Cycle Assessment of Eutectic Freeze Crystallization and Evaporative Crystallization for the Treatment of Saline Wastewater," *Desalination*, vol. 306, pp. 17–23, 2012.

[32] E. Drioli, G. Di Profio and E. Curcio, *Membrane-Assisted Crystallization Technology*, Imperial College Press, London, 2015.

[33] B. Ohs, M. Falkenberg and M. Wessling, "Optimizing Hybrid Membrane-Pressure Swing Adsorption for Biogenic Hydrogen Reovery," *Chemical Engineering Journal*, vol. 364, pp. 452–461, 2019.

[34] T. Hu, H. Zhou, H. Peng and H. Jiang, "Nitrogen Production by Efficiently Removing Oxygen from Air Using a Perovskite Hollow-Fiber Membrane with Porous Catalytic Layer," *Frontiers in Chemistry*, vol., 2018, doi: 10.3389/fchem.2018.00329.

[35] A. F. Ismail and T. Matsuura, *Membrane Separation Processes: Theories, Processes, and Solutions*, Elsevier, Amsterdam, 2021.

[36] P. H. Au-Yeung, S. M. Resnick, P. M. Witt, T. C. Frank, F. A. Donate and L. A. Robbins, "Horizontal Reactive Distillation for Multicomponent Chiral Resolution," *AIChE Journal*, vol. 59, no. 7, pp. 2603–2620, 2013.

[37] W. Kujawski and Ł. Zieliński, "Bioethanol – One of the Renewable Energy Sources," *Environmental Protection Engineering*, vol. 32, no. 1, pp. 143–150, 2006.

Bibliography

Chapter 7

M. F. Doherty, Z. T. Fidkowski, M. F. Malone and R. Taylor, "Distillation," Section 13 in *Perry's Chemical Engineers' Handbook*, 9th ed., D. W. Green and M. Z. Southard, Eds., McGraw-Hill, New York, 2018.

H. Z. Kister, P. M. Mathius, D. E. Steinmeyer, W. Roy Penney, V. S. Monical and J. R. Fair, "Equipment for Distillation, Gas Absorption, Phase Dispersion, and Phase Separation," Section 14, in *Perry's Chemical Engineers' Handbook*, 9th ed., D. W. Green and M. Z. Southard, Eds., McGraw-Hill, New York, 2018.

T. J. Mathew, R. Tumbalam Gooty, M. Tawarmalani and R. Agrawal, "Quickly Assess Distillation Columns," *Chemical Engineering Progress*, vol. vol. 116, no. 12, pp. 27–34, 2020.

H. Z. Kister, *Distillation Operation*, McGraw-Hill, New York, 1990.

H. Z. Kister, *Distillation Design*, McGraw-Hill, New York, 1992.

H. Z. Kister, *Distillation Troubleshooting*, Wiley-AIChE, Hoboken, New Jersey, 2006.

R. F. Strigle Jr., *Random Packings and Packed Towers*, Gulf Publishing Co., Houston, 1987.

W. L. Luyben, *Distillation Design and Control Using Aspen^{TM} Simulation*, 2nd ed., Wiley, Hoboken, New Jersey, 2013.

AIChE Equipment Testing Procedures Committee, "Evaluating Distillation Column Performance," *Chemical Engineering Progress*, vol. 109, no. 6, pp. 27–35, 2013.

M. Pilling and B. S. Holden, "Back to Basics: Choosing Trays and Packings for Distillation," *Chemical Engineering Progress*, vol. 105, no. 9, pp. 44–50, 2009.

B. S. Holden, P. H. Au-Yeung and T. W. Kajden, "Watch Out for Trapped Components in Towers," *Chemical Processing*, vol. 75, no. 5, pp. 38–41, 2012.

Z. Xu, D. R. Summers, T. Cai, B. Holden, R. Olsson, L. Pless, M. Schultes, F. Seibert and S. X. Xu, *AIChE Equipment Testing Procedure – Trayed & Packed Columns: A Guide to Performance Evaluation*, 3rd ed., AIChE-Wiley, Hoboken, New Jersey, 2014.

K. T. Chuang and K. Nandakumar, *Distillation–Tray Columns: Design*, Academic Press, Cambridge, Massachusetts, 2000.

A. Gorak and E. Sorensen, Eds., *Distillation: Fundamentals and Principles*, Academic Press, Cambridge, Massachusetts, 2014.

A. Gorak and H. Schoenmakers, Eds., *Distillation: Operation and Applications*, Academic Press, Cambridge, Massachusetts, 2014.

A. Gorak and Z. Olujic, Eds., *Distillation: Equipment and Processes*, Academic Press, Cambridge, Massachusetts, 2014.

L. A. Robbins, *Distillation Control, Optimization, and Tuning: Fundamentals and Strategies*, CRC Press, Boca Raton, Florida, 2011.

C. L. Smith, *Distillation Control: An Engineering Perspective*, Wiley, Hoboken, New Jersey, 2012.

C. L. Smith, *Control of Batch Processes*, Wiley-AIChE, Hoboken, New Jersey, 2014.

U. M. Diwekar, *Batch Distillation: Simulation, Optimal Design, and Control*, 2nd ed., CRC Press, Boca Raton, Florida, 2016.

M. Mujtabe, *Batch Distillation: Design and Operation, Series on Chemical Engineering*, Vol. 3, R. T. Yang, Ed., Imperial College Press, London, 2004.

K. Arvid Berglund, "Artisan Distilling: A Guide for Small Distillers," Department of Chemical Engineering and Materials Science, Michigan State University, East Lansing, Michigan, 2004.

K. McCarley, "Finding the Capacity of a Distillation Column," *Chemical Engineering Progress*, vol. 117, no. 12, pp. 23–28, 2021.

https://doi.org/10.1515/9783110695052-020

Chapter 8

M. F. Doherty, Z. T. Fidkowski, M. F. Malone and R. Taylor, Section 13, "Distillation," in *Perry's Chemical Engineers' Handbook*, 9th ed., D. W. Green and M. Z. Southard, Eds., McGraw-Hill, New York, 2019.

M. F. Doherty and M. F. Malone, *Conceptual Design of Distillation Systems*, McGraw-Hill, New York, 2001.

S. Widagdo and W. D. Seider, "Azeotropic Distillation," *AIChE Journal*, vol. 42, no. 1, pp. 96–130, 1996.

L. Berg and A.-I. Yeh, "The Unusual Behavior of Extractive Distillation – Reversing the Volatility of the Acetone-Isopropyl Ether System," *AIChE Journal*, vol. 31, no. 3, pp. 504–506, 1985.

V. Gerbaud, I. Rodriguez-Donis, L. Hegely, P. Lang, F. Denes and X. You, "Review of Extractive Distillation. Process Design, Operation, Optimization and Control," *Chemical Engineering Research and Design*, vol. 141, pp. 229–271, 2019.

J. P. Knapp and M. F. Doherty, "Minimum Entrainer Flows for Extractive Distillation: A Bifurcation Theoretic Approach," *AIChE Journal*, vol. 40, no. 2, pp. 243–268, 2004.

W. L. Luyben and I.-L. Chien, *Design and Control of Distillation Systems for Separating Azeotropes*, Wiley, Hoboken, New Jersey, 2010.

J. P. Knapp and M. F. Doherty, "A New Pressure-Swing Distillation Process for Separating Homogeneous Azeotropic Mixtures," *Industrial & Engineering Chemistry Research*, vol. 31, no. 1, pp. 346–357, 1992.

T. C. Frank, "Break Azeotropes with Pressure-Sensitive Distillation," *Chemical Engineering Progress*, vol. 93, no. 4, pp. 52–63, 1997.

Chapter 9

T. C. Lo and M. H. I. Baird, in *Handbook of Solvent Extraction*, C. Hanson, Ed., Wiley, Hoboken, New Jersey, 1983.

J. D. Thornton, Ed., *Science and Practice of Liquid-Liquid Extraction*, Vol. 1 Phase Equilibria; Mass Transfer and Interfacial Phenomena; Extractor Hydrodynamics, Selection, and Design, Oxford University Press, Oxford, England 1992.

J. D. Thornton, Ed., *Science and Practice of Liquid-Liquid Extraction*, Vol. 2 Process Chemistry and Extraction Operations in the Hydrometallurgical, Nuclear, Pharmaceutical, and Food Industries, Oxford University Press, Oxford, England, 1992.

J. C. Godfrey and M. J. Slater, *Liquid-Liquid Extraction Equipment*, Wiley, Hoboken, New Jersey, 1994.

J. Rydberg, *Solvent Extraction Principles and Practice*, 2nd ed., CRC Press, Boca Raton, Florida, 2004.

T. C. Frank, B. S. Holden and A. F. Seibert, Section 15, "Liquid-Liquid Extraction and Other Liquid-Liquid Operations and Equipment," in *Perry's Chemical Engineers' Handbook*, 9th ed., D. W. Green and M. Z. Southard, Eds., McGraw-Hill, New York, 2018.

T. C. Frank, L. Dahuron, B. S. Holden, W. D. Prince, A. F. Seibert and L. C. Wilson, Section 15, "Liquid-Liquid Extraction and Other Liquid-Liquid Operations and Equipment," in *Perry's Chemical Engineers' Handbook*, 8th ed., D. W. Green, Ed., McGraw-Hill, New York, 2007.

V. H. Shah, V. Pham, P. A. Larsen, S. Biswas and T. C. Frank, "Liquid-liquid Extraction for Recovering Low Margin Chemicals: Thinking Beyond the Partition Ratio," *Industrial & Engineering Chemistry Research*, vol. 55, no. 6, pp. 1731–1739, 2016.

Chapter 10

M. Douglas Levan, G. Carta, J. A. Ritter and K. S. Walton, "Adsorption and Ion Exchange," Section 16, in *Perry's Chemical Engineers Handbook*, 9th ed., D. W. Green and M. Z. Southard, Eds., McGraw-Hill, New York, 2018.

C. Tien, *Introduction to Adsorption: Basics, Analysis, and Applications*, Elsevier, Amsterdam, 2019.

O. Kopsidas, Scale up of Adsorption in Fixed-Bed Column Systems, Univ. Piraeus, Piraeus, Greece, 2016.

A. Dąbrowski, "Adsorption – From Theory to Practice," *Advances in Colloid and Interfacial Science*, vol. 93, no. 1-3, pp. 135–224, 2001.

A. Gabelman, "Adsorption Basics: Part 1," *Chemical Engineering Progress*, vol. 113, no. 7, pp. 48–53, 2017.

A. Gabelman, "Adsorption Basics: Part 2," *Chemical Engineering Progress*, vol. 113, no. 8, pp. 38–45, 2017.

B. W. Pynnonen, S. D. Feist, Y. Hasabnis, D. R. Albers and T. C. Frank, "Quickly Evaluate SMB Chromatography for Your Application," *Chemical Processing*, vol. 73, no. 5, pp. 29–33, 2010.

S. D. Feist, Y. Hasabnis, B. W. Pynnonen and T. C. Frank, "SMB Chromatography Design Using Profile Advancement Factors, Miniplant Data, and Rate-Based Process Simulation," *AIChE Journal*, vol. 55, no. 11, pp. 2848–2860, 2009.

S. Sircar, "Pressure Swing Adsorption," *Industrial & Engineering Chemistry Research*, vol. 41, no. 6, pp. 1389–1392, 2002.

C. Voss, "Applications of Pressure Swing Adsorption Technology," *Adsorption*, vol. 11, pp. 527–529, 2005.

C. A. Grande, "Advances in Pressure Swing Adsorption for Gas Separation," *Chemical Engineering*, Review Article 2012, Article ID 982934, 13 pages, 2012, doi: 10.5402/2012/982934.

Chapter 11

A. S. Myerson, D. Erdemir and A. Y. Lee, Eds., *Handbook of Industrial Crystallization*, 3rd ed., Cambridge University Press, Cambridge, England, 2019.

W. J. Genck, B. Albin, F. A. Baczek, D. S. Dickey, C. G. Gilbert, T. Herrera, T. J. Laros, W. Li, P. McCurdie, J. K. McGillicuddy, T. P. McNulty, C. G. Moyers, F. Schoenbrunn, T. W. Wisdom and W. Chen, "Liquid-Solid Operations and Equipment," Section 18, in *Perry's Chemical Engineers' Handbook*, 9th ed., D. W. Green and M. Z. Southard, Eds., McGraw-Hill, New York, 2018.

W. J. Genck, "A Clearer View of Crystallizers," *Chemical Engineering Magazine*, pp. 28–32, July 2011.

W. J. Genck, "Guidelines for Crystallizer Selection and Operation," *Chemical Engineering Progress*, vol. 100, no. 10, pp. 26–32, 2004.

C. Wibowo and K. M. Ng, *Conceptual Design of Crystallization Processes*, De Gruyter, Berlin/Boston, 2021.

N. Yazdanpanah and K. N. Zoltan, Eds., *The Handbook of Continuous Crystallization*, Royal Society of Chemistry, London, 2020.

W. Beckmann, Ed., *Crystallization: Basic Concepts and Industrial Applications*, Wiley-VCH, Hoboken, New Jersey, 2013.

N. C. S. Kee, L. M. Goh, E. Rusli, G. He, V. Bhamidi, R. B. H. Tan, P. J. A. Kenis, C. F. Zukoski and R. D. Braatz, "Design of Crystallization Processes from Laboratory Research and Development to Manufacturing Scale," Parts I and II, *American Pharmaceutical Review*, vol. 11, no. 5 and 6, 2008.

C. G. Moyers Jr. and R. W. Rousseau, "Crystallization Operations," Chapter 11, in *Handbook of Separation Process Technology*, R. W. Rousseau, Ed., Wiley-Interscience, Hoboken, New Jersey, 1987.

G. J. Sloan and A. R. McGhie, "Techniques of Melt Crystallization," in *Techniques of Chemistry Series*, Vol. XIX, A. Weissberger and W. H. Saunders Jr., Eds., Wiley-Interscience, Hoboken, New Jersey, 1988.

A. D. Randolph and M. A. Larson, *Theory of Particulate Processes: Analysis and Techniques of Continuous Crystallization*, 2nd ed., Academic Press, Cambridge, Massachusetts, 1988.

D. Ramkrishna, *Population Balances: Theory and Applications to Particulate Systems in Engineering*, Academic Press, Cambridge, Massachusetts, 2000.

J. F. Lutsko, "How Crystals Form: A Theory of Nucleation Pathways," *Science Advances*, vol. 5, article eaav7399, 2019.

D. Q. Kern, *Process Heat Transfer*, McGraw-Hill, New York, 1950.

A. Marie Flynn, T. Akashige and L. Theodore, *Kern's Process Heat Transfer*, 2nd ed., Wiley, Hoboken, New Jersey, 2018.

Chapter 12

H. Ludwig, *Reverse Osmosis Seawater Desalination*, Springer, Heildelberg, 2022.

A. Ismail and T. Matsuura, *Membrane Separation Processes: Theories, Processes, and Solutions*, Elsevier, Amsterdam, 2021.

L. W. Jye and A. Fauzi Ismail, *Nanofiltration Membranes: Synthesis, Characterization, and Applications*, CRC Press, Boca Raton, Florida, 2019.

V. Gude, Ed., *Sustainable Desalination Handbook*, Elsevier, Amsterdam, 2018.

K. Nath, *Membrane Separation Processes*, 2nd ed., PHI Learning, Delhi, 2017.

H. Lutz, *Ultrafiltration for Bioprocessing*, Elsevier, Amsterdam, 2015.

M. Cheryan, *Ultrafiltration and Microfiltration Handbook*, 2nd ed., CRC Press, Boca Raton, Florida, 1998.

L. M. Robeson, "The Upper Bound Revisited," *Journal of Membrane Science*, vol. 320, pp. 390–400, 2008.

Chapter 13

G. Subramanian, Ed., *Bioseparation and Bioprocessing*, 2 volume set, 2nd ed., Wiley-VCH, Hoboken, New Jersey, 2007.

D. Forciniti, *Industrial Bioseparations: Principles and Practice*, Wiley-Blackwell, Hoboken, New Jersey, 2008.

P. G. Jessop, D. J. Heldebrant, L. Xiaowang, C. A. Eckert and C. L. Liotta, "Reversible Nonpolar-to-Polar Solvent," *Nature*, vol. 436, pp. 1102, 2005.

P. Pollet, C. A. Eckert and C. L. Liotta, "Switchable Solvents," *Chemical Science*, vol. 2, no. 4, pp. 609–614, 2011.

M. McHugh and V. Krukonis, *Supercritical Fluid Extraction: Principles and Practice*, 2nd ed., Butterworth-Heinemann, Oxford, England, 1994.

J. L. Martinez, Ed., *Supercritical Fluid Extraction of Nutraceuticals and Bioactive Compounds*, CRC Press, Boca Raton, Florida, 2008.

S. T. Patel, N. K. Prajapati and K. M. Joshi, *Chirality in Practical Utility: Introduction to Chiral Chemical Compounds*, LAP LAMBERT Academic Publishing, London, 2014.

H. Lorenz and A. Seidel-Morgenstern, "Processes To Separate Enantiomers," *Angewandte Chemie*, vol. 53, no. 5, pp. 1218–1250, 2014.

M. B. Hocking, *Handbook of Chemical Technology and Pollution Control*, 3rd ed., Elsevier, Amsterdam, 2006.

Chapter 14

J. M. Douglas, "A Hierarchical Decision Procedure for Process Synthesis," *AIChE Journal*, vol. 31, no. 3, pp. 3530362, 1985.

J. M. Douglas, *Conceptual Design of Chemical Processes*, McGraw-Hill, New York, 1988.

J. D. Seader, E. J. Henley and D. K. Roper, *Separation Process Principles with Applications Using Process Simulators*, 4th ed., Wiley, Hoboken, New Jersey, 2015.

A. B. de Haan, H. Burak Eral and B. Schuur, "Separation Method Selection," Chapter 12, in *Industrial Separation Processes: Fundamentals*, De Gruyter, Berlin/Boston, 2020.

F. J. Keil, "Process Intensification," *Reviews in Chemical Engineering*, 2017, doi.org/10.1515/revce-2017-0085.

A. Lucia, A. Amale and R. Taylor, "Energy Efficient Hybrid Separation Processes," *Industrial & Engineering Chemistry Research*, vol. 45, no. 25, pp. 8319–8328, 2006.

Z. Jiang and R. Agrawal, "Process Intensification in Multicomponent Distillation: A Review of Recent Advancements," *Chemical Engineering Research and Design*, vol. 147, pp. 122–145, 2019.

C. Wibowo and K. M. Ng, "Unified Approach for Synthesizing Crystallization-Based Separation Processes," *AIChE Journal*, vol. 46, no. 7, pp. 1400–1421, 2000.

J. G. Speight, *The Refinery of the Future*, 2nd ed., Elsevier, Amsterdam, 2021.

D. W. Mackinnon, "The Nature and Nurture of Creative Talent," *American Psychologist*, vol. 17, no. 7, pp. 484–495, 1962.

E. P. Torrance, *Creativity*, National Education Association, Washington, D.C., 1963.

G. Thompson and M. A. Lordan, "A Review of Creativity Principles Applied to Engineering Design," *Proceedings of the Institution of Mechanical Engineers, Part E: Journal of Process Mechanical Engineering*, vol. 213, no. 1, pp. 17–31, 1999.

T. Arciszewski, *Inventive Engineering: Knowledge and Skills for Creative Engineers*, CRC Press, 2016.

P. Livotov and V. Petrov, *TRIZ: Innovation and Inventive Problem Solving Handbook*, TriS Europe – Innovation Knowledge Company, Freiburg im Breisgau, Germany, 2015.

J. Cleese, *Creativity: A Short and Cheerful Guide*, Crown Publishing, New York, 2020.

R. Jolly, "What's the Hurry," *London Business School Review*, vol. 26, no. 3, pp. 22, 2015.

A. E. Herman, *Fixed: How to Perfect the Fine Art of Problem Solving*, Harper Wave, New York, 2021.

A. B. VanGundy Jr., *Techniques of Structured Problem Solving*, 2nd ed., Van Nostrand Reinhold, New York, 1988.

K. Knight, *Mind Mapping: Improve Memory, Concentration, Communication, Organization, Creativity, and Time Management (Mental Performance)*, CreateSpace Independent Publishing Platform, Scotts Valley, California, 2012.

Index

https://doi.org/10.1515/9783110695052-021

www.ingramcontent.com/pod-product-compliance
Lightning Source LLC
Chambersburg PA
CBHW080119220326
41598CB00032B/4893